T0336932

EMOTIONS AND AFFECT IN HUMAN FACTORS AND HUMAN–COMPUTER INTERACTION

EMOTIONS AND AFFECT IN HUMAN FACTORS AND HUMAN–COMPUTER INTERACTION

Edited by

MYOUNGHOON JEON

Michigan Technological University, Houghton, MI, United States

Academic Press is an imprint of Elsevier
125 London Wall, London EC2Y 5AS, United Kingdom
525 B Street, Suite 1800, San Diego, CA 92101-4495, United States
50 Hampshire Street, 5th Floor, Cambridge, MA 02139, United States
The Boulevard, Langford Lane, Kidlington, Oxford OX5 1GB, United Kingdom

Copyright © 2017 Elsevier Inc. All rights reserved.

No part of this publication may be reproduced or transmitted in any form or by any means, electronic or mechanical, including photocopying, recording, or any information storage and retrieval system, without permission in writing from the publisher. Details on how to seek permission, further information about the Publisher's permissions policies and our arrangements with organizations such as the Copyright Clearance Center and the Copyright Licensing Agency, can be found at our website: www.elsevier.com/permissions.

This book and the individual contributions contained in it are protected under copyright by the Publisher (other than as may be noted herein).

Notices
Knowledge and best practice in this field are constantly changing. As new research and experience broaden our understanding, changes in research methods, professional practices, or medical treatment may become necessary.

Practitioners and researchers must always rely on their own experience and knowledge in evaluating and using any information, methods, compounds, or experiments described herein. In using such information or methods they should be mindful of their own safety and the safety of others, including parties for whom they have a professional responsibility.

To the fullest extent of the law, neither the Publisher nor the authors, contributors, or editors, assume any liability for any injury and/or damage to persons or property as a matter of products liability, negligence or otherwise, or from any use or operation of any methods, products, instructions, or ideas contained in the material herein.

Library of Congress Cataloging-in-Publication Data
A catalog record for this book is available from the Library of Congress

British Library Cataloguing-in-Publication Data
A catalogue record for this book is available from the British Library

ISBN: 978-0-12-801851-4

For information on all Academic Press publications visit our website at https://www.elsevier.com/books-and-journals

Working together to grow libraries in developing countries

www.elsevier.com • www.bookaid.org

Publisher: Nikki Levy
Acquisition Editor: Emily Ekle
Editorial Project Manager: Barbara Makinster
Production Project Manager: Nicky Carter
Designer: Matthew Limbert

Typeset by Thomson Digital

Contents

3. Mood Effects on Cognition: Affective Influences on the Content and Process of Information Processing and Behavior

JOSEPH P. FORGAS

4. Cross-Cultural Similarities and Differences in Affective Processing and Expression

JAMES A. RUSSELL

5. On the Moral Implications and Restrictions Surrounding Affective Computing

ANTHONY F. BEAVERS, JUSTIN P. SLATTERY

II

FRAMEWORKS OF AFFECTIVE SCIENCES IN HUMAN FACTORS AND HUMAN–COMPUTER INTERACTION

13. On Computational Models of Emotion Regulation and Their Applications Within HCI

TIBOR BOSSE

IV

APPLICATIONS: CASE STUDIES AND APPLIED EXAMPLES

14. Evolution of Emotion Driven Design

OYA DEMIRBILEK

List of Contributors

Jacob Aday Northern Michigan University, Marquette, MI, United States

Robert K. Atkinson Arizona State University, Tempe, AZ, United States

Mustafa Baydogan Bogazici University, Istanbul, Turkey

Anthony F. Beavers The University of Evansville, Evansville, IN, United States

Jenay M. Beer University of South Carolina, Columbia, SC, United States

Tibor Bosse VU University Amsterdam, Amsterdam, The Netherlands

Winslow Burleson New York University, New York, NY, United States

Rafael A. Calvo School of Electrical and Information Engineering, University of Sydney, Sydney, NSW, Australia

Joshua M. Carlson Northern Michigan University, Marquette, MI, United States

Andrea Cavallaro Queen Mary University of London, London, United Kingdom

Maria Elena Chavez-Echeagaray Arizona State University, Tempe, AZ, United States

David Cherry Clemson University, Clemson, SC, United States

Shaundra B. Daily University of Florida, Gainesville, FL, United States

Shelby S. Darnell Clemson University, Clemson, SC, United States

Oya Demirbilek University of New South Wales, Sydney, NSW, Australia

Stephen H. Fairclough Liverpool John Moores University, Liverpool, United Kingdom

Paul A. Fishwick University of Texas at Dallas, Richardson, TX, United States

Joseph P. Forgas University of New South Wales, Sydney, NSW, Australia

Andrea Gaggioli Catholic University of Sacred Heart; Istituto Auxologico Italiano, Milan, Italy

Javier Gonzalez-Sanchez Arizona State University, Tempe, AZ, United States

Hatice Gunes University of Cambridge, Cambridge, United Kingdom

Peter A. Hancock University of Central Florida, Orlando, FL, United States

Bruce Hanington Carnegie Mellon University, Pittsburgh, PA, United States

Eva Hudlicka Psychometrix Associates, College of Information and Computer Sciences, University of Massachusetts-Amherst, Amherst, MA, United States

Joseph Isaac University of Florida, Gainesville, FL, United States

Melva T. James Clemson University, Clemson, SC; Massachusetts Institute of Technology, Lexington, MA, United States

Myounghoon Jeon Michigan Technological University, Houghton, MI, United States

Karina R. Liles University of South Carolina, Columbia, SC, United States

Seyedeh Maryam Fakhrhosseini Michigan Technological University, Houghton, MI, United States

Tal Oron-Gilad Ben-Gurion University of the Negev, Be'er Sheva, Israel

Sujan Pakala University of South Carolina, Columbia, SC, United States

Dorian Peters School of Electrical and Information Engineering, University of Sydney, Sydney, NSW, Australia

John J. Porter, III Clemson University, Clemson, SC, United States

Andreas Riener Johannes Kepler University Linz, Linz, Austria; Ingolstadt University of Applied Sciences, Ingolstadt, Germany

Giuseppe Riva Catholic University of Sacred Heart; Istituto Auxologico Italiano, Milan, Italy

Will Rizer Northern Michigan University, Marquette, MI, United States

Tania Roy Clemson University, Clemson, SC, United States

James A. Russell Boston College, Chestnut Hill, MA, United States

Evangelos Sariyanidi Queen Mary University of London, London, United Kingdom

Justin P. Slattery Indiana University, Bloomington, IN, United States

Xian Wu University of South Carolina, Columbia, SC, United States

Toshimasa Yamanaka University of Tsukuba, Tsukuba, Japan

Preface

THE ORGANIZATION OF THE HANDBOOK

This handbook aims to serve as a "field guide" in conducting affect-related research and design projects in H/F and HCI, by introducing necessary concepts, approaches, methods, applications, and emerging areas so that readers can apply those ingredients to their own area. On the other hand, since this handbook deals with comprehensive topics, I hope that experts in this domain can also utilize this handbook as a necessary reference or a teaching material. The handbook is organized in the following way.

PART I: FOUNDATIONS OF AFFECTIVE SCIENCES

Part I provides a theoretical background of Affective Sciences. More and more research on emotions and affect has recently appeared in applied H/F and HCI settings, whereas considerable research has not been based on affect theory or mechanism. Engineers and designers can learn and apply these psychological theories and mechanisms to account for their affect-related research and can develop their own domain-specific theory. In Chapter 1 Jeon presents an overview of Affective Sciences in H/F and HCI. This is a compact version of the handbook. In Chapter 2 Aday, Rizer, and Carlson introduce the neural mechanisms associated with affective phenomena, such as appraisal, reactivity, feeling, and regulation with a focus on the role of amygdala. Then, Forgas provides an extensive review on the affective effects on cognitive processes—content of thinking and quality of information processing with recent empirical studies (Chapter 3). In Chapter 4 Russell proposes a new conceptual framework of affect about commonalities and differences among individuals and cultures, highlighting subjective experience and facial expressions of emotion. Beavers and Slattery discuss a moral dilemma and ethical issues, which arise, specifically from "simulated affect" of machines in affective computing. This chapter should serve as a good starting point about other ethical issues regarding affect induction, detection, data storage and access, and intervention for affect regulation, which are critical components in affect-related projects in H/F and HCI.

PART II: FRAMEWORKS OF AFFECTIVE SCIENCES IN HUMAN FACTORS AND HCI

In Part II, representative (most widely cited, discussed, and currently followed) approaches to Affective Sciences in H/F and HCI are introduced, including Emotional Design, Hedonomics, Kansei Engineering, and Affective Computing. Chapters cover each approach's background, goal, development, characteristics, applications, and future directions. Readers will be able to learn how these approaches are similar to each other and how they are distinct from each other. Based on that, readers can get some sense of what type of approach they should take for their research domain. This part begins with Hanington's Chapter 6 about design and emotion; its history, theories, methodologies, and approaches. This chapter also supplies a bird's eye view about this part. Next, Oron-Gilad and Hancock reveal the evolution of human factors from traditional Ergonomics into Hedonomics in Chapter 7. They emphasize to catch up the tremendous progress in the domain and change their focus from functional usability to hedonomic fulfillment. In Chapter 8 Yamanaka tells us about philosophical background of the Kansei approach and relationships among Kansei, Kansei science, and Kansei Engineering. He finishes up his chapter by demonstrating a design example which Kansei design procedure has been applied for. Daily and her coworkers wrap up the history, core components (i.e., affect sensing and generation), current applications, and future trends of Affective Computing. They also include interesting topics on ethical consideration.

PART III: METHODOLOGIES: INTRODUCTION AND EVALUATION OF TECHNIQUES

To conduct affect research, systematic knowledge about practical methods is necessary. Part III introduces ample methodologies, such as affect induction, measurement, and detection techniques. Also, regulation strategies and computational models for training and intervention system development are included. Readers can learn the pros and cons of each method and how to choose or combine those methods for their own research. Both Chapters 10 and 11 are written to offer a boot camp for novice researchers so that they are able to follow the steps. In Chapter 10, FakhrHosseini and Jeon demonstrate diverse affect induction methods used in actual research settings—listening to music, writing experiences, reading passages, watching videos, or photos, etc. Then, Gonzalez-Sanchez and his coworkers describe sensing technologies (brain–computer interfaces, facial expression detection, eye-tracking systems, physiological sensors, body language recognition, and writing patterns) and detail their technical know-how from data collection to analysis in Chapter 11. In a similar line, Sariyanidi, Gunes, and Cavallaro discuss facial expression more in depth in Chapter 12. Specifically, they focus on registration and representation, which are important both in cognitive sciences and computer science. Even without understanding the algorithms, readers can readily follow their ideas and get the gist. This part closes with computational models of affect regulation by Bosse (Chapter 13). He covers emotion modeling and its application to regulation processes, which can be used in a variety of application settings.

PART IV: APPLICATIONS: CASE STUDIES AND APPLIED EXAMPLES

Application domains are selected based on the amount of existing affect research and documented literature in a specific domain. However, these are just a few selected areas, not exhaustive. Chapters in this part discuss various H/F and HCI application areas, including product design, human–robot interaction, healthcare, and driving. Core research topics, design practices and case studies, theoretical and practical issues, and design guidelines are discussed. Demirbilek in Chapter 14 provides abundant product design examples, which include emotional sensory modalities—look (color), feel (touch), and sound. She also delineates positive emotional triggers for delightful experiences, such as fun, cuteness, and familiarity. In Chapter 15 Beer and her coworkers give an overview about affective human–robot interaction. They define both humans and robots as social entities and explore the existing applications and future possibilities of social robots for aging, companionship, and education. Then, in Chapter 16 Hudlicka discusses the state-of-the-art in modeling emotion–cognition interaction and the application of these models for understanding the mechanisms mediating psychopathology and therapeutic action, and for facilitating more accurate assessment and diagnosis of affective disorders. Changing gear, Jeon shows lacking of the driving models that have emotional components and presents recent empirical research outcomes about the effects of various affective states on driving in Chapter 17.

PART V: EMERGING AREAS

The last part briefly touches emerging interdisciplinary areas related to Affective Sciences in H/F and HCI, including Positive Technology, Subliminal Perception, Physiological Computing, and Aesthetic Computing. There are new conferences, workshops, journal special issues, or new books for these emerging areas. The presence of all these relevant disciplines supports the importance of affective approaches to H/F and HCI. Chapters in this part describe their background, goal, research topics, and how they are relevant to Affective Sciences. This part provides a new perspective to understand a big picture of Affective Sciences. Gaggioli and coworkers start the part with Chapter 18, by depicting how positive psychology has sought happiness and well-being, which has brought about positive technology and positive computing. They offer theoretical background, applications, and future research agenda of these approaches to human happiness and well-being. In Chapter 19 Riener questions if we can provide additional information to users while they are not aware of that perception, which is called, "subliminal perception." He shows compelling examples and potentials of subliminal approaches for different sensory channels, specifically in the automotive domain. In Chapter 20 Fairclough provides an overview about physiological computing in which physiological data from the brain and body are transformed into input control to inform software adaptation. He explains how a dynamic user representation can be used as a form of input control and guides a process of intelligent adaptation and this adaptation can enhance human–computer interaction ultimately. In the last chapter of this handbook, Fishwick answers to core questions in aesthetic computing: "How can embodied cognition be situated within formal languages?" and "How can embodied cognition result in novel computer interfaces for formal languages?" He challenges us to build a new generation of human–computer interfaces that are informed by embodied principles and use these principles as design elements for interacting with formal languages.

As readers can see, I tried to offer multiple perspectives on the topic for a holistic approach to Affective Sciences in H/F and HCI. It was a long but rewarding journey to edit this handbook. I have taught the course with the same title as this handbook, but I myself have learned a lot more about the domain while editing this handbook. By no means, this handbook addresses all relevant topics of Affective Sciences in H/F and HCI. However, this volume should help frame useful, timely questions about the roles of emotions and affect in interactions between people and technology. Yes, it cannot answer them just yet. In the next turn, this book could be further updated based on more empirical research studies. The approach outlined in this handbook is expected to begin to close the existing gap between the traditional affect research and the emerging field of affective design and affective computing.

I hope that readers in this topic—scholars, students, designers, engineers, and practitioners, can benefit from reading diverse perspectives and applications, and ultimately integrating multiple approaches into their own work.

Myounghoon Jeon

Acknowledgments

This handbook would not have been possible without the contribution of all the authors, who have generously shared their invaluable views and research outcomes to make this book. I deeply appreciate their time and effort for their excellent work.

I would like to thank Emily Ekle, Timothy Bennett, Barbara Makinster, and Nicky Carter at Elsevier for suggesting the edits in this handbook and managing the publication processes. I also thank Dr. Bruce Seely, Dr. Susan Amato-Henderson, and Dr. Min Song for their practical advice and strong support. Finally, I am extremely grateful to my academic advisors—Dr. Kwang Hee Han and Dr. Bruce Walker, my students in the Mind Music Machine Lab, and my family.

PART I

FOUNDATIONS OF AFFECTIVE SCIENCES

CHAPTER

1

Emotions and Affect in Human Factors and Human–Computer Interaction: Taxonomy, Theories, Approaches, and Methods

Myounghoon Jeon
Michigan Technological University, Houghton, MI, United States

INTRODUCTION

Emotions and affect are known as a motivating and guiding force in our attention and perception (Izard, 1993). Psychologists even accept that it is impossible for people to have a thought or perform an action without engaging their emotional system, at least unconsciously (Nass et al., 2005). However, despite their importance and prevalence in everyday life, emotions and affect have had fewer chances to be a dominant topic of human factors (HF) and human–computer interaction (HCI) for a long time. Given that the information processing approach of traditional Cognitive Sciences has been a dominant paradigm of HF and HCI, there has not been much room for emotions and affect in the mainstream of human–machine system research until recently. However, affective elements allow for a systems approach (Czaja and Nair, 2006) and a more holistic view to understanding the human–machine system. For example, affective states have been known to play a major role in influencing all aspects of situation assessment and belief information, from cue identification and extraction to situation classification, and decision selection

Emotions and Affect in Human Factors and Human–Computer Interaction. http://dx.doi.org/10.1016/B978-0-12-801851-4.00001-X
Copyright © 2017 Elsevier Inc. All rights reserved.

(Hudlicka, 2003). Therefore, embracing emotions and affect is expected to enrich human–machine system research even more.

There is a growing realization of this need to acknowledge emotions and affect in the design challenges of complex sociotechnical systems (Lee, 2006) and this is high time to look back on research on emotions. First, this trend has been reflected in publications in special issues of Ergonomics (Helander and Tham, 2003) and the *International Journal of Human-Computer Studies* (Hudlicka, 2003), the birth of IEEE Transactions on Affective Computing (Picard, 2010), books (Calvo et al., 2014; Norman, 2004; Picard, 1997), and book chapters (e.g., Lee, 2006). Thus, we have much more sources and resources than before in terms of theory, framework, and approach. Second, the traditional cognitive paradigm is being shifted. A recent new wave of embodied cognition—for example, special issues of embodied cognition (Davis and Markman, 2012) and embodied interaction (Marshall et al., 2013)—is good justification we need to include emotions and affect in human–machine system research. In addition, the development of physiological computing or aesthetic computing (Fishwick, 2006) is closely related to Affective Sciences and strongly encourages researchers to integrate emotional elements in their research. Finally, with the rapid advances of sensing technology, emotional research has appeared to be easier and more feasible. In fact, recent CHI conferences have had successive workshops and tutorials about affect detection techniques. Based on this background, this handbook tries to outline the big picture of emotions and affect research in HF/HCI. Providing taxonomy as a field guide would be, especially beneficial to domain novices and students, by listing and specifying the emerging frameworks of emotions and affect research in HF/HCI.

DEFINITION AND TAXONOMY

Affect research includes several relevant constructs, such as emotions, feelings, moods, and affect which are distinct, but frequently used as interchangeable terms. Specifying the terms would facilitate further discussion and clarify their usage. Emotions generally refer to the physiological response of the brain and body to threats and opportunities, whereas feelings are the conscious mental representation and interpretation of that bodily response (Damasio, 1994, 2001). Hence, feelings are known to follow emotions evolutionarily and experientially. Emotions have a salient cause, occur and diminish quickly and thus are relatively intense and clear cognitive contents (e.g., anger or fear). In contrast, moods are less intense, more diffuse, and more enduring and thus unclear to the individual experiencing them (e.g., just like feeling good or feeling bad) (Bodenhausen et al., 1994; Forgas, 1995). Although moods are more subtle, they often exert a more enduring effect on behavior (Forgas, 2002).

Affect is sometimes referred to as something weaker than emotions (Slovic et al., 2004). However, researchers currently agree that affect denotes any type of affective experience (Forgas, 1995; Hudlicka, 2003; Lottridge et al., 2011; Mayer, 1986). In this chapter, I would distinguish affect from unchangeable long-term characteristics, such as personality traits or temperament. Here, focus is on dynamically changeable affective states depending on contexts. Of course, where appropriate, I follow the traditional usage of the terms (e.g., the mood-congruent effects).

Sometimes, emotions can be categorized as background emotions, basic emotions (e.g., Ekman's emotion set), or social emotions (Damasio, 1999). Also, depending on the relationship between the source of the affect and the task-relevancy, integral affect (related to the task) and incidental affect (not related to the task) are differentiated (Bodenhausen, 1993). The most salient distinction among affect researchers is discrete versus dimensional approaches. For example, Ekman (1992) suggested basic emotions (happiness, sadness, anger, fear, surprise, and disgust) as a universal emotion set across the world. Similarly, Plutchik (1994) proposed a basic emotion set (happiness, sadness, fearfulness, anger, and tenderness). On the other hand, a common dimensional approach to understanding the experience of emotions is the circumplex model, in which emotions are arranged in a circle around the intersections of two core dimensions of affect (Russell, 1980). The circumplex model maps emotions according to their valence, indicating how negative or positive they are, and their activation, indicating how arousing they are. There is no clear cut answer about which approach is better. It depends on the research goal, the task, situations, and measurement methods.

MECHANISMS AND THEORIES

Relationship Between Affect and Cognition

Traditionally, psychological sciences have suggested various theories on the relationship among emotion, physiological arousal, and cognition [e.g., the James-Lange theory (James, 1884), the Cannon-Bard theory (Cannon, 1927), and the two-factor theory (Schacter and Singer, 1962)]. One of the unresolved issues is whether affective processes should be considered as a part of the cognitive representational system or as an entirely separate mental faculty (Fiedler, 1988; Hilgard, 1980; Salovey and Mayer, 1990). Researchers have shown with empirical findings that affect can be aroused without the participation of cognitive processes and thus, can function independently (Clore et al., 1994; LeDoux, 1996; Niedenthal, 1990; Schwarz and Clore, 1988; Strack et al., 1988; Zajonc, 1984). Therefore, it is important to assume that cognition and affect are distinct and dynamically interact with each other (Forgas, 1995).

Emotional Functions and Neural Mechanisms

The evolutionarily older part of the brain, such as brainstem and cerebellum and the newer cerebral cortex are separated by the limbic system. The brain structures in the limbic system are known especially important for controlling basic motivations, including emotions (Gazzaniga et al., 2010), while affect-related processing in the human brain is distributed across the brainstem, limbic, paralimbic, and neocortical regions. Even though some researchers consider affective processes as states of nonspecific autonomic arousal (e.g., Mandler, 1985), affective states take the form of patterns collaborated within and across the autonomic, endocrine, and motor systems (Derryberry and Tucker, 1992). Rosen and Levenson (2009) classified emotional functions into four categories: appraisal, reactivity, regulation, and understanding. Each of them has been addressed within the HF/HCI research domain.

The most well-examined emotional function is *appraisal*, which includes processing and interpreting external (e.g., visual, auditory, other perceptions, etc.) and internal (thoughts) stimuli. It is widely accepted that the amygdala is the most critical element of the appraisal system. The amygdala plays a role regarding responses to stimuli predicting potential harm (Bechara et al., 1995; LeDoux, 1992) or recognition of negative facial expressions (Adolphs and Tranel, 2004; Williams et al., 2005). The frontal lobes (medial and orbital portions) are also involved in emotional appraisal (Rolls, 2004; Sotres-Bayon et al., 2006).

Affect sometimes comes prior to, and directs, cognitive processes (Zajonc, 1984) and this is supported by the presence of two separate appraisal pathways. Information reaches the amygdala along two separate pathways (Armony and LeDoux, 1997). The first path is a "quick and dirty" system that processes sensory information nearly instantaneously. Sensory information travels quickly through the thalamus to the amygdala for priority processing. The second pathway is somewhat slower, but it leads to more deliberate and thorough evaluations. In this case, sensory material travels from the thalamus to the sensory cortex, where the information is scrutinized in greater depth before it is passed along to the amygdala. In brief, the fast system prepares people to quickly respond and the slower system confirms the threat. These two separate appraisal systems have been applied and conceptualized in the HF/HCI domain. For example, research has proposed that people apprehend reality and risk by two interactive, parallel processing systems: the experiential system and the analytic system (Epstein, 1994; Slovic et al., 2004). The experiential system is intuitive, fast, mostly automatic, and not very accessible to conscious awareness. In contrast, the analytic system is a deliberative, analytical system that functions by way of established rules of logic and evidence, such as probability calculation. This distinction is echoed as "primary" and "secondary" emotional processes (Damasio, 1994), or "experiential" and "reactive" cognition (Norman, 1993).

Reactivity includes psychological and physiological responses to emotional stimuli. This activity can be generated by subcortical regions (e.g., hypothalamic and brainstem structures and striatum) even without the modulation of higher cortical input (Panksepp, 1998). However, higher cortical regions (e.g., anterior cingulated) also play an important role in control of emotional reactivity. Again, the amygdala seems to organize behavioral and autonomic emotional reactions (Armony and LeDoux, 1997). Assistive technology research for people with autism has mainly addressed this reactivity issue—for example, affective prosody learning (Jeon and Rayan, 2011) or affective interaction with robots (Feil-Seifer and Mataric, 2008; Michaud and Theberge-Turmel, 2002). Emotional reactions are typically adjusted to match situational demands by a process, called emotion *regulation*. In terms of reappraisal (i.e., reexamining its meaning), studies have shown increased activation in dorsalateral and ventrolateral frontal structures and decreased activation in the hypothalamus and amygdala compared to unregulated emotion (Ochsner and Gross, 2005). In contrast, studies of emotion suppression have shown increased activation in the amygdala (Goldin et al., 2008). Regarding emotion regulation research in HF/HCI, see Bouchard et al. (2011) for learning and training in the army and Harris and Nass (2011), Hudlicka and McNeese (2002), Jeon (2012) for dynamic environments, such as driving.

Emotional *understanding* refers to our ability to recognize emotions in ourselves and in others, and is necessary for empathy. The amygdala is the most important structure in identifying emotion in both feeling own emotions and in others. Since cognitive processes also play an important role in empathy, parietal lobes and several parts of prefrontal cortex are included in this process (Decety and Jackson, 2004). Emotional understanding and empathy is one of the main topics in affective computing research (e.g., el Kaliouby, 2005).

Theories of Affective Effects

Psychological sciences have provided a number of affect theories and mechanisms to account for affective effects. However, applied researchers have tended to focus more on "how" (e.g., detection methods) to deal with emotions and affect, and they have picked a phenomenological approach to affect. Therefore, more theory-driven affect research with a balanced view of "what" (e.g., which emotion is important in a specific domain) and "why" (mechanisms) is required in the HF/HCI domain.

Affective Effects on Attention and Information Processing

With respect to the relationship between affect and attention, there have been numerous studies that show positive affect promotes a greater focus on global processing and negative affect promotes a greater focus on local processing. (Basso et al., 1996; Derryberry and Reed, 1998; Derryberry and

Tucker, 1994; Easterbrook, 1959; Fredrickson and Branigan, 2005; Gasper and Clore, 2002). Theorists have postulated several hypotheses to explain this tendency. A capacity explanation assumes that positive states activate a larger network of associations than negative states, thereby reducing the resources available for effortful processing (Mackie and Worth, 1989; Worth and Mackie, 1987), which means that positive affect leads to reduced cognitive processing. On the other hand, a motivational explanation assumes that people avoid expending effort on tasks that are not enjoyable to maintain their current positive states (Isen, 1987; Wegener et al., 1995). In the same line, the information provided by positive affect might signal that one's goal has already been achieved (Martin et al., 1993) or that further processing is unnecessary (Clore et al., 1994). However, there is no evidence that negative affect elicits more extensive processing than positive affect (Gasper and Clore, 2002). The affect-as-information approach can provide an alternative explanation (Schwarz and Clore, 1983, 1988, 1996) in which affective feelings are considered as consciously accessible information from ongoing, nonconscious appraisals. However, differences between positive and neutral states in global processing are not explained by this affect-as-information approach because people generally tend to be in positive moods and a global bias is considered normative. The broaden-and-build theory may be an alternative mechanism (Fredrickson, 1998, 2001), which predicts that the cognitive consequences of positive states are distinct from those evident in neutral states. According to the broaden-and-build theory: (1) positive emotions may broaden the scopes of attention, cognition, and action, widening the array of percepts, thoughts, and actions, and (2) people can build a variety of personal resources from positive emotional states, including physical, social, intellectual, and psychological resources. However, more longitudinal research is required to support the build hypothesis. The easing of inhibitory control might alter the quality of attention, resulting in a shift from a narrow focused state to a broader and more diffuse attentional focus. Positive states, by loosening the reins on inhibitory control, may result in a fundamental change in the breadth of attentional allocation to both external visual and internal representational spaces (Rowe et al., 2007). Nevertheless, there has been no underlying mechanism beyond those conceptual discussions, such as the neurological underpinnings of those effects.

Affective Effects on Judgment and Decision-Making

Considerable research has examined whether affect influences how people estimate the likelihood of future events (Johnson and Tversky, 1983) and how people actually select among different options. Research has generally shown mood-congruent effects in judgment and decision-making: participants in positive moods estimated positive events as more likely than participants in negative moods and vice versa. There are a couple of hypothetical mechanisms to explain the affective effects on judgment and decision-making. The availability (or accessibility) heuristic (MacLeod and Campbell, 1992; Tversky and Kahneman, 1973, 1974) denotes the process by which participants form estimates of likelihood based on how easily they can retrieve instances from memory. Similar to the availability heuristic, the affect-priming model also suggests that affect can indirectly inform judgments by facilitating access to related cognitive categories (Bower, 1981; Forgas, 1995, 2006; Isen, 1987). A number of studies supported this hypothesis based on mood-congruent memory facilitation (Derry and Kuiper, 1981; Greenberg and Beck, 1989). The affect-as-information hypothesis also accounts for the mood-congruent judgment. Clore and Huntsinger (2007) proposed that participants use the information delivered by affective states strategically during the judgment process. However, the affect-as-information model shows limited effects depending on the circumstances. For example, the mood influenced global judgments about overall life satisfaction (i.e., heuristic processing) but had no effect on judgments about specific life domains (i.e., direct access processing) (Schwarz et al., 1987).

Specific Affective Effects on Cognitive Processes

Despite general affective effects described previously, recent research has shown that a specific approach for each affect is needed beyond the valence dimension (positive vs. negative). Readers might notice that the affective effects in the attention level would not necessarily agree with the effects in the judgment or decision making level. For instance, in the attention level, positive affect seems to reduce the resources available for effortful processing, but in the judgment or decision making level, positive affect seems to result in positive estimates, which does not always correspond to the attention level case. Furthermore, research shows specified affective effects on each level. For attentional processing, anger may facilitate attentional circuits best suited for combat (Blanchard and Blanchard, 1988), fear may enable circuits best suited for evaluating danger and flight (Lang et al., 2000), disgust may be linked to circuits involved in expelling harmful bodily substances (Berridge, 2003), and joy and surprise may be linked to processing information in a global and fluent manner (Fredrickson, 2003). For judgment and decision-making, students with increased anxiety for a statistics exam showed an increase in risk perception for the exam but not for other tasks (Constans and Mathews, 1993). Even within the same valence, different affective states specifically influence judgments. A study found that happy participants made faster lexical decisions to happiness-related words, but not to other positive words (Niedenthal and Setterlund, 1994). Research also found that fear increased risk estimates, but anger reduced risk estimates (Lerner and Keltner, 2000). Sometimes, the different effects of negative affect can be explained by cognitive appraisal mechanisms (e.g.,

Ellsworth and Smith, 1988; Keltner et al., 1993). Ellsworth and Smith (1988) proposed eight appraisal dimensions that are important in differentiating emotional experience: pleasantness, anticipated effort, attentional activity, certainty, responsibility, control, legitimacy, and perceived obstacle. These cognitive appraisals are involved in particular affect because appraisals have important implications for the individual's well-being and the affect associated with those appraisals prepares the person to adaptively cope with those implications (Ellsworth and Smith, 1988). Subsequent studies have validated this cognitive appraisal mechanism with empirical data. To illustrate, Lerner and Keltner (2000) showed that fear is associated with uncertainty and situational control, but anger is associated with certainty and individual control. However, cognitive appraisals do not always work for every case. As discussed, affect often works prior to cognition or independently of cognitive processes.

All of these mechanisms and theories are good theoretical ingredients that affect-related HF/HCI researchers can apply to their own research and develop further. To develop a healthier theory, researchers need to identify a more elaborated common language, which has a clear relationship with existing domain constructs, including workload (Jeon et al., 2014), situation awareness (Jeon, 2012), automation (Lee, 2006), etc. By attempting to actively apply existing theories and to search for a robust new theory, this budding research field may be able to attain a suitable framework for potentially fruitful new avenues of research.

APPROACHES TO AFFECTIVE SCIENCES IN HF/HCI

Among many approaches to emotional and affective components in HF/HCI, four different approaches were selected in this handbook based on their pervasiveness and scholarly impact—Emotional design (Norman, 2004), Hedonomics (Hancock et al., 2005; Helander, 2002), Kansei Engineering (Nagamachi, 1995, 2002), and Affective Computing (Picard, 1997). Moreover, each of them has a different origin of area—design, ergonomics, mass production, and computing. Of course, these four might neither be mutually exclusive nor exhaustive. Nowadays, researchers in each area interchangeably use each other's methods.

Emotional Design

"Is usability alone sufficient?"

As he confessed in his book, *Emotional Design* (Norman, 2004), Don Norman did not take emotions into account when he wrote, *The Design of Everyday Things* (Norman, 1988), which is widely used in HCI classes over the world. To him, *Emotional Design* is a product of his "*re*cognition" about the importance of emotions and aesthetics in design, so that he can overcome a "usable but ugly" (pp. 8) situation, which designers would face when they were to follow Norman's prescriptions before emotional design. Therefore, from the first sight it seems that Norman betrays himself by returning to visual attractiveness and trashing usability that he has enthusiastically propagated. However, a closer look reveals that he has finally found the importance of balance between the different levels of design. He classified design into three different levels: visceral, behavioral, and reflective designs, which are equally important in design.

The visceral level wins in most cases. Simply to say, visceral design means that "attractive things work better" just as in nature. In fact, this has been supported by empirical research (e.g., Kätsyri et al., 2012). It is about symmetry and pretty appearance, and may not be about great art, but more enjoyable experience. The visceral design is dominated by physical features, such as look, feel, and sound. In other words, it is about immediate emotional impact, which traditional market research and visual designers have been involved with. When it comes to aesthetics theory, however, this is not equivalent to visual attractiveness. Given that the meaning of aesthetics has dramatically changed since Marcel Duchamp in the 20th century (i.e., from *visual attractiveness* to *shock*), the concept of digital aesthetics has not settled down yet (Tractinsky, 2004).

The behavioral design is all about "use," which HF and usability people (including Norman) have been actively involved with. Here, performance matters. Norman included four components of good behavioral design, such as function, understandability, usability, and physical feel. The first step is to understand how people will use a product, which is human-centered design. Norman's solution (or methodology) for this level is designers' "clever observations" (pp. 72) instead of having a conversation with users, such as focus groups, questionnaires, or surveys because these methods are poor at extracting their actual behaviors.

The reflective design is about "meaning" of a product or its use (e.g., message, culture, personal remembrances, or self-image, etc.), which seems compatible with contemporary user experience notions, such as value-sensitive design (Friedman et al., 2006). Norman acknowledges that beauty also comes from the reflective level (vs. attractiveness is mainly a visceral-level phenomenon). To him, product aesthetics is below the surface, coming from conscious reflection and experience, influenced by knowledge, learning, and culture. More importantly, the reflective level often determines an individual's overall impression of a product and influences the relationship between the user and the product. In conclusion, even though Norman has much room to be misunderstood, he does not solely focus on visual attractiveness of product appearance.

In (*Emotional Design*), Norman did not provide any specific methodologies that designers can pick up and apply to their emotional design. Instead, he provided many foods for thoughts, serving as an evangelist of emotional design. Further, he described his notions on affective computing and thus, he would embrace emotionally *adaptive* systems in his emotional design concept beyond premade products. That may also imply that product design for all (or general people, that is, nomothetic approach of generalization) can evolve into design for an individual (i.e., idiographic approach of specification) (Thomae, 1999). Given his influence, indeed, his work on emotional design has made researchers and practitioners look back on their design goals and processes and blew the age of *user experience*. It was also a very successful attempt to provide a general audience with a novel view about emotion and technology.

Practical recommendation:

- Balance between the three levels of design: visceral, behavioral, and reflective design.

Hedonomics

"How to design pleasurable products?"

At the similar time to Norman, but separately, Hancock (Hancock et al., 2005) and Helander (Helander, 2002; Helander and Tham, 2003) tried to make an agenda for "affective design" in the Human Factors and Ergonomics community. They coined the term, "hedonomics" in several venues and guest-edited a special issue of *Ergonomics* on Hedonomics and affective design. Hedonomics stems from the Greek "hedone" (i.e., pleasure) and "nomos" (i.e., laws, principles). Their direction is very clear in that they want to move from increasing performance and preventing pain, to promoting pleasure in addition. They contended that HF design has to move from identifying commonalities across all people to identifying an individual-based design, which is echoed in emotional design (Norman, 2004). That might imply design research should direct from a nomothetic approach to an idiographic approach.

Helander would embrace the design of products, human–computer interfaces, CSCW, and work tasks in the realm of hedonomics. Taking a Maslow's hierarchical needs model (1970) as an analogy, they created a hierarchy of ergonomic and hedonomic needs. From the bottom, three levels including safety, functionality, and usability are referred to as ergonomics and top two levels including pleasurable experience and individuation are referred to as hedonomics (Helander and Khalid, 2006). Helander and Tham (2003) prioritized the research issue in affective HF design as theory formation and affect measurement. Regarding theory, Helander proposed to start from activity theory (Leontjev, 1978) in that simple, operational, and intellectual activities are mapped onto affect, emotions, and sentiments. Measurement methods in hedonomics are delineated more in (Helander and Khalid, 2006) later on, and overlapped much with Kansei Engineering and Affective Computing as described in the next sections.

In terms of the practical application of hedonomic design, Jordan (1998) seems to serve a good role. In his paper, he simply adopted a semistructured interview technique to survey pleasure and displeasure in product use. He extracted pleasurable feelings: security, confidence, pride, excitement, satisfaction, entertainment, freedom, and nostalgia; and displeasurable feelings: aggression, feeling cheated, resignation, frustration, contempt, anxiety, and annoyance. He also analyzed when and from what users experience these feelings. In both pleasurable and displeasurable cases, "during use" was reported as the highest, compared to before or after use or other times, which supports the importance of Norman's behavioral design. Additionally, users experienced displeasure most with poor usability, whereas they experienced pleasure most with good features. He also showed that pleasure experiences can further influence users' behaviors. He cautiously suggested that users might use a product more because of the pleasure and it might serve as a benchmark (or negative benchmark) for a future purchase decision and be related to brand loyalty. One of the important points is that product pleasure can be independent of task pleasure, which implies design can make pleasurable products even associated with inherently disliked tasks. Finally, Jordan suggested developing a survey tool that can measure various feelings, such as pride or excitement in addition to simple user satisfaction.

Hedonomics seems to be a fresh attempt in Human Factors and Ergonomics society, and the innate evolution of the discipline. Especially, from the beginning it sought to build its theoretical foundation and practical methods, which is desirable. Based on this notion, designers might design a pleasurable product based on the analysis of pleasurable or displeasurable points. However, it might not promptly respond to users' dynamic situations unless the product contains dynamic adaptability, which is more discussed in Section "Affective Computing."

Practical recommendation:

- Analyze pleasurable and displeasurable experiences of the products.
- Take the whole life cycle of the product into account.

Kansei Engineering

"How to design a product that fits to its target image (kansei)?"

Kansei Engineering (Nagamachi, 2002) or Sensibility Ergonomics (Lee, 1999) is defined as translating the customer's kansei (in Japanese, consumer's feeling and image in mind) into the product design field including product mechanical functions. Kansei Engineering refers to a specific design methodology. In contrast to using general taxonomies for emotions (e.g., Ekman, 1992; Plutchik, 1994), they construct a domain-specific affect dimension in Kansei Engineering. The "surveyed" kansei are transferred to physical traits, and these are transformed to the design. As the word "surveyed" reveals, "kansei" is people's aggregated feeling or image (e.g., looks graceful or intelligent) about products rather than an individual's own emotional response (e.g., sad or angry). That research tradition emerged first from Eastern Asia in 1970s, but nowadays it is pervasive over the world. For example, researchers from 25 countries attended the KEER 2010 (Kansei Engineering and Emotion Research International Conference, 2010) in Paris. It can be applied to the designs of anything, such as costume, cosmetics, cars, home appliances, office/construction machines, etc. (Nagamachi, 1995).

Sometimes, the Kansei Engineering process can be done by an expert system where the kansei keywords are input and these words are recognized by the system database. Then, the keywords are matched to the image database and calculated by an inference engine to find the best-fit design, which is shown on the display. The Kansei Engineering system has a flow from the kansei (keywords) to the design domain, called "forward kansei engineering." The hybrid kansei engineering also has a backward flow from the design sketch with the kansei keywords.

Kansei Engineering researchers have built even a virtual space (e.g., VIrtual Kansei Engineering, "VIKE" or "HousMall") to fit to customers' kansei and customers can investigate the appropriateness of a design for their kansei (e.g., Matsubara and Nagamachi, 1996). This concept has already been popular in Japan and is expected to be further linked to the use of 3D printers, nowadays.

There have been plenty of examples of the application of those methods to product design (Jeon et al., 2008; Kleiss, 2008), facial feature extraction (Park et al., 2002), web design (Sun, 2002), sound design (Jeon et al., 2004; Lee et al., 2003), and graphical mobility (Lim and Han, 2002).

Kansei Engineering has developed its distinct and robust methodologies, which have yielded many successful design outputs in real settings. As a result, it is now pervasive over the world and has got many followers. However, it seems to focus on a static image (kansei) of the product (or object centered) based on average data, rather than designing a personalized (or subject-centered) product.

Practical recommendation:

- Find out optimal kansei of the product and design products to fit to kansei.

Affective Computing

"How to design systems that can detect users' affect and adapt to it on the fly?"

Rosalind Picard (1997) coined the term, "affective computing," which is defined as computing that relates to, arises from, or influences emotions. The initial goal of affective computing is to design a computer system that at least recognizes and expresses affect. Then, the ultimate goal is to design further so that the computer can have emotions and use them in making decisions. As she describes in her invited paper, "Affective computing: From laughter to IEEE" in the first issue of the journal, IEEE Affective Computing (Picard, 2010), at first, it was considered just as an absurd dream. Note that she did not propose that the computer could measure human's affective state directly, but rather proposed to measure observable functions of such states. She considered affect recognition as a dynamic pattern recognition problem and wanted to model an individual's or group of individuals' affective states or mixture of such states based on machine learning or neural network algorithms. When it comes to display of recognized affective states, she proposed an "affective symmetry." Just as we can see ourselves via a small image in video teleconferencing, computers that read emotions can also show us what they are reading. This interactive affect display notion is supported by the affect-as-information hypothesis as discussed—that is, even just awareness of one's affective state or the source of the state can make the person less influenced by those affective states (Tiedens and Linton, 2001).

Application domains of affective computing that Picard envisioned and developed include computer-assisted learning, perceptual information retrieval, arts and entertainment, and human health, across which affective computing researchers are currently working. Specifically, Picard and her affective computing group at MIT developed various affective wearable computers for people with autism. Her first step to aid people with autism was to sense a person's affective-cognitive state. For example, Expression Glasses discriminates facial expressions of interest or surprise from those of confusion or dissatisfactions, allowing students to communicate feedback anonymously to the teacher in real time, without having to shift attention from trying to understand the teacher (Scheirer et al., 1999). The galvactivator is a skin-conductance sensing glove that converts level of skin conductance to the brightness of a glowing LED (Picard and Scheirer, 2001). StartleCam is a wearable camera system that saves video information based on a physiological response, such as the skin conductance arousal response, tagging the data with information about whether or not it was exciting (Healey and Picard, 1998). Then, as a framework for machine perception and mental state recognition, her coworkers, el Kaliouby (2005) developed a computational model of mindreading. This framework combines bottom-up vision-based processing of the face with top-down predictions of mental state models. This effort attempted to integrate multiple modalities, such as face and posture, to infer affective states, such as the level of engagement (e.g., interest vs. boredom). The other approach was to develop technologies that enhance empathizing. For instance, the affective social quotient project was to develop assistive technologies for autism using physical input devices, namely four dolls (stuffed dwarfs), which appeared to be happy, angry, sad, or surprised (Blocher and Picard, 2002). The system plays short digital videos that embody one of the four emotions, and then encourages the child to choose the dwarf that went with the appropriate emotion.

Affective Computing and Kansei Engineering have a similar philosophical tenet; we can detect or predict our "internal emotional state" or "feelings about objects" by a mathematical or computational approach. However, while Kansei Engineering seeks to find out commonalities of kansei of target customers, Affective Computing seeks to find out and adapt to an individual's idiosyncratic emotional elements. Compared to other approaches, Affective Computing looks more dynamic in terms of interaction with artifacts depending on a user's context (Table 1.1).

TABLE 1.1 Comparisons of Four Approaches

	Emotional design	**Hedonomics**	**Kansei engineering**	**Affective computing**
Background domain	UI/product design/ usability	Ergonomics	Mass production (car manufacturer)	Interactive Computing
Goals	Balance of visceral, behavioral, and reflective design	Design of pleasurable products	Design of a product fitting to customers' kansei	Design of a system to detect a user's affect and adapt to it
Methods	Emphasize "observation," not specified at the outset → all of the related methods	Analysis of pleasurable and displeasurable experiences of product use → all of the related methods	Kansei engineering	Data fusion of multiple sensors and machine learning
Approaches	Nomothetic → idiographic	Nomothetic → idiographic	Nomothetic	Idiographic → nomothetic
Adaptability	Static → dynamic	Static	Static	Dynamic

Practical recommendation:

- Design a system that can detect a user's affective state and adaptively respond to that state.
- For better detection, fuse sensing data from multiple modalities.
- Develop a system that can empathize with a user.
- Design a system that can have emotions and use them in decision-making.

METHODOLOGIES

Affect Induction Techniques

To investigate the effects of affect on sociotechnical contexts, researchers often need to induce users' emotional states. In traditional psychological research, they have had participants look at photos (Lang et al., 1990) or watch film clips (e.g., Fredrickson and Branigan, 2005), read scenarios or stories (e.g., Johnson and Tversky, 1983; Raghunathan and Pham, 1999), listen to music (e.g., Jefferies et al., 2008; Rowe et al., 2007), and write down their past experience (e.g., Gasper and Clore, 2002). These various induction methods can also be applied to HF/HCI research. However, in psychological lab research, participants were usually required to fill out a short questionnaire or conduct a simple judgment task immediately after affect induction. For research that requires a complicated or long-lasting task, researchers should cautiously select and apply their induction method. The other consideration is whether researchers are interested in integral (relevant to task) affect or incidental (irrelevant to task) affect. For example, in driving research, as integral affect induction, researchers have embedded frustrating or angry events on the driving scenarios (e.g., Jeon, 2012; Lee, 2010). As incidental affect, they have participants read emotional paragraphs and write down their emotional experience (Jeon, 2012; Jeon et al., 2011; Jeon and Zhang, 2013).

HCI research requests a system to induce more dynamic emotions to users and to respond to those users' affective states on the fly. To illustrate, researchers have slightly distorted users' facial expression on the mirror to examine affective effects on their preference of clothes and decision-making in purchase (Yoshida et al., 2013). Another experiment (Nass et al., 2005) showed that when in-vehicle voice emotion matched the driver's emotional state (e.g., energetic to happy and subdued to upset), drivers had fewer accidents and attended more to the road (actual and perceived), and even spoke more with the car. However, given that friendly interfaces or anthropomorphism (Rogers et al., 2011) has not always been successful (e.g., Microsoft's "at home with bob" or "clippy"), more case studies are still required.

Affect Measure and Detection Techniques

The ability to dynamically detect users' affective states is crucial in predicting their behaviors and performance, and exerting adaptive interactions. During the past decade, there has been a proliferation of research on emotion/affect detection. For example, a recent survey on affect detection (Zeng et al., 2009) included 160 references even though it addressed only very recent research. Researchers have applied a variety of methods to measure and detect users' affective states in each of their domains. Nowadays, they borrow and combine each other's methods or even create new ones (see more methods, e.g., Boehner et al., 2007; Helander and Khalid, 2006; Hudlicka, 2003; Isbister et al., 2006; Mauss and Robinson, 2009): self-report (for a dimensional framework), autonomic measures (e.g., electrodermal gland and cardiovascular for a dimensional framework), startle response magnitude (e.g., eye blink, EMG, etc. for valence), brain states (e.g., EEG, fMRI, PET for approach-avoidance related states), and behavior measure (e.g., vocal characteristics for arousal, facial behavior for valence, and whole-body behavior for discrete emotions), diagnostic tasks and expert observer evaluation (or knowledge-based assessment). In addition, HF/HCI researchers have employed novel methods, such as using body movements (Crane and Gross, 2007), mouse and keyboard inputs (e.g., Zimmermann et al., 2003), steering wheel grip intensity (e.g., Oehl et al., 2007), etc.

Each technique has its own advantages and disadvantages (Table 1.2). Facial detection is not intrusive and highly accurate in discrete emotional detection. However, it has some practical issues, such as light, head movement, angle, etc. in real environments. Speech detection is also nonintrusive and gets closer to natural language with higher accuracy rate. Nonetheless, music, conversation, or phone calls might interfere with voice commands and speech detection. If users are alone, they might not say anything or are not likely to initiate conversation with the system.

Researchers have tried to use more and more physiological sensors. However, it is certainly intrusive. The mapping between sensing data and a specific emotional state is not yet robust. There are also some practical issues observed in actual research, such as hairy skin, different size of body part, sweating in the summer.

A considerable number of affect detection applications have used combinations of these methods in a range of domains, including learning, agents, robots, games, and engaging people with autism (e.g., see *International Journal of Human-Computer Studie*, 59 special issue, Hudlicka, 2003). For example, Hudlicka and McNeese (2002) integrated a variety of user assessment methods involving knowledge-based assessments, self-reports, diagnostic tasks, and physiological sensing to detect a pilot's affective state. For a more accurate affect assessment, they also included personality traits and personal history (e.g., training, recent events) as

TABLE 1.2 Comparisons of Affect Detection Methods

	Intrusiveness	Accuracy	Robustness	Continuity (time resolution)	Cost
Self-report	No	Very good	No	No (pre–post test)	No
Heart rate/respiration rate	Intrusive	Good	Robust	Yes	Low
EEG/fMRI	Very intrusive	Not good	Not robust	Yes	Very high
Facial expression	No	Very good	Very robust	Yes	Develop-mental cost
Speech recognition	No	Very good	Very robust	Yes	Develop-mental cost
Body posture/motion	No	Not good	Not robust	Yes	Develop-mental cost

variables. Recently, researchers combine multiple sensor data and determine affective states using machine-learning algorithms (e.g., Gaponova and Benke, 2013).

ISSUES AND FUTURE WORKS

Despite active research in emotions and affect in HF/HCI, a number of theoretical and practical issues and research questions still remain unanswered.

First of all, constructing a robust, generic affect research model is required. In Cognitive Sciences, for example, there are several standardized computational models (e.g., Human Processor Model, ACT-R, EPIC, SOAR, etc.). In contrast, there are not many standardized models in Affective Sciences. With respect to the affect dimension, creating domain-specific affect taxonomy just as in Kansei Engineering seems to be promising in terms of its accuracy. Then, it needs to be answered whether researchers should construct their own affect taxonomy for every research. Meanwhile, it is also required to form the clear relationship between affect and other core concepts in HF/HCI, such as workload, situation awareness, automation, etc. "How affect is involved in such processes" or "whether affect is an independent construct of each of them" should be addressed based on empirical research. Emerging areas, such as embodied cognition or embodied interaction is closely related to Affective Sciences. Thus, understanding those areas together would also be helpful to understand Affective Sciences better and expand its boundary.

One of the primary issues related to the practical application of Affective Sciences include induction methodologies. For example, "what is the most appropriate affect induction for a certain research domain? and why?" Or "is it even ethically acceptable at all?" Another question involves, "How can we guarantee that the induced affective states in the research setting are equivalent to affective states in real world?" Situations are similar to affect measurement or detection. Some people are not very expressive, but reserved by nature. Other people still show aversion about machine's control (they might not want an artificial system to detect their emotional states and take control of them). Another issue includes security and privacy about emotional data. In the age of Internet of Things, people are likely to be concerned about data storage and further utilization even though they are told that the data are not going to be used for any other purposes. Implementation of the display of detected results is also a big design space. Furthermore, HF/HCI researchers might want to design a system to help people regulate their affective states or mitigate affective effects on their task. "How can we improve performance by enhancing emotional experience?" would be another core question in HF/HCI. See (Lottridge et al., 2011) for more related discussions.

CONCLUSIONS

This chapter described Affective Sciences in Human Factors and Human–Computer Interaction, including taxonomy, theories, historical approaches and repertoire, and various methods, followed by discussions on theoretical and practical issues. To advance the field of Affective Sciences further, first of all, a common ground is required in terms of terminology and standardized methodologies. As shown through the chapter, there is no overarching mechanism or theory that can account for complicated affective phenomena. However, there are considerable raw ingredients that HF/HCI researchers can borrow and make their own mechanisms on top of them. Historically, there have been several approaches to Affective Sciences in HF/HCI. Their background, goals, and methods are different. However, all of them clearly show that emotions and affect are not peripheral, but one of the necessary considerations in HF/HCI design and try to integrate all the methodologies for successful outcomes. This handbook is expected to begin to close the existing gap between the traditional HF/HCI research and the emerging field of Affective Sciences research, and to contribute to each. The balanced viewpoint and a more systematic approach may provide a legitimate framework of HF/HCI research.

References

Adolphs, R., Tranel, D., 2004. Impaired judgments of sadness but not happiness following bilateral amygdala damage. J. Cogn. Neurosci. 16 (3), 453–462.
Armony, J.L., LeDoux, J.E., 1997. How the brain processes emotional information. Ann. NY Acad. Sci. 821, 259–270.
Basso, M.R., Schefft, B.K., Ris, M.D., Dember, W.N., 1996. Mood and global-local visual processing. J. Intl. Neuropsychol. Soc. 2, 249–255.
Bechara, A., Tranel, D., Damasio, H., Adolphs, R., Rockland, C., Damasio, A.R., 1995. Double dissociation of conditioning and declarative knowledge relative to the amygdala and hippocampus in humans. Science 269 (5227), 1115–1118.
Berridge, K., 2003. Pleasures of the brain. Brain Cogn. 52, 106–128.
Blanchard, D.C., Blanchard, R.J., 1988. Ethoexperimental approaches to the biology of emotion. Annu. Rev. Psychol. 39, 43–68.
Blocher, K., Picard, R.W., 2002. Affective social quest: emotion recognition therapy for autistic children. In: Dautenhahn, K., Bond, A., Canamero, L., Edmonds, B. (Eds.), Socially Intelligent Agents—Creating Relationships With Computers and Robots. Kluwer Academic Publishers, The Netherlands, (Chapter 16).
Bodenhausen, G.V., 1993. Emotion, arousal, and stereotypic judgment: a heuristic model of affect and sterotyping. In: Mackie, D., Hamilton, D. (Eds.), Affect, Cognition and Stereotyping: Interactive Processes in Intergroup Perception. Academic Press, CA, pp. 13–37.
Bodenhausen, G.V., Sheppard, L.A., Kramer, G.P., 1994. Negative affect and social judgment: the differential impact of anger and sadness. Eur. J. Soc. Psychol. 24, 45–62.
Boehner, K., DePaula, R., Dourish, P., Sengers, P., 2007. How emotion is made and measured. Int. J. Hum. Comput. Stud. 65 (4), 275–291.
Bouchard, S., Guitard, T., Bernier, F., Robillard, G., 2011. Virtual reality and the training of military personnel to cope with acute stressors. In: Brahnam, S., Jain, L.C. (Eds.), Advanced Computational Intelligence Paradigms in Healthcare 6, SCI 337. Springer-Verlag, Berlin Heidelberg, pp. 109–128, (Chapter 6).
Bower, G.H., 1981. Mood and memory. Am. Psychol. 36 (2), 129–148.
Calvo, R.A., D'Mello, S., Gratch, J., Kappas, A. (Eds.), 2014. The Oxford Handbook of Affective Computing. Oxford University Press, USA, New York, NY.
Cannon, W.B., 1927. The James-Lange theory of emotions: a critical examination and an alternative theory. Am. J. Psychol. 39, 106–124.
Clore, G.L., Huntsinger, J.R., 2007. How emotions inform judgment and regulate thought. Trends Cogn. Sci. 11, 393–399.
Clore, G.L., Schwarz, N., Conway, M., 1994. Affective causes and consequences of social information processing. In: Wyer, R.S., Srull, T.K. (Eds.), Handbook of Social Cognition. second ed. Erlbaum, Hillsdale, NJ, pp. 323–341.
Constans, J.I., Mathews, A.M., 1993. Mood and the subjective risk of future events. Cogn. Emot. 7, 545–560.
Crane, E., Gross, M., 2007. Motion capture and emotion: affect detection in whole body movement. In: Paiva, A., Prada, R., Picard, R.W. (Eds.), ACII 2007, LNCS 4738. Springer-Verlag, Berlin Heidelberg, pp. 95–101.
Czaja, S.J., Nair, S.N., 2006. Human factors engineering and systems design. In: Salvendy, G. (Ed.), Handbook of Human Factors and Ergonomics. John Wiley & Sons, NJ, pp. 32–49.
Damasio, A.R., 1994. Descartes' Error. Avon Books, New York.
Damasio, A.R., 1999. The Feeling of What Happens: Body and Emotion in the Making of Consciousness. Harcourt, New York.
Damasio, A.R., 2001. Fundamental feelings. Nature 413 (6858), 781.
Davis, J.I., Markman, A.B., 2012. Embodied cognition as a practical paradigm. Special Issue Top. Cogn. Sci. 4, 685–793.

Decety, J., Jackson, P.L., 2004. The functional architecture of human empathy. Behav. Cogn. Neurosci. Rev. 3 (2), 71–100.
Derry, P.A., Kuiper, N.A., 1981. Schematic processing and self-reference in clinical depression. J. Abnorm. Psychol. 90, 286–297.
Derryberry, D., Reed, M.A., 1998. Anxiety and attentional focusing: trait, state and hemispheric influences. Pers. Individ. Dif. 25, 745–761.
Derryberry, D., Tucker, D.M., 1992. Neural mechanisms of emotion. J. Consult. Clin. Psychol. 60 (3), 329–338.
Derryberry, D., Tucker, D.M., 1994. Motivating the focus of attention. In: Niedenthal, P.M., Kitayama, S. (Eds.), The Heart's Eye: Emotional Influences in Perception and Attention. Academic Press, San Diego, CA, pp. 167–196.
Easterbrook, J.A., 1959. The effect of emotion on cue utilization and the organization of behavior. Psychol. Rev. 66, 183–201.
Ekman, P., 1992. An argument for basic emotions. Cogn. Emot. 6, 169–200.
el Kaliouby, R., 2005. Mind-Reading Machines: Automated Inference of Complex Mental States. Computer Laboratory, University of Cambridge.
Ellsworth, P.C., Smith, C.A., 1988. From appraisal to emotion: differences among unpleasant feelings. Motiv. Emot. 12 (3), 271–302.
Epstein, S., 1994. Integration of the cognitive and the psychodynamic unconscious. Am. Psychol. 49, 709–724.
Feil-Seifer, D., Mataric, M., 2008. Robot-assisted therapy for children with autism spectrum disorders. IDC Proceedings—Workshop on Special Needs. June 11–13, Chicago, IL.
Fiedler, K., 1988. Emotional mood, cognitive style, and behaviour regulation. In: Fiedler, K., Forgas, J. (Eds.), Affect, Cognition and Social Behavior. Hogrefe, Gottingen, Germany, pp. 100–119.
Fishwick, P.A. (Ed.), 2006. Aesthetic Computing. MIT Press, MA.
Forgas, J.P., 1995. Mood and judgment: the affect infusion model (AIM). Psychol. Bull. 117 (1), 39–66.
Forgas, J.P., 2002. Feeling and doing: affective influences on interpersonal behavior. Pyschol. Inq. 13 (1), 1–28.
Forgas, J.P. (Ed.), 2006. Affect in Social Thinking and Behavior. Psychology Press, New York.
Fredrickson, B.L., 1998. What good are positive emotions? Rev. Gen. Psychol. 2, 300–319.
Fredrickson, B.L., 2001. The role of positive emotions in positive psychology: the broaden-and-build theory of positive emotions. Am. Psychol. 56, 218–226.
Fredrickson, B.L., 2003. The value of positive emotions. Am. Sci. 91, 330–335.
Fredrickson, B.L., Branigan, C., 2005. Positive emotions broaden the scope of attention and thought-action repertoires. Cogn. Emot. 19 (3), 313–332.
Friedman, B., Kahn, P.H.J., Borning, A., 2006. Value sensitive design and information systems. In: Galletta, D., Zhang, P. (Eds.), Human-Computer Interaction in Management Information Systems: Applications. M. E. Sharpe, Armonk, NY, pp. 348–372, (Chapter 16).
Gaponova, I., Benke, V., 2013. Robust emotion assessment from physiological data. Proceedings of the International Conference on Automotive User Interfaces and Vehicular Applications (AutomotiveUI'13). Eindhoven, The Netherlands.
Gasper, K., Clore, G.L., 2002. Attending to the big picture: mood and global versus local processing of visual information. Psychol. Sci. 13 (1), 34–40.
Gazzaniga, M., Heatherton, T., Halpern, D., 2010. Psychological Science. W.W. Norton & Company, New York.
Goldin, P.R., McRae, K., Ramel, W., Gross, J.J., 2008. The neutral bases of emotion regulation: reappraisal and suppression of negative emotion. Biol. Psychiatry 63 (6), 577–586.
Greenberg, M.S., Beck, A.T., 1989. Depression versus anxiety: a test of the content-specificity hypothesis. J. Abnorm. Psychol. 98, 9–13.
Hancock, P., Pepe, A.A., Murphy, L.L., 2005. Hedonomics: the power of positive and pleasurable ergonomics. Ergon. Des. 13 (1), 8–14.

Harris, H., Nass, K., 2011. Emotion regulation for frustrating driving contexts. Proceedings of the ACM SIGCHI Conference on Human Factors in Computing Systems (CHI 2011). Vancouver, BC, Canada.

Healey, J., Picard, R.W., 1998. StartleCam: a cybernetic wearable camera. Proceedings of the International Symposium on Wearable Computers. October 19–20, Pittsburgh, pp. 42–29.

Helander, M.G., 2002. Hedonomics—affective human factors design. Proceedings of the Human Factors and Ergonomics Society 46th Annual Meeting. pp. 978–982.

Helander, M.G., Khalid, H.M., 2006. Affective and pleasurable design. In: Salvendy, G. (Ed.), Handbook of Human Factors and Ergonomics. John Wiley & Sons, Inc, NJ.

Helander, M.G., Tham, M.P., 2003. Hedonomics: affective human factors design. Ergonomics 46 (13-14), 1269–1272.

Hilgard, E.R., 1980. The trilogy of mind: cognition, affection, and conation. J. Hist. Behav. Sci. 16, 107–117.

Hudlicka, E., 2003. To feel or not to feel: the role of affect in human-computer interaction. Int. J Hum. Comput. Stud. 59, 1–32.

Hudlicka, E., McNeese, M.D., 2002. Assessment of user affective and belief states for interface adaptation: application to an air force pilot task. User Model. User Adapt. Interact. 12, 1–47.

Isbister, K., Höök, K., Sharp, M., Laaksolahti, J., 2006. The sensual evaluation instrument: developing an affective evaluation tool. In: Grinter, R., Rodden, T., Aoki, P., Cutrell, E., Jeffries, R., Olson, G., (Eds.), Proceedings of the SIGCHI Conference on Human Factors in Computing Systems. Montreéal, Queébec, Canada, April 22–27, 2006, ACM, New York, pp. 1163–1172.

Isen, A.M., 1987. Positive affect, cognitive process, and social behavior. In: Berkowitz, L. (Ed.), Advances in Experimental Social Psychology. Academic press, San Diego, pp. 203–253.

Izard, C.E., 1993. Four systems for emotion activation: cognitive and noncognitive processes. Psychol. Rev. 100 (1), 68–90.

James, W., 1884. What is an emotion? Mind 9, 188–205.

Jefferies, L.N., Smilek, D., Eich, E., Enns, J.T., 2008. Emotional valence and arousal interact in attentional control. Psychol. Sci. 19 (3), 290–295.

Jeon, M., 2012. Effects of affective States on Driver Situation Awareness and Adaptive Mitigation Interfaces: Focused on Anger. Doctoral Dissertation. Georgia Institute of Technology, Atlanta, GA.

Jeon, M., Heo, U., Ahn, J.H., Kim, J., 2008. Emotional palette: affective user experience elements for product design according to user segmentation. Proceedings of the 6th International Conference of Cognitive Science. Seoul, Korea.

Jeon, M., Lee, J.H., Kim, Y.E., Han, K.H., 2004. Analysis of musical features and affective words for affection-based music search system. Proceedings of Korean Conference on Cognitive Science. Seoul, Korea.

Jeon, M., Rayan, I.A., 2011. The effect of physical embodiment of an animal robot on affective prosody recognition. Proceedings of the 14th International Conference on Human-Computer Interaction. Orlando, FL.

Jeon, M., Walker, B.N., Yim, J.-B., 2014. Effects of specific emotions on subjective judgment, driving performance, and perceived workload. Transp. Res. Part F Traffic Psychol. Behav. 24, 197–209.

Jeon, M., Yim, J., Walker, B.N., 2011. An angry driver is not the same as a fearful driver: different effects of specific negative emotions on risk perception, driving performance, and workload. Proceedings of the Third International Conference on Automotive User Interfaces and Interactive Vehicular Applications (AutomotiveUI11). Salzburg, Austria.

Jeon, M., Zhang, W., 2013. Sadder but wiser? Effects of negative emotions on risk perception, driving performance, and perceived workload. Proceedings of the Human Factors and Ergonomics Society International Annual Meeting (HFES2013), vol. 57, No 1. September 30–October 4, San Diego, CA, pp. 1849–1853.

Johnson, E.J., Tversky, A., 1983. Affect, generalization, and the perception of risk. J. Pers. Soc. Psychol. 45, 20–31.
Jordan, P.W., 1998. Human factors for pleasure in product use. Appl. Ergon. 29 (1), 25–33.
Kätsyri, J., Ravaja, N., Salminen, M., 2012. Aesthetic images modulate emotional responses to reading news messages on a small screen: a psychophysiological investigation. Int. J. Hum. Comput. Stud. 70, 72–87.
Keltner, D., Ellsworth, P.C., Edwards, K., 1993. Beyond simple pessimism: effects of sadness and anger on social perception. J. Pers. Soc. Psychol. 64, 740–752.
Kleiss, J.A., 2008. Characterizing and differentiating the semantic qualities of auditory tones for products. Proceedings of the Human Factors and Ergonomic Society 52nd Annual Meeting. NY, USA.
Lang, P.J., Bradley, M.M., Cuthbert, B.N., 1990. Emotion, attention, and the startle reflex. Psychol. Rev. 97 (3), 377–395.
Lang, P.J., Davis, M., Ohman, A., 2000. Fear and anxiety: animal models and human cognitive psychophysiology. J. Affect. Disord. 61, 137–159.
LeDoux, J., 1996. The Emotioinal Brain, The Mysterious Underpinnings of Emotional Life. Simon & Schuster, New York.
LeDoux, J.E., 1992. Brain mechanisms of emotion and emotional learning. Curr. Opin. Neurobiol. 2 (2), 191–197.
Lee, J.D., 2006. Affect, attention, and automation. In: Kramer, A., Wiegmann, D., Kirlik, A. (Eds.), Attention: From Theory to Practice. Oxford University Press, New York.
Lee, J.H., Jeon, M., Han, K.H., 2003. The sound sensibility dimension for auditory displays. Proceedings of the 2003 International Conference on Cognitive Science. Sidney, Australia.
Lee, K., 1999. Sensibility Ergonomics and human sensibility. Proceedings of the International Workshop on Harmonized Technology With Human Life. Takarazuka, Japan.
Lee, Y., 2010. Measuring drivers' frustration in a driving simulator. Proceedings of the Human Factors and Ergonomics Society 54th Annual Meeting. CA, USA.
Leontjev, A.N., 1978. Activity, Consciousness and Personality. Prentice-Hall, London.
Lerner, J.S., Keltner, D., 2000. Beyond valence: toward a model of emotion-specific influences on judgment and choice. Cogn. Emot. 14 (4), 473–493.
Lim, E., Han, K.H., 2002. The affective effect of mobility. Proceedings of the Korean Conference on Cognitive Science. Seoul, Korea.
Lottridge, D., Chignell, M., Jovicic, A., 2011. Affective interaction: understanding, evaluating, and designing for human emotion. DeLucia, P.R. (Ed.), Reviews of Human Factors and Ergonomics, vol. 7, pp. 197–237, (Chapter 5).
Mackie, D.M., Worth, L.T., 1989. Processing deficits and the mediation of positive affect in persuasion. J. Pers. Soc. Psychol. 57, 27–40.
MacLeod, C., Campbell, L., 1992. Memory accessibility and probability judgments: an experimental evaluation of the availability heuristics. J. Pers. Soc. Psychol. 63 (6), 890–902.
Mandler, G., 1985. Mind and Body: Psychology of Emotion and Stress. Norton, New York.
Marshall, P., Antle, A., van den Hoven, E., Rogers, Y., 2013. Special issue on the theory and practice of embodied interaction in HCI and interaction design. ACM Trans. Comput. Hum. Interact. (TOCHI) 2 (1), 20, (Article 1-8).
Martin, L.L., Ward, D.W., Achee, J.W., Wyer, R.S., 1993. Mood as input: people have to interpret the motivational implications of their moods. J. Pers. Soc. Psychol. 64, 317–326.
Maslow, A.H., 1970. Motivatoin and Personality, second ed. Harper and Row, New York.
Matsubara, Y., Nagamachi, M., 1996. Kansei virtual reality technology and evaluation on kitchen design. Manufacturing Agility Hybrid Automation I. pp. 81-84.
Mauss, I.B., Robinson, M.D., 2009. Measures of emotion: a review. Cogn. Emot. 23 (2), 209–237.
Mayer, J.D., 1986. How mood influences cognition. Sharkey, N.E. (Ed.), Advances in Cognitive Science, vol. 1, Ellis Horwood, Chicester, England, pp. 290–314.

Michaud, F., Theberge-Turmel, C., 2002. Mobile robotic toys and autism. In: Dautenhahn, K., Bond, A.H., Canamero, L., Edmonds, B. (Eds.), Socially Intelligent Agents—Creating Relationships With Computers and Robots. Springer, pp. 125–132.
Nagamachi, M., 1995. Kansei engineering: a new ergonomic consumer-oriented technology for product development. Int. J. Ind. Ergon. 15 (1), 3–11.
Nagamachi, M., 2002. Kansei Engineering as a powerful consumer-oriented technology for product development. Appl. Ergon. 33, 289–294.
Nass, C., Jonsson, I.-M., Harris, H., Reaves, B., Endo, J., Brave, S., Takayama, L., 2005. Improving automotive safety by pairing driver emotion and car voice emotion. Proceedings of the SIGCHI Conference on Human Factors in Computing Systems (CHI05). Portland, Oregon, USA.
Niedenthal, P.M., 1990. Implicit perception of affective information. J. Exp. Soc. Psychol. 26, 505–527.
Niedenthal, P.M., Setterlund, M.B., 1994. Emotion congruence in perception. Pers. Soc. Psychol. Bull. 20, 401–411.
Norman, D.A., 1988. The Design of Everyday Things. Basic Books, New York.
Norman, D.A., 1993. Things That Make Us Smart. Addison-Wesley Publishing Company, Boston, MA.
Norman, D.A., 2004. Emotional Design. Basic Books, New York.
Ochsner, K.N., Gross, J.J., 2005. The cognitive control of emotion. Trends Cogn. Sci. 9 (5), 242–249.
Oehl, M., Siebert, F.W., Tews, T.-K., Höger, R., Pfister, H.-R., 2007. Improving human-machine interaction—a noninvasive approach to detect emotions in car drivers. In: Jacko, J.A. (Ed.), Human-Computer Interaction, Part III, HCII 2011, LNCS 6763. Springer-Verlag, Berlin Heidelberg, pp. 577–585.
Panksepp, J., 1998. Affective Neuroscience. Oxford, New York.
Park, S., Han, J., Chung, C., 2002. The analysis of the facial features and affective keywords. Korean J. Cogn. Sci. 13 (3), 1–10.
Picard, R., 1997. Affective Computing. MIT Press, Cambridge.
Picard, R., 2010. Affective computing: from laughter to IEEE. IEEE Trans. Affect. Comput. 1 (1), 11–17.
Picard, R.W., Scheirer, J., 2001. The galvactivator: a glove that senses and communicates skin conductivity. Paper Presented at the International Conference on Human-Computer Interaction, August. New Orleans, pp. 1538–1542.
Plutchik, R., 1994. The Psychology and Biology of Emotion. Harper Collins, NY.
Raghunathan, R., Pham, M.T., 1999. All negative moods are not equal: motivational influences of anxiety and sadness on decision making. Organ. Behav. Hum. Decis. Process. 79, 56–77.
Rogers, Y., Shartp, H., Preece, J., 2011. Interaction Design: Beyond Human-Computer Interaction. John Wiley, Chichester.
Rolls, E.T., 2004. The functions of the orbitofrontal cortex. Brain Cogn. 55 (1), 11–29.
Rosen, H.J., Levenson, R.W., 2009. The emotional brain: combining insights from patients and basic science. Neurocase 15 (3), 173–181.
Rowe, G., Hirsh, J.B., Anderson, A.K., 2007. Positive affect increases the breadth of attentional selection. Proc. Natl. Acad. Sci. USA 104 (1), 383–388.
Russell, J.A., 1980. A circumplex model of affect. J. Pers. Soc. Psychol. 39, 1161–1178.
Salovey, P., Mayer, J.D., 1990. Emotional intelligence. Imagin. Cogn. Pers. 9, 185–211.
Schacter, S., Singer, J.E., 1962. Cognitive, social, and psychological determinants of emotional state. Physiol. Rev. 69, 379–399.
Scheirer, J., Fernandez, R., Picard, R.W., 1999. Expression glasses: a wearable device for facial expression recognition. Proceedings of the Conference on Human Factors in Computing Systems (CHI'99). May 15–20. ACM, Pittsburgh, PA, New York, pp. 262–263.

Schwarz, N., Clore, G.L., 1983. Mood, misattribution, and judgments of well-being: informative and directive functions of affective states. J. Pers. Soc. Psychol. 45, 513–523.

Schwarz, N., Clore, G.L. (Eds.), 1988. How Do I Feel About It? The Informative Function of Affective States. C.J. Hogrefe, Toronto, Ontario.

Schwarz, N., Clore, G.L., 1996. Feelings and phenomenal experience. In: Higgins, E.T., Kruglanski, A. (Eds.), Social Psychology: A Handbook of Basic Principles. Guilford Press, New York, pp. 433–465.

Schwarz, N., Strack, F., Kommer, D., Wagner, D., 1987. Soccer, rooms and the quality of your life: mood effects on judgments of satisfaction with life in general and with specific life domains. Eur. J. Soc. Psychol. 17, 69–79.

Slovic, P., Finucane, M., Peters, E., MacGregor, D.G., 2004. Risk as analysis and risk as feelings: some thoughts about affect, reason, risk, and rationality. Risk Anal. 24 (2), 1–12.

Sotres-Bayon, F., Cain, C.K., LeDoux, J.E., 2006. Brain mechanisms of fear extinction: historical perspectives on the contribution of prefrontal cortex. Biol. Psychiatry 60 (4), 329–336.

Strack, F., Martin, L.L., Stepper, S., 1988. Inhibiting and facilitating conditions of the human smile: a nonobtrusive test of the facial feedback hypothesis. J. Pers. Soc. Psychol. 54, 768–777.

Sun, J., 2002. The affective effect of color and layout in web design. Masters Thesis, Yonsei University, Seoul, Korea.

Thomae, H., 1999. The nomothetic-idiographic issue: some roots and recent trends. Int. J. Group Tensions 28 (1), 187–215.

Tiedens, L.Z., Linton, S., 2001. Judgment under emotional certainty and uncertainty: the effects of specific emotions on information processing. J. Pers. Soc. Psychol. 81, 973–988.

Tractinsky, N., 2004. Toward the study of aesthetics in information technology. Proceedings of the 25th International Conference on Information Systems.

Tversky, A., Kahneman, D., 1973. Availability: a heuristic for judging frequency and probability. Cogn. Psychol. 5, 207–232.

Tversky, A., Kahneman, D., 1974. Judgment under uncertainty: heuristics and biases. Science 185, 1124–1131.

Wegener, D.T., Petty, R.E., Smith, S.M., 1995. Positive mood can increase or decrease message scrutiny: the hedonic contingency view of mood and message processing. J. Pers. Soc. Psychol. 69, 5–15.

Williams, M.A., McGlone, F., Abbott, D., Mattingley, J.B., 2005. Differential amygdala responses to happy and fearful facial expressions depend on selective attention. Neuroimage 24 (2), 417–425.

Worth, L.T., Mackie, D.M., 1987. Cognitive mediation of positive affect in persuasion. Soc. Cogn. 5, 76–94.

Yoshida, S., Tanikawa, T., Sakurai, S., Hirose, M., Narumi, T., 2013. Manipulation of an emotional experience by real-time deformed facial feedback. AH'13 Proceedings of the 4th Augmented Human International Conference. pp. 35–42.

Zajonc, R., 1984. On the primacy of affect. Am. Psychol. 39, 117–124.

Zeng, Z., Pantic, M., Roisman, G.I., Huang, T.S., 2009. A survey of affective recognition methods: audio, visual, and spontaneous expressions. IEEE Trans. Pattern Anal. Mach. Intell. 31 (1), 39–58.

Zimmermann, P., Guttormsen, S., Danuser, B., Gomez, P., 2003. Affective computing—a rationale for measuring mood with mouse and keyboard. Int. J. Occup. Saf. Ergo. 9 (4), 539–551.

CHAPTER

2

Neural Mechanisms of Emotions and Affect

Jacob Aday, Will Rizer, Joshua M. Carlson
Northern Michigan University, Marquette, MI, United States

INTRODUCTION

Imagine that you and your friend have decided to skydive for the first time. You are on the plane as it ascends to altitude—13,000 ft. above the ground—and now you are starting to wonder what you have gotten yourself into! What if the shoot doesn't open? What if there is a rip in the shoot? What if the straps are not working appropriately? Your hands are clammy, your heart is racing, your mouth is dry, you have butterflies in your stomach, and you are feeling very anxious. Before you know it, you are falling. The air rushes past your face. It is incredibly loud. Then suddenly, your attention is drawn to your friend falling beside you. They are smiling and enthusiastically screaming with joy. You begin to realize that everything is going to be ok—it is safe—and now you too are feeling a euphoric rush of excitement and joy! As the shoot opens, and you slowly descend, you start to appreciate the amazing view—it is one that you will never forget. Once you land your heart starts to slow down, but you can't stop smiling and sharing your experience with your friend and others who made the jump. Eventually, the euphoria gives way to a pleasant state of calmness.

We would probably all agree that what was just described can be considered an emotional event or experience. However, defining what exactly an emotion is can be tricky. Emotion theorists debate about what types of events are considered emotions. For the purposes of this chapter, we will define *emotions* as internal states elicited by events or stimuli that significantly impact an individual's needs and initiate behaviors to fulfill these needs (Schirmer, 2014).

Emotions and Affect in Human Factors and Human–Computer Interaction. http://dx.doi.org/10.1016/B978-0-12-801851-4.00002-1
Copyright © 2017 Elsevier Inc. All rights reserved.

Emotional events, such as the skydiving example described previously, are multidimensional phenomena, which include the following components: (1) appraisal of the event for emotional significance, (2) response to the emotional stimulus (e.g., autonomic and muscular activity), and (3) a subjective experience or feeling (James, 1884). Assessment of your safety prior to jumping from a plane would be an example of a type of emotional appraisal. However, it should be noted that appraisal processes are not necessarily assumed to require conscious deliberation and many evaluations of emotional significance occur below the level of conscious awareness. Regardless the level of processing, emotional appraisals elicit functional responses that prepare or aid the organism for the stimulus/event. As in the case of falling to your imminent death, emotional arousal generally elicits activation of the sympathetic branch of the autonomic nervous system including accelerated heart rate, respiration, perspiration, and adrenaline release among other responses (Ekman, 1992; Lang et al., 1993). Sympathetic activity readies bodily resources to prepare the individual for action. According to James (1884) and others, what we experience as emotional feelings are in fact the brain's perception of the body as well as the representation of emotion-elicited changes to the body (e.g., autonomic activity, muscle tension, and facial expressions). That is, it is because of your clammy hands, racing heart, dry mouth, "butterflies" in your stomach, and the body's ability to detect these physiological responses that you feel anxious in the skydiving example. Thus, emotional processing contains appraisal, response, and feeling components. In this chapter, we will discuss the brain systems supporting these three components of emotion.

Although we are only using skydiving as an example of an emotional event and it may or may not be an experience that you would willingly partake in, researchers have used the skydiving model to study both stress/anxiety (Chatterton et al., 1997) as well as euphoria/joy (Carlson et al., 2012b). Skydiving research has shown that on average people tend to feel most anxious immediately prior to skydiving and on the other hand, feel most euphoric immediately after landing (Carlson et al., 2012b). As suggested in the description of the skydive, research suggests that skydive-elicited stress results in increased heart rate, but also increased stress hormone (i.e., cortisol) release (Carlson et al., 2012b; Chatterton et al., 1997) and activation of the immune system (Breen et al., 2016). Researchers have also been able to study the neural correlates of skydive-elicited anxiety and skydive-elicited euphoria. These studies suggest that an amygdala-based system is associated with greater levels of anxiety and cortisol release in anticipation of skydiving (Mujica-Parodi et al., 2014), while the medial orbitofrontal cortex (OFC) is related to feelings of joy and euphoria upon landing (Carlson et al., 2015a). In this chapter, we will further discuss the amygdala and its role in emotional processing. We will also discuss the medial OFC more in the section on emotional feelings. It should be noted that although extreme real-world emotional events, such as a skydive have been used to study emotion and its neural correlates, most neuroscience research on emotion has involved highly controlled laboratory studies using a variety of more mild emotional stimuli, such as facial expressions, affective images, monetary gains and losses, electric shocks, as well as a number of other stimuli that can be highly controlled. Most of the affective neuroscience research discussed in this chapter will be based on highly controlled and relatively mild emotional stimuli. However, there is evidence that these laboratory measures of neural reactivity are strongly related to real-world emotional reactivity (Mujica-Parodi et al., 2014).

Although we commonly associate emotions with their feelings, the functional significance of emotion is often overlooked. For most affective scientists, the functional role that emotions play in guiding behavior is central to understanding emotion. For example, consider emotional facial expressions. Why do we (and other species including canines, apes, rodents, and others) have facial expressions? Research by Paul Ekman and his coworkers has shown that across the globe facial expressions of fear, disgust, happiness, sadness, anger, and surprise are universally expressed in the same way (Ekman et al., 1969). Why? There must be a functional or adaptive value to these different muscle movements depending on the environmental context. Relatively recent research has compared the facial expressions of fear and disgust, which appear to have opposing muscle movements and perhaps opposing functional goals (Susskind et al., 2008). Fearful facial expressions are characterized by enlarged eye-whites and dilated pupils, flared nostrils, and an open mouth. On the other hand, disgusted facial expressions are characterized by squinted or partially closed eyes, wrinkled nose, and puckered lips. It appears that these facial features are not random, but rather allow for increased versus decreased sensory acquisition for fearful and disgusted facial expressions, respectively. In particular, widened eyes result in a wider visual field while flared nostrils and an open mouth allow for enhanced respiration and air intake (Susskind et al., 2008). From a functional perspective, these different facial expressions result in emotion-appropriate advantages. That is, in states of fear it makes sense to increase sensory intake to better understand where the potential threat is located and to be able to detect other cues related to the nature of the threat. On the other hand, if there is a disgusting stimulus in the environment, we don't want to smell, taste, or look at it! The idea that emotional expressions have adaptive value and functional significance is not new (Darwin, 1872), but it is still central to how affective scientists think about emotion.

In this chapter, we will focus on the functional influence of emotion in guiding what information we attend to in our environment, what events we remember, how we come to make decisions, and other influences of emotion on cognition. In the skydiving example, our attention was automatically drawn to our friend's facial expression of joy. Other facial expressions and emotional stimuli, especially those that signal potential danger (e.g., weapons or dangerous animals), capture our attention and research suggests that these emotional stimuli will capture our attention even if we are not consciously aware of them (Beaver et al., 2005; Blanchette, 2006; Carlson and Reinke, 2008, 2014; Carlson et al., 2009a; Cooper and Langton, 2006; Fox, 2002; Fox and Damjanovic, 2006; Koster et al., 2004; Mogg and Bradley, 1999, 2002; Ohman et al., 2001; Pourtois et al., 2004; van Honk et al., 2001). However, once an emotional stimulus captures our attention, it is more likely than other stimuli to reach a state of conscious awareness (Anderson and Phelps, 2001). Perhaps because we attended to or are more conscious of emotional stimuli and events in our lives, we are also more likely to remember these events than boring or mundane events (Kensinger and Corkin, 2004). Indeed, you are unlikely to remember what you had for lunch last year on this date (unless it was a very tasty or sickening lunch!), but an emotional event, such as a first-time skydive, will likely be permanently engrained into your long-term memory. Thus, emotional reactions have important functional roles in guiding behavior including how we perceive the world and what we remember. The neural systems involved in the various functional roles of emotion will be explored later in this chapter.

Although emotional responses play an adaptive functional role in aiding behaviors, such as attention, memory, and decision-making, the expression (or overexpression) of emotion in certain contexts can be disadvantageous. For example, if you and your friend share a joke, it is appropriate to smile and laugh robustly. However, if you share this joke during a lecture or public movie screening, it would be inappropriate to laugh out loud and disrupt others in the audience. Even if you find the joke to be incredibly funny, most individuals can regulate their expression of emotion in these types of situations and not laugh. Similarly, if you are worried about an upcoming exam, worry can be advantageous, if it leads to increased preparation or studying for the exam. Yet, an appropriate level of emotion regulation helps us to not be overcome by worry. Thus, another important aspect of emotional processing involves the regulation of emotion. The neural systems involved in emotion regulation appear to be (at least partially) distinct from those involved in emotional appraisal, reactivity, and feeling (Ochsner and Gross, 2005).

Looking ahead to the remainder of this chapter, we will review the neural mechanisms underlying emotion. We will first discuss the various methods that affective neuroscientists use to study the neural mechanisms of emotion and affect. Our focus will be on the neural systems involved in the appraisal of emotion, the response to emotion, and the feeling of emotion. We will also address the neural systems involved in the regulation of emotion. Given the range of emotions that we feel and range of circumstances in which we feel them, this chapter is necessarily going to offer an abbreviated discussion of the neural mechanisms of emotion. Nevertheless, we will cover many important neural mechanisms of emotions and affect. Some emotion theorists believe that our emotions fall into discrete categories including fear, anger, sadness, joy/happiness, disgust, and surprise (Ekman, 1992). Others believe that emotions exist along continuous dimensions, such as unpleasant/negative to pleasant/positive and calm to excited/aroused (Bradley and Lang, 1994; Bradley et al., 2001). Given that fear/negative and happy/positive emotions have received a great deal of research attention, these will be the model systems that this chapter uses to study the neural mechanisms of emotion.

TECHNIQUES USED IN THE STUDY OF AFFECTIVE NEUROSCIENCE

Brief Overview of the Nervous System

When neuroscientists study emotion, they work off the assumption that the brain plays a central role in the production and regulation of emotion. At the most basic level, the brain can be divided into three general regions: the forebrain, midbrain, and hindbrain. The forebrain consists of the cerebrum, basal ganglia, hippocampus, and amygdala. The cerebrum, or cerebral cortex, is two wrinkled hemispheres consisting of six layers of cells that can be further divided into four lobes: frontal, temporal, parietal, and occipital (Fig. 2.1). Each lobe contains neurons tasked with particular functions. The frontal lobe is associated with short-term memory, motor and action planning, and a number of other higher-level cognitive processes; the temporal lobe focuses on hearing, learning and memory, and

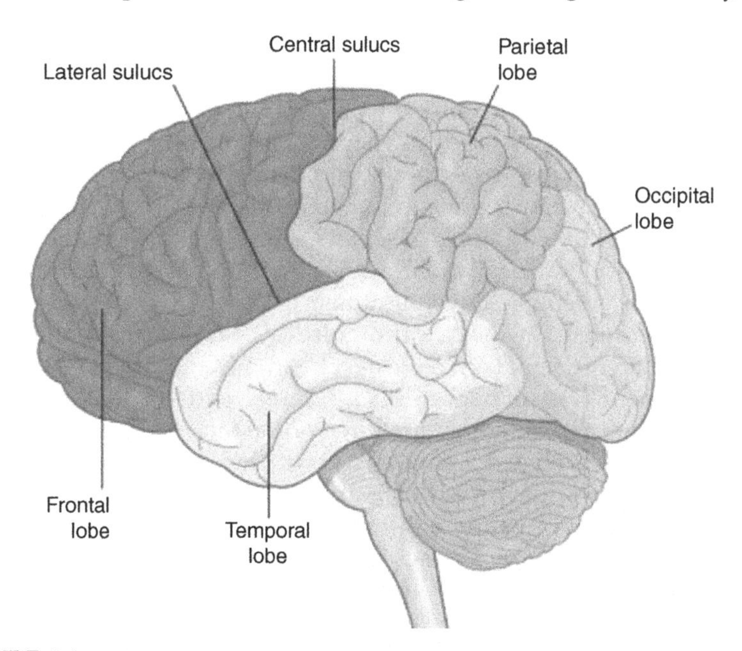

FIGURE 2.1 **A brain with basic divisions highlighting the four lobes of the cerebrum and the cerebellum.**

higher-level vision; the parietal lobe functions as the site of somatic sensation, body image, and extrapersonal space functions; lastly, the occipital lobe deals with vision (Kandel et al., 2012). The midbrain is responsible for regulating basic motor and sensory responses, such as reflexive eye movements triggered by visual and auditory stimuli. The hindbrain contains three structures: medulla, pons, and cerebrum. The medulla is responsible for the vital functions of the human body, such as heart rate, breathing, and digestion. The pons serves, in part, as a relay to carry information about movement from the cerebrum to the spinal cord and cerebellum, which modulates movement and motor learning (Kandel et al., 2012). Although these ways dividing the brain are accurate, even further subdivisions exist and will be discussed in this text.

As the brain is not easily accessible for study—especially, the human brain—technologies have been developed to image the brain and record signals related to its function. The smaller cellular units that make up the brain, called neurons and glia, are responsible for producing detectable signals related to emotion and other behaviors. Neurons communicate with each other via chemical and electrical mechanisms and are the cells mainly responsible for affective processing, while glia cells are support structures that play a homeostatic role. It is neurons that are the focus of many of the techniques utilized in studying affective neuroscience. Neurons consist of a cell body, or soma, which has protrusions called dendrites that receive information from other neurons. Neurons contain a single protrusion that sends information called an axon, which contains presynaptic terminals that release chemicals to communicate with other cells. Neurons typically communicate with each other by releasing stored chemical messengers, called neurotransmitters, from their presynaptic terminals in a process called synaptic transmission. In synaptic transmission, neurotransmitters bind to appropriate receptors at the dendrites or cell body of the next neuron at a junction known as the synapse. When enough excitatory neurotransmitters bind to a cell's receptors, it results in a depolarization in the neuron's membrane potential, which leads to an electrical impulse (i.e., the action potential) traveling down the axon. Upon reaching the presynaptic terminals, this electrical impulse causes a release in neurotransmitters, sending the signal on to the next neuron (Kandel et al., 2012).

Different neurotransmitters exist in different parts of the brain; oxytocin, dopamine, and serotonin are three of the major neurotransmitters utilized by emotion-related brain areas. Oxytocin is produced by neurons in the hypothalamus that project their axons to the pituitary gland, amygdala, hippocampus, striatum, and anterior cingulate. Functions related to oxytocin release include feelings of comradery or love associated with formation of social bonds and recognition of emotions and empathy (Insel, 2010). Neurons that synthesize dopamine have been identified in the substantia nigra and ventral tegmental area. Dopamine is thought to play a role in processing positive emotional stimuli, rewards, and motivation

(Wise, 2004). Additionally, recent research has linked the dopamine system (or at least certain populations of dopamine neurons) to aversive emotional processing as well (Kravitz et al., 2012). Serotonin is derived from the essential amino acid tryptophan, which is ingested through food and passes through the blood–brain barrier where serotonin can then be synthesized in the neurons of the raphe nucleus. From the raphe nucleus, serotonergic neurons project to many cortical and subcortical structures. Serotonin is involved in the improvement of negative feelings; this has been demonstrated by increasing the amount of serotonin at the postsynaptic neuron (Harmer et al., 2004; Mohammad-Zadeh et al., 2008). The brain uses these chemical messengers to communicate between neurons allowing for the processing of stimuli and these neurotransmitters are necessary for the changes in membrane potential of the neuron.

Techniques

Many neuroimaging techniques utilize the physiology of the brain and associated natural phenomena to detect signals produced by the brain. Before such technologies were developed, one way to determine brain function was to examine people (or animal subjects) with brain damage. Often referred to as lesion studies, the justification for studying brain-damaged individuals is that once an area of the brain is damaged, it may lose its normal function and thus we can compare data collected from lesioned individuals to healthy controls. Human patients can acquire brain damage in many different ways, such as stroke, trauma, disease, or other pathological conditions. Regardless of the method of damage, functional deficits may follow. In the case of stroke, it has been observed that 40% of poststroke patients can be classified as depressed. Furthermore, evidence from stroke patients has demonstrated a possible laterality for crying and overt sadness to the left hemisphere (Bogousslavsky, 2003).

One of neuroscience's most famous lesion patients, Phineas Gage, experienced extreme emotional impairments after forceful damage to the brain. Phineas Gage was a railroad foreman whose responsibility was to detonate explosives to clear the way for new rails. One day, Phineas tamped a hole filled with explosives that had not been properly filled with sand first—resulting in an explosion that sent a 190 cm long tamping rod through his face, skull, and brain. Phineas survived the incident, but not without acquiring extreme behavioral changes, such as aggressive outbursts and other changes to his personality that made him unemployable. Unfortunately, Phineas died without an autopsy performed; however, the recovery of his skull has led researchers to believe the damage suffered was localized to the medial portions of the left frontal lobe (Damasio et al., 1994). This example demonstrates that localized brain damage results in localized behavioral impairments. We will discuss other examples of what the lesion technique has taught us about the emotional brain throughout this chapter.

Luckily, we no longer need to depend on tragedy to study the brain. Engineers have developed technologies that allow us to noninvasively study the function of the human brain. All neuroimaging devices aim to detect signals produced by the brain. Usually this means detecting a particular physiological event or phenomenon that informs our larger understanding of brain function. The most common phenomena examined are the electrical activity of neurons and the associated delivery of oxygen to active neurons. A direct way to record the changes of a neuron's electrical potential is to use intracranial depth electrodes. Depth electrode recording is highly invasive; therefore, recording the electrical activity of neurons is typically done using electroencephalography (EEG).

EEG records the oscillations of local field potentials of neuronal masses detected at the scalp using electrodes. The tissue and bone between the electrode and brain cause the signal to be weak and difficult to localize, as the head can conduct electrical signals across the scalp surface to locations distal to the generator. The EEG activity recorded at the scalp surface consists mainly of the summed postsynaptic potentials of many neurons (called a neuronal mass) aligned in the same direction and firing synchronously; this pattern is called palisade (Lopez da Silva and van Rotterdam, 2005). EEG can only detect the activity of neuronal masses in a palisade pattern, but even this signal is extremely weak and is amplified so that physiological processes are monitored, but not distorted (Teplan, 2002). The bodies of neurons whose activity can be detected at the scalp can be thought of as oscillators and their activity as oscillations. Electrical oscillations detected on the human scalp using EEG have a frequency range from minutes for a single cycle to 600 Hz. The frequency range has been sorted into bins (or frequency bands) that are described using Greek letters: delta (0.05–4 Hz), theta (4–8 Hz), alpha (8–12 Hz), beta (12–30 Hz), and gamma (>30 Hz). The frequency bands are not functionally defined, as different bodies of neurons can produce the same oscillations. Furthermore, there is still the inverse problem of determining an electrical signal's source after it travels through an inhomogeneous volume conductor (i.e., the head; Buzsáki and Draguhn, 2004). Although a single frequency band cannot be attributed to emotional processing, some research has found evidence that certain emotional stimuli, such as angry faces can cause event-related increases in alpha and beta bands (Güntekin and Basar, 2007).

EEG data can be thought of as a mixture of signals produced by neurons of the brain that are in a palisade pattern in addition to signals unrelated to brain activity. Given what is known about EEG data collected at the scalp and the underlying generators, the data can be analyzed in a number of different ways. A few methods for analyzing EEG data are: independent component analysis (ICA), principle component analysis (PCA), frequency analysis, and event-related potential (ERP). The most common of these is ERP (also referred to as evoked potentials), although PCA and ICA can be used on ERPs.

In the ERP methodology, a subject is repeatedly presented with stimuli throughout the experiment and stimulus presentation times are flagged in the raw EEG data. The data around these flags are segmented by a computer program into approximately one second segments. The time-locked segments are then averaged together. The assumption is that the averaged data represents the brain signals that are common across stimulus presentations while eliminating the randomly distributed noise (e.g., inconsistent brain activity and environmental noise). After averaging, a waveform is left that has positive and negative peaks and troughs of varying latencies called components. Each component in the waveform is either positive or negative and can be labeled according to polarity (positive or negative) and order (1, 2, 3, etc.) or latency (in ms). For example, a positive component at 100 ms poststimulus could be labeled as P1, order based, or P100, latency based (Key et al., 2005).

Some well-known ERP components elicited by visual stimuli can be modulated when the images have emotional content: P1, N1, N170, and P2. The posterior P1 component is modulated most effectively by the repeated presentation of emotional faces, more so than other emotional images (Hajcak et al., 2012). Modulation of the central N1 component may represent early processes in the evaluation of emotional stimuli as it is modulated by both pleasant and unpleasant stimuli (Weinberg and Hajcak, 2010). Simple line drawings of faces that were evaluated as being neutral, pleasant, or unpleasant have also been shown to modulate the central P2 component (Begleiter et al., 1979; Hajcak et al., 2012). Another commonly studied ERP component in affective science is the N170, which is modulated by emotional faces (Blau et al., 2007). When evaluating ERP studies, it is important to keep in mind that the scalp distribution of components may indicate a different generator is responsible for the production of said component. For instance, a visual P1 may occur at occipital (posterior) electrode sites while auditory P1 occurs at temporal sites. Nonetheless, ERPs contain useful information as to the time-course and resources used in neuronal processing.

EEG is used to look at the electrical activity of certain populations of neurons. When a neuron fires, the membrane potential changes and must be returned to the resting potential (Bean, 2007). To return to its resting potential, the neuron needs energy in the form of adenosine triphosphate (ATP). ATP is produced in the brain by the oxidation of glucose, which requires oxygen (Sokoloff et al., 1977). Scientists have developed multiple ways to observe activity related to the delivery of oxygen to the brain via cerebral blood flow (CBF). Here we will review the most commonly used of these methods—functional magnetic resonance imaging (fMRI).

Functional MRI can be used to detect signals related to brain function due to the coupling of neurons with blood vessels (known as neurovascular coupling), the paramagnetic properties of deoxygenated hemoglobin, and a property held by the nuclei of some atoms called nuclear magnetic resonance (NMR). As previously described, it has been determined that neurons are coupled with blood vessels and although the density of vascularization may be different between cortical layers, the cortical surface is covered with blood vessels with just hundreds of microns between them (Nonaka et al., 2002). This potential for blood to be delivered to the whole brain at 1 × 1 mm resolution means that fMRI has very good spatial resolution. However, CBF can take up to 5 s to reach an area of activation—meaning CBF measures have low temporal resolution (Magri et al., 2012). For these reasons fMRI is often considered complimentary to EEG, since EEG has high temporal resolution and low spatial resolution.

Functional MRI utilizes the phenomena of NMR that is held by some atomic nuclei. By using a number of different magnetic fields and a radiofrequency coil for stimulation and detection, fMRI is able to disturb and record changes in the magnetic resonance of atomic nuclei with NMR. One such atom is hydrogen, which is ubiquitous in the human brain as it is one of the atoms that form the water molecule and water makes up 73% of the human brain (Mitchell et al., 1945). When hydrogen atoms are placed in the static magnetic field, most begin to precess (or rotate) around an axis parallel to the magnetic field, in a low-energy state. Using the radiofrequency coil, energy can be added to the atom making it take on a high-energy state that precesses antiparallel to the axis of the magnetic field in a process known as excitation. Once excitation is ended, a process called relaxation occurs in which energy is released as the molecules return to the low-energy state. The detection of this released energy is the basis of the MR signal (Heuttel et al., 2009).

Functional MRI detects a blood-oxygen-level dependent (BOLD) signal. When a neuron fires, its membrane potential changes and must be returned to the resting potential (Bean, 2007). This requires energy in the form of ATP. Research has shown that the brain needs six oxygen molecules to oxidize one glucose molecule (Sokoloff et al., 1977). Oxygen molecules are carried in the blood by hemoglobin, and thus hemoglobin can be oxygenated (HbO) or deoxygenated (HbR). Research has shown that the magnetic properties of hemoglobin are different depending on the oxygenation of the hemoglobin. Specifically, it was found that HbO is diamagnetic and HbR is paramagnetic. The fact that HbR is paramagnetic means that it can affect a nearby magnetic field and thus affect the precession of nearby nuclei with NMR (Heuttel et al., 2009). One group of researchers was able to determine that the effect on the MR image of HbR was to decrease the signal, making the image appear darker in areas of HbR concentration (Ogawa et al., 1990). When a neuron fires it utilizes the oxygen in local

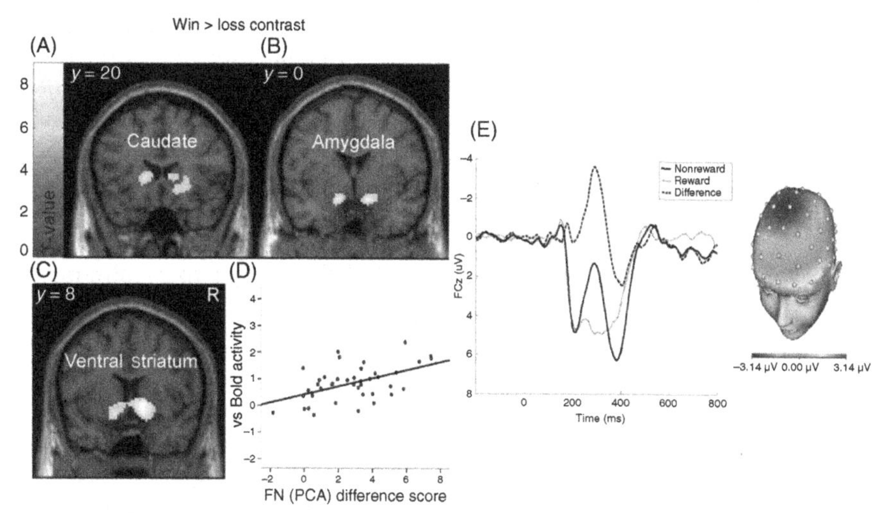

FIGURE 2.2 (A–D) Areas of increased functional magnetic resonance imaging (fMRI) blood-oxygen-level dependent *(BOLD)* signal for monetary wins relative to losses in a simple gambling task. (E) ERPs for rewards and losses as well as difference waves for FCz along with the EEG topography during the simple gambling task. *Source: Recreated from Carlson, J.M., Foti, D., Mujica-Parodi, L.R., Harmon-Jones, E., Hajcak, G., 2011a. Ventral striatal and medial prefrontal BOLD activation is correlated with reward-related electrocortical activity: a combined ERP and fMRI study.* Neuroimage *57 (4), 1608–1616.*

HbO, resulting in HbR. CBF then restores HbO to the active site (Malonek and Grinvald, 1996). Therefore, the increase in MR signal is not due to an increase in oxygen but a decrease in HbR, which was suppressing the signal. This is the basis of the BOLD signal (Heuttel et al., 2009). BOLD fMRI is often considered the gold standard for neuroimaging and has made a considerable impact on affective science since its inception (Fig. 2.2).

One of the first fMRI studies of emotion made a discovery in line with earlier theories and animals studies of emotional processing (King and Meyer, 1958; MacLean, 1952). In particular, this early fMRI study found increased amygdala activity when presenting fearful and happy faces compared to neutral ones, implying that it is activated in response to emotionally salient stimuli (Breiter et al., 1996). Since this early finding, the amygdala has been at the center of fMRI and structural MRI research into the neural correlates of emotion. One of the initial structural MRI studies was used to identify bilateral amygdala damage in one patient (Nahm et al., 1993).

Functional MRI and EEG are the most common imaging modalities and many affective imaging studies use these tools. However, other methods exist that can help inform the relationship between affect and brain function. Some other common methods are magnetoencephalogram (MEG), near-infrared spectroscopy (NIRS), and positron emission tomography (PET). Both NIRS and PET image signals are related to CBF, while MEG detects magnetic fields associated with electrical fields generated by the brain. NIRS measures CBF and hemoglobin oxygenation by using two wavelengths of light between 650 and 950 nm emitted by lasers that are absorbed by HbO and HbR, while some of the scattered light is recorded using detectors also placed on the scalp (Orbig et al., 2000). The intensity of light at the detector can be converted to optical density and then hemoglobin concentrations (Fekete et al., 2011a, 2011b; Hupert et al., 2009). PET is invasive as it requires the injection of a radioactive tracer. Radioactive tracers for brain metabolism and neurotransmission can be used with PET. As the radioactive tracers break down, they emit gamma rays that can be detected; increases in radiation are correlated with increases in brain activity (Heuttel et al., 2009). Although PET, NIRS, and MEG are not as commonly used as EEG or fMRI, they still provide important information on how the brain processes emotion. For instance, MEG measures magnetic fields that are associated with electrical fields generated by the brain and PET can examine processes associated with neurotransmitter activity (Cuffin and Cohen, 1979; Magistretti et al., 1999). Ultimately, when affective scientists use imaging tools they must understand the signals related to the physiological process they wish to learn more about to understand the relationship between brain function and emotion.

Reviewed here were various techniques for imaging the brain. It should be noted that neuroimaging doesn't mean to "take pictures." Even structural MRI is based off the distribution of hydrogen atoms in water and not an image of the gray or white matter itself. What neuroimaging methods collect are signals emitted by physiological processes of the brain that, with a proper understanding, can lead to informed scientific discovery. In the realm of affective science, EEG and fMRI have dominated the field whether it be for cost effectiveness and ease of use (EEG) or robustness of data and localization capabilities (fMRI). An understanding of these techniques and their signals allows researchers to better understand their use in affective science and make their own decisions regarding the validity of findings.[a]

EMOTIONAL APPRAISAL AND RECOGNITION

Early theories on the neural mechanisms of emotion focused on structures believed to be phylogenetically (evolutionarily) older as emotion is also believed to be phylogenetically older. Many of these structures

[a]For the computer scientist wishing to integrate emotion-related brain signals and computing technology, an understanding of the brain-produced signals associated with affective behavior will be necessary for brain–computer interfacing. Furthermore, the types of stimuli used in affective science research may also be useful in developing methods for emotional communication in human–computer interactions and computer-based human interactions.

were subcortical (i.e., beneath the neocortex). Two early models of emotional brain systems were the "Papez circuit" (Papez, 1937) and "limbic system" (MacLean, 1949, 1952). The limbic system consists of the cortex adjacent to the olfactory tract, pyriform area, hippocampus complex, parasplenial gyrus, cingulate gyrus, subcallosal gyrus, amygdala, hypothalamus, and anterior thalamus (MacLean, 1952). MacLean used known anatomical connections between brain regions to identify the structures of his limbic system that were thought to be involved in emotion processing. The positioning of this system, particularly the hippocampus, between phylogenetically older and newer structures, was proposed to be optimal for associating visceral sensation and perceptions of the physical characteristics of external stimuli (MacLean, 1949). Today, the brain systems as described by Papez and MacLean are not considered to be accurate correlates of emotional processing. Nevertheless, at the heart of the limbic system is the amygdala, which has been known to be important for detecting the salience of environmental stimuli since early research by Kluver and Bucy (1939), who lesioned the amygdala in monkeys and found that these animals no longer showed fearful responses toward dangerous stimuli, such as snakes and more dominant monkeys.

The amygdala is a complex subcortical structure consisting of more than a dozen subnuclei situated in the anterior portion of the medial temporal lobe. These subnuclei can be grouped into three main complexes. The largest of these complexes, the lateral nuclear complex, receives inputs from all sensory systems and is important for mediating emotion-related behaviors (Fanselow and LeDoux, 1999). The central complex contains the medial nucleus, which is the primary output region for emotional responses (LeDoux et al., 1988). Lastly, the smallest group is the cortical nucleus whose main inputs are from the olfactory bulb and olfactory cortex (McDonald, 1998). Since the amygdala is one of the earliest (nonsensory) structures to receive sensory input and receives inputs from all sensory modalities, it has been the focus of many studies examining the appraisal and recognition of emotional stimuli. The outputs of the amygdala extend to regions involved in autonomic responses and regions involved in perception, memory, cognition, and emotion (LeDoux, 2007). Many of these connections are related to the amygdala's role in emotional reactivity, which will be discussed in more detail in the following section. While the amygdala was initially studied for its role in fear conditioning, and was believed to be mostly involved in fear/negative emotion, more recent research suggests that the amygdala is a general salience or relevance detector (Sander et al., 2003). We will first focus on the convergence of sensory information in the amygdala and then discuss the role of the amygdala, as well as other structures, in emotional appraisal and recognition.

Convergence of Sensory Information in the Amygdala

As mentioned previously, the lateral nucleus is the primary input region into the amygdala and receives environmental information from all sensory modalities (Fanselow and LeDoux, 1999). There is evidence that sensory information reaches the amygdala through two routes: one for quick, but less discriminative responses, and one for slower more discriminative responses (LeDoux, 1996). According to LeDoux (1996), the slower and more discriminative route involves sensory information being relayed from the sensory receptors, to the sensory thalamus, to the sensory cortex, and then to the amygdala. This route is referred to as the "high-road." The quicker, dirtier, route starts the same, but involves direct projections from the sensory thalamus (or other subcortical structures) to the amygdala, bypassing the sensory cortex. This route is referred to as the "low-road." The information relayed to the amygdala by the sensory thalamus has not been processed by the sensory cortex, and is therefore of lower resolution. In bypassing the sensory cortex, low-road information is thought to be quicker and allow the amygdala to initiate fast responses to important stimuli based off of less detailed low-level sensory information. Thus, the low-road acts to initiate an immediate "better safe than sorry" response. In contrast, high-road information receives higher resolution processing by the cortex resulting in a later, but more appropriate response (LeDoux, 1996).

Early evidence that the amygdala receives information both from the sensory cortices and sensory thalamus comes from animal lesion studies. Researchers conditioned a rabbit to fear one tone by pairing it with a shock. A very similar tone was not paired with a shock. After repeated conditioning, the rabbit would display an increased heart rate to the conditioned tone, but not the similar tone. Following conditioning, the auditory cortex was lesioned and testing the fear response to the tone resumed. Once the rabbits were lesioned, they were unable to make the discrimination between the 2 tones and instead displayed fear responses to both stimuli (Schneiderman et al., 1974). This finding suggests that the cortex is needed only to make fine sensory discriminations, but the amygdala can still perform fear learning based off of the low-resolution sensory information sent from the thalamus (LeDoux, 1996). Similar results have been reported in rodent models measuring freezing behavior in fear conditioning paradigms (Phillips and LeDoux, 1992). Collectively, these results suggest that damage to the sensory cortex results in less specific fear learning. Yet, if the amygdala and sensory thalamus are still intact, less discriminative fear responses can still be learned—presumably through the low-road.

Research in humans has also provided evidence supporting these two routes to the amygdala. Patient G.Y. has cortical blindness, which is blindness as a result of damage to the visual cortex, rather than ocular damage. Despite the lack of a conscious visual experience (i.e., blindness), G.Y. is still able to discriminate between emotion-relevant and neutral visual stimuli. Brain activity was recorded with fMRI while G.Y. was asked to make discriminations of visual stimuli in their blind hemifield. The fMRI results indicated that the amygdala and superior colliculus (a subcortical visual structure in the midbrain) were activated in response to the presentation of emotional stimuli in the blind hemifield. On the other hand, when emotional images were presented to the intact hemifield, areas of visual cortex were active. These results provide support for a subcortical pathway that bypassed the damaged visual cortex to provide the amygdala with low-resolution information about emotionally salient stimuli in the environment (Morris et al., 2001).

To approximate blindsight in healthy subjects, a "masking" procedure can be utilized to prevent conscious-level processing of a presented image. One group used this technique by presenting happy and fearful faces for 33 ms then immediately presenting a neutral face in the same spot for 167 ms; participants only reported seeing the neutral face. Performing this task in conjunction with fMRI has revealed that the amygdala is more active in response to masked fearful versus masked happy faces, again suggesting the amygdala plays a role in detecting important stimuli without the need for conscious awareness (Whalen et al., 1998). Findings of amygdala activity to masked threats have subsequently been replicated numerous times (Armony et al., 2005; Carlson et al., 2010; Carlson et al., 2009b; Liddell et al., 2005; Morris et al., 1998, 1999; Rauch et al., 2000; Sheline et al., 2001; Suslow et al., 2006; Whalen et al., 2004; Williams et al., 2009; Williams et al., 2005). Furthermore, masking studies have shown that amygdala activation to nonconscious visual threat is associated with activation in subcortical visual structures, such as the superior colliculus and pulvinar nucleus of the thalamus (Liddell et al., 2005; Morris et al., 1999). Taken together, the evidence from animal lesion, blindsight, and masking studies provides substantial support for a fast, but low-resolution, subcortical "low-road" to the amygdala (LeDoux, 1996; Liddell et al., 2005; Morris et al., 2001; Pourtois et al., 2013; Tamietto and de Gelder, 2010; Whalen et al., 1998).

While evidence has mounted in support of a subcortical route to the amygdala, it is not universally accepted that this route is necessarily responsible for faster, less detailed, processing of emotional information. One recent theory that takes direct opposition against the two-route theory is the multiple wave hypothesis (Pessoa and Adolphs, 2010). Pessoa and Adolphs (2010) argue that evidence for a subcortical pathway that processes emotional stimuli faster than other stimuli is not sufficient for all aspects of emotional processing. Furthermore, given the speed of cortical processing, the necessity for a fast subcortical pathway is unclear. From the low-road perspective, the pulvinar is thought to relay low-resolution visual information to the amygdala. However, some evidence suggests that the pulvinar may process high-resolution information (Greene and Oliva, 2009; Pessoa and Adolphs, 2010; Pessoa and Ungerleider, 2004). In opposition to the high and low-roads to the amygdala, the multiple wave theory proposes that many different routes exist for information to reach the amygdala. While information does come from the pulvinar, in the context of the multiple wave theory the pulvinar plays an important role in distributing high-resolution stimulus information to the amygdala.

In sum, there is not a singular accepted theory as to how sensory information reaches the amygdala, or what particular level of detail is relayed via different sensory routes to the amygdala. Nevertheless, the theories reviewed here acknowledge that both cortical and subcortical structures project to the amygdala. The lateral amygdala receives sensory inputs from somatosensory, visual, auditory, taste, and olfactory sensory cortices as well as associated thalamic inputs (LeDoux, 1996, 2007). Despite an ongoing debate as to the number of routes and the exact nature of the information relayed by these routes, there is general consensus that the amygdala receives a vast amount of sensory information, which makes it perfectly anatomically suited for a role in emotional appraisal.

Emotional Appraisal and Recognition Processing Within the Amygdala

Evidence that the amygdala is involved in emotional appraisal and recognition comes from studies of patients with amygdala damage. Perhaps the most well-known person with amygdala damage, and also one of neurology's most famous patients, is known as patient S.M. S.M.'s amygdala damage was due to Urbach–Weithe disease, a condition that causes nearly complete bilateral destruction of the amygdala, while sparing nearby hippocampus and cortical structures (Adolphs et al., 1994). Aside from her amygdala damage, S.M. lives a fairly normal life; she has been married, a normal IQ, a high school education, and is a single-mother of three healthy children. Her main symptom is that she has little to no capacity to show or feel the basic emotion of fear and thus has impairments in avoiding situations which are dangerous and is perhaps overtrusting and overfriendly in her social interactions. Due to her rare condition, she has been the subject of extensive research on the amygdala and its role in emotional appraisal.

S.M.'s condition revealed an important role of the amygdala in the processing of facial affect (Adolphs et al., 1994). S.M. and 12 brain-damaged controls with similar IQ were shown facial expressions of six basic emotions (happiness, surprise, fear, anger, disgust, sadness), as well as neutral faces, and asked to rate each face according to several emotional adjectives. S.M.'s assessments differed dramatically from controls, consistently rating expressions of fear, anger, and surprise as less intense than any of the brain-damaged controls. S.M. also showed severe recognition impairment specific to fear: her ratings of fearful faces correlated less with normal ratings than those of any brain-damaged control. In contrast, her recognition of the unique identity of faces was fully preserved: she was able to learn new faces and perfectly identify familiar faces. The results suggest dissociation between neural processing of facial identity and facial affect and that they are likely maintained by anatomically separate neural systems. These findings in the patient S.M. have been replicated in a larger sample of patients with varying degrees of amygdala damage. All but three of the patients in this study had been previously reported on, but had yet to be tested under the same experimental procedures. The patients with amygdala damage rated negative emotions (fear, anger, and disgust) lower than controls. Furthermore, the ratings from amygdala-damaged patients were again less correlated with the ratings of normal controls than the ratings of brain-damaged controls. The impaired recognition of fear intensity in the amygdala-damaged group suggests that it is involved in detecting emotional signals of threat or harm (Adolphs et al., 1999).

Additional evidence points to a role for the amygdala in processing information from the eyes when recognizing emotional facial expressions. While S.M.'s injury and subsequent functional losses are the result of amygdala damage, it has been determined that her inability to detect fear is the result of not looking at the eyes of faces. S.M. and a control group were both presented with approximately 3000 faces: half of which were fearful and half of which were happy. In each trial, only certain portions of the face were revealed. The participants identified whether the faces expressed happiness or fear. S.M. required nearly twice as much of the face to be exposed as controls to identify a fearful face; however, her recognition of happy faces was normal. A closer examination of the data revealed that S.M. failed to make use of information from the eye region whereas controls consistently used this information. Interestingly, using the same experimental set up, but changing the task to identifying gender, revealed that S.M. does use information from the eye region in the same way healthy controls do if the task is gender recognition. To further test the theory that the eyes may be necessary for normal emotion recognition, the researchers digitally erased the eyes from whole face images and had S.M. and the control subjects try to recognize the expressed emotion. Removing the eyes from the images did not affect S.M.'s recognition accuracy, but the control subjects had a decrease in accuracy. Subsequent tests using eye tracking revealed that S.M. spent less time looking at the eye region in emotional faces than healthy controls. The results from this series of experiments in the patient S.M. indicate that the ability to recognize fear requires attention to the eye region and this process is mediated by the amygdala (Adolphs et al., 2005).

Early research on the amygdala focused primarily on its role in fear. However, it is now understood that the amygdala is involved in the processing of other affectively salient stimuli. For example, according to the multiple wave theory the amygdala is involved in processing abstract dimensions of information, such as salience, ambiguity, unpredictability, and other aspects of biological importance (Pessoa and Adolphs, 2010). In general, the amygdala has been shown to activate for both negatively valent emotional stimuli, such as bad odors, aversive images, and threatening images as well as positively valent emotional stimuli, such as images of pleasant animals, babies, nudes, appetizing food, and monetary reward (Carlson et al., 2011a; Hamann et al., 2002; Redouté et al., 2000; Walter et al., 2008; Zald and Pardo, 2000). The amygdala is preferentially involved in the processing of fearful faces, but also activates in response to other socially relevant stimuli, such as faces in general, faces with directional gaze, and the processing of one's own name (Kawashima et al., 1999; N'Diaye et al., 2009; Portas et al., 2000; Sander et al., 2003; Wicker et al., 2003b). Even emotionally neutral, but unusual (e.g., surrealist films), stimuli activate the amygdala (Hamann et al., 2002). To explicitly test the hypothesis that the amygdala is involved in the evaluation of motivationally relevant stimuli, researchers had participants make evaluations of both positive and negative famous names while amygdala activity was recorded with fMRI (e.g., Adolph Hitler, Paris Hilton, George Clooney, etc.; Cunningham et al., 2008). Amygdala activity was found in conditions where evaluations were made of all names and when making positive or negative evaluations of positive or negative names, respectively. Collectively, these findings dispute early theories that the amygdala was involved only in fear processing. Converging research now suggests that the amygdala is involved in stimulus relevance evaluation based on the organism's needs (Cunningham and Brosch, 2012; Hamann et al., 2002).

Emotional Appraisal and Recognition Processing Beyond the Amygdala

Areas of the cortex also play a role in the recognition of emotions. Recognizing emotions, especially in other people, is an important aspect of social behavior. Without the ability to recognize that another human can feel pain and joy similar to oneself, a person would have a hard time fitting in with society. Many experiments examining the ability to recognize emotions, including experiments described here, use facial expressions as stimuli to determine whether someone can recognize an emotion. The emotions communicated by faces are communicated by the changeable aspects of the face, such as the eyes or mouth (Bassili, 1979). For instance, fearful faces can be characterized by a widening of the eyes, flaring nostrils, and gaping mouth (Susskind et al., 2008). Research suggests that the ability to recognize a facial identity and the ability to recognize the changeable aspects of a face involve separate neural systems. Face identification is related to the lateral fusiform gyrus whereas recognition of facial expression is related to the superior temporal sulcus (STS; Haxby et al., 2000; Hoffman and Haxby, 2000). A study using MEG found that responses in the STS occurred along with responses in the amygdala when judging the expression of an emotional face compared to a neutral face (Streit et al., 1999). Further evidence that a network involving the STS and amygdala is involved in the recognition of emotion in faces comes from fMRI and lesion studies (Campbell et al., 1990; Lahnakoski et al., 2012).

The STS has connections to many different areas of the brain involved in different dimensions of social cognition. Of particular interest to this chapter are the connections from the STS to the amygdala. Researchers presented healthy subjects with videos that depicted different social features (faces, bodies, biological motion, goal-oriented action, emotion, social interaction, pain, and speech) while they underwent fMRI scans. The authors found that the anterior STS and the amygdala formed a network sensitive to speech and the communication of emotions through facial expressions and/or body language (Lahnakoski et al., 2012). In rhesus monkeys, it has also been found that ablation of the STS leads to deficits in the ability to discriminate the direction of eye gaze, a result that is also found in humans with prosopagnosia (i.e., the inability to recognize faces; Campbell et al., 1990).

Damage to the somatosensory cortex can also lead to an inability to recognize emotion in others even when the amygdala is intact. One group of researchers examined 108 patients with brain damage to assess possible regions of cortex associated with deficits in the ability to recognize emotion. In two experiments, the brain-damaged patients along with controls were shown 6 blocks of the same 36 emotional faces (6 faces each of happiness, surprise, fear, anger, disgust, and sadness) and asked to rate the intensity of the emotion in one experiment and match the expression with the written name of the expression in the second experiment. In a third experiment, participants sorted a subset of 18 facial expressions (3 of each emotion) into emotion categories that were similar. The researchers divided the group in half based on intensity scores in experiment one and found that those in the bottom half (more impaired) had localized damage to the somatosensory cortex. When the groups were further divided into the top and bottom 25% of intensity rankings, localized somatosensory damage was found again. Examination of performance as a function of lesion location in the second and third experiments further revealed the necessity for intact somatosensory cortex for normal recognition of all emotions. A post hoc analysis of medical records found a correlation between impairment in emotion recognition and somatic sensation (Adolphs et al., 2000). Based on these results and results from other studies of STS function, the authors concluded that visual and somatosensory representations of emotions and the white matter connecting regions of the brain that support the representation of this information allow one to recognize the mental state of another by simulating the observed state in oneself (Adolphs et al., 1994; Adolphs et al., 2000; Adolphs et al., 2005; Campbell et al., 1990; Haxby et al., 2000; Lahnakoski et al., 2012).

Although the amygdala has been extensively studied in the context of negative emotions, it also plays an important role in positive emotions and can be thought of as a general salience detector (Sander et al., 2003). The brain appears to also have a region that is primarily associated with reward processing—the nucleus accumbens of the ventral striatum (Knutson et al., 2001a; Kravitz et al., 2012; Olds and Milner, 1954). Early studies on reward and pleasure in the brain used a technique of electrical self-stimulation in rats (Olds, 1958; Olds and Milner, 1954). This technique involves placing an electrode into the brain of a rat. The wires leading into the electrode hang from the ceiling of a Skinner box to allow free movement by the animal. The Skinner box, a controlled environment used in animal behavior experiments, contains a lever that controls the electricity sent to the subject's brain and the subject is free to push the lever. Olds and Milner (1954) used this technique, implanting electrodes into different brain regions and then correlating bar presses and extinction of bar presses to electrode placement. The results indicated that electrode implantation in the septal region, beneath the corpus callosum, resulted in the highest amount of bar presses. Furthermore, the results of self-stimulation were comparable with those of studies using typical reinforcers, such as food. A modern take on this study has used optogenetic techniques to stimulate dopaminergic activity in the striatum of rats (Kravitz et al., 2012). In this study, a rat was placed in a Skinner box with two levers, one of which did nothing and the other turned on a laser that, depending on experimental condition, would either stimulate excitatory D1 receptors or inhibitory D2 receptors. The results indicate that activation of D1 receptors increases the probability that the rat will respond again, whereas D2 activation results in the animal moving away from the lever and decreased probability of response (Kravitz et al., 2012). Thus, the striatum and dopamine may be involved in the detection of both rewarding and punishing stimuli.

An increase in BOLD signal has been found in the nucleus accumbens/ventral striatum in anticipation of increased reward using fMRI (Knutson et al., 2001a). In a seminal study, researchers presented participants with a task, during fMRI scans, in which a cue either signaled a small or larger monetary reward (gain $0.20, $1, $5), a small or larger monetary loss (lose $0.20, $1, $5), or no monetary loss or gain. During conditions in which the cue stimulus signaled loss or gain, the subjects responded to a target appearing on screen for 160–260 ms. If they pressed the button while the square was on screen they either avoided the loss or received more money. The results of fMRI analysis indicated greater BOLD signals in the nucleus accumbens during anticipation of reward compared to no reward. An additional fMRI study revealed that the ventral striatum is related to predicted monetary reward versus nonreward outcomes (Knutson et al., 2001b). Similar fMRI activation in the ventral striatum has been observed in anticipation of other positive/rewarding stimuli such as erotic movie clips (Greenberg et al., 2015). Collectively, these findings provide evidence that the nucleus accumbens is active during the anticipation of reward, meaning it is involved in the appraisal and recognition of rewarding stimuli.

Appraisal of emotional stimuli and accurate recognition of the emotion are important aspects of survival as well as the more mundane aspects of daily life. In the case of S.M., damage to the amygdala resulted in an inability to recognize fear in faces, show normal fear responses, and a tendency to be more trustworthy and friendly than the normal person (Adolphs et al., 1998; Bechara et al., 1995). Thus, there are very real consequences as a result of damage to the amygdala. In addition to the amygdala, different cortical regions are also necessary for appraising emotional stimuli and recognizing emotions. The somatosensory cortex appears to be necessary for recognizing other's emotional expressions by simulating the feelings of others in ourselves (Adolphs et al., 2000). The STS is connected to the amygdala, which is important for processing facial expression and body language (Haxby et al., 2000). The nucleus accumbens also plays a role in the evaluation of emotional stimuli as it is involved in predicting rewarding outcomes (Knutson et al., 2001b). The amygdala is an important structure in the appraisal of environmental stimuli for emotional significance. However, other structures are also involved. For instance, the cortically generated N1 ERP component can be modulated by both positive and negative stimuli (Weinberg and Hajcak, 2010). Thus, there are multiple and distributed regions in the brain for detecting and appraising the significance of emotional input and we have only introduced a few of these structures here.

EMOTIONAL REACTIVITY: IMMEDIATE AND LONG-TERM FUNCTIONS

As mentioned in the introduction to this chapter, affective scientists believe that the functional role emotions play in guiding behavior is central to understanding emotion. The example discussed in the introduction focuses on emotional facial expressions and how these universally expressed contractions of facial muscles are not random movements, which only serve as a means of nonverbal communication, but also as a functional modulation of sensory processing mediated by overt actions on the sensory organs themselves. Fearful faces are characterized by exposed eye-whites, dilated pupils, flared nostrils, and an open mouth, which allows for a wider visual field and enhanced respiration (Susskind et al., 2008). On the other hand, disgust facial expressions are characterized by partially closed eyes, wrinkled nose, and puckered lips, which results in the opposite effect: decreased visual field and decreased air intake. Other facial and bodily expressions serve emotion-specific functions (Darwin, 1872). Angry expressions focus attention at the source of one's anger. Bodily postures of fear keep threat away and minimize bodily harm. Thus, from a functional perspective, emotional expressions result in emotion-appropriate advantages. However, facial and bodily expressions are just one of many reactive responses that occur during an emotional event. Although these expressions are observable to the naked eye, many emotional responses cannot be directly observed.

As discussed in the section on emotional appraisal, the amygdala is critically implicated in the processing of emotionally salient stimuli (e.g., fearful facial expressions) and is responsive to emotional stimuli that are both consciously (Breiter et al., 1996; Morris et al., 1996) and nonconsciously processed (Liddell et al., 2005; Whalen et al., 1998). These human neuroimaging studies build upon earlier work by Joseph LeDoux (1996) implicating the rodent amygdala in detecting and responding to threats via a quick and dirty route through the sensory thalamus as well as a slower, more detailed route through the sensory cortex. In short, the amygdala receives a vast amount of sensory (afferent) input across sensory modalities (LeDoux, 2007; Zald, 2003) making it aptly suited to evaluate and appraise incoming information for emotional salience.

Just as the amygdala is anatomically suited for emotional appraisal, it also displays a number of outgoing (efferent) connections (Amaral and Price, 1984), which make it aptly suited for initiating or modulating emotional responses. Primary outputs from the amygdala include the hypothalamus, brainstem, prefrontal cortex (PFC), hippocampus, striatum, and sensory regions. These outputs play an important role in emotional reactivity including the modulation of hormone release, autonomic nervous system activation, attention, decision-making, memory formation, and perceptual processing to name a few (Davis and Whalen, 2001; LeDoux, 2007). Given the great variety of emotional responses that can occur and appear to, at least in part, be mediated by the amygdala, we will focus our discussion on an exemplary response of immediate functional significance and an exemplary response of long-term functional significance—attention and memory, respectively.

Emotional Attention

Our sensory systems are continuously bombarded with vast amounts of sensory information which compete for our attention and access to consciousness. We simply cannot process every piece of sensory information and thus certain sensory stimuli are prioritized over others. The emotional salience of a stimulus is an important factor leading to prioritized processing. In laboratory settings, one of the more common methods of studying emotional influences on attention is the dot-probe task (MacLeod and Mathews, 1988). Trials contain two stimuli simultaneously presented to the left and right of a central fixation point. One stimulus is emotionally relevant and the other is neutral (or of a different emotion type). These two stimuli are then followed by a single "dot" probe appearing randomly at either the left or right location. The rationale is that emotional stimuli automatically capture attention and therefore reaction times should be faster for probes presented in the location occupied by emotional stimuli (referred to as congruent trials) compared to when probes are presented in the location occupied by neutral stimuli (referred to as incongruent trials). Indeed, research suggests that a variety of emotional images, especially those conveying threat, capture our attention. This includes both social stimuli (e.g., threatening facial expressions) and nonsocial stimuli, such as weapons as well as poisonous and predatory animals (Beaver et al., 2005; Blanchette, 2006; Carlson et al., 2009a; Carlson and Reinke, 2008, 2014; Cooper and Langton, 2006; Fox, 2002; Fox and Damjanovic, 2006; Koster et al., 2004; Mogg and Bradley, 1999, 2002; Ohman et al., 2001; Pourtois et al., 2004; van Honk et al., 2001).

It should be noted that although most research on the neural system for emotional attention has used fearful faces as the emotionally relevant attention-grabbing stimulus, positive emotional stimuli, such as happy faces and other affiliative expressions have been shown to capture our attention as well (Elam et al., 2010; Kreta et al., 2016; Torrence et al., 2017). Furthermore, much neuroimaging research has focused on the visual domain, but other sensory stimuli carrying emotional significance, including auditory stimuli, have been found to capture attention (Brosch et al., 2008; Gerdes et al., 2014). Therefore, despite the fact that much work on emotional attention has used visual signals of threat, the capture of attention by emotional stimuli goes beyond visual threat signals including other emotions and sensory domains.

How does the brain go about prioritizing emotional stimuli or biasing our attention toward emotionally salient stimuli? As you may suspect, the amygdala plays an important role. We already know that the amygdala receives diverse sensory inputs and it projects to a number of regions important for modulating emotional reactivity including the PFC and sensory cortices. These connections alone suggest that the amygdala plays an important role in modulating attention. However, the amygdala's involvement in modulating attention is further supported by empirical research. Amygdala activity is increased during the dot-probe task, which, as described previously, measures emotional attention (Carlson et al., 2009b; Monk et al., 2008). Furthermore, studies of patients with amygdala damage indicate that the amygdala is necessary for emotional prioritization

(Anderson and Phelps, 2001; Bach et al., 2014). For example, in the attentional blink paradigm, individuals are presented a list of visual stimuli in rapid succession one after the other and asked to identify target stimuli (e.g., words presented in a particular text color). The "attentional blink" refers to a phenomenon, which occurs when two targets are presented in close temporal proximity to one another and the second target is undetected as if the individual has blinked. It is believed that when our attentional resources are being utilized (in this case focusing on the first target presented) there is a period of time when subsequent stimuli are not detected. Interestingly, however, if the second target contains emotion-related information (e.g., the word "death") then the attentional blink is usually overcome and the second target is detected (Anderson and Phelps, 2001). Thus, emotional stimuli automatically capture our attention. The ability of emotional stimuli to overcome the attentional blink has been found to be dependent upon an intact amygdala (Anderson and Phelps, 2001). Collectively, there is both anatomical and empirical evidence to support the role of the amygdala as an important neural structure in the prioritization of emotional stimuli in attention.

Thus far, the evidence discussed indicates that the amygdala plays a critical role in the allocation of visuospatial attention to emotionally salient stimuli, but what exactly does the amygdala do? Evidence from nonhuman primates suggests that neurons in the amygdala code for stimulus value/salience (Peck et al., 2013), which is consistent with the role of the amygdala in emotional appraisal reviewed in the previous section. Furthermore, neurons in the nonhuman primate amygdala code for spatial location (Peck et al., 2013). These two properties of the amygdala are essential for prioritizing sensory processing at a particular location in one's environment.

After the amygdala has coded for the salience and spatial location of sensory information (Peck et al., 2013), this signal is relayed to the PFC (Amaral and Price, 1984). This amygdala–prefrontal pathway appears to be a central mechanism for the attentional capture by emotional stimuli. At a structural level, variability in the integrity of the amygdala–prefrontal white matter pathway has been found to correlate with attention bias scores in the dot-probe task (Carlson et al., 2013, 2014). At a functional level, the strength of the coupling between amygdala and PFC activity is correlated both with attention bias scores as well as the integrity of the amygdala–prefrontal white matter pathway (Carlson et al., 2013). Within the PFC, the anterior cingulate appears to be the primary region interacting with the amygdala to direct attentional resources toward emotionally relevant stimuli within the environment. Activity within the anterior cingulate is elevated during the capture of attention in the dot-probe task (Armony and Dolan, 2002; Carlson et al., 2009b; Price et al., 2014) and greater anterior cingulate volume correlates with greater attentional bias in the dot-probe task (Carlson et al., 2012a).

The anterior cingulate, and the PFC more generally, are thought to be involved in cognitive control processes including conflict resolution (Botvinick et al., 1999; Botvinick et al., 2004; Etkin et al., 2006; Horga and Maia, 2012; van Gaal et al., 2012). Consistent with this broader role, the anterior cingulate detects multiple stimuli competing for attention and resolves this conflict by selecting the most salient stimulus for preferential access to attention. The amygdala biases this prefrontal monitoring system to favor emotion-related information (Carlson et al., 2013, 2014). An additional role of the anterior cingulate in affective attention comes from findings that activity in this region is greater during incongruent compared to congruent trials in the dot-probe task. Given the typical delay in reaction times during incongruent trials, this finding suggests that the anterior cingulate regulates the duration of attentional engagement by emotional stimuli (Fu et al., 2015; Price et al., 2014). As is the case with the amygdala (Adolphs et al., 2005; Anderson and Phelps, 2001; Bach et al., 2014; Vuilleumier et al., 2004), damage to the PFC (including the anterior cingulate) results in impaired attention to emotion (Wolf et al., 2014).

The neural system for directing attention to emotionally salient stimuli includes additional structures beyond the amygdala and PFC. Of particular importance are the sensory cortices themselves. Indeed, it has been long known that visual emotional stimuli increase visual cortical processing above and beyond nonemotional visual stimuli (Lang et al., 1998). In addition to the broad increase in visual cortical processing elicited by emotional stimuli, is a location-specific enhancement in visual cortical processing for emotional stimuli. For example, if you detect a spider in the lower right portion of your visual field, there will be a selective increase in the regions of visual cortex responsible for processing this area of the visual field (i.e., upper left quadrant of visual cortex). This location-specific enhancement of the visual cortex serves to prioritize the processing of stimuli at this spatial location and can therefore be thought to represent the "spotlight" of attentional capture. Evidence for a location-specific enhancement of visual cortical processing comes from ERP and fMRI studies using the dot-probe task (Carlson et al., 2011c; Carlson and Reinke, 2010). Remember, neurons in the amygdala code for spatial location (Peck et al., 2013) and project back to sensory cortex (Adolphs, 2004; Vuilleumier, 2005; Vuilleumier et al., 2004); providing an anatomical substrate by which location-specific sensory processing can be enhanced. The importance of this amygdala-to-sensory cortex feedback mechanism is demonstrated in neuroimaging studies of patients with amygdala damage. In contrast to the enhanced visual cortical processing observed in healthy individuals by emotional stimuli, those with amygdala damage display visual cortical processing that is similar to nonemotional stimuli (Vuilleumier et al., 2004). In addition to direct feedback from the amygdala-to-sensory cortex, there is an indirect route through the nucleus basalis in the basal forebrain. The amygdala projects to acetylcholine producing neurons in the nucleus basalis, which in turn project diffusely throughout the cortex including the sensory cortices. This cholinergic system has been implicated in increased alertness and affective attention (Carlson et al., 2012a; Holland and Gallagher, 1999; Peck and Salzman, 2014). In sum, the amygdala and the PFC appear to be critical components of a broader neural network, including sensory regions, which are involved in the prioritized processing of emotional stimuli and the spatial location in which they occupy.

Emotional Learning and Memory

This amygdala-centered brain system for emotional attention is important for the immediate prioritization and subsequent processing of emotional stimuli. However, additional mechanisms are involved in a more long-term prioritization of stimuli that have been linked to emotional outcomes or emotional life events. For example, if there is an electrical outlet at your house or workplace that is not working appropriately and emits an electrical shock when something is plugged in, it would be beneficial for you to avoid plugging items into this outlet. If you were to plug in your laptop or phone charger into the outlet (conditioned stimulus) and get shocked (unconditioned stimulus) you will likely withdraw from the outlet and emit a startle response (unconditioned response to the shock). The brain has a system in place to associate this outlet with danger and ultimately motivate you to avoid this particular outlet in the future (conditioned response). Even if you did not get shocked by the outlet, but observed someone else get shocked or were simply told about this relationship, the brain learns the same stimulus-response relationship and the outcome is the same.

In the case of fear learning, the amygdala plays a critically important role. Sensory information related to the unconditioned stimulus and the conditioned stimulus converges in the amygdala (lateral nucleus). Through the coactivation of these convergent pathways structural changes occur at the synaptic level, which result in the conditioned stimulus eliciting a conditioned fear response. Across a number of mammalian species, damage to the amygdala has been shown to abolish fear learning. Initial research in rodents demonstrated that amygdala damage results in not only the inability to acquire new fear associations, but also the inability to express previously learned fear associations (Davis, 1992a, 1992b; Iwata et al., 1986; LeDoux, 1996). This work in rodents has subsequently been replicated in humans. In a seminal study, human patients with unilateral temporal lobectomy (including removal of the amygdala) showed impairment in fear conditioning (LaBar et al., 1995). In particular, the researchers presented squares of different colors (e.g., blue and green) on a computer screen and a certain colored square would occasionally be copresented with an electric shock. Healthy participants without amygdala damage would develop a conditioned skin conductance response (SCR) (a measure of sympathetic nervous system activity associated with emotional arousal) to the conditioned stimulus (e.g., blue square) in the absence of the unconditioned stimulus (e.g., electric shock) suggesting that an association between the two stimuli had been formed. On the other hand, although patients with amygdala damage displayed an increase in skin conductance to the electric shock (no different than healthy controls), they did not display a SCR to the conditioned stimulus (e.g., blue square) in the absence of the unconditioned stimulus (e.g., electric shock) suggesting that no association between the two stimuli had been formed and that the amygdala is necessary for this type of fear learning. Subsequent neuroimaging research indicates that the amygdala is active in healthy individuals during fear learning (Armony and Dolan, 2002; Knight et al., 2005; LaBar et al., 1998; Morris et al., 1998; Phelps, 2004). In rodent models, fear learning has been more precisely localized to the lateral nucleus of the amygdala (LeDoux, 2007). Thus, the amygdala is important for the formation and storage of new fear associations.

Research also suggests that emotional learning affects cortically generated ERPs. Researchers tested the differences between pre- and post-conditioning ERPs after pairing unpleasant faces with shock (conditioned stimulus) and pleasant faces without shock (control stimulus; Wong et al., 1997). The results revealed significant differences between the conditioned and control stimuli on P1, N1, and P3 ERP components. The most pronounced effects were on the P3, an ERP component related to higher-level cognitive processes, such as attention and memory (Polich, 2007). These results provide evidence that emotional learning has widespread effects beyond the amygdala that can be detected in cortical neurons.

Think back to our example of the faulty electrical outlet. Is the immediately adjacent outlet safe? What about other outlets in the room? What about outlets in other buildings? The brain's ability to identify the precise nature of a threat (or other emotional association) and to appropriately generalize this association is another important aspect of emotional learning. Indeed, abnormalities in fear generalization are at the heart of affective disorders, such as generalized anxiety disorder (GAD), which is characterized by an overgeneralized threat response. Human neuroimaging research suggests that the ability to generalize conditioned fear appropriately is less dependent upon the amygdala, but rather a neural system including the insula, caudate nucleus, anterior cingulate, ventral medial PFC, somatosensory cortex, and a number of other regions (Dunsmoor et al., 2011; Greenberg et al., 2013a; Lissek et al., 2014). In order to study this generalization of emotional learning, researchers have paired an otherwise neutral stimulus, such as rectangle of a particular size (conditioned stimulus), with a shock (unconditioned stimulus). Intermixed with the conditioned–unconditioned stimulus pairings are generalization stimuli. These generalization stimuli are perceptually similar to the conditioned stimulus, but distinct. For example, in one study researchers paired a rectangle with an electric shock and created generalization stimuli that were ±20, ±40, and ±60% in width (Greenberg et al., 2013a). Using fMRI, researchers then looked at which areas of the brain were maximally activated (or deactivated) by the conditioned stimulus and displayed graded gradually decreasing (or increasing) activity to the generalization stimuli as they became less similar to the conditioned stimulus (i.e., CS to ±20% to ±40% to ±60%). The results suggest that the amygdala quickly habituates in this context; however, structures, such as the anterior insula, dorsal anterior cingulate, caudate nucleus, supplementary motor area, ventral medial PFC, and somatosensory cortex display generalization gradients (Greenberg et al., 2013a). Structures within this broad generalization circuit are functionally coupled with the amygdala, and therefore implicate a role of the amygdala in mediating this widespread system for fear generalization (Greenberg et al., 2013b). Interestingly, within this system there is an overgeneralization of conditioned fear in the ventral medial PFC, somatosensory cortex, and ventral tegmental area (a cluster of dopamine producing neurons in the midbrain) in patients with GAD (Cha et al., 2014a,b; Greenberg et al., 2013b). These results suggest that emotional learning involves structures beyond the amygdala—especially for sharp discriminations between appropriate and inappropriate generalization of learned emotional associations.

Environmental context is another important factor in emotional learning. For example, you see a man holding a gun. Is the gun a threat? It depends on the context. If you are at a track meet and the man holding the gun is starting the race, then no. If you are at a military funeral with a 21-gun salute, then no. However, if you are alone in a dark alley and approached by a stranger with a gun, then the answer is likely yes! In these examples, context dictates the level of threat or salience associated with the emotional stimulus. Researchers using rodent models have found that contextual fear learning is mediated by a pathway from the hippocampus to the amygdala. In particular, placing an animal into a testing box where it previously received a shock elicits freezing behavior (a conditioned response) even before the conditioned (tone) or unconditioned (shock) stimuli are presented. This does not occur for testing environments in which an animal has not been shocked. Animals with amygdala damage display no memory for the conditioned stimulus or the context (Phillips and LeDoux, 1992). On the other hand, if the hippocampus is damaged the animal maintains the memory for the conditioned stimulus, but not the contextual memory. Thus, the hippocampus appears to communicate contextual information to the amygdala to regulate the emotional response.

We do not need to directly experience an aversive outcome to learn a fear association. Simply observing another individual get shocked when using our faulty outlet will lead us to learn about the association between this outlet and shock. Learning by watching others is referred to as observational learning. The medial PFC has been linked to the understanding of one's own emotional state and the emotional state of others (Amodio and Frith, 2006). As previously mentioned, the medial PFC has rich reciprocal connections with the amygdala (Amaral and Price, 1984) and it appears that through these connections the medial PFC communicates the conditioned stimulus–unconditioned stimulus relationship to the amygdala as the neural basis of emotional learning by observation (Olsson et al., 2007). Thus, whether we directly experience an aversive outcome or observe it, the emotional associations that are formed involve the amygdala as well as the medial PFC in the case of observational learning.

So far, we have discussed emotional learning that results in automatic emotional responses. For example, pairing a blue square with an electric shock elicits a conditioned SCR and the ability to develop this skin conductance is dependent upon the amygdala. However, patients with amygdala damage *know* that the blue square results in a shock. That is, when asked about the relationship between the blue square and the electric shock these patients have no problem telling researchers the association between the two (LaBar et al., 1995). Thus, there is a distinction between explicitly "knowing" the association between the conditioned and unconditioned stimuli and the ability to form an implicitly learned emotional response. In contrast to amygdala damage, damage to the hippocampus impairs the ability to form a declarative memory about the association between the blue square and shock, but does not impair the ability to form the conditioned emotional response (Bechara et al., 1995). These results suggest the amygdala is necessary for fear conditioning, but what role (if any) does it play in the formation of memories that we explicitly recall and describe verbally?

As with other types of emotional learning, animal models have been used to identify the neural correlates involved in long-term declarative memories involving emotion. One task that has been used in rodent models is the Morris water maze. In this task, a rat is placed in a circular shaped pool of murky water. Within the pool is a submerged platform on which the rat can safely stand. Initially, the rat does not know the location of the platform and will swim around the maze until it accidentally comes across the platform. After several trials, the rat learns the location of the platform and will immediately swim to that location on all subsequent trials. Damage to the hippocampus leads to the rat displaying no memory of learning the location of the platform (i.e., the rat swims randomly throughout the maze). If a state of emotional arousal is induced prior to learning, the animal will show enhanced memory for the location of the platform (i.e., takes less time to reach the platform). If the amygdala is damaged, the animal still remembers the location of the platform, but the facilitation of learning typically observed under conditions of emotional arousal is abolished (McGaugh, 2002, 2004; McGaugh et al., 2002; Packard et al., 1994). Thus, the amygdala is not necessary for the formation of long-term declarative memories, but the enhancement of memory by emotion is dependent upon the amygdala. It appears the amygdala modulates the encoding of long-term declarative memory through activation of neuromodulatory systems, such as the norepinephrine system (Cahill and McGaugh, 1996; Cahill et al., 1994). This amygdala-dependent enhancement of declarative memory is also true in humans—patients with amygdala damage do not display the typical enhancement of long-term memory by emotion (Adolphs et al., 1997; Cahill et al., 1995). Furthermore, neuroimaging research suggests that amygdala activity during an emotional event predicts later memory for this event (Cahill et al., 1996; Canli et al., 2000). The amygdala influences the hippocampus during the initial encoding or formation of the emotional event (Dolcos et al., 2004) and is also active during the recall of an emotional memory (Dolcos et al., 2005). This amygdala-mediated enhancement of emotional memories is driven by arousal rather than valence (Kensinger and Corkin, 2004) suggesting that the amygdala broadly enhances explicit memory for emotional salient events.

It is not only important to learn the emotional significance of environmental stimuli, but also to be able to disassociate environmental stimuli from emotional outcomes when this relationship no longer holds. For example, if the faulty outlet from the earlier example gets fixed and works appropriately, it is no longer adaptive to avoid this outlet as it is no longer associated with an emotional outcome. Indeed, the brain has a mechanism in place for this type of learning or unlearning. In laboratory settings, this type of learning is assessed by measuring the extent to which conditioned responses (e.g., freezing) are elicited by conditioned stimuli (e.g., tones) when the conditioned stimulus is no longer paired with the unconditioned stimulus (e.g., shock) over repeated trials. This process is referred to as extinction. Research using both rodent (Quirk et al., 2006) and human (Phelps et al., 2004) models suggests that the medial PFC plays an important role in extinction learning. In particular, the connection between the amygdala and medial PFC has been found to be important for both the acquisition (Burgos-Robles et al., 2009) and extinction (Vouimba and Maroun, 2011) stages of emotional memory. Thus, the amygdala stores emotional memories and through connections with the medial PFC these memories are weakened or become extinct when the learned association is no longer predictive.

In summary, the amygdala projects to a number of regions, which makes it optimally suited for its role in emotional responding. Primary outputs from the amygdala include the hypothalamus, brainstem, PFC,

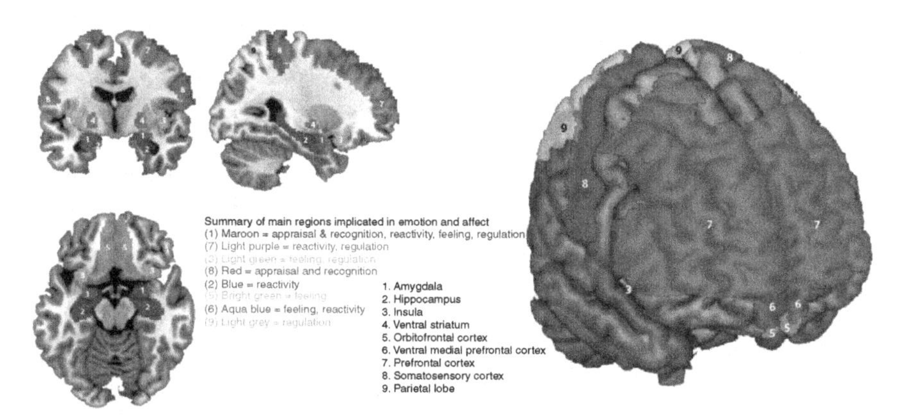

FIGURE 2.3 Illustrated are the primary areas of the brain introduced in this chapter as being associated with emotional appraisal, reactivity, feeling, and regulation. Note that the amygdala (1) is associated with appraisal, reactivity, feeling, and regulation processes.

hippocampus, striatum, and sensory regions, which result in the modulation of the following responses: hormone release, autonomic nervous system activation, attention, decision-making, memory formation, and perceptual processing (Fig. 2.3). Attention and memory are two primary reactive behaviors modulated by emotion that have short-term and long-term adaptive value, respectively. Through connections with the PFC, sensory cortices, and nucleus basalis cholinergic modulatory center the amygdala mediates a rapid orienting of spatial attention and immediate preferential processing of emotionally relevant stimuli. When new emotional associations are formed through learning, the amygdala appears to be the site of implicit emotional memories, while connections between the amygdala and hippocampus are important for the emotional enhancement of explicit memory. Although we have focused our discussion of emotional learning on the amygdala, other brain regions, such as the ventral striatum (Day and Carelli, 2007) and insula (Kiefer et al., 1982) have been shown to be important for other types of emotional learning including reward and disgust, respectively. In short, the amygdala is part of the brain system which initiates a number of reactive responses to emotionally arousing stimuli including enhanced attention and memory.

EMOTIONAL FEELINGS AND THEIR FUNCTIONS

In this section, we discuss what occurs in the brain when experiencing a subjective emotional feeling. Up to this point, we have reviewed the neural processes involved leading up to the experience of an emotion, but what structures are necessary for the "feeling" of an emotion?

As you may have guessed, the answer is going to vary depending on the emotional state and features of the emotion-inducing stimuli. There has been exponential innovation in neuroimaging techniques in recent years (Donoghue, 2002) and recent interest in studying feelings scientifically. Therefore, it is likely we are just beginning to scratch the surface of how emotional feelings are represented and produced by different structures and circuits in the brain. Nevertheless, significant strides have been made in our understanding of the complex relationship between neural processes and emotional feelings.

The Role of the Amygdala in Emotional Feelings

As discussed in detail in the preceding sections, the amygdala is an important subcortical structure involved in the appraisal of and response to emotional stimuli. The amygdala has connections with many other areas of the brain. Important for its role in emotional feelings, are connections to regions that regulate the activity of the autonomic nervous system and regions involved in hormonal control (Buijs and Van Eden, 2000; LeDoux et al., 1988). Initial research on the amygdala implicated its tremendous role in the emotion of fear. Individuals with amygdala damage can explicitly identify dangerous stimuli, but their peripheral physiology shows no signs of arousal nor do they feel any subjective fear (LaBar et al., 1995).

In a study with patient S.M., researchers directly confronted her with fear-inducing stimuli and observed her behavior while also monitoring her subjective state (Feinstein et al., 2011). The researchers chose three situations that were capable of inducing fear but held little to no risk of direct harm. They studied her while visiting an exotic pet store with snakes and spiders (she claimed to "hate" snakes and spiders and "tries to avoid them," this suggests the cognitive aspect of fear is intact), walking through a haunted house, and watching film clips of scary movies (e.g., *The Blair Witch Project, The Shining, The Silence of the Lambs)*. At no point did S.M. show any behavioral or subjective indications of fear. Likewise, 3 months of real-life experience sampling and self-report questionnaires on a life history full of traumatic events revealed S.M. repeatedly showed an absence of overt fear behaviors and an overall diminished experience of fear. She has been robbed at gunpoint and knifepoint, almost killed in a domestic violence incident, and has been the subject of several death threats. The disproportionate number of traumatic events in S.M.'s life can presumably be tied to her inability to experience fear when entering situations that may be dangerous or when danger is looming. Despite her lack of fear, S.M. is able to display other basic emotions and experience their respective feelings. These findings suggest that the amygdala plays an emotion-specific role in the experience of fear.

The most revealing finding about S.M.'s amygdala damage comes from Feinstein et al. (2013). The researchers were looking to replicate and extend findings from animal studies that showed that the amygdala detects CO_2 and acidosis to produce fear behaviors. Due to S.M.'s amygdala damage, they did not expect CO_2 inhalation to produce a fear response. To their great surprise, when S.M. inhaled the CO_2 she reported fear and even experienced a panic attack. To the researchers' knowledge, this was the first time S.M. experienced fear since her childhood, which makes sense as the amygdala develops primarily during adolescence (Feinstein et al., 2011). The researchers replicated these findings with two other individuals with Urbach–Weithe disease who also reported experiencing panic attacks after inhalation of 35% CO_2.

These findings raise the question of what is different about CO_2 compared to the other stimuli that failed to induce fear or panic. One possibility posited by Feinstein et al. (2013) is that the amygdala is a brain region critical for triggering a state of fear when an individual encounters threatening stimuli in the external environment, but not when the fear is triggered interoceptively—directly through internal stimulation. Fear-inducing stimuli, other than CO_2, are exteroceptive in nature—environmental stimuli mainly processed through the visual and auditory pathways that project to the amygdala. In contrast, CO_2 acts internally at acid-activated receptors and causes an assortment of physiological changes. Structures in the brainstem are likely responsible for interoceptive fear-inducing stimuli, not the amygdala.

This conclusion is consistent with our knowledge of the functional neuroanatomy of the amygdala. S.M.'s sensory and association cortices, necessary for representing external stimuli, are intact, as is the brainstem and hypothalamic circuitry responsible for initiating the autonomic and hormonal fear response. S.M.'s amygdala damage essentially disconnects these two components, making it unlikely and perhaps impossible, for sensory stimuli to trigger fear responses.

To conclude, these results indicate that the amygdala is necessary for fear and panic initiated by external threats in the environment, but not fear triggered internally by CO_2. Since patients with bilateral amygdala damage were still able to experience fear when induced interoceptively, we can assume that the amygdala plays a crucial role in initiating fear responses, *but it is not necessarily where the subjective feeling of fear occurs.* Patients with amygdala damage are able to explicitly identify dangerous and threatening stimuli, but show no signs of peripheral or subjective fear to external stimuli.

The Role of the Insula in Emotional Feelings

If the amygdala is not the locus of emotional feeling, what part of the brain is? There is extensive research suggesting that the insula may be where this process occurs (Craig, 2009). As its size and complex cytoarchitecture indicate, the insula performs a number of diverse functions (Chikama et al., 1997). It is located in a junction between the frontal, temporal, and parietal lobes and is not visible from an outside view of the brain. The insula is richly interconnected with many cortical and subcortical structures including the amygdala (Amaral and Price, 1984), allowing for high-level representations of affective state (Singer et al., 2009). It is interconnected with subcortical structures like the brainstem, amygdala, and basal ganglia, supporting its role in representing autonomic, emotional, and visceral states. The insula also has dense connections with regions in the frontal lobe that support decision-making like the PFC and anterior cingulate cortex (ACC).

The anterior insular cortex (AIC) is a multisensory hub that plays a significant role in emotion-specific processing (Lamm and Singer, 2010). The anteroventral aspect of the AIC is thought to play a role in olfactory and autonomic function. The anterodorsal and medial sections facilitate our sense of taste by forming the primary gustatory cortex (GC). Researchers are able to elicit activity in the primary GC by comparing disgust-relevant with neutral stimuli, suggesting a connection between our sense of taste and the emotion of disgust (Wicker et al., 2003a). Another negatively valenced emotion the insula seems to be associated with is fear. Schienle et al. (2002) found the insula to be active when comparing fear inducing with neutral stimuli and in anticipation or expectation of something bad happening. The most posterior part of the AIC establishes a multisensory integration area that receives somatosensory, visual, and auditory information.

Initial research on the AIC indicated its role was to provide representations of the body that serve as the basis for all subjective feelings from the body and potentially emotional awareness. Functional MRI studies have found that AIC activity tracks states of awareness for one's physiological state (Craig, 2002; Critchley et al., 2004) as well as reported affective experience or feelings (Carlson et al., 2011b; Singer et al., 2009). Furthermore, these processes share an overlapping common substrate within the AIC (Zaki et al., 2012). The body and brain are constantly sending signals back and forth and the AIC seems to be the region responsible for representing bodily states. There is now meta-analytic evidence that suggests that the AIC plays a fundamental role in human awareness (Craig, 2009). Compelling evidence suggests the AIC contains the neurological substrate for the evolved ability of humans to be aware of themselves, others, and the environment. Functional MRI studies report activation of the AIC in response to a wide range of stimuli (e.g., sensing one's heartbeat, feeling pain and touch, social exclusion, sexual arousal, etc. see Craig, 2009 for review), all of which involve conscious feelings. Consciousness is thought to be made of a mental and physical component. The insula is active in response to seemingly all bodily reactions that reach consciousness as well as all conscious feelings and appraisals. Consequently, the AIC should be considered as a potential neural correlate of consciousness (Craig, 2009).

Other studies have shown the AIC to be selectively activated in response to subjects viewing pictures of their own face or body (Devue et al., 2007). In one such study, participants performed a task in which they were to indicate the real appearance of themselves and of a gender-matched close colleague among intact and altered pictures of faces and bodies. They found the AIC and right frontal cortex to be the main regions specifically involved in visual self-recognition compared with visual processing of other highly familiar persons. Additionally, the right AIC in concert with the right ACC appear to play a role in the integration of information about oneself independently of the stimulus domain (Devue et al., 2007). These results led the authors to suggest that the right AIC and ACC "could give rise to an abstract representation of oneself that could possibly participate in maintaining a sense of self."

The Role of the Orbitofrontal Cortex in Emotional Feelings

The OFC is a subregion of the PFC and frontal lobe associated with reward processing and subjective feelings of pleasantness (Kringelbach et al., 2003). The topography of the OFC can be broken down into three main subregions: (1) medial OFC (mOFC), (2) lateral OFC (lOFC), and (3) pars orbitalis. There is now meta-analytic evidence suggesting the mOFC activates in response to a wide range of pleasant or rewarding stimuli, whereas the lOFC is more sensitive to punishers and negatively valenced stimuli (Kringelbach and Rolls, 2004). Additionally, there is research showing that greater expression of positive emotion is associated with larger medial, but not lateral, OFC volumes (Welborn et al., 2009). Evidence from postmortem studies (Rajkowska, 2000), literature reviews (Lorenzetti et al., 2009), and meta-analyses (Kempton et al. 2011; Koolschijn et al., 2009) shows that individuals with major depressive disorder (MDD) and schizophrenia (Baaré et al., 1999; Gur et al. 2000; Liao et al. 2015) have reduced mOFC brain volumes. This commonality can be explained through their shared symptom of anhedonia or lack of positive affect. Collectively, these findings suggest that the mOFC is linked with reward processing and reduced mOFC volume is related to disorders characterized by blunted positive affect.

In an unorthodox, but inventive experimental design, Carlson et al. (2015a) collected self-reported ratings of euphoria after participants had just touched ground from their first ever skydiving experience and correlated the findings with MRI measures. What the researchers found was that individuals' level of skydive-induced euphoria was uniquely linked to the volume of their left mOFC, such that, greater mOFC volumes predicted greater euphoria. These findings complement the work done on patients with MDD and schizophrenia, both of which are characterized by anhedonia and reduced volume of the mOFC (Cardoner et al., 2007; Liao et al., 2015). Another activity humans engage in to experience pleasure is listening to music, which can trigger highly euphoric experiences and even "musical chills." Similar to skydiving, the type of music that elicits such chills is person specific and is linked to activity in the mOFC, further suggesting a link between subjective feelings of euphoria and the mOFC (Blood and Zatorre, 2001).

Brief Overview of Emotional Feelings in Decision-Making

When making decisions, individuals use both cognitive and emotional processes to assess the incentive value of the options presented to them. Cognitive processes alone can become overloaded when presented with complex and conflicting choices. When this happens, your brain may use somatic markers to help make a decision (Damasio et al., 1996). According to Damasio's somatic marker hypothesis, somatic markers are associations between stimuli and the physiological affective state they induce, which nonconsciously influence our behavior based on past experience with that stimulus. Somatic markers produce a feeling state that biases individuals toward certain behaviors; this is advantageous since you can then make a quick decision rather than relying on logic-based, but time-consuming, strategies. The brain is constantly sending new signals to the body when processing incoming sensory information. The body in turn sends signals back to the brain, relaying the physiological changes. The perception of these bodily changes are transmitted to the cortex and experienced as a subjective emotion. These physiological changes become associated with different incoming stimuli, and the valence of this association is one factor that is immediately considered when making a decision. Somatic markers therefore bias behavior toward more advantageous options, simplifying the decision-making process.

There is extensive research that indicates that the ventral medial prefrontal cortex (vmPFC) is one area critical to utilizing somatic markers. Individuals with damage to their vmPFC show marked changes in their ability to use past emotional and social experiences in their decision-making (Damasio et al., 1996). This illustrates that emotion is vital to sound decision-making, contrary to the centuries-old notion that emotion hinders reason. Patients with vmPFC damage have severe defects in their daily functioning that are not related to disturbances in pertinent knowledge, intellectual ability, language, basic working memory, or basic attention (Damasio et al., 1996). One study in particular legitimized Damasio's somatic marker hypothesis and the role of the vmPFC in utilizing emotion in decision-making (Damasio et al., 1996). The researchers had healthy controls and patients with vmPFC damage play a game known as the Iowa Gambling Task. In the task, participants are to choose cards from various decks that contain unique amounts of monetary rewards and losses. The game is too demanding to consciously keep track of which decks have more rewards and losses than others. However, individuals without brain damage are able to implicitly start choosing from decks that give them higher rewards than losses. Patients with damage localized to the vmPFC, are unable to do this and continue picking from "bad" decks; this implies that they were unable to form an emotional association when picking from "good" and "bad" decks. These patients are still able to generate emotional feelings, but are unable to use them later on in guiding their decisions. These results corroborate the defects shown in their daily lives and support the role of the vmPFC in guiding decision-making through the integration of emotional feelings.

As you can see, the "feeling" of an emotion is produced by an incredibly complex system that is subject to internal and external influences at every stage of the process. The amygdala is required for initiating the feeling of fear when encountering a threat in the environment. However, studies utilizing CO_2 inhalation show that individuals with amygdala damage can still feel fear in some contexts. The insula seems to be where these "feelings" occur and can be considered as a potential neural correlate of consciousness, since we see it active in response to any physical or mental feeling. Activity in other regions of the brain, such as the OFC, influences the feeling states that ultimately occur in the insula. Emotional feelings serve an important function. For example, not being able to associate an incoming stimulus with a previous emotional encounter is incredibly debilitating, as evidenced by patients with vmPFC damage who are unable to form these connections. Without the fast-acting effects of emotion, we would rely on very time-consuming, logic-based strategies to choose our behaviors. Subjective emotional experience provides the basis for "gut feelings," which immediately bias our behavior toward certain outcomes based upon innate and conditioned learning.

EMOTION REGULATION

Emotion regulation is the application of strategies that increase, maintain, or decrease the intensity, duration, and/or quality of an emotion (Gyurak et al., 2011). Emotions serve to motivate behavior and being able to regulate your emotions is advantageous because it allows you to engage in behaviors that may not provide an immediate reward. Being able to regulate your emotional reaction to an event also allows you to manipulate the likelihood of you engaging in behaviors associated with that emotion. For example, downregulating the emotion of anger to decrease its intensity makes you less likely to engage in behaviors associated with anger, such as fighting or aggression. Research on the neural mechanisms involved in emotion regulation has focused primarily on attempts to downregulate negatively valenced emotions. To date, emotion regulation strategies have been largely grouped into three different classes: (1) attentional control, (2) cognitive reappraisal, and (3) response modulation (Webb et al., 2012). Surprisingly, researchers have had little success in reliably differentiating which brain structures are specifically implicated with each strategy (Giuliani et al., 2011; Kühn et al., 2011). There seems to be much overlap in the neural resources recruited by these varied strategies.

Emotion Regulation Strategies

The three major emotion regulation strategies all work to modulate the intensity, duration, and/or quality of an emotion. Attentional control involves focusing one's attention away from an emotion-eliciting stimulus. Studies on attentional control have focused primarily on responses to pain and have found that various methods of shifting attention or distraction (simply being asked to "think of something else") diminish the averseness of pain (Tracey et al., 2002). Cognitive reappraisal involves reframing the problem or thinking about it in a different way. This has been explored in various ways including asking participants to imagine a context in which an emotional event or experience would be neutral (Ochsner et al., 2002), in which it retreats into the distance (Davis et al., 2011), and in which the participants act as objective, scientific observers without being emotionally involved (Goldin et al., 2008). Response modulation involves altering emotion expressive behavior. For example, when you attempt to suppress tears when you are sad to prevent yourself from becoming even sadder, you are using response modulation. This strategy may be effective in the short term, but research shows that frequent expressive suppression may be dysfunctional as it can result in diminished control of emotion, interpersonal functioning, memory, well-being, and greater depressive symptomatology (Gross and John, 2003).

Neural Correlates of Emotion Regulation

As mentioned previously, research focused on differentiating neural structures associated with different emotion regulation strategies has been largely unclear or not replicable (Giuliani et al., 2011; Kühn et al., 2011). However, there do seem to be some structures implicated in all three emotion regulation strategies. This overlap is found primarily within the frontal, temporal, and parietal lobes.

Successful downregulation of negative emotion using cognitive reappraisal is associated with decreased activity in the amygdala (Goldin et al., 2008; Ochsner et al., 2002, 2004). This is consistent with our knowledge of its pervasive role in emotional appraisal and reactivity. The insula is another region implicated in emotion regulation. Similar to the amygdala, fMRI activation of this structure is typically reduced when individuals effectively downregulate emotion using cognitive reappraisal (Carlson and Mujica-Parodi, 2010; Goldin et al., 2008; Herwig et al., 2007). This makes sense due to its role in monitoring body physiology; effective downregulation means you should have less peripheral arousal and thus, less activity in the insula. In general, this pattern of activity in the amygdala and insula makes sense—if you downregulate an emotion, there should be decreased activity in brain areas associated with emotional responding and feeling.

The temporal lobe's role in emotion regulation is primarily due to its connections with the amygdala; however, the temporal pole (TP) also appears to be involved. The TP is a region in the temporal lobe situated roughly between the OFC and the amygdala that receives and sends connections to both regions. There has been relatively little research on this region, but it is thought to potentially play a role in emotion regulation by incorporating emotional and social information (Olson et al., 2007). Humans and monkeys with TP damage show profound changes in personality, emotion regulation, and social behavior (Gorno-Tempini et al., 2004; Olson et al., 2007).

The frontal lobe also contains several regions implicated in emotion regulation. Research has shown that activity in the ventrolateral prefrontal cortex (vlPFC) is correlated with reduced negative emotional experience during cognitive reappraisal of aversive images (Wager et al., 2008). Several fMRI studies have noted increased activity in other regions of the PFC when participants are instructed to deploy cognitive strategies that reduce negative emotional experience, including the dorsolateral prefrontal cortex (dlPFC) and the dorsomedial prefrontal cortex (dmPFC; Ochsner and Gross, 2005). More specifically, recent research has found that the right vlPFC is activated when participants exert attentional control over their emotions and when they engage in cognitive reappraisal (Bishop et al., 2004; Ochsner and Gross, 2005). It seems that the vlPFC and ACC are implicated in almost all imaging studies of emotion regulation, and regardless of whether the participants are trying to increase or decrease the intensity of an emotion (Ochsner et al., 2004). These frontal regions incorporate emotional and social information and are able to influence emotional reactions through their connections to the amygdala (Ochsner and Gross, 2005). Evidence suggests that control processes initiated in the frontal lobe serve to inhibit amygdala function. In particular, the PFC and amygdala are negatively coupled during emotion regulation (Banks et al., 2007) and successful emotion regulation is mediated by this pathway (Wager et al., 2008). Collectively, these findings suggest a strong role of the frontal lobe, specifically the PFC, in emotion regulation.

Another region implicated in the various emotion regulation strategies is the parietal lobe. The parietal lobe's role in emotion regulation likely has to do with its role in maintaining attention. All of the emotion regulation strategies recruit some level of attention to modify emotional responses. The posterior P300 ERP component, localized to the parietal lobe, is elicited by emotional stimuli and more recent ERP research indicates that in addition to an increased P300, emotional stimuli are associated with a more sustained positivity in the ERP, known as the late positive potential (LPP; Hajcak et al., 2010). Converging evidence suggests that increases in the LPP are directly related to the emotional salience of the eliciting stimulus, such that highly arousing pleasant and unpleasant stimuli, such as erotica and threat scenes, result in corresponding increases in the LPP (Schupp et al., 2004). The LPP is also implicated when participants are instructed to regulate their emotions. Studies that have participants reinterpret unpleasant images in a less negative way have shown that reductions in the LPP are correlated with reductions in self-reported emotional experience following reappraisal (Hajcak and Nieuwenhuis, 2006). Furthermore, fMRI research indicates that the intraparietal sulcus, the inferior parietal lobe, and dorsal premotor cortex are richly interconnected and form what's known as the frontoparietal attention network (Ptak, 2011). These close functional and structural interconnections with other areas involved in emotion regulation support the notion that parietal activation helps direct and maintain attention toward emotion regulation goals.

As you can see, the various emotion regulation strategies recruit a vast number of overlapping brain structures. This large distribution of regions makes sense given the importance of emotion regulation in our emotional, physical, and social well-being. Engaging in emotion regulation allows us to utilize our cognitive abilities when forming an emotional reaction. This can be highly advantageous as we can incorporate more information into our decisions—such as planning, emotional intelligence, social norms—and not just rely on past stimulus-reward (or threat) associations. Emotion regulation strategies have been generally grouped into three categories: (1) attentional control, (2) cognitive reappraisal, and (3) response modulation. Most of the research on emotion regulation has focused on attempts to downregulate negative emotions and numerous studies have found that effective downregulation of negative emotions is associated with increased PFC activity and decreased insula and amygdala activity. These results corroborate our knowledge of the PFC's role in cognitive control and the insula and amygdala's relationship in feeling negative emotional states.

INDIVIDUAL DIFFERENCES

> No two persons are born exactly alike; but each differs from the other in natural endowments, one being suited for one occupation and the other for another.
>
> —*Plato*

Most psychological and neuroscience research aims at finding general differences *between* two groups of individuals or between experimental conditions. However, it is also critical to understand what is different *within* certain populations of people that distinguish them from others; this is the study of individual differences. In any study, variation exists between individuals and insight into these differences can help explain and predict behavior. Within the brain, every single person has a unique composition of cytoarchitecture that gives rise to each of our unique experiences. These neural differences between individuals can be viewed as differences in degree rather than any fundamental structural difference. For example, like a face, individuals' brains are typically composed of the same components, barring major deficits during embryological development. Most faces are composed of eyes, eyebrows, lips, cheeks, and a nose. Even though almost every face is made of the same components, each is distinguishable in its own unique way through slight modulations giving it its own unique identity. Likewise, almost every brain is composed of largely the same components, but different features have been modulated to varying degrees to produce individual differences. This section will explore a small number of examples of individual differences in the brain that seem to be linked to characteristic behaviors and psychological disorders.

Anxiety

Anxiety is one area of research where we are beginning to attain a better understanding of how individual differences in brain structure and function contribute to behavior. (Almost) everybody feels anxious at times; it is only when anxiety disrupts an individual's ability to live a normal life that it is classified as a psychological disorder. It has been long established that the amygdala plays a critical role in fear responses and anxiety (Kluver and Bucy, 1939). However, this initial understanding has been refined and we now have a much better understanding of how different regions within the amygdala and its subnuclei produce fear and anxiety. There has been extensive research done, on both humans and rodents, that suggests the central nucleus of the amygdala is most important in responding to immediate threats and the bed nucleus of the stria terminalis (BNST) seems to be involved in more prolonged threats and in anticipation to threat, more akin to anxiety (Fox et al., 2015).

One study found increasing BNST activity as the subjects imagined a spider advancing closer and closer to their foot (Mobbs et al., 2010). In another study, participants were shown a fake reading of their SCR and told they would receive a shock when it reached a certain threshold. The researchers found increased BNST activation as their fake SCR approached the threshold (Somerville et al., 2010). In a Pacman-like game where capture was associated with an electric shock, researchers found increased

BNST activity as the computer closed in on the subject's avatar (Mobbs et al., 2007). In all of the aforementioned studies, researchers found increased BNST activity in both anxious and healthy populations. However, we see even greater activations of the BNST when looking specifically at individuals with different types of anxiety disorders. Notable examples of this increased BNST activation in anxious populations include spider phobics during anticipation of phobia-relevant images (Fox et al., 2015), posttraumatic stress disorder (PTSD) sufferers when viewing combat-relevant scenes (Morey et al., 2009), and in patients with GAD during a gambling task (Yassa et al., 2012). These experiments and many others support the idea that variability in BNST activity may distinguish those with anxiety disorders from the general population.

As mentioned earlier in the chapter, individuals with clinical anxiety also display abnormal patterns of brain activity in the vmPFC, somatosensory cortex, and ventral tegmental area during the generalization of conditioned fear (Cha et al., 2014a,b; Greenberg et al., 2013b), which have been found to vary on an individual-by-individual level in relation to symptom severity. Thus, in addition to the BNST a number of neural regions beyond those discussed in this chapter have been shown to vary in relation to symptoms of anxiety.

Depression

Like anxiety, depression is a psychological disorder that is accompanied by individual differences in brain structure and function. Depression has a very diverse set of possible symptoms, which presents a major problem when trying to localize variations in brain structure and activity to the disease. Some individuals diagnosed with depression may only have some, or as little as one, of the symptoms another person diagnosed with the same disease has, since the DSM-V only requires 5/9 depressive symptoms for a diagnosis (American Psychiatric Association, 2013). However, depression is generally characterized by a persistent feeling of sadness, lack of interest, anhedonia, and commonly various somatic complaints.

The ACC, and more broadly the mPFC, is a region commonly implicated in depression and other mood disorders. Drevets et al. (2008) demonstrated that the mean gray matter volume of the ACC is abnormally reduced in patients with MDD and bipolar disorder, regardless of mood state. Interestingly, they found that this abnormality was associated with a reduction in glia cells, with no equivalent loss of neurons. Although MDD is associated with reduced gray matter in the ACC, numerous researchers have actually found increased activity in this region in depressed patients compared to controls. In studies of remitted MDD patients, researchers have found that ACC metabolism increases during depressive relapses (Hasler et al., 2008; Neumeister et al., 2004). Successful remission is associated with decreased activity in the ACC, regardless if the depression is treated with antidepressants (Drevets et al., 2002), deep brain stimulation (Mayberg et al., 2005), or electroconvulsive therapy (ECT; Nobler et al., 2001).

Another region heavily implicated in the depression literature is the ventral striatum (Ubeda-Bañon et al., 2007). As mentioned earlier, it has been shown to play an important role in reward processing and mediating the motivational effects of emotionally significant stimuli (Cardinal et al., 2002). Compared to normal subjects, depressed individuals show significantly less ventral striatal and mPFC fMRI activity in response to positive stimuli (Epstein et al., 2006). Converging evidence from ERPs suggests that the reward-related positivity—a positive amplitude deflection occuring 250-350 ms post-stimulus onset at central frontal electrode cites—is a potential biomarker for depression (Proudfit, 2015). The reward-related positivity is maximally elicited by rewarding stimuli and has been found to correlate with ventral striatal and mPFC activity in a combined ERP and fMRI study (Carlson et al., 2011a) and correlates with the volume of the midbrain (Carlson et al., 2015c). The reward-related positivity ERP has been found to be blunted in depressed individuals (Bress et al., 2013) and has been linked to blunted ventral striatal activation, specifically in anhedonic depression (Foti et al., 2014). Thus, the decreased activation in the ventral striatum is correlated with decreased interest/pleasure in and performance of activities. Additional evidence from rodent studies has found that long-term exposure to various stressors decreases dopamine release in the ventral striatum and mPFC (Scheggi et al., 2002). Grieve et al. (2013) found decreased gray matter volume in the mPFC in MDD participants when compared to control groups. Other research has shown that additional individual differences further mediate this relationship as lower mPFC volume in depression has been found in males, but less so in females (Carlson et al., 2015b). These results are part of a growing literature implicating frontolimbic dysfunction in depression.

Overall, we see that individual differences in brain structure and activity lie at the heart of the behavioral differences observed between individuals. In any sample, variation can be found between individuals and this variation is the focus of study when exploring individual differences. The individual differences presented in this section are not exhaustive and are subject to other internal and external influences, such as personality, culture, age, gender, etc. The take-home point is that there will always be some variation between individuals in a population or group and this variation will influence how they feel and behave. This has been alluded to in other sections of this chapter—patients with MDD and schizophrenia are characterized by reduced mOFC volumes and mOFC volumes predict how euphoric a person will feel when skydiving. No two brains are formed completely alike or completely different and thus, no two people are completely alike or completely different.

BRIEF OVERVIEW OF OTHER EMOTIONS

So far we have focused mainly on fear and joy as examples of how the brain processes emotion. We have provided evidence that the amygdala is part of a neural system associated with detecting emotional salience and mediating the emotional response to a wide variety of emotional stimuli. However, specific categories of emotions appear to exist and for some of these emotions the brain appears to have discrete regions dedicated to their functions. Emotions can influence behaviors meant to support inclusion in a social group. Often this means maintaining or increasing physical or emotional proximity to the group (Farrell and Alberts, 2002; Levine et al., 1987). Other times the organism wishes to increase the distance between itself and some unpleasant stimulus. In the first case, researchers have identified the behaviors associated with sadness as being meant to express distress and despair so that they can receive attention from a social group or significant social partner (Farrell and Alberts, 2002; Levine et al., 1987). In the second case, expressions of disgust are elicited by displeasure with something and decrease contact with the eliciting stimulus both in the short term, but also in the future, particularly as it pertains to food sources (Rozin and Fallon, 1987). In this section, we will examine the neural correlates of sadness and disgust.

Sadness

Sadness is viewed by some researchers to be a two-stage process in which the initial behaviors include vocalizations (e.g., crying, whining, whimpering, etc.) in an attempt to restore contact with a social group. The second stage includes withdrawal behaviors potentially meant to reduce further strain on the organism's standing with the social group or communicate "hurt" feelings (Michalak et al., 2009). Exclusion from a social group or removal of an important social partner, such as a break-up or death, is often related to sadness. Functional MRI studies examining the neural correlates of sadness have attempted to replicate social exclusion in experimental settings. One widely tested hypothesis is that the neural correlates of social pain (e.g., sadness from social exclusion) are the same as those involved in physical pain. Studies using fMRI have found increased activation in the ACC in response to physical pain (Coghill et al., 2003).

To create the negative emotional experience of social exclusion in an experimental setting, researchers developed a game for participants to play called "Cyber Ball" (Eisenberger et al., 2003). In this experiment, participants were told that they would play a digital game of catch with two other participants, while in reality they were playing a game of catch with a computer program. The experiment had one condition where participants were told there was an issue and they wouldn't be able to play, but could watch the others play (watching condition). In another condition, participants were allowed to play with the other "participants" (inclusion condition). In a final condition, the ball would be thrown to the subject a few times, but then the other two participants would only throw to each other (social exclusion condition). The results of the fMRI scans indicated that the ACC became significantly more active in the exclusion and watching conditions relative to the inclusion condition. Additionally, there was greater activity in the right ventral prefrontal cortex (VPFC) in the exclusion condition compared to the inclusion condition. Further analysis revealed that increased activation in the VPFC was associated with decreased activation in the ACC—suggesting that the VLPFC mitigates the negative feelings associated with social exclusion (Eisenberger et al., 2003). The regulatory relationship between the VPFC and ACC for social pain detected by Eisenberger et al. (2003) is similar to that known to help regulate physical pain (Foltz and White, 1968; Eisenberger and Lieberman, 2004).

Although Cyber Ball studies provide evidence in support of the idea that emotional and physical pain shares a common neural substrate, namely the ACC, a recent meta-analysis has brought this understanding into question. The authors examined the fMRI results of 244 participants across a number of Cyber Ball studies and determined that there was no activation in the pain network (Cacioppo et al., 2013). The authors suggest that the results found in the original Cyber Ball studies may have been the by-product of low statistical power due to small sample sizes. Furthermore, rejection in Cyber Ball comes from strangers and not close significant others and therefore may not be eliciting social exclusion per se. Although the quantitative meta-analysis was unable to support claims that social and physical pain uses the same neural network, there are multiple studies supporting the initial Cyber Ball findings including some with relatively high sample sizes (Eisenberger et al., 2003, 2006, 2007; Kawamoto et al., 2012; MacDonald and Leary, 2005). Thus, although it was initially believed that the ACC was the neural locus of sadness, mixed results and potentially underpowered statistical analyses have called this view point into question.

Disgust

Behaviors related to sadness typically result in maintaining or decreasing physical or emotional proximity, whereas disgust behaviors communicate dissatisfaction with some stimulus in an effort to avoid it. Sadness and disgust are both negative emotions, but the distinct observable behaviors and motivated outcomes related to these emotions indicate that they are likely mediated by distinct neural substrates. Cyber Ball studies have revealed the ACC as a region associated with sadness and fMRI studies using disgust-related stimuli have implicated the insula (Wicker et al., 2003a).

External behaviors related to emotions are beneficial to the individual organism and can be used to communicate information regarding an organism's feelings about something in the environment. In the case of disgust, these behaviors can benefit the group evolutionarily (e.g., making a disgusted face after drinking out of a dirty puddle may help others in the social group avoid the dirty and potentially toxic water). Some theories on the importance of disgust view it as mostly digestion related, meaning disgust behaviors communicate toxicity in objects and potential objects of consumption within the environment (Rozin and Fallon, 1987). Indeed, it has been found that the amygdala and OFC are activated when coming in contact with aversive gustatory stimuli (Zald et al., 1998). In monkeys, a region of the brain known as the gustatory cortex (GC; homolog of the gustatory region of the human insula) shows increased neuronal activity to pleasant and unpleasant tastes regardless of satiation (Rolls et al., 1988). Furthermore, ablation of the GC in rats leads to a failure to learn aversion to unpleasant tastes at the same rate as control organisms (Kiefer and Orr, 1992). Functional MRI studies in humans have indicated that the insula is the primary neural substrate of disgust (Phillips et al., 1997; Wicker et al., 2003a).

To detect the neural substrates of disgust in humans, one group presented subjects with disgusted, fearful, and neutral faces while undergoing fMRI (Phillips et al., 1997). The results of comparing BOLD signals produced when viewing fearful to viewing neutral faces indicated activation of the amygdala as found in many other studies, but disgust did not cause significant changes in amygdala activation. The comparison of disgust to neutral faces did reveal significant changes in insula activation. These results were some of the first to demonstrate different neural substrates for fear and disgust. Furthermore, the results of this study found activation in the anterior insula similar to that found in studies where subjects taste aversive substances, suggesting the insula is involved in feeling and recognizing disgust (Yaxley et al., 1988). In another experiment, participants underwent fMRI scans while being shown disgusted faces in some conditions and being presented with unpleasant odors in others (Wicker et al., 2003a). The analysis of fMRI BOLD signal indicated that the same regions involved in seeing disgusted faces were also involved in feeling disgust related to unpleasant odors. Thus, as has been discussed with other emotions, the recognition of disgust in others relies on regions in the brain that are active while feeling disgust subjectively.

Although emotions, such as sadness and disgust are both considered negative emotions, they have unique neural substrates that produce behaviors with specific outcomes. Sadness is thought to be the result of exclusion from a social group (Farrell and Alberts, 2002; Levine et al., 1987).

Phrases like "hurt feelings" in reference to sadness are not uncommon in English speaking countries (Eisenberger et al., 2006). The results of fMRI studies attempting to create hurt feelings via social exclusion in the Cyber Ball game have revealed that the ACC may underlie the negative feelings associated with both physical and social pain (Eisenberger et al., 2003, 2006, 2007; Kawamoto et al., 2012; MacDonald and Leary, 2005). While the ACC may play a role in sadness, the GC/insula plays a role in disgust. Subjective feelings of disgust and displays of disgust are important for the survival of an individual organism and the group by communicating toxicity in the environment (Rozin and Fallon, 1987). Functional MRI studies have provided evidence suggesting that the same brain regions involved in feeling disgust are involved in recognizing disgust in others (Phillips et al., 1998). We have reviewed structures, such as the amygdala and nucleus accumbens, that are involved in more general aspects of emotional processing like salience detection and reward processing, respectively. Here, we used the examples of sadness and disgust to demonstrate how some specific emotions and related behaviors appear to have both common and unique neural correlates.

CONCLUSIONS

In this chapter, we provided a brief introduction into the neural mechanisms thought to underlie emotion. The research used to make these claims utilized a multitude of techniques in both human and animal subjects. These techniques focused on the loss of function following localized brain damage, as well as neuroimaging measures of electrical and BOLD activity in EEG/ERP and fMRI, respectively. Much of this chapter discussed one particular brain structure—the amygdala. Since the early work by Kluver and Bucy (1939) and its inclusion in the limbic system (MacLean, 1952), the amygdala has been prominently associated with emotion. Both animal and human research has found that the amygdala is critically involved in fear learning and a variety of other fear-related processes. However, after reading this chapter, it should be clear that although the amygdala is a major component of the emotional brain, it is by itself not *the* emotional brain. Furthermore, we hope the reader appreciates the fact that although the amygdala is tightly linked to fear-related processes, it is not a fear-only region, but plays a larger role in evaluating emotional salience more broadly.

Emotions are believed to serve evolutionarily adaptive functions and are thought to contain multiple processing stages including appraisal, reactivity, and feeling. There are a number of different types of emotional appraisals and associated appraisal regions/networks in the brain including the amygdala, superior temporal sulcus, somatosensory cortex, and ventral striatum among others. There are also a number of different emotional reactions including enhanced attention to and memory of emotional stimuli/events. The neural network for emotional attention appears to be centered on the amygdala, but also includes the PFC, sensory cortices, and diffuse modulatory centers, such as the cholinergic basal forebrain. The neural substrate for implicit emotional memories appears to be the amygdala. Through its interactions with the hippocampus as well as the insula, anterior cingulate, vmPFC, caudate, and other regions, context and generalization factors add to emotional learning. The amygdala also modulates explicit memory through its influence on the hippocampus. Other emotional reactions mediated by the amygdala include autonomic, hormonal, and muscular activity. These responses in particular have direct effects on the body, which are detected and appear to be relayed to the anterior insula. There is accumulating evidence that the anterior insula serves as a hub for many processes involving conscious awareness—including the representation of subjective emotional feeling states. Areas involved in emotional reactivity and feeling decrease in activity when emotion is successfully regulated, while activity in areas of the regulatory PFC increases. Although structures, such as the amygdala, are involved in appraising and reacting to emotionally salient stimuli in general, certain emotions, such as sadness and disgust appear to selectively recruit the anterior cingulate and the insula (gustatory portion), respectively. In sum, emotion is not a singular process and the emotional brain is not a singular region. Rather, emotions are complex multistage phenomena and the neural representation is likewise distributed and related to the stage of emotional processing.

References

Adolphs, R., 2004. Emotional vision. Nat. Neurosci. 7 (11), 1167–1168.

Adolphs, R., Tranel, D., Damasio, H., Damasio, A., 1994. Impaired recognition of emotion in facial expressions following bilateral damage to the human amygdala. Nature 372 (6507), 669–672.

Adolphs, R., Cahill, L., Schul, R., Babinsky, R., 1997. Impaired declarative memory for emotional material following bilateral amygdala damage in humans. Learn. Mem. 4 (3), 291–300.

Adolphs, R., Tranel, D., Damasio, A.R., 1998. The human amygdala in social judgment. Nature 393 (6684), 470–474.

Adolphs, R., Tranel, D., Hamann, S., Young, A.W., Calder, A.J., Phelps, E.A., et al., 1999. Recognition of facial emotion in nine individuals with bilateral amygdala damage. Neuropsychologia 37 (10), 1111–1117.

Adolphs, R., Damasio, H., Tranel, D., Cooper, G., Damasio, A.R., 2000. A role for somatosensory cortices in the visual recognition of emotion as revealed by three-dimensional lesion mapping. J. Neurosci. 20 (7), 2683–2690, http://www.jneurosci.org/.

Adolphs, R., Gosselin, F., Buchanan, T.W., Tranel, D., Schyns, P., Damasio, A.R., 2005. A mechanism for impaired fear recognition after amygdala damage. Nature 433 (7021), 68–72.

Amaral, D.G., Price, J.L., 1984. Amygdalo-cortical projections in the monkey (*Macaca fascicularis*). J. Comp. Neurol. 230 (4), 465–496.

American Psychiatric Association, 2013. Diagnostic and Statistical Manual of Mental Disorders, fifth ed. American Psychiatric Association, Washington DC.
Amodio, D.M., Frith, C.D., 2006. Meeting of minds: the medial frontal cortex and social cognition. Nat. Rev. Neurosci. 7 (4), 268–277.
Anderson, A.K., Phelps, E.A., 2001. Lesions of the human amygdala impair enhanced perception of emotionally salient events. Nature 411 (6835), 305–309.
Armony, J.L., Dolan, R.J., 2002. Modulation of spatial attention by fear-conditioned stimuli: an event-related fMRI study. Neuropsychologia 40 (7), 817–826.
Armony, J.L., Corbo, V., Clement, M.H., Brunet, A., 2005. Amygdala response in patients with acute PTSD to masked and unmasked emotional facial expressions. Am. J. Psychiatry 162 (10), 1961–1963.
Baaré, W.F., Hulshoff Pol, H.E., Hijman, R., Th. Mali, W.P., Viergever, M.A., Kahn, R.S., 1999. Volumetric analysis of frontal lobe regions in schizophrenia: relation to cognitive function and symptomatology. Biol. Psychiatry 45 (12), 1597–1605.
Bach, D.R., Hurlemann, R., Dolan, R.J., 2014. Impaired threat prioritisation after selective bilateral amygdala lesions. Cortex 63C, 206–213.
Banks, S.J., Eddy, K.T., Angstadt, M., Nathan, P.J., Phan, K.L., 2007. Amygdala-frontal connectivity during emotion regulation. Soc. Cogn. Affect. Neurosci. 2 (4), 303–312.
Bassili, J.N., 1979. Emotion recognition: the role of facial movement and the relative importance of upper and lower areas of the face. J. Pers. Soc. Psychol. 37 (11), 2049–2058.
Bean, B.P., 2007. The action potential in mammalian central neurons. Nat. Rev. Neurosci. 8 (6), 451–465.
Beaver, J.D., Mogg, K., Bradley, B.P., 2005. Emotional conditioning to masked stimuli and modulation of visuospatial attention. Emotion 5 (1), 67–79.
Bechara, A., Tranel, D., Damasio, H., Adolphs, R., Rockland, C., Damasio, A.R., 1995. Double dissociation of conditioning and declarative knowledge relative to the amygdala and hippocampus in humans. Science 269 (5227), 1115–1118.
Begleiter, H., Porjesz, B., Garozzo, R., 1979. Visual evoked potentials and affective ratings of semantic stimuli. In: Begleiter, H. (Ed.), Evoked Brain Potentials and Behavior. Springer, New York, pp. 127–141.
Bishop, S., Duncan, J., Brett, M., Lawrence, A.D., 2004. Prefrontal cortical function and anxiety: controlling attention to threat-related stimuli. Nat. Neurosci. 7 (2), 184–188.
Blanchette, I., 2006. Snakes, spiders, guns, and syringes: how specific are evolutionary constraints on the detection of threatening stimuli? QJ Exp. Psychol. 59 (8), 1484–1504.
Blau, V.C., Maurer, U., Tottenham, N., McCandliss, B.D., 2007. The face-specific N170 component is modulated by emotional facial expression. Behav. Brain Funct. 3 (7).
Blood, A.J., Zatorre, R.J., 2001. Intensely pleasurable responses to music correlate with activity in brain regions implicated in reward and emotion. Proc. Natl. Acad. Sci. 98 (20), 11818–11823.
Bogousslavsky, J., 2003. William Feinberg lecture 2002: emotions, mood, and behavior after stroke. Stroke 34, 1046–1050.
Botvinick, M., Nystrom, L.E., Fissell, K., Carter, C.S., Cohen, J.D., 1999. Conflict monitoring versus selection-for-action in anterior cingulate cortex. Nature 402 (6758), 179–181.
Botvinick, M.M., Cohen, J.D., Carter, C.S., 2004. Conflict monitoring and anterior cingulate cortex: an update. Trends Cogn. Sci. 8 (12), 539–546.
Bradley, M.M., Lang, P.J., 1994. Measuring emotion: the self-assessment manikin and the semantic differential. J. Behav. Ther. Exp. Psychiatry 25 (1), 49–59.
Bradley, M.M., Codispoti, M., Cuthbert, B.N., Lang, P.J., 2001. Emotion and motivation I: defensive and appetitive reactions in picture processing. Emotion 1 (3), 276–298.
Breen, M.S., Beliakova-Bethell, N., Mujica-Parodi, L.R., Carlson, J.M., Ensign, W.Y., Woelk, C.H., Rana, B.K., 2016. Acute psychological stress induces short-term variable immune response. Brain Behav. Immun. 53, 172–182.
Breiter, H.C., Etcoff, N.L., Whalen, P.J., Kennedy, W.A., Rauch, S.L., Buckner, R.L., et al., 1996. Response and habituation of the human amygdala during visual processing of facial expression. Neuron 17 (5), 875–887.

Bress, J.N., Foti, D., Kotov, R., Klein, D.N., Hajcak, G., 2013. Blunted neural response to rewards prospectively predicts depression in adolescent girls. Psychophysiology 50 (1), 74–81.
Brosch, T., Grandjean, D., Sander, D., Scherer, K.R., 2008. Behold the voice of wrath: cross-modal modulation of visual attention by anger prosody. Cognition 106 (3), 1497–1503.
Buijs, R., Van Eden, C., 2000. The integration of stress by the hypothalamus, amygdala and prefrontal cortex: balance between the autonomic nervous system and the neuroendocrine system. Prog. Brain Res. 126, 117–132.
Burgos-Robles, A., Vidal-Gonzalez, I., Quirk, G.J., 2009. Sustained conditioned responses in prelimbic prefrontal neurons are correlated with fear expression and extinction failure. J. Neurosci. 29 (26), 8474–8482.
Buzsáki, G., Draguhn, A., 2004. Neuronal oscillations in cortical networks. Science 304 (5679), 1926–1929.
Cacioppo, S., Frum, C., Asp, E., Weiss, R.M., Lewis, J.W., Cacioppo, J.T., 2013. A quantitative meta-analysis of functional imaging studies of social rejection. Sci. Rep. 3 (2027).
Cahill, L., McGaugh, J.L., 1996. The neurobiology of memory for emotional events: adrenergic activation and the amygdala. Proc. West Pharmacol. Soc. 39, 81–84.
Cahill, L., Prins, B., Weber, M., McGaugh, J.L., 1994. Beta-adrenergic activation and memory for emotional events. Nature 371 (6499), 702–704.
Cahill, L., Babinsky, R., Markowitsch, H.J., McGaugh, J.L., 1995. The amygdala and emotional memory. Nature 377 (6547), 295–296.
Cahill, L., Haier, R.J., Fallon, J., Alkire, M.T., Tang, C., Keator, D., et al., 1996. Amygdala activity at encoding correlated with long-term, free recall of emotional information. Proc. Natl. Acad. Sci. USA 93 (15), 8016–8021.
Campbell, R., Heywood, C.A., Cowey, A., Regard, M., Landis, T., 1990. Sensitivity to eye gaze in prosopagnosic patients and monkeys with superior temporal sulcus ablation. Neuropsychologia 28 (11), 1123–1142.
Canli, T., Zhao, Z., Brewer, J., Gabrieli, J.D., Cahill, L., 2000. Event-related activation in the human amygdala associates with later memory for individual emotional experience. J. Neurosci. 20 (19), RC99.
Cardinal, R.N., Parkinson, J.A., Hall, J., Everitt, B.J., 2002. Emotion and motivation: the role of the amygdala, ventral striatum, and prefrontal cortex. Neurosci. Biobehav. Rev. 26 (3), 321–352.
Cardoner, N., Soriano-Mas, C., Pujol, J., Alonso, P., Harrison, B.J., Deus, J., et al., 2007. Brain structural correlates of depressive comorbidity in obsessive–compulsive disorder. Neuroimage 38 (3), 413–421.
Carlson, J.M., Mujica-Parodi, L.R., 2010. A disposition to reappraise decreases anterior insula reactivity during anxious anticipation. Biol. Psychol. 85 (3), 383–385.
Carlson, J.M., Reinke, K.S., 2008. Masked fearful faces modulate the orienting of covert spatial attention. Emotion 8 (4), 522–529.
Carlson, J.M., Reinke, K.S., 2010. Spatial attention-related modulation of the N170 by backward masked fearful faces. Brain Cogn. 73 (1), 20–27.
Carlson, J.M., Reinke, K.S., 2014. Attending to the fear in your eyes: facilitated orienting and delayed disengagement. Cogn. Emot. 28 (8), 1398–1406.
Carlson, J.M., Fee, A.L., Reinke, K.S., 2009a. Backward masked snakes and guns modulate spatial attention. Evol. Psychol. 7 (4), 527–537.
Carlson, J.M., Reinke, K.S., Habib, R., 2009b. A left amygdala mediated network for rapid orienting to masked fearful faces. Neuropsychologia 47 (5), 1386–1389.
Carlson, J.M., Greenberg, T., Mujica-Parodi, L.R., 2010. Blind rage? Heightened anger is associated with altered amygdala responses to masked and unmasked fearful faces. Psychiatry Res. 182, 281–283.
Carlson, J.M., Foti, D., Mujica-Parodi, L.R., Harmon-Jones, E., Hajcak, G., 2011a. Ventral striatal and medial prefrontal BOLD activation is correlated with reward-related electrocortical activity: a combined ERP and fMRI study. Neuroimage 57 (4), 1608–1616.

Carlson, J.M., Greenberg, T., Rubin, D., Mujica-Parodi, L.R., 2011b. Feeling anxious: anticipatory amygdalo-insular response predicts the feeling of anxious anticipation. Soc. Cogn. Affect. Neurosci. 6, 74–81.

Carlson, J.M., Reinke, K.S., LaMontagne, P.J., Habib, R., 2011c. Backward masked fearful faces enhance contralateral occipital cortical activity for visual targets within the spotlight of attention. Soc. Cogn. Affect. Neurosci. 6 (5), 639–645.

Carlson, J.M., Beacher, F., Reinke, K.S., Habib, R., Harmon-Jones, E., Mujica-Parodi, L.R., Hajcak, G., 2012a. Nonconscious attention bias to threat is correlated with anterior cingulate cortex gray matter volume: A voxel-based morphometry result and replication. Neuroimage 59 (2), 1713–1718.

Carlson, J.M., Dikecligil, G.N., Greenberg, T., Mujica-Parodi, L.R., 2012b. Trait reappraisal is associated with resilience to acute psychological stress. J. Res. Pers. 46 (5), 609–613.

Carlson, J.M., Cha, J., Mujica-Parodi, L.R., 2013. Functional and structural amygdala—anterior cingulate connectivity correlates with attentional bias to masked fearful faces. Cortex 49 (9), 2595–2600.

Carlson, J.M., Cha, J., Harmon-Jones, E., Mujica-Parodi, L.R., Hajcak, G., 2014. Influence of the BDNF genotype on amygdalo-prefrontal white matter microstructure is linked to nonconscious attention bias to threat. Cereb. Cortex 24 (9), 2249–2257.

Carlson, J.M., Cha, J., Fekete, T., Greenberg, T., Mujica-Parodi, L.R., 2015a. Left medial orbitofrontal cortex volume correlates with skydive-elicited euphoric experience. Brain Struct. Funct. 21, 4269–4279.

Carlson, J.M., Depetro, E., Maxwell, J., Harmon-Jones, E., Hajcak, G., 2015b. Gender moderates the association between dorsal medial prefrontal cortex volume and depressive symptoms in a subclinical sample. Psychiatry Res. 233 (2), 285–288.

Carlson, J.M., Foti, D., Harmon-Jones, E., Proudfit, G.H., 2015c. Midbrain volume predicts fMRI and ERP measures of reward reactivity. Brain Struct. Funct. 220 (3), 1861–1866.

Cha, J., Carlson, J.M., Dedora, D.J., Greenberg, T., Proudfit, G.H., Mujica-Parodi, L.R., 2014a. Hyper-reactive human ventral tegmental area and aberrant mesocorticolimbic connectivity in overgeneralization of fear in generalized anxiety disorder. J. Neurosci. 34 (17), 5855–5860.

Cha, J., Greenberg, T., Carlson, J.M., DeDora, D.J., Hajcak, G., Mujica-Parodi, L.R., 2014b. Circuit-wide structural and functional measures predict ventromedial prefrontal fear generalization: implications for generalized anxiety disorder. J. Neurosci. 34 (11), 4043–4053.

Chatterton, Jr., R.T., Vogelsong, K.M., Lu, Y.C., Hudgens, G.A., 1997. Hormonal responses to psychological stress in men preparing for skydiving. J. Clin. Endocrinol. Metab. 82 (8), 2503–2509.

Chikama, M., McFarland, N., Amaral, D., Haber, S., 1997. Insular cortical projections to functional regions of the striatum correlate with cortical cytoarchitectonic organization in the primate. J. Neurosci. 17 (24), 9686–9705.

Coghill, R.C., McHaffie, J.G., Yen, Y.F., 2003. Neural correlates of interindividual differences in the subjective experience of pain. Proc. Natl. Acad. Sci. 100 (14), 8538–8542.

Cooper, R.M., Langton, S.R., 2006. Attentional bias to angry faces using the dot-probe task? It depends when you look for it. Behav. Res. Ther. 44 (9), 1321–1329.

Craig, A.D., 2002. How do you feel? Interoception: the sense of the physiological condition of the body. Nat. Rev. Neurosci. 3 (8), 655–666.

Craig, A.D., 2009. How do you feel—now? The anterior insula and human awareness. Nat. Rev. Neurosci. 10 (1), 59–70.

Critchley, H.D., Wiens, S., Rotshtein, P., Ohman, A., Dolan, R.J., 2004. Neural systems supporting interoceptive awareness. Nat. Neurosci. 7 (2), 189–195.

Cuffin, B.N., Cohen, D., 1979. Comparison of the magnetoencephalogram and electroencephalogram. Electroencephalogra. Clin. Neurophysiol. 47 (2), 132–146.

Cunningham, W.A., Brosch, T., 2012. Motivational salience amygdala tuning from traits, needs, values, and goals. Curr. Dir. Psychol. Sci. 21 (1), 54–59.

Cunningham, W.A., Van Bavel, J.J., Johnsen, I.R., 2008. Affective flexibility evaluative processing goals shape amygdala activity. Psychol. Sci. 19 (2), 152–160.

Damasio, H., Grabowski, T., Frank, R., Galaburda, A.M., Damasio, A.R., 1994. The return of Phineas Gage: clues about the brain from the skull of a famous patient. Science 264 (5162), 1102–1105.

Damasio, A.R., Everitt, B.J., Bishop, D., 1996. The somatic marker hypothesis and the possible functions of the prefrontal cortex [and discussion]. Philos. Trans. R Soc. B Biol. Sci. 351 (1346), 1413–1420.

Darwin, C., 1872. The Expression of the Emotions in Man and Animals. John Murray, London.

Davis, M., 1992a. The role of the amygdala in fear-potentiated startle: implications for animal models of anxiety. Trends Pharmacol. Sci 13 (1), 35–41.

Davis, M., 1992b. The role of the amygdala in fear and anxiety. Annu. Rev. Neurosci. 15, 353–375.

Davis, M., Whalen, P.J., 2001. The amygdala: vigilance and emotion. Mol. Psychiatry 6 (1), 13–34.

Davis, J.I., Gross, J.J., Ochsner, K.N., 2011. Psychological distance and emotional experience: what you see is what you get. Emotion 11 (2), 438–444.

Day, J.J., Carelli, R.M., 2007. The nucleus accumbens and Pavlovian reward learning. Neuroscientist 13 (2), 148–159.

Devue, C., Collette, F., Balteau, E., Degueldre, C., Luxen, A., Maquet, P., Brédart, S., 2007. Here I am: the cortical correlates of visual self-recognition. Brain Res. 1143, 169–182.

Dolcos, F., LaBar, K.S., Cabeza, R., 2004. Interaction between the amygdala and the medial temporal lobe memory system predicts better memory for emotional events. Neuron 42 (5), 855–863.

Dolcos, F., LaBar, K.S., Cabeza, R., 2005. Remembering one year later: role of the amygdala and the medial temporal lobe memory system in retrieving emotional memories. Proc. Natl. Acad. Sci. USA 102 (7), 2626–2631.

Donoghue, J.P., 2002. Connecting cortex to machines: recent advances in brain interfaces. Nat. Neurosci. 5, 1085–1088.

Drevets, W.C., Bogers, W., Raichle, M.E., 2002. Functional anatomical correlates of antidepressant drug treatment assessed using PET measures of regional glucose metabolism. Eur. Neuropsychopharmacol. 12 (6), 527–544.

Drevets, W.C., Savitz, J., Trimble, M., 2008. The subgenual anterior cingulate cortex in mood disorders. CNS Spectr. 13 (8), 663.

Dunsmoor, J.E., Prince, S.E., Murty, V.P., Kragel, P.A., LaBar, K.S., 2011. Neurobehavioral mechanisms of human fear generalization. Neuroimage 55 (4), 1878–1888.

Eisenberger, N.I., Lieberman, M.D., 2004. Why rejection hurts: a common neural alarm system for physical and social pain. Trends Cogn. Sci. 8 (7), 294–300.

Eisenberger, N.I., Lieberman, M.D., Williams, K.D., 2003. Does rejection hurt? An fMRI study of social exclusion. Science 302 (5643), 290–292.

Eisenberger, N.I., Jarcho, J.M., Lieberman, M.D., Naliboff, B.D., 2006. An experimental study of shared sensitivity to physical pain and social rejection. Pain 126 (1), 132–138.

Eisenberger, N.I., Gable, S.L., Lieberman, M.D., 2007. Functional magnetic resonance imaging responses relate to differences in real-world social experience. Emotion 7 (4), 745–754.

Ekman, P., 1992. Are there basic emotions? Psychol. Rev. 99 (3), 550–553.

Ekman, P., Sorenson, E.R., Friesen, W.V., 1969. Pan-cultural elements in facial displays of emotion. Science 164 (3875), 86–88.

Elam, K.K., Carlson, J.M., DiLalla, L.F., Reinke, K.S., 2010. Emotional faces capture spatial attention in 5-year-old children. Evol. Psychol. 8 (4), 754–767.

Epstein, J., Pan, H., Kocsis, J.H., Yang, Y., Butler, T., Chusid, J., Hochberg, H., Murrough, J., Strohmayer, E., Stern, E., Silbersweig, D.A., 2006. Lack of ventral striatal response to positive stimuli in depressed versus normal subjects. Am. J. Psychiatry 163 (10), 1784–1790.

Etkin, A., Egner, T., Peraza, D.M., Kandel, E.R., Hirsch, J., 2006. Resolving emotional conflict: a role for the rostral anterior cingulate cortex in modulating activity in the amygdala. Neuron 52 (6), 1121–11121.

Fanselow, M.S., LeDoux, J.E., 1999. Why we think plasticity underlying Pavlovian fear conditioning occurs in the basolateral amygdala. Neuron 23 (2), 229–232.

Farrell, W.J., Alberts, J.R., 2002. Stimulus control of maternal responsiveness to Norway rat (*Rattus norvegicus*) pup ultrasonic vocalizations. J. Comp. Psychol. 116 (3), 297.

Feinstein, J.S., Adolphs, R., Damasio, A., Tranel, D., 2011. The human amygdala and the induction and experience of fear. Curr. Biol. 21 (1), 34–38.

Feinstein, J.S., Buzza, C., Hurlemann, R., Follmer, R.L., Dahdaleh, N.S., Coryell, W.H., Wemmie, J.A., 2013. Fear and panic in humans with bilateral amygdala damage. Nat. Neurosci. 16 (3), 270–272.

Fekete, T., Rubin, D., Carlson, J.M., Mujica-Parodi, L.R., 2011a. The NIRS analysis package: noise reduction and statistical inference. PLoS One 6 (9), e24322.

Fekete, T., Rubin, D., Carlson, J.M., Mujica-Parodi, L.R., 2011b. A stand-alone method for anatomical localization of NIRS measurements. Neuroimage 56 (4), 2080–2088.

Foltz, E.L., White, L.E., 1968. The role of rostral cingulumotomy in" pain" relief. Int. J. Neurol. 6 (3–4), 353.

Foti, D., Carlson, J.M., Sauder, C.L., Proudfit, G.H., 2014. Reward dysfunction in major depression: multimodal neuroimaging evidence for refining the melancholic phenotype. Neuroimage 101, 50–58.

Fox, E., 2002. Processing emotional facial expressions: the role of anxiety and awareness. Cogn. Affect. Behav. Neurosci. 2 (1), 52–63.

Fox, E., Damjanovic, L., 2006. The eyes are sufficient to produce a threat superiority effect. Emotion 6 (3), 534–539.

Fox, A.S., Oler, J.A., Tromp, D.P., Fudge, J.L., Kalin, N.H., 2015. Extending the amygdala in theories of threat processing. Trends Neurosci. 38 (5), 319–329.

Fu, X., Taber-Thomas, B.C., Perez-Edgar, K., 2015. Frontolimbic functioning during threat-related attention: relations to early behavioral inhibition and anxiety in children. Biol. Psychol 122, 98–109.

Gerdes, A.B., Wieser, M.J., Alpers, G.W., 2014. Emotional pictures and sounds: a review of multimodal interactions of emotion cues in multiple domains. Front. Psychol. 5, 1351.

Giuliani, N.R., Drabant, E.M., Bhatnagar, R., Gross, J.J., 2011. Emotion regulation and brain plasticity: expressive suppression use predicts anterior insula volume. NeuroImage 58 (1), 10–15.

Goldin, P.R., McRae, K., Ramel, W., Gross, J.J., 2008. The neural bases of emotion regulation: reappraisal and suppression of negative emotion. Biol. Psychiatry 63 (6), 577–586.

Gorno-Tempini, M.L., Rankin, K.P., Woolley, J.D., Rosen, H.J., Phengrasamy, L., Miller, B.L., 2004. Cognitive and behavioral profile in a case of right anterior temporal lobe neurodegeneration. Cortex 40 (4–5), 631–644.

Greenberg, T., Carlson, J.M., Cha, J., Hajcak, G., Mujica-Parodi, L.R., 2013a. Neural reactivity tracks fear generalization gradients. Biol. Psychol. 92 (1), 2–8.

Greenberg, T., Carlson, J.M., Cha, J., Hajcak, G., Mujica-Parodi, L.R., 2013b. Ventromedial prefrontal cortex reactivity is altered in generalized anxiety disorder during fear generalization. Depress Anxiety 30 (3), 242–250.

Greenberg, T., Carlson, J.M., Rubin, D., Cha, J., Mujica-Parodi, L., 2015. Anticipation of high arousal aversive and positive movie clips engages common and distinct neural substrates. Soc. Cogn. Affect. Neurosci. 10 (4), 605–611.

Greene, M.R., Oliva, A., 2009. The briefest of glances: the time course of natural scene understanding. Psychol. Sci. 20 (4), 464–472.

Grieve, S.M., Korgaonkar, M.S., Koslow, S.H., Gordon, E., Williams, L.M., 2013. Widespread reductions in gray matter volume in depression. Neuroimage Clin. 3, 332–339.

Gross, J.J., John, O.P., 2003. Individual differences in two emotion regulation processes: implications for affect, relationships, and well-being. J. Pers. Soc. Psychol. 85 (2), 348–362.

Güntekin, B., Basar, E., 2007. Emotional face expressions are differentiated with brain oscillations. Int. J. Psychophysiol. 64 (1), 91–100.

Gur, R.E., Cowell, P.E., Latshaw, A., Turetsky, B.I., Grossman, R.I., Arnold, S.E., Gur, R.C., 2000. Reduced dorsal and orbital prefrontal gray matter volumes in schizophrenia. Arch. Gen. Psychiatry 57 (8), 761.

Gyurak, A., Gross, J.J., Etkin, A., 2011. Explicit and implicit emotion regulation: a dual-process framework. Cogn. Emot. 25 (3), 400–412.

Hajcak, G., Nieuwenhuis, S., 2006. Reappraisal modulates the electrocortical response to unpleasant pictures. Cogn. Affect. Behav. Neurosci. 6 (4), 291–297.

Hajcak, G., MacNamara, A., Olvet, D.M., 2010. Event-related potentials, emotion, and emotion regulation: an integrative review. Dev. Neuropsychol. 35 (2), 129–155.

Hajcak, G., Weinberg, A., MacNamara, A., Foti, D., 2012. ERPs and the study of emotion. The Oxford Handbook of Event-Related Potential Components. pp. 441–474.

Hamann, S.B., Ely, T.D., Hoffman, J.M., Kilts, C.D., 2002. Ecstasy and agony: activation of the human amygdala in positive and negative emotion. Psychol. Sci. 13 (2), 135–141.

Harmer, C.J., Shelley, N.C., Cowen, P.J., Goodwin, G.M., 2004. Increased positive versus negative affective perception in healthy volunteers following selective serotonin and norepinephrine reuptake inhibition. Am. J. Psychiatry 161 (7), 1256–1263.

Hasler, G., Fromm, S., Carlson, P.J., Luckenbaugh, D.A., Waldeck, T., Geraci, M., Roiser, J., Neumeister, A., Meyers, N., Charney, D., Drevets, W., 2008. Neural response to catecholamine depletion in unmedicated subjects with major depressive disorder in remission and healthy subjects. Arch. Gen. Psychiatry 65 (5), 521–531.

Haxby, J.V., Hoffman, E.A., Gobbini, M.I., 2000. The distributed human neural system for face perception. Trends Cogn. Sci. 4 (6), 223–233.

Herwig, U., Baumgartner, T., Kaffenberger, T., Bruhl, A., Kottlow, M., Schreiter-Gasser, U., et al., 2007. Modulation of anticipatory emotion and perception processing by cognitive control. Neuroimage 37 (2), 652–662.

Heuttel, S.A., Song, A.W., McCarthy, G., 2009. Functional Magnetic Resonance Imaging, second ed. Sinauer Associates, Sunderland.

Hoffman, E., Haxby, J., 2000. Distinct representations of eye gaze and identity in the distributed human neural system for face perception. Nat. Neurosci. 3 (1), 80–84.

Holland, P.C., Gallagher, M., 1999. Amygdala circuitry in attentional and representational processes. Trends Cogn. Sci. 3 (2), 65–73.

Horga, G., Maia, T.V., 2012. Conscious and unconscious processes in cognitive control: a theoretical perspective and a novel empirical approach. Front. Hum. Neurosci. 6, 199.

Hupert, T.J., Diamond, S.G., Franceschini, M.A., Boas, D.A., 2009. HomER: a review of time-series analysis methods for near-infrared spectroscopy of the brain. Appl. Opt. 48 (10), D280–D298.

Insel, T.R., 2010. The challenge of translation in neuroscience: a review of oxytocin, vasopressin, and affiliative behavior. Neuron 65 (6), 768–779.

Iwata, J., LeDoux, J.E., Meeley, M.P., Arneric, S., Reis, D.J., 1986. Intrinsic neurons in the amygdaloid field projected to by the medial geniculate body mediate emotional responses conditioned to acoustic stimuli. Brain Res. 383 (1–2), 195–214.

James, W., 1884. What is an emotion? Mind 9, 188–205.

Kandel, E.R., Schwartz, J.H., Jessell, T.M., Siegelbaum, S.A., Hudspeth, A.J., 2012. Principles of Neural Science, fifth ed. McGraw-Hill, New York.

Kawamoto, T., Onoda, K., Nakashima, K.I., Nittono, H., Yamaguchi, S., Ura, M., 2012. Is dorsal anterior cingulate cortex activation in response to social exclusion due to expectancy violation? An fMRI study. Front. Evol. Neurosci. 4 (11).

Kawashima, R., Sugiura, M., Kato, T., Nakamura, A., Hatano, K., Ito, K., et al., 1999. The human amygdala plays an important role in gaze monitoring. Brain 122 (4), 779–783.

Kempton, M.J., Salvador, Z., Munafo, M.R., Geddes, J.R., Simmons, A., Frangou, S., Williams, S.C., 2011. Structural neuroimaging studies in major depressive disorder. Meta-analysis and comparison with bipolar disorder. Arch. Gen. Psychiatry 68 (7), 675–690.

Kensinger, E.A., Corkin, S., 2004. Two routes to emotional memory: distinct neural processes for valence and arousal. Proc. Natl. Acad. Sci. USA 101 (9), 3310–3315.
Key, A.P., Dove, G.O., Maguire, M.J., 2005. Linking brainwaves to the brain: an ERP primer. Dev. Neuropsychol. 27 (2), 183–215.
Kiefer, S.W., Orr, M.R., 1992. Taste avoidance, but not aversion, learning in rats lacking gustatory cortex. Behav. Neurosci, 106 (1), 140–146.
Kiefer, S.W., Rusiniak, K.W., Garcia, J., 1982. Flavor-illness aversions: gustatory neocortex ablations disrupt taste but not taste-potentiated odor cues. J. Comp. Physiol. Psychol. 96 (4), 540–548.
King, F.A., Meyer, P.M., 1958. Effects of amygdaloid lesions upon septal hyperemotionality in the rat. Science 128 (3325), 655–656.
Kluver, H., Bucy, P.C., 1939. Preliminary analysis of functions of the temporal lobes in monkeys. Arch. Neurol. Psychiatry 42 (6), 979.
Knight, D.C., Nguyen, H.T., Bandettini, P.A., 2005. The role of the human amygdala in the production of conditioned fear responses. Neuroimage 26 (4), 1193–1200.
Knutson, B., Adams, C.M., Fong, G.W., Hommer, D., 2001a. Anticipation of increasing monetary reward selectively recruits nucleus accumbens. J. Neurosci. 21 (16), RC159, http://www.jneurosci.org/.
Knutson, B., Fong, G.W., Adams, C.M., Varner, J.L., Hommer, D., 2001b. Dissociation of reward anticipation and outcome with event-related fMRI. Neuroreport 12 (17), 3683–3687.
Koolschijn, P., Van Haren, N., Lensvelt-Mulders, G., Pol, H.H., Kahn, R., 2009. Brain volume abnormalities in major depressive disorder: a meta-analysis of magnetic resonance imaging studies. NeuroImage 47, S152.
Koster, E.H., Crombez, G., Verschuere, B., De Houwer, J., 2004. Selective attention to threat in the dot probe paradigm: differentiating vigilance and difficulty to disengage. Behav. Res. Ther. 42 (10), 1183–1192.
Kravitz, A.V., Tye, L.D., Kreitzer, A.C., 2012. Distinct roles for direct and indirect pathway striatal neurons in reinforcement. Nat. Neurosci. 15 (6), 816–818.
Kreta, M.E., Jaasmab, L., Biondac, T., Wijnend, J.G., 2016. Bonobos (*Pan paniscus*) show an attentional bias toward conspecifics' emotions. Proc. Natl. Acad. Sci. USA 113, 3761–3766.
Kringelbach, M., Rolls, E., 2004. The functional neuroanatomy of the human orbitofrontal cortex: evidence from neuroimaging and neuropsychology. Prog. Neurobiol. 72 (5), 341–372.
Kringelbach, M., O'Doherty, J., Rolls, E., Andrews, C., 2003. Activation of the human orbitofrontal cortex to a liquid food stimulus is correlated with its subjective pleasantness. Cereb. Cortex 13 (10), 1064–1071.
Kühn, S., Gallinat, J., Brass, M., 2011. Keep calm and carry on": structural correlates of expressive suppression of emotions. PLoS One 6 (1), e16569.
LaBar, K.S., LeDoux, J.E., Spencer, D.D., Phelps, E.A., 1995. Impaired fear conditioning following unilateral temporal lobectomy in humans. J. Neurosci. 15 (10), 6846–6855.
LaBar, K.S., Gatenby, J.C., Gore, J.C., LeDoux, J.E., Phelps, E.A., 1998. Human amygdala activation during conditioned fear acquisition and extinction: a mixed-trial fMRI study. Neuron 20 (5), 937–945.
Lahnakoski, J.M., Glerean, E., Salmi, J., Jääskeläinen, I.P., Sams, M., Hari, R., Nummenmaa, L., 2012. Naturalistic fMRI mapping reveals superior temporal sulcus as hub for the distributed brain network for social perception. Front. Hum. Neurosci. 6 (14), 233.
Lamm, C., Singer, T., 2010. The role of anterior insular cortex in social emotions. Brain Struct. Funct. 214 (5–6), 579–591.
Lang, P.J., Greenwald, M.K., Bradley, M.M., Hamm, A.O., 1993. Looking at pictures: affective, facial, visceral, and behavioral reactions. Psychophysiology 30 (3), 261–273.
Lang, P.J., Bradley, M.M., Fitzsimmons, J.R., Cuthbert, B.N., Scott, J.D., Moulder, B., Nangia, V., 1998. Emotional arousal and activation of the visual cortex: an fMRI analysis. Psychophysiology 35 (2), 199–210.

LeDoux, J.E., 1996. The Emotional Brain the Mysterious Underpinnings of Emotional Life. Weidenfeld and Nicholson, London.

LeDoux, J., 2007. The amygdala. Curr. Biol. 17 (20), R868–R874.

LeDoux, J.E., Iwata, J., Cicchetti, P.R.D.J., Reis, D.J., 1988. Different projections of the central amygdaloid nucleus mediate autonomic and behavioral correlates of conditioned fear. J. Neurosci. 8 (7), 2517–2529, http://www.jneurosci.org/.

Levine, S., Wiener, S.G., Coe, C.L., Bayart, F.E., Hayashi, K.T., 1987. Primate vocalization: a psychobiological approach. Child Dev. 58 (6), 1408–1419.

Liao, J., Yan, H., Liu, Q., Yan, J., Zhang, L., Jiang, S., Zhang, X., Dong, Z., Yang, W., Cai, L., Wang, Y., Zimeng, L., Tian, L., Zhang, D., Wang, F., 2015. Reduced paralimbic system gray matter volume in schizophrenia: correlations with clinical variables, symptomatology and cognitive function. J. Psychiatric Res. 65, 80–86.

Liddell, B.J., Brown, K.J., Kemp, A.H., Barton, M.J., Das, P., Peduto, A., et al., 2005. A direct brainstem-amygdala-cortical 'alarm' system for subliminal signals of fear. Neuroimage 24 (1), 235–243.

Lissek, S., Bradford, D.E., Alvarez, R.P., Burton, P., Espensen-Sturges, T., Reynolds, R.C., Grillon, C., 2014. Neural substrates of classically conditioned fear-generalization in humans: a parametric fMRI study. Soc. Cogn. Affect. Neurosci. 9 (8), 1134–1142.

Lopez da Silva, F., van Rotterdam, A., 2005. Biophysical aspects of EEG and magnetoencephalographic generation. In: Niedermeyer, E., Lopes da Silva, F. (Eds.), Electroencephalography: Basic Principles, Clinical Applications and Related Fields. fifth ed. Lippincott, Williams & Wilkins, New York, pp. 107–110.

Lorenzetti, V., Allen, N.B., Fornito, A., Yücel, M., 2009. Structural brain abnormalities in major depressive disorder: a selective review of recent MRI studies. J. Affect. Disord. 117 (1–2), 1–17.

MacDonald, G., Leary, M.R., 2005. Why does social exclusion hurt? The relationship between social and physical pain. Psychol. Bull. 131 (2), 202–223.

MacLean, P.D., 1949. Psychosomatic disease and the "visceral brain": recent developments bearing on the Papez theory of emotion. Psychosom. Med. 11 (6), 338–353.

MacLean, P.D., 1952. Some psychiatric implications of physiological studies on frontotemporal portion of the limbic system (visceral brain). Electroencephalogr. Clin. Neurophysiol. 4 (4), 407–418.

MacLeod, C., Mathews, A., 1988. Anxiety and the allocation of attention to threat. QJ Exp. Psychol. A 40 (4), 653–670.

Magistretti, P.J., Pellerin, L., Rothman, D.L., Shulman, R.G., 1999. Energy on demand. Science 283 (5401), 496.

Magri, C., Schridde, U., Murayama, Y., Panzeri, S., Logothetis, N.K., 2012. The amplitude and timing of the BOLD signal reflects the relationship between local field potential power at different frequencies. J. Neurosci. 32 (4), 1395–1407.

Malonek, D., Grinvald, A., 1996. Interactions between electrical activity and cortical microcirculation revealed by imaging spectroscopy: implications for functional brain mapping. Science 272 (5261), 551–554.

Mayberg, H.S., Lozano, A.M., Voon, V., McNeely, H.E., Seminowicz, D., Hamani, C., Schwalb, J.M., Kennedy, S.H., 2005. Deep brain stimulation for treatment-resistant depression. Neuron 45 (5), 651–660.

McDonald, A.J., 1998. Cortical pathways to the mammalian amygdala. Prog. Neurobiol. 55 (3), 257–332.

McGaugh, J.L., 2002. Memory consolidation and the amygdala: a systems perspective. Trends Neurosci. 25 (9), 456.

McGaugh, J.L., 2004. The amygdala modulates the consolidation of memories of emotionally arousing experiences. Annu. Rev. Neurosci. 27, 1–28.

McGaugh, J.L., McIntyre, C.K., Power, A.E., 2002. Amygdala modulation of memory consolidation: interaction with other brain systems. Neurobiol. Learn Mem. 78 (3), 539–552.

Michalak, J., Troje, N.F., Fischer, J., Vollmar, P., Heidenreich, T., Schulte, D., 2009. Embodiment of sadness and depression—gait patterns associated with dysphoric mood. Psychosom. Med. 71 (5), 580–587.

Mitchell, H.H., Hamilton, T.S., Steggerda, F.R., Bean, H.W., 1945. The chemical composition of the human body and its bearing on the biochemistry of growth. J. Biol. Chem. 158, 625–637.

Mobbs, D., Petrovic, P., Marchant, J.L., Hassabis, D., Weiskopf, N., Seymour, B., Raymond, J.B., Frith, C.D., 2007. When fear is near: threat imminence elicits prefrontal-periaqueductal gray shifts in humans. Science 317 (5841), 1079–1083.

Mobbs, D., Yu, R., Rowe, J.B., Eich, H., FeldmanHall, O., Dalgleish, T., 2010. Neural activity associated with monitoring the oscillating threat value of a tarantula. Proc. Natl. Acad. Sci. 107 (47), 20582–20586.

Mogg, K., Bradley, B.P., 1999. Some methodological issues in assessing attentional biases for threatening faces in anxiety: a replication study using a modified version of the probe detection task. Behav. Res. Ther. 37 (6), 595–604.

Mogg, K., Bradley, B.P., 2002. Selective orienting of attention to masked threat faces in social anxiety. Behav. Res. Ther. 40 (12), 1403–1414.

Mohammad-Zadeh, L.F., Moses, L., Gwaltney-Brant, S.M., 2008. Serotonin: A review. J. Vet. Pharmacol. Ther. 31 (3), 187–199.

Monk, C.S., Telzer, E.H., Mogg, K., Bradley, B.P., Mai, X., Louro, H.M., et al., 2008. Amygdala and ventrolateral prefrontal cortex activation to masked angry faces in children and adolescents with generalized anxiety disorder. Arch. Gen. Psychiatry 65 (5), 568–576.

Morey, R.A., Dolcos, F., Petty, C.M., Cooper, D.A., Hayes, J.P., LaBar, K.S., McCarthy, G., 2009. The role of trauma-related distractors on neural systems for working memory and emotion processing in posttraumatic stress disorder. J. Psychiatric Res. 43 (8), 809–817.

Morris, J.S., Frith, C.D., Perrett, D.I., Rowland, D., Young, A.W., Calder, A.J., Dolan, R.J., 1996. A differential neural response in the human amygdala to fearful and happy facial expressions. Nature 383 (6603), 812–815.

Morris, J.S., Ohman, A., Dolan, R.J., 1998. Conscious and unconscious emotional learning in the human amygdala. Nature 393 (6684), 467–470.

Morris, J.S., Ohman, A., Dolan, R.J., 1999. A subcortical pathway to the right amygdala mediating "unseen" fear. Proc. Natl. Acad. Sci. USA 96 (4), 1680–1685.

Morris, J.S., DeGelder, B., Weiskrantz, L., Dolan, R.J., 2001. Differential extrageniculostriate and amygdala responses to presentation of emotional faces in a cortically blind field. Brain 124 (Pt 6), 1241–1252.

Mujica-Parodi, L.R., Carlson, J.M., Cha, J., Rubin, D., 2014. The fine line between 'brave' and 'reckless': amygdala reactivity and regulation predict recognition of risk. Neuroimage 103, 1–9.

Nahm, F.K., Tranel, D., Damasio, H., Damasio, A.R., 1993. Cross-modal associations and the human amygdala. Neuropsychologia 31 (8), 727–744.

N'Diaye, K., Sander, D., Vuilleumier, P., 2009. Self-relevance processing in the human amygdala: gaze direction, facial expression, and emotion intensity. Emotion 9 (6), 798–806.

Neumeister, A., Nugent, A.C., Waldeck, T., Geraci, M., Schwarz, M., Bonne, O., Baine, E., Luckenbaugh, E., Herscovitch, P., Charney, D., Drevets, W., 2004. Neural and behavioral responses to tryptophan depletion in unmedicated patients with remitted major depressive disorder and controls. Arch. Gen. Psychiatry 61 (8), 765–773.

Nobler, M.S., Oquendo, M.A., Kegeles, L.S., Malone, K.M., Campbell, C., Sackeim, H.A., Mann, J.J., 2001. Decreased regional brain metabolism after ECT. Am. J. Psychiatry 158 (2), 305–308.

Nonaka, H., Akima, M., Hatori, T., Nagayama, T., Zhang, Z., Ihara, F., 2002. The microvasculature of the human cerebellar meninges. Acta Neuropathologica 104 (6), 608–614.

Ochsner, K.N., Gross, J.J., 2005. The cognitive control of emotion. Trends Cogn. Sci. 9 (5), 242–249.

Ochsner, K.N., Bunge, S.A., Gross, J.J., Gabrieli, J.D., 2002. Rethinking feelings: an fMRI study of the cognitive regulation of emotion. J. Cogn. Neurosci. 14 (8), 1215–1229.

Ochsner, K.N., Ray, R.D., Cooper, J.C., Robertson, E.R., Chopra, S., Gabrieli, J.D., Gross, J.J., 2004. For better or for worse: neural systems supporting the cognitive down- and up-regulation of negative emotion. NeuroImage 23 (2), 483–499.

Ogawa, S., Lee, T.M., Nayak, A.S., Glynn, P., 1990. Oxygenation-sensitive contrast in magnetic resonance image of rodent brain in high magnetic fields. Magn. Reson. Med. 14 (1), 68–78.

Ohman, A., Flykt, A., Esteves, F., 2001. Emotion drives attention: detecting the snake in the grass. J. Exp. Psychol. Gen. 130 (3), 466–478.

Olds, J., 1958. Self-stimulation of the brain its use to study local effects of hunger, sex, and drugs. Science 127 (3294), 315–324.

Olds, J., Milner, P., 1954. Positive reinforcement produced by electrical stimulation of septal area and other regions of rat brain. J. Comp. Physiol. Psychol. 47 (6), 419–427.

Olson, I.R., Plotzker, A., Ezzyat, Y., 2007. The enigmatic temporal pole: a review of findings on social and emotional processing. Brain 130 (7), 1718–1731.

Olsson, A., Nearing, K.I., Phelps, E.A., 2007. Learning fears by observing others: the neural systems of social fear transmission. Soc. Cogn. Affect. Neurosci. 2 (1), 3–11.

Orbig, H., Wenzel, R., Kohl, M., Horst, S., Wobst, P., Steinbrink, J., et al., 2000. Near-infrared spectroscopy: does it function in functional activation studies of the adult brain? Int. Psychophysiol. 35 (2), 125–142.

Packard, M.G., Cahill, L., McGaugh, J.L., 1994. Amygdala modulation of hippocampal-dependent and caudate nucleus-dependent memory processes. Proc. Natl. Acad Sci. USA 91 (18), 8477–8481.

Papez, J.W., 1937. A proposed mechanism of emotion. Arch. Neurol. Psychiatry 38 (4), 725–743.

Peck, C.J., Salzman, C.D., 2014. The amygdala and basal forebrain as a pathway for motivationally guided attention. J. Neurosci. 34 (41), 13757–13767.

Peck, C.J., Lau, B., Salzman, C.D., 2013. The primate amygdala combines information about space and value. Nat. Neurosci. 16 (3), 340–348.

Pessoa, L., Adolphs, R., 2010. Emotion processing and the amygdala: from a 'low road' to 'many roads' of evaluating biological significance. Nat. Rev. Neurosci. 11 (11), 773–783.

Pessoa, L., Ungerleider, L.G., 2004. Neural correlates of change detection and change blindness in a working memory task. Cereb. Cortex 14 (5), 511–520.

Phelps, E.A., 2004. Human emotion and memory: interactions of the amygdala and hippocampal complex. Curr. Opin. Neurobiol. 14 (2), 198–202.

Phelps, E.A., Delgado, M.R., Nearing, K.I., LeDoux, J.E., 2004. Extinction learning in humans: role of the amygdala and vmPFC. Neuron 43 (6), 897–905.

Phillips, R.G., LeDoux, J.E., 1992. Differential contribution of amygdala and hippocampus to cued and contextual fear conditioning. Behav. Neurosci. 106 (2), 274–285.

Phillips, M.L., Young, A.W., Scott, S.K., Calder, A.J., Andrew, C., Giampietro, V., Gray, J.A., 1998. Neural responses to facial and vocal expressions of fear and disgust. Proc. Royal Soc. B: Biol. Sci. 265 (1408), 1809–1817.

Phillips, M.L., Young, A.W., Senior, C., Brammer, M., Andrew, C., Calder, A.J., et al., 1997. A specific neural substrate for perceiving facial expressions of disgust. Nature 389 (6650), 495–498.

Polich, J., 2007. Updating P300: an integrative theory of P3a and P3b. Clin. Neurophysiol. 118, 2128–2148.

Portas, C.M., Krakow, K., Allen, P., Josephs, O., Armony, J.L., Frith, C.D., 2000. Auditory processing across the sleep-wake cycle: simultaneous EEG and fMRI monitoring in humans. Neuron 28 (3), 991–999.

Pourtois, G., Grandjean, D., Sander, D., Vuilleumier, P., 2004. Electrophysiological correlates of rapid spatial orienting towards fearful faces. Cereb. Cortex 14 (6), 619–633.

Price, R.B., Siegle, G.J., Silk, J.S., Ladouceur, C.D., McFarland, A., Dahl, R.E., Ryan, N.D., 2014. Looking under the hood of the dot-probe task: an fMRI study in anxious youth. Depress Anxiety 31 (3), 178–187.

Pourtois, G., Schettino, A., Vuilleumier, P., 2013. Brain mechanisms for emotional influences on perception and attention: what is magic and what is not. Biol. Psychol. 92 (3), 492–512.
Proudfit, G.H., 2015. The reward positivity: from basic research on reward to a biomarker for depression. Psychophysiology 52 (4), 449–459.
Ptak, R., 2011. The frontoparietal attention network of the human brain: action, saliency, and a priority map of the environment. Neuroscientist 18 (5), 502–515.
Quirk, G.J., Garcia, R., Gonzalez-Lima, F., 2006. Prefrontal mechanisms in extinction of conditioned fear. Biol. Psychiatry 60 (4), 337–343.
Rajkowska, G., 2000. Postmortem studies in mood disorders indicate altered numbers of neurons and glial cells. Biol. Psychiatry 48 (8), 766–777.
Rauch, S.L., Whalen, P.J., Shin, L.M., McInerney, S.C., Macklin, M.L., Lasko, N.B., et al., 2000. Exaggerated amygdala response to masked facial stimuli in posttraumatic stress disorder: a functional MRI study. Biol. Psychiatry 47 (9), 769–776.
Redouté, J., Stoléru, S., Grégoire, M.C., Costes, N., Cinotti, L., Lavenne, F., et al., 2000. Brain processing of visual sexual stimuli in human males. Hum. Brain Mapp. 11 (3), 162–177, http://dx.doi.org/10.1002/1097-0193(200011)11:3%3C162::AID-HBM30%3E3.0.CO;2-A.
Rolls, E.T., Scott, T.R., Sienkiewicz, Z.J., Yaxley, S., 1988. The responsiveness of neurones in the frontal opercular gustatory cortex of the macaque monkey is independent of hunger. J. Physiol. 397 (1), 1–12.
Rozin, P., Fallon, A.E., 1987. A perspective on disgust. Psychol. Rev. 94 (1), 23–41.
Sander, D., Grafman, J., Zalla, T., 2003. The human amygdala: an evolved system for relevance detection. Rev. Neurosci. 14 (4), 303–316.
Scheggi, S., Leggio, B., Masi, F., Grappi, S., Gambarana, C., Nanni, G., Rauggi, R., De Montis, M.G., 2002. Selective modifications in the nucleus accumbens of dopamine synaptic transmission in rats exposed to chronic stress. J. Neurochem. 83 (4), 895–903.
Schienle, A., Stark, R., Walter, B., Blecker, C., Ott, U., Kirsch, P., Vaitl, D., 2002. The insula is not specifically involved in disgust processing: an fMRI study. Neuroreport 13 (16), 2023–2026.
Schirmer, A., 2014. Emotion. SAGE Publications, Los Angeles, CA.
Schneiderman, N., Francis, J., Sampson, L.D., Schwaber, J.S., 1974. CNS integration of learned cardiovascular behavior. In: DiCara, L. (Ed.), Limbic and Autonomic Nervous Systems Research. Springer, New York, pp. 277–309.
Schupp, H., Cuthbert, B., Bradley, M., Hillman, C., Hamm, A., Lang, P., 2004. Brain processes in emotional perception: motivated attention. Cogn. Emot. 18 (5), 593–611.
Sheline, Y.I., Barch, D.M., Donnelly, J.M., Ollinger, J.M., Snyder, A.Z., Mintun, M.A., 2001. Increased amygdala response to masked emotional faces in depressed subjects resolves with antidepressant treatment: an fMRI study. Biol. Psychiatry 50 (9), 651–658.
Singer, T., Critchley, H.D., Preuschoff, K., 2009. A common role of insula in feelings, empathy and uncertainty. Trends Cogn. Sci. 13 (8), 334–340.
Sokoloff, L., Reivich, M., Kennedy, C., Des Roisers, M.H., Patlak, C.S., Pettigrew, K.D., Sakurada, O., Shinohara, M., 1977. The 14C-2-deoxyglucose method for the measurement of local cerebral glucose utilization. Theory, procedure, and normal values in the conscious anaesthetized rat. J. Neurochem. 28 (5), 897–916.
Somerville, L.H., Whalen, P.J., Kelley, W.M., 2010. Human bed nucleus of the stria terminalis indexes hypervigilant threat monitoring. Biol. Psychiatry 68 (5), 416–424.
Streit, M., Ioannides, A.A., Liu, L., Wölwer, W., Dammers, J., Gross, J., et al., 1999. Neurophysiological correlates of the recognition of facial expressions of emotion as revealed by magnetoencephalography. Cogn. Brain Res. 7 (4), 481–491.
Suslow, T., Ohrmann, P., Bauer, J., Rauch, A.V., Schwindt, W., Arolt, V., et al., 2006. Amygdala activation during masked presentation of emotional faces predicts conscious detection of threat-related faces. Brain Cogn. 61 (3), 243–248.
Susskind, J.M., Lee, D.H., Cusi, A., Feiman, R., Grabski, W., Anderson, A.K., 2008. Expressing fear enhances sensory acquisition. Nat. Neurosci. 11 (7), 843–850.

Tamietto, M., de Gelder, B., 2010. Neural bases of the non-conscious perception of emotional signals. Nat. Rev. Neurosci. 11 (10), 697–709.

Teplan, M., 2002. Fundamentals of EEG measurement. Meas. Sci. Rev. 2 (2), 1–11, http://www.measurement.sk/.

Torrence, R.D., Wylie, E., Carlson, J.M., 2017. The Time-Course for the Capture and Hold of Visuospatial Attention by Fearful and Happy Faces. J. Nonverbal Behav., 1–15.

Tracey, I., Ploghaus, A., Gati, J., Clare, S., Smith, S., Menon, R., Matthews, P., 2002. Imaging attentional modulation of pain in the periaqueductal gray in humans. J. Neurosci. 22, 2748–2752.

Ubeda-Bañon, I., Novejarque, A., Mohedano-Moriano, A., Pro-Sistiaga, P., de la Rosa-Prieto, C., Insausti, R., Martinez-Garcia, F., Lanuza, E., Martinez-Marcos, A., 2007. Projections from the posterolateral olfactory amygdala to the ventral striatum: neural basis for reinforcing properties of chemical stimuli. BMC Neurosci. 8 (1), 1.

van Gaal, S., de Lange, F.P., Cohen, M.X., 2012. The role of consciousness in cognitive control and decision making. Front. Hum. Neurosci. 6, 121.

van Honk, J., Tuiten, A., de Haan, E., van den Hout, M., Stam, H., 2001. Attentional biases for angry faces: relationships to trait anger and anxiety. Cogn. Emot. 15 (3), 279–297.

Vouimba, R.M., Maroun, M., 2011. Learning-induced changes in mPFC-BLA connections after fear conditioning, extinction, and reinstatement of fear. Neuropsychopharmacology 36 (11), 2276–2285.

Vuilleumier, P., 2005. How brains beware: neural mechanisms of emotional attention. Trends Cogn. Sci. 9 (12), 585–594.

Vuilleumier, P., Richardson, M.P., Armony, J.L., Driver, J., Dolan, R.J., 2004. Distant influences of amygdala lesion on visual cortical activation during emotional face processing. Nat. Neurosci. 7 (11), 1271–1278.

Wager, T.D., Davidson, M.L., Hughes, B.L., Lindquist, M.A., Ochsner, K.N., 2008. Prefrontal-subcortical pathways mediating successful emotion regulation. Neuron 59 (6), 1037–1050.

Walter, M., Bermpohl, F., Mouras, H., Schiltz, K., Tempelmann, C., Rotte, M., et al., 2008. Distinguishing specific sexual and general emotional effects in fMRI-subcortical and cortical arousal during erotic picture viewing. Neuroimage 40 (4), 1482–1494.

Webb, T.L., Miles, E., Sheeran, P., 2012. Dealing with feeling: a meta-analysis of the effectiveness of strategies derived from the process model of emotion regulation. Psychol. Bull. 138 (4), 775–808.

Weinberg, A., Hajcak, G., 2010. Beyond good and evil: the time-course of neutral activity elicited by specific picture subtypes. Emotion 10 (6), 767–782.

Welborn, B.L., Papademetris, X., Reis, D.L., Rajeevan, N., Bloise, S.M., Gray, J.R., 2009. Variation in orbitofrontal cortex volume: relation to sex, emotion regulation and affect. Soc. Cogn. Affect. Neurosci. 4 (4), 328–339.

Whalen, P.J., Rauch, S.L., Etcoff, N.L., McInerney, S.C., Lee, M.B., Jenike, M.A., 1998. Masked presentations of emotional facial expressions modulate amygdala activity without explicit knowledge. J. Neurosci. 18 (1), 411–418.

Whalen, P.J., Kagan, J., Cook, R.G., Davis, F.C., Kim, H., Polis, S., et al., 2004. Human amygdala responsivity to masked fearful eye whites. Science 306 (5704), 2061.

Wicker, B., Keysers, C., Plailly, J., Royet, J.P., Gallese, V., Rizzolatti, G., 2003a. Both of us disgusted in My insula: the common neural basis of seeing and feeling disgust. Neuron 40 (3), 655–664.

Wicker, B., Perrett, D.I., Baron-Cohen, S., Decety, J., 2003b. Being the target of another's emotion: a PET study. Neuropsychologia 41 (2), 139–146.

Williams, L.M., Barton, M.J., Kemp, A.H., Liddell, B.J., Peduto, A., Gordon, E., Bryant, R.A., 2005. Distinct amygdala-autonomic arousal profiles in response to fear signals in healthy males and females. Neuroimage 28 (3), 618–626.

Williams, L.E., Bargh, J.A., Nocera, C.C., Gray, J.R., 2009. The unconscious regulation of emotion: nonconscious reappraisal goals modulate emotional reactivity. Emotion 9 (6), 847–854.

Wise, R.A., 2004. Dopamine, learning, and motivation. Nat. Rev. Neurosci. 5 (6), 483–494.
Wolf, R.C., Philippi, C.L., Motzkin, J.C., Baskaya, M.K., Koenigs, M., 2014. Ventromedial prefrontal cortex mediates visual attention during facial emotion recognition. Brain 137 (Pt 6), 1772–1780.
Wong, P.S., Bernat, E., Bunce, S., Shevrin, H., 1997. Brain indices of nonconscious associative learning. Consciousness Cogn. 6 (4), 519–544.
Yassa, M.A., Hazlett, R.L., Stark, C.E., Hoehn-Saric, R., 2012. Functional MRI of the amygdala and bed nucleus of the stria terminalis during conditions of uncertainty in generalized anxiety disorder. J. Psychiatric Res. 46 (8), 1045–1052.
Yaxley, S., Rolls, E.T., Sienkiewicz, Z.J., 1988. The responsiveness of neurons in the insular gustatory cortex of the macaque monkey is independent of hunger. Physiol. Behav. 42 (3), 223–229.
Zaki, J., Davis, J.I., Ochsner, K.N., 2012. Overlapping activity in anterior insula during interoception and emotional experience. Neuroimage 62 (1), 493–499.
Zald, D.H., 2003. The human amygdala and the emotional evaluation of sensory stimuli. Brain Res. Brain Res. Rev. 41 (1), 88–123.
Zald, D.H., Pardo, J.V., 2000. Functional neuroimaging of the olfactory system in humans. Int. J. Psychophysiol. 36 (2), 165–181.
Zald, D.H., Lee, J.T., Fluegel, K.W., Pardo, J.V., 1998. Aversive gustatory stimulation activates limbic circuits in humans. Brain 121 (6), 1143–1154.

CHAPTER

3

Mood Effects on Cognition: Affective Influences on the Content and Process of Information Processing and Behavior

Joseph P. Forgas

University of New South Wales, Sydney, NSW, Australia

INTRODUCTION

How does affect influence our thinking and behavior? This is an age-old question that philosophers, writers, and artists have struggled with since time immemorial. In the last few decades there have been major advances in the experimental study of the ways that mild, everyday affective states, or moods, may influence the way people process social information—the way they think, remember, and judge the social world around them. Research on the influence of affect on cognition also contributes to the age-old quest to understand the relationship between the rational and the emotional aspects of human nature. These mood-induced influences on the content and process of thinking have very important implications for everyday interpersonal behavior in general, and for human–computer interactions (HCI) in particular.

The evidence increasingly suggests that an evolutionary process shaped the development of all affective responses. Thus, temporary experiences of happiness and sadness, in addition to having a positive or negative hedonic quality, also appear to function as useful signals, spontaneously triggering different information processing strategies that appear to be

Emotions and Affect in Human Factors and Human–Computer Interaction. http://dx.doi.org/10.1016/B978-0-12-801851-4.00003-3
Copyright © 2017 Elsevier Inc. All rights reserved.

highly adaptive to the requirements of different social situations. In this way, positive and negative moods may assist people by recruiting information and processing strategies that are most appropriate to deal with a given situation. In addition, positive or negative affective states also influence the way we access and use information stored in memory (mood congruence). A growing number of studies emphasize the important applied aspects of mood effects on cognition that are particularly relevant in human factors and HCI, as well as in applied settings, such as legal, clinical, forensic, and educational psychology where concern with human performance is crucial.

Understanding the delicate interplay between feeling and thinking or affect and cognition is one of the most important tasks for psychological research. This chapter reviews recent research documenting the multiple roles that moods play in influencing both the *content* and the *process* of cognition. After a brief introduction reviewing early work and theories exploring the links between mood and cognition, the chapter is divided into two main parts. First, research documenting the way moods influence the *content and valence* of cognition is reviewed, focusing on mood congruence in cognition and behavior. The second part of the chapter presents evidence for the *processing effects* of moods, showing that mood states influence the quality of information processing as well. The chapter concludes with a discussion of the theoretical and applied implications of this work, and future prospects for these lines of inquiry are considered.

We need to emphasize at the outset that this review deals with mood effects rather than the influence of more distinct and intense emotions. We may define moods as "relatively low-intensity, diffuse, subconscious, and enduring affective states that have no salient antecedent cause and therefore little cognitive content" (Forgas, 2006, pp. 6–7). Distinct emotions in contrast are more intense, conscious, and short-lived experiences (e.g., fear, anger, or disgust). Moods tend to have relatively more robust, reliable, and enduring cognitive consequences, and the research reported here largely focused on the effects of mild, nonspecific positive and negative moods on thinking and behavior, although more specific states, such as anger have also been studied (Unkelbach et al., 2008).

BACKGROUND

Human beings are a remarkably moody species. Fluctuating positive and negative affective states accompany, underlie, and color everything we think and do, and our thoughts and behaviors are often determined by spontaneous affective reactions. Since the dawn of Western civilization, a long list of writers and philosophers have explored the role of moods in the way we think, remember, and interact with our environment. It is all the more surprising that empirical research on how moods influence cognition is a relatively recent phenomenon. Apart from some early exceptions (Rapaport, 1961; Razran, 1940), concentrated empirical research on this phenomenon in psychology is but a few decades old, perhaps because the affective nature of human beings has long been considered secondary and inferior to the study of rational thinking (Adolphs and Damasio, 2001; Hilgard, 1980).

Of the two major paradigms that dominated the brief history of our discipline (behaviorism and cognitivism), neither assigned much importance to the study of affective states or moods. Radical behaviorists considered all mental events, such as moods beyond the scope of scientific psychology. The emerging cognitive paradigm in the 1960s was largely directed at the study of cold, affectless mental processes, and initially had little interest in the study of affect and moods. In contrast, research since the 1980s has shown that moods play a central role in how information about the world is represented, and affect determines the cognitive representation of many of our social experiences (Forgas, 1979).

Early evidence. Although radical behaviorists generally showed little interest in exploring the nature of mood effects, Watson's research with Little Albert may be viewed as an early demonstration of affect congruence in judgments (Watson, 1929; Watson and Rayner, 1920). These studies showed that evaluations of a neutral stimulus, such as a furry rabbit, became more negative after it has been associated with threatening stimuli, such as a loud noise. Watson thought that most complex affective reactions are acquired in a similar manner throughout life due to cumulative stimulus associations. In another early mood study, Razran (1940) showed that people evaluated sociopolitical messages more favorably when in a good rather than in a bad mood, induced either by a free lunch or aversive smells. This work also provides an early demonstration of mood congruence (Bousfield, 1950). In another pioneering study, Feshbach and Singer (1957) induced negative affect in subjects through electric shocks and then instructed some of them to suppress their fear. Fearful subjects' evaluations of another person were more negative, and ironically, this effect was even greater when subjects were trying to suppress their fear (Wegner, 1994). Feshbach and Singer (1957) explained this in terms of the psychodynamic mechanism of projection, suggesting that "suppression of fear facilitates the tendency to project fear onto another social object" (p.286). Mood-congruent effects on evaluative judgments were also found by Byrne and Clore (1970) and Clore and Byrne (1974) using a classical conditioning approach. They used pleasant or unpleasant environments (the unconditioned stimuli) to elicit good or bad moods (the unconditioned response), and then assessed evaluations of a person encountered in this environment (the conditioned stimulus; Gouaux, 1971; Gouaux and Summers, 1973; Griffitt, 1970). These early studies provided convergent evidence for the existence of a mood-congruent pattern in thinking and judgments, but did not as yet offer a clear explanation for the psychological mechanisms responsible for these effects. Nevertheless, these early studies paved the way for the emergence of more focused research on mood congruence in thinking and judgments in the 1980s.

THEORIES OF AFFECT CONGRUENCE

As we have seen, it has long been recognized that affective states may influence the content and valence of thinking, although psychological explanations for this phenomenon have remained scarce until recently. Several early studies demonstrated such informational effects, that is, ways that positive and negative moods may influence the *content* of cognition. Three main theories accounting for mood congruence will be reviewed: (1) *associative network* theories emphasizing memory processes (Bower, 1981; Bower and Forgas, 2000), (2) *affect-as-information* theory relying on inferential processes (Clore et al., 2001; Clore and Storbeck, 2006; Schwarz and Clore, 1983), and (3) an integrative *Affect Infusion Model* (AIM) (Forgas, 1995, 2002).

EMPHASIS ON MEMORY: THE ASSOCIATIVE NETWORK THEORY

Bower (1981) proposed a comprehensive model that assumes that moods are linked to an associative network of memory representations. A mood state may thus automatically prime or activate representations linked to that mood in the past, which in turn are more likely to be used in subsequent constructive cognitive tasks. Several experiments found support for such *affective priming*. For example, happy or sad people were more likely to recall mood-congruent details from their childhood and also remembered better mood-congruent rather than mood-incongruent events that occurred in the past few weeks (Bower, 1981). Mood congruence was also observed in how people interpreted ongoing social behaviors (Forgas et al., 1984) and formed impressions of others (Forgas and Bower, 1987).

Subsequent work found however that mood congruence is also subject to several boundary conditions (Blaney, 1986; Bower, 1987; Singer and Salovey, 1988). Mood-congruence in memory and judgments is most reliable (1) when moods are intense (Bower and Mayer, 1985), (2) meaningful (Bower, 1991), (3) when the cognitive task is self-referential (Blaney, 1986), and (4) when open, elaborate thinking (or constructive processing) is used. In particular, tasks requiring constructive and more elaborate processing, such as associations, inferences, impression formation, and interpersonal behaviors are most likely to show mood-congruent effects (e.g., Bower and Forgas, 2000; Fiedler, 1990; Mayer et al., 1992). The reason for this is that an open, elaborate processing strategy amplifies the opportunities for affectively primed incidental memories and associations to become incorporated into a newly constructed cognitive response. Tasks that can be performed automatically, relying on preexisting reactions require little or no constructive processing, and show less mood congruence. For example, tasks involving the simple recognition or reproduction of existing reactions (Forgas, 1995, 2002, 2006) are impervious to mood effects, because narrow and targeted thinking offers little opportunity for affectively primed information to be incorporated in a response.

MISATTRIBUTION: THE AFFECT-AS-INFORMATION THEORY

An interesting alternative theoretical approach seeks to explain mood congruence by suggesting that "rather than computing a judgment on the basis of recalled features of a target, individuals may ask themselves: 'how do I feel about it?' [and] in doing so, they may mistake feelings due to a preexisting state as a reaction to the target" (Schwarz, 1990, p. 529; see also Schwarz and Clore, 1983; Clore and Storbeck, 2006). In essence, this model suggests that people misattribute a pre-existing mood state as indicative about their reaction to an unrelated target. The model is closely derived from research on misattribution and judgmental heuristics. However, its predictions are often empirically indistinguishable from those derived from earlier conditioning models assumes that blind associations—evaluative conditioning—between a preexisting mood state and another, unrelated stimulus (e.g., Clore and Byrne, 1974). Evidence shows that people mainly rely on their mood as a simple and convenient heuristic cue to infer their evaluative reactions when "the task is of little personal relevance, when little other information is available, when problems are too complex to be solved systematically, and when time or attentional resources are limited" (Fiedler, 2001, p. 175). In contrast, if the task is of high personal relevance and there are cognitive resources available, then affective priming is the most likely strategy resulting in mood congruence.

The affect-as-information model has been supported by studies that showed, for example, that mood induced by good or bad weather can significantly influence judgments on a variety of unexpected and unfamiliar opinion survey questions in a telephone interview (Schwarz and Clore, 1983). In another study, Forgas and Moylan (1987) found mood congruence in survey responses of almost 1000 subjects who completed a questionnaire after they had seen funny or sad films at the cinema. As in the earlier study by Schwarz and Clore (1983), respondents presumably had little time, interest, motivation, or capacity to engage in elaborate constructive processing in such a survey situation requiring rapid responses, and so relied on their mood as a simple and convenient heuristic shortcut to infer their evaluative reactions. There are also some important limitations of this model. As the informational value of a mood state is not fixed but rather depends on the situational context (Martin, 2000), such mood effects may also be highly context-specific. Furthermore, the affect-as-information model mostly applies to valenced evaluative judgments, and may have difficulty accounting for mood congruence in attention, learning, and memory. In one sense misattributing mood to an unrelated target is probably the exception rather than the norm in real-life mood effects on cognition.

TOWARD AN INTEGRATION: THE AFFECT INFUSION MODEL

The AIM (Forgas, 1995, 2002) suggests that mood effects on cognition should depend on the kind of information processing strategy used. The model identifies four processing strategies that vary in terms of (1) their openness and *constructiveness* and (2) the degree of *effort* exerted in seeking a solution:

1. The first, *direct access* strategy involves the simple and direct retrieval of an already stored preexisting response. This is most likely when the task is highly familiar and there is no reason to engage in more elaborate thinking (e.g., retrieving a friend's mobile number). As this is a low-effort and a non- constructive processing strategy, affect infusion should not occur.
2. The second, *motivated processing* strategy refers to effortful, yet highly selective and targeted thinking that is dominated by a specific motivational objective (e.g., making a good impression in a job interview). This strategy is dominated by a powerful motivational objective and involves little open, constructive processing and therefore should be impervious to affect infusion and may even produce mood-incongruent effects (Clark and Isen, 1982; Sedikides, 1994).
3. *Heuristic processing* refers to constructive but truncated, low-effort processing, which is likely to be adopted when time and personal resources such as motivation, interest, attention, and working-memory capacity are scarce (e.g., evaluating your friend's new company car). Heuristic processing may result in mood congruence when affect is used as a heuristic cue as predicted by the affect-as-information

model (Schwarz and Clore, 1983; see also Clore et al., 2001; Clore and Storbeck, 2006).

4. Finally, *substantive processing* involves open, constructive thinking, and is used whenever the task is new and demanding and there are no ready-made direct access responses or motivational goals available to guide the response. Substantive processing is most likely to produce memory-based affect infusion into cognition as mood may selectively prime or enhance the accessibility of mood-congruent thoughts, memories, and interpretations (Forgas, 1994, 1999a,b).

Further, the AIM also identifies a range of contextual variables related to the task, the person, and the situation that jointly determine processing choices (Forgas, 2002; Smith and Petty, 1995), and the model also recognizes that affect itself can influence processing choices (Bless and Fiedler, 2006). The key prediction of the AIM is the *absence* of affect infusion when direct access or motivated processing is used, and the presence of affect infusion during heuristic and substantive processing. Affect infusion is most likely in the course of constructive processing that involves the substantial transformation rather than the mere reproduction of existing information. Such processing requires a relatively open information search strategy and a significant degree of generative elaboration of the available stimulus details. Thus, affect "will influence cognitive processes to the extent that the cognitive task involves the active generation of new information as opposed to the passive conservation of information given" (Fiedler, 1990, pp. 2–3). The implications of this model have now been explored and supported in a number of the experiments considered below. In particular, mood congruence in cognition turns out to be *greater* when *more extensive* and *elaborate* processing is required to deal with a more complex, demanding task (Forgas, 2002; Sedikides, 1995). Conversely, affect infusion is curtailed whenever open, constructive processing is suppressed.

EVIDENCE FOR MOOD CONGRUENCE IN THINKING AND BEHAVIOR

Mood Congruence in Memory and Attention

Mood congruency in memory was one of the first reliable effects demonstrated. Several studies found that people are better at retrieving both early and recent autobiographical memories that happen to match their prevailing mood (Bower, 1981; Miranda and Kihlstrom, 2005). In a similar manner, depressed patients preferentially remember aversive experiences and negative information (Direnfeld and Roberts, 2006). Implicit tests of memory provide particularly good evidence of mood congruence. For example, depressed people complete more word stems (e.g., *can-*) with negative rather than positive words they had studied earlier (e.g., *cancer* vs. *candy*; Ruiz-Caballero and Gonzalez, 1994). Person memory also is subject to mood congruent effects: happy and sad people selectively remembered more positive and negative details respectively about people they had read about (Forgas and Bower, 1987).

These mood-congruent memory effects occur because of the selective activation of an affect-related associative base, resulting in mood-congruent information receiving greater attention and more extensive processing and encoding (Bower, 1981). That is, people spend longer reading and learning about mood-congruent material, integrating it into a richer network of primed associations, and as a result, they are better able to remember such information later on (Bower and Forgas, 2000). There is also growing evidence for mood congruence at the attention stage. In one recent inattentional blindness study (Becker and Leinenger, 2011), mood selectively influenced participants' attentional filter, increasing the chance to notice unexpected faces that carried a mood-congruent rather than a mood-incongruent emotional expression. Other research also demonstrated that positive mood leads to an attentional bias toward rewarding words (Tamir and Robinson, 2007), and broadened attention to positive images (Wadlinger and Isaacowitz, 2006). In contrast, depressed patients tend to pay greater attention to negative information (Koster et al., 2005), and show better learning and memory for depressive, negative words (Watkins et al., 1992), and negative facial expressions (Gilboa-Schechtman et al., 2002).

However, selective focus on mood-congruent information does not necessarily lead to an escalating cycle of positivity or negativity. Sad people eventually may escape the vicious circle of focusing on and remembering negative information by means of deliberately employing *mood-incongruent* attention and memory. Consistent with the hypothesis of such motivational *mood repair* (Isen, 1985), Josephson et al. (1996) showed that after initially retrieving negative memories, nondepressed participants in a negative mood deliberately shifted to retrieving positive memories in order to lift their mood (Detweiler-Bedell and Salovey, 2003; Heimpel et al., 2002).

Mood-State Dependent Memory

In addition to mood-congruence described earlier, mood has another significant influence on memory by selectively facilitating the retrieval of information that has been learnt in a matching rather than a nonmatching mood. Such *mood-state dependent memory* may play a particularly important role in the memory deficits found in patients with alcoholic blackout, chronic depression, dissociative identity and other psychiatric disorders (Goodwin, 1974; Reus et al., 1979; Schacter and Kihlstrom, 1989).

However, the evidence suggests that these effects are rather subtle (Bower and Mayer, 1989; Kihlstrom, 1989; Leight and Ellis, 1981), and there are several moderating factors that influence their occurrence. Constructive tasks, such as free recall are more sensitive to mood-dependent memory effects than are reproductive tasks such as recognition as also suggested by the AIM described earlier (Bower, 1992; Eich, 1995a; Fiedler, 1990; Forgas, 1995; Kenealy, 1997).

The effects are most reliable when people engage in highly constructive processing and generate their own events to be remembered and their own retrieval cues rather than when they are confronted with fixed materials and predetermined retrieval cues (Beck and McBee, 1995; Eich and Metcalfe, 1989). It seems that the more a person needs to rely on self-constructed information, the more likely that memory for corresponding events will be mood-dependent. Eich et al. (1994) confirmed this, reporting that mood dependence effects were markedly greater when the recalled events were self-generated. Recall was consistently better when encoding mood and retrieval mood were matched rather than different, and this effect pattern was obtained with different mood induction methods (Eich et al., 1994; Eich, 1995b). Similar mood dependence in memory was demonstrated in bipolar patients (Eich et al., 1997). Mood-dependent memory is also enhanced when the intensity, authenticity, or distinctiveness of encoding, and retrieval moods is high rather than low (Eich, 1995a; Eich and Macaulay, 2000; Eich and Metcalfe, 1989; Ucros, 1989).

Given that individual differences in personality play an important part in mood-congruent memory (Bower and Forgas, 2000; Smith and Petty, 1995), such factors may also moderate mood-state dependent memory. Thus, mood-state dependent memory is less likely to occur in experiments that employ simple, irrelevant tasks, such as list-learning experiments, and when the mood induction is weak and not particularly distinctive to be effective as a retrieval cue. In terms of the affect infusion model (Forgas, 1995, 2002), the higher the level of constructive processing and affect infusion that occurs both at the encoding and at the retrieval stages, the more likely that mood-dependence can be demonstrated.

Mood Congruence in Inferences and Associations

In terms of the associative network model, the selective priming of mood-consistent materials in memory can have a particularly marked influence on how complex or ambiguous information is interpreted (Bower and Forgas, 2000; Clark and Waddell, 1983). For example, people generated more mood-congruent ideas when daydreaming or free associating to projective thematic apperception test (TAT) pictures, and happy subjects generated more positive than negative associations to words, such as *life* (e.g., *love* and *freedom* vs. *struggle* and *death*) than did sad subjects (Bower, 1981).

The selective priming and activation of mood-congruent constructs can have particularly strong effects on constructive social judgments, such as perceptions of faces (Forgas and East, 2008a; Gilboa-Schechtman et al., 2002; Schiffenbauer, 1974), impressions of people (Forgas and Bower, 1987), and self-perceptions (Sedikides, 1995). These associative effects on inferences and associations are again diminished when the targets to be judged are more simple and clear-cut (Forgas, 1994, 1995), confirming that open, constructive processing is crucial for mood congruence to occur.

Mood Congruence in Social Judgments

Consistent with the affect infusion model, several studies have found that the more people need to think in order to compute a judgment, the greater the likelihood that affectively primed ideas will influence the outcome. For example, mood had a greater influence on judgments about unusual, complex characters that require more constructive and elaborate processing than on judgments of simple, typical targets (Forgas, 1992). Mood also had a greater influence on judgments about unusual, badly matched couples than on typical, well-matched couples (Forgas, 1993).

Judgments about one's real-life partners showed similar mood congruence (Forgas, 1994). Mood significantly influenced the evaluation of one's partner and relationship conflicts, and paradoxically, these effects were stronger for judgments about complex, difficult conflicts that required more constructive processing, confirming that affect infusion into social judgments depends on the processing strategy recruited by the task at hand. Some personality characteristics, such as trait anxiety, may moderate such mood congruence effects on judgments, as highly anxious people are less likely to process information in an open, constructive manner (Ciarrochi and Forgas, 1999). Affect intensity may be another important trait moderator of mood congruence effects, as people who scored high on measures assessing openness to feelings showed greater mood congruence (Ciarrochi and Forgas, 2000).

Moods also exert an important influence on self-related judgments (Sedikides, 1995). Students in a positive mood were more likely to claim credit for success in a recent exam, and made more internal and stable attributions for their high test scores, but were less willing to assume personal responsibility for failure. Those in a negative mood engaged in more self-deprecating attributions and blamed themselves more for failure and took less credit for success (Forgas et al., 1990). These findings were replicated in a study by Detweiler-Bedell and Detweiler-Bedell (2006), who concluded that consistent with the AIM, "constructive processing accompanying most self-judgments is critical in producing mood-congruent perceptions of personal success" (p. 196).

Social judgments about the self also show similar mood effects. Sedikides (1995) also found support for the AIM, reporting that well-rehearsed "central" conceptions of the self were processed more automatically and less constructively and thus were less influenced by mood than were "peripheral" self-conceptions that required more substantive processing and showed stronger mood congruence. Individual differences in self-esteem may also influence affect infusion into self-judgments, as mood-congruent effects on self-related memories were stronger for low rather than high self-esteem people (Smith and Petty, 1995). These findings are in line with the principle that low self-esteem persons have a less clearly defined and less stable self-concept that is more likely to be influenced by prevailing mood (Brown and Mankowski, 1993).

Consistent with the AIM, these results show that low self-esteem is linked to the more open and constructive processing of information about the self, increasing the scope for mood-related associations to influence the outcome. Other work suggests that mood congruence may sometimes be a self-limiting process and can be spontaneously corrected as a result of shifting to the motivated processing strategy. For example, initially mood-congruent thoughts can be spontaneously reversed over time (Sedikides, 1994). Further research by Forgas and Ciarrochi (2002) replicated these results and found that the spontaneous reversal of negative self-judgments was strongest in people with high self-esteem, suggesting the operation of a homeostatic process of mood management.

Mood-Congruence in Social Behaviors

Social interaction is a complex and cognitively demanding process, and planning strategic social behaviors necessarily requires some degree of constructive, open information processing (Heider, 1958), suggesting that moods may also produce distinct behavioral effects. Positive mood, by priming positive evaluations and inferences, should elicit more optimistic, positive, confident, and cooperative behaviors, whereas negative mood may produce more avoidant, defensive, and unfriendly behaviors. In one experiment, happy or sad mood was induced in people before they engaged in a strategic negotiation task (Forgas, 1998a). Those in a happy mood employed more trusting, optimistic, and cooperative negotiating strategies, and achieved better outcomes, while those in a negative mood were more pessimistic and competitive in their negotiating moves. Other experiments examined the effects of induced mood on the way people formulate and use verbal requests (Forgas, 1999a). These studies found that due to more optimistic inferences about the receptiveness/willingness of the persons receiving the request, positive mood resulted in more confident and less polite request formulations. In contrast, negative affect triggered a more cautious, polite, and elaborate requesting strategy as a

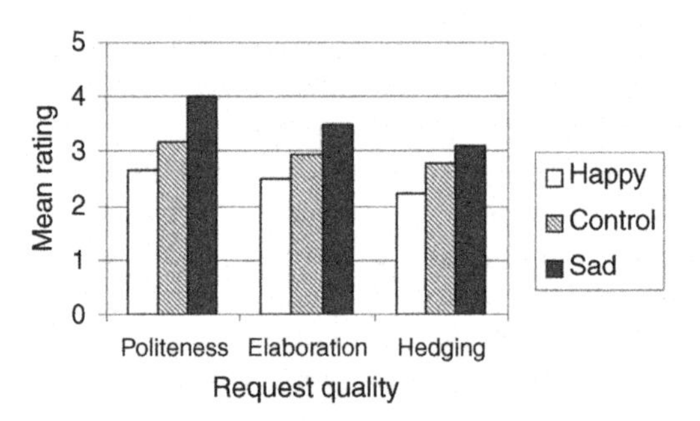

FIGURE 3.1 Mood effects on naturally produced requests. Positive mood increases, and negative mood decreases the degree of politeness, elaboration, and hedging in strategic communications. *Source: After Forgas, J.P., 1999b. Feeling and speaking: mood effects on verbal communication strategies. Person. Soc. Psychol. Bull. 25, 850–863.*

result of rather pessimistic inferences regarding the request's chance of success (Fig. 3.1).

A further unobtrusive exploration of these mechanisms showed that moods also influence how people *respond* to an impromptu request (Forgas, 1998b). Mood was induced by leaving folders containing mood-inducing materials (pictures as well as text) on empty library desks. After occupying the desks and examining the mood induction materials, students received an unexpected polite or impolite request from a confederate asking for paper needed to complete an essay. Results revealed a clear mood-congruent response pattern: negative mood resulted in less compliance and more critical, negative evaluations of the request and requester, whereas positive mood yielded the opposite results. Again, the effects were stronger when the request was formulated in an unusual and impolite way and therefore recruited more substantive processing.

Some strategic interpersonal behaviors, such as *self-disclosure,* are critical for the development and maintenance of intimate relationships, for mental health, and for social adjustment. It seems that by facilitating mood-congruent associations and inferences about a conversational partner, affective states can directly influence people's preferred self-disclosure strategies (Forgas, 2011a). Several recent experiments found that, consistent with the predicted mood congruence effects, those in a positive mood are more confident and optimistic and preferred to disclose information that was more intimate, more varied, more abstract, and more positive than was the case for people in a neutral mood. Interestingly, negative affect had exactly the opposite effect and resulted in less intimate, less varied, and less abstract disclosure (Fig. 3.2), and this pattern was even stronger when the conversational partner reciprocated with a high degree of disclosure.

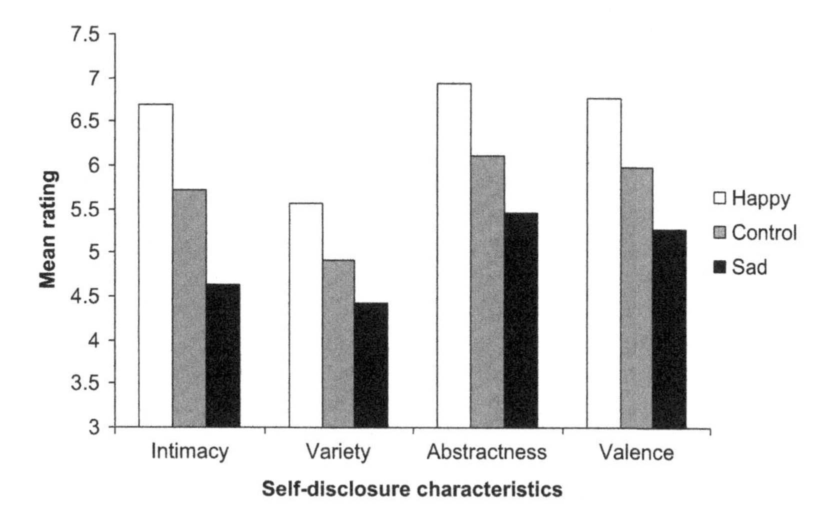

FIGURE 3.2 **The effects of positive, neutral, and negative mood on the intimacy, variety, abstractness, and valence of self-disclosing messages.** Mean number of target items seen in a shop recalled as a function of the mood (happy vs. sad) induced by the weather.

Thus, these experiments provide convergent evidence that temporary fluctuations in mood can produce marked changes in the quality, valence, and reciprocity of strategic interpersonal behaviors, such as self-disclosure, suggesting that mood congruence is likely to occur in the context of many other unscripted and unpredictable strategic interpersonal behaviors. When considered jointly, the evidence shows that transient moods play an important informative function, influencing the content and valence of memory, attention, associations, inferences, judgments, and social behaviors in a predominantly mood-congruent way.

However, these effects are dependent on the information processing strategy adopted, with open, constructive processing more likely to be influenced by moods than are other kinds of processing strategies (Forgas, 1995, 2002). When such substantive processing is used, affective priming appears to be the most likely mechanism responsible for mood congruence effects (Bower, 1981), while some evaluative judgments made under suboptimal processing conditions may also become mood congruent as a result of the heuristic affect-as-information mechanism. The overall pattern of results seems consistent with the AIM, suggesting that mood congruence is unlikely when a task can be performed using simple, well-rehearsed direct access or motivated processing, as there is little opportunity for moods to influence cognition. According to the AIM, mood congruence is most likely when individuals engage in substantive, constructive processing.

Mood Effects on Information Processing Strategies

We shall now turn to the second major influence that mood has on cognition. The evidence surveyed so far clearly shows that mood states can have a significant *informational* influence on the content and valence of cognition, producing mood-congruent effects on memory, attention, associations, judgments, and social behaviors. In addition to influencing cognitive content, that is, *what* people think, moods may also influence the process of cognition, that is, *how* people think. This section will review evidence for the information processing consequences of moods. In the past few decades, a growing number of studies suggested that people experiencing a positive mood rely on a more superficial and less effortful information processing strategy (Forgas, 2013, 2015).

Those in a good mood were consistently found to reach decisions more quickly, used less information, avoided systematic and demanding thinking, and, ironically, appeared more confident about their decisions. In contrast, negative mood apparently triggered a more effortful, systematic, analytic, and vigilant processing style (Clark and Isen, 1982; Isen, 1984, 1987; Schwarz, 1990). Nevertheless, more recent studies show that positive mood sometimes produces distinct processing advantages. For instance, happy people tend to adopt a more creative, open, and inclusive thinking style, use broader cognitive categories, show greater mental flexibility, and perform better on secondary tasks (Bless and Fiedler, 2006; Fiedler, 2001; Fredrickson, 2009; Hertel and Fiedler, 1994; Isen and Daubman, 1984).

How can we explain these mood-induced processing differences? Early explanations emphasized the *motivational* consequences of good and bad moods. According to the *mood maintenance/mood repair* hypothesis, those in a positive mood may be motivated to maintain this rewarding state by avoiding effortful activity, such as elaborate information processing. In contrast, a negative mood should motivate people to engage in more vigilant, effortful information processing as an adaptive strategy to relieve their aversive state (Clark and Isen, 1982; Isen, 1984, 1987). More recently, several studies also showed that the cognitive consequences of affective states may depend on whether the mood state is high or low in approach motivational intensity. For example, positive affect with low specific approach motivation toward a goal seems to broaden cognitive categorization and attention, but positive affect linked to high approach motivation towards a particular goal tends to narrow cognitive categorization (Fredrickson, 2009).

A somewhat different *cognitive tuning* account (Schwarz, 1990) argues that positive and negative moods have a fundamental signaling/tuning function, informing the person whether a relaxed, effort-minimizing (positive mood) or a vigilant, effortful (negative mood) processing style is required. Both these models emphasize a functionalist/evolutionary view regarding moods as fulfilling adaptive functions (Forgas et al., 2008).

Yet another theory focuses on the impact of moods on *information processing capacity*, suggesting that mood states may influence processing style because they take up scarce processing capacity. Curiously, both positive mood (Isen, 1984) and negative mood (Ellis and Ashbrook, 1988) are hypothesized to reduce processing capacity. However, as positive and negative mood clearly promote qualitatively different processing styles, it is unlikely that the conflicting capacity reduction explanations put forward by Isen (1984) and Ellis and Ashbrook (1988) are both correct.

THE ASSIMILATION/ACCOMMODATION MODEL

Most of the earlier explanations of mood effects on information processing suggest that moods influence processing style by altering the degree of motivation, vigilance, and effort exerted. However, this view has been challenged by some experiments demonstrating that positive mood does not necessarily impair processing effort, as performance on simultaneously presented secondary tasks was not impaired (Fiedler, 2001; Hertel and Fiedler, 1994). A recent integrative theory, Bless and Fiedler's (2006) assimilation/accommodation model suggests that the fundamental, evolutionary significance of moods lies not so much in regulating processing effort, but rather, in triggering equally effortful but qualitatively different *processing styles*. The model identifies two complementary adaptive functions, *assimilation* and *accommodation* (Piaget, 1954). Assimilation means to impose internalized structures onto the external world, whereas accommodation means to modify internal structures in accordance with external constraints.

Thus, according to the model, "the adaptive function of positive mood is to facilitate assimilation, whereas the role of negative mood is to strengthen accommodation functions" (Bless and Fiedler, 2006; p. 66). Several lines of evidence now support the assimilative/accommodative processing dichotomy. For example, those in a positive mood tend to use broader, more assimilative cognitive categories (Fredrickson, 2009; Isen, 1984), sorted stimuli into fewer and more inclusive groups (Isen and Daubman, 1984), and classified behavioral descriptions into fewer and more inclusive types (Bless et al., 1992a). Positive affect also recruited more assimilative and abstract representations in language choices, as happy people generated more abstract and general event descriptions than did sad participants (Beukeboom, 2003), and were more likely to retrieve a generic rather than specific representation of a persuasive message (Bless et al., 1992b). Negative mood was also found to improve people's ability to detect ambiguous, unclear verbal messages, and to conform more closely to Grice's conversational postulates (Koch et al., 2013; Matovic et al., 2014). Similar mood-induced effects on processing style were found with nonverbal tasks. For example, happy mood resulted in greater focus on the global rather than the local features of geometric patterns (Gasper and Clore, 2002; Sinclair, 1988).

How can we best explain these pervasive mood-induced differences in processing style? Bless and Fiedler (2006) suggest that moods perform an adaptive function essentially preparing us to respond to different environmental challenges. Positive mood indicates that the situation is safe and familiar, and that existing knowledge can be relied upon. In contrast, negative mood functions like a mild alarm signal, indicating that the situation is novel and unfamiliar, and that the careful monitoring of new, external information is required.

There is supporting evidence suggesting that positive affect increases, and negative affect decreases the tendency to rely on internal knowledge rather than external information in cognitive tasks, resulting in a selective memory bias for self-generated information (Bless et al., 1992b; Fiedler et al., 2003). The theory thus predicts that *both* positive and negative mood can produce processing advantages albeit in response to different situations requiring different processing styles. Given the almost exclusive emphasis on the benefits of positive affect in our culture, this is an important message with some intriguing real-life implications. Numerous studies now suggest that negative mood can produce definite processing advantages in situations when the careful and detailed monitoring of new, external information is required, as we shall see later.

MOOD EFFECT ON MEMORY PERFORMANCE

If negative mood indeed recruits a more accommodative, externally focused processing style, then it should result in improved memory for incidentally encountered information. This is one key area where the processing consequences of good or bad moods have been explored. In one experiment, happy or sad subjects read a variety of essays advocating alternative positions on public policy issues. Later, their cued recall memory of the essays was assessed (Forgas, 1998b, Exp. 3). Results showed that those in a negative mood remembered the details of the essays significantly better than those in a happy mood, consistent with negative mood promoting more externally focused, accommodative thinking.

This effect was subsequently further explored in a field experiment, when happy or sad shoppers (on sunny or rainy days, respectively) saw a variety of small objects displayed on the check-out counter of a local news agency. After leaving the store, they were asked to recall and

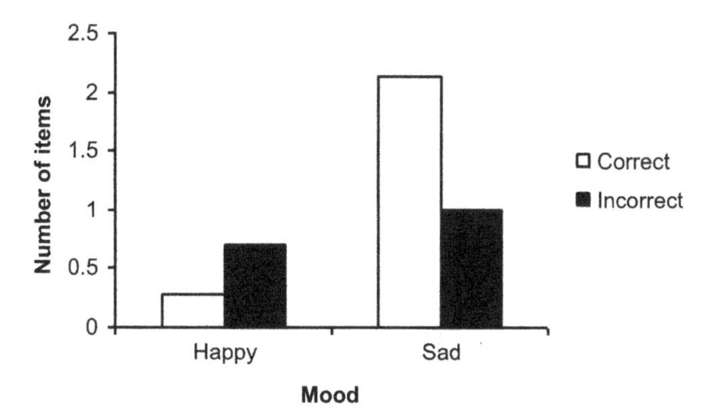

FIGURE 3.3 **The interaction between mood and the presence or absence of misleading information on eye-witness memory.** Positive mood increased and negative mood decreased the tendency to incorporate false, misleading details into eye-witness reports (false alarms).

recognize the objects they had seen on the check-out counter. It turned out that mood, induced by the weather, had a significant effect. Those in a negative mood (on rainy days) had significantly better memory for what they had seen in the shop than did happy people (on sunny days), confirming that mood states have a subtle but reliable memory effect, and negative mood actually improves memory for incidentally encountered information (Fig. 3.3).

A series of further experiments explored mood effects on eye-witness memory, predicting that, due to promoting more assimilative thinking (Isen, 1987), positive affect should increase, and negative affect should decrease, the tendency of eye-witnesses to incorporate false details into their memories (Forgas et al., 2005). In one study (Forgas et al., 2005, Exp. 1), participants viewed pictures of a car crash (negative event) and a wedding party (positive event). After 1 h, they received a mood induction (recalling happy or sad events from their past) and answered questions about the initially viewed scenes that either contained or did not contain misleading, false information. After a further 45-min interval, the accuracy of their eye-witness memory for the two scenes was tested. As predicted, positive mood increased, and negative mood decreased the amount of false, misleading information incorporated (assimilated) into their eye-witness memories. In contrast, negative mood almost completely eliminated this "misinformation effect," improving eyewitness memory as confirmed by a signal detection analysis.

In a second, more realistic experiment students witnessed a staged 5-min aggressive encounter between a lecturer and a female intruder (Forgas et al., 2005, Exp. 2). One week later, while in a happy or sad mood, they received a questionnaire that either did or did not contain planted,

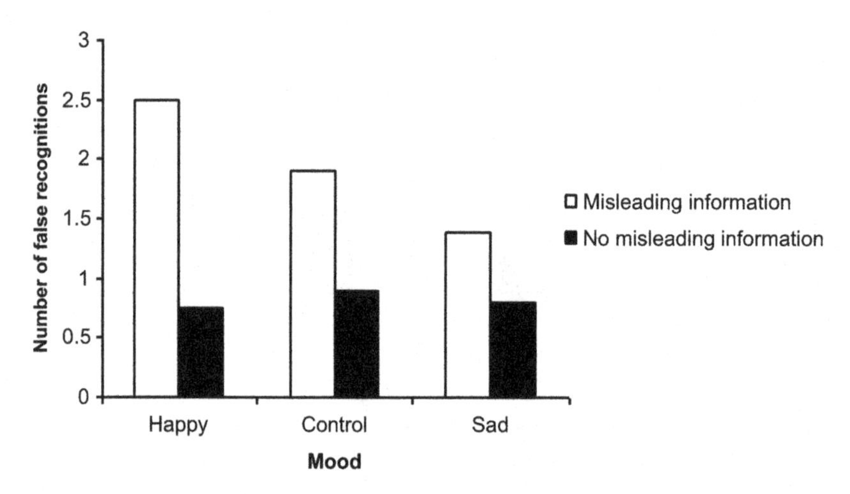

FIGURE 3.4 **Mood effects on the tendency to incorporate misleading information into eyewitness memory (Experiment 2).** Negative mood reduced, and positive mood increased eyewitness distortions due to misleading information (false alarms) *Source: After Forgas, J.P., Vargas, P., Laham, S., 2005. Mood effects on eyewitness memory: affective influences on susceptibility to misinformation. J. Exp. Soc. Psychol. 41, 574–588.*

misleading information. After a further interval, their eye-witness memory was assessed. Those in a happy mood when exposed to misleading information were more likely to assimilate false details into their memory. In contrast, negative mood eliminated this source of error in eye-witness memory, consistent with negative mood recruiting more accommodative processing and thus improving subject's ability to discriminate between correct and misleading details (Fig. 3.4).

In a further experiment in this series, participants saw videotapes showing (1) a robbery and (2) a wedding scene. After a 45-min interval, they received an audio-visual mood induction and completed a short questionnaire that either contained or did not contain misleading information about the events. Additionally, some were instructed to "disregard and control their affective states." Exposure to misleading information reduced eye-witness accuracy most when people were in a happy rather than a sad mood. However, direct instructions to control one's affect proved ineffective to reduce this mood effect.

Conceptually, similar results were reported by Storbeck and Clore (2011), who found that "individuals in a negative mood were significantly less likely to show false memory effects than those in positive moods" (p. 785). These authors explain their findings in terms of the affect-as-information mechanism. These experiments offer convergent evidence that negative moods recruit more accommodative thinking and therefore can improve memory performance by means of reducing susceptibility to misleading information. Paradoxically, happy mood *reduced* eye-witness accuracy yet *increased* subjective confidence, suggesting that judges were unaware of the processing consequences of their mood states.

MOOD EFFECTS ON JUDGMENTAL ACCURACY

Is it possible that mood states, through their influence on processing style, may also improve or impair the accuracy of our social judgments? For example, can good or bad mood influence the common tendency for people to form evaluative judgments based on their first impressions? One recent experiment examined mood effects on this "primacy effect," which occurs because people pay disproportionate attention to early rather than later information when forming impressions (Forgas, 2011b). After an autobiographical mood induction (recalling happy or sad past events), participants formed impressions about a character (Jim) described either in an introvert–extrovert or an extrovert–introvert sequence. As primacy effects occur because of the assimilative processing of later information, the subsequent impression formation judgments revealed that positive mood significantly increased the primacy effect by recruiting more top-down, assimilative processing. In contrast, negative mood, by recruiting a stimulus-based, accommodative processing style, almost eliminated the primacy effect.

Social perceivers are also often influenced by the relative salience or fluency of the targets. More visible and easier to process targets are often perceived as more important and influential than are less visible and fluent targets. It was found in a recent study that moods can also influence these judgmental effects (Forgas, 2015). In this experiment, perceivers had to form impressions about individuals based on a recorded conversation between them. Visual salience and fluency was manipulated by showing one target in a large, color picture and the other target in a small, black and white picture, and these manipulations were counterbalanced. The visual salience of the photos had a significant influence on impressions, with the more visible targets judged as more important and influential. However, mood again significantly mediated this effect: positive mood increased, and negative mood reduced salience effects on impressions (Forgas, 2015).

Many common judgmental errors in everyday life occur because people are imperfect and often inattentive information processors. For example, the *fundamental attribution error* (FAE) or *correspondence bias* refers to the pervasive tendency by people to infer intentionality and internal causation and underestimate the impact of situational constraints and forces when making judgments about the behavior of others (Gilbert and Malone, 1995). This error occurs because people focus on central and salient information, that is, the actor, whereas they ignore equally relevant but less salient information about external influences on the actor (Gilbert and Malone, 1995).

As negative mood promotes vigilant, detail-oriented processing, it should reduce the incidence of the FAE by directing greater attention to external influences on actors. This prediction was tested in one experiment (Forgas, 1998c) where happy or sad subjects read an essay and made attributions about its writer advocating a popular or unpopular position (for or against nuclear testing). The writer's position was described as either assigned (implies external causation) or freely chosen (implies internal causation). Results showed that happy persons were more likely and sad people were less likely than controls to commit the FAE by incorrectly inferring an internally caused attitude based on a coerced essay. Such mood-induced differences in judgmental accuracy do occur in real life. In a field study (Forgas, 1998c), happy or sad participants (after watching happy or sad movies) read essays and made attributions about writers advocating popular positions (pro recycling) or unpopular positions (contra recycling).

Again, positive affect increased and negative affect decreased the tendency to mistakenly infer internally caused attitudes based on coerced essays. In a further study, recall of the essays was additionally assessed as an index of processing style (Forgas, 1998c, Exp. 3). Negative mood again reduced and positive mood increased the incidence of the FAE. Recall memory data confirmed that those in a negative mood remembered more details, indicating enhanced accommodative processing. Furthermore, a mediation analysis showed that this mood-induced difference in processing style significantly mediated the observed mood effects on the incidence of the FAE. We should note, however, that negative mood only improves judgmental accuracy when relevant stimulus information is actually available. Ambady and Gray (2002) found that in the absence of diagnostic details, "sadness impairs [judgmental] accuracy precisely by promoting a more deliberative information processing style" (p. 947).

MOOD EFFECTS ON SKEPTICISM AND THE DETECTION OF DECEPTION

Many messages, such as most interpersonal communications, are by their very nature ambiguous and not open to objective validation. Much of our knowledge about the world is based on such second-hand information we receive from others. Only some claims (such as "urban myths") can potentially be evaluated against objective evidence, although such testing is usually not practicable. One of the most important cognitive tasks people face in everyday life is to decide whether to trust and accept, or distrust and reject social information. Rejecting valid information (excessive skepticism) is just as dangerous as accepting invalid information (gullibility).

What determines whether the information we come across in everyday life is judged true or false? There is some recent evidence that by recruiting assimilative or accommodative processing, mood states may significantly influence skepticism and gullibility (Forgas and East, 2008a,b). For example, one study asked happy or sad participants to judge the probable truth of a number of urban legends and rumors. Positive mood promoted greater gullibility for novel and unfamiliar claims, whereas negative mood promoted skepticism, which is consistent with the more externally focused, attentive, and detail-oriented accommodative thinking style. In another experiment, participants' recognition memory was tested 2 weeks after initial exposure to true and false statements taken from a trivia game. Only sad participants were able to correctly distinguish between the true and false claims they had seen previously. In contrast, happy participants tended to rate all previously seen and thus familiar statements as true (in essence, a fluency effect).

This pattern suggests that happy mood produced reliance on the "what is familiar is true" heuristic, whereas negative mood conferred a clear cognitive advantage improving judges' ability to accurately remember the truth or untruth of the statements. Unlike many "urban myths," interpersonal communications are often intrinsically ambiguous and have no objective truth value (Heider, 1958). Accepting or rejecting such messages is particularly problematic, yet critically important for effective social interaction. It turns out that mood effects on processing style may also influence people's tendency to accept or reject interpersonal communications as genuine. People in a negative mood were significantly less likely and those in a positive mood were more likely to accept various facial expressions communicating feelings as authentic (Forgas and East, 2008a). Taking this line of reasoning one step further, does mood, through its effect on processing styles, influence people's ability to detect deception? In one study, happy or sad participants watched videotaped interrogations of suspects accused of theft who were either guilty or not guilty of this offence (Forgas and East, 2008b). Surprisingly, those in a more happy mood were more gullible, as they accepted more denials as true. In contrast, sad mood resulted in more guilty judgments, and actually improved the participants' ability to correctly identify targets as deceptive (guilty) or honest, consistent with a more accommodative processing style (Fig. 3.5). These experiments offer convergent evidence that negative mood increases skepticism, and may significantly improve people's ability to accurately detect deception.

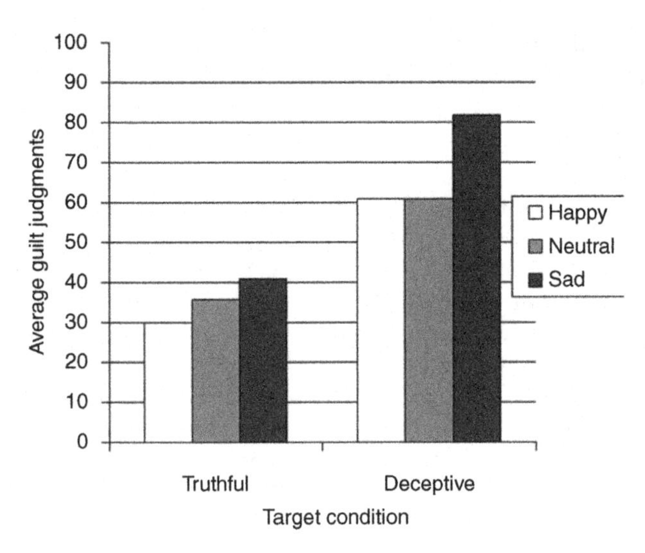

FIGURE 3.5 The effects of mood and the target's veracity (truthful, deceptive) on judgments of guilt of targets accused of committing a theft (average percentage of targets judged guilty in each condition. *Source: After Forgas, J.P., East, R., 2008. On being happy and gullible: mood effects on skepticism and the detection of deception. Exp. Soc. Psychol. 44, 1362–1367.*

MOOD EFFECTS ON STEREOTYPING

Assimilative processing in positive mood should promote, and accommodative processing in negative mood should reduce the use of preexisting knowledge structures, such as stereotypes. In several studies, Bodenhausen (1993) and Bodenhausen et al. (1994) found that happy participants relied more on ethnic stereotypes when evaluating a student accused of misconduct, whereas negative mood reduced this tendency. Generally speaking, sad individuals tend to pay greater attention to specific, individuating information when forming impressions of other people (Bless et al., 1996).

Similar effects were demonstrated in a recent experiment where happy or sad subjects had to form impressions about the quality and other aspects of a brief philosophical essay allegedly written by a middle-aged male academic (stereotypical author) or by a young, alternative-looking female writer (atypical author). Results showed that happy mood increased the judges' tendency to be influenced by irrelevant stereotypical information about the age and gender of the author. In contrast, negative mood eliminated this effect (Forgas, 2011c). Again, this pattern is entirely consistent with the predicted assimilative versus accommodative processing style recruited by good or bad moods, respectively (Fig. 3.6).

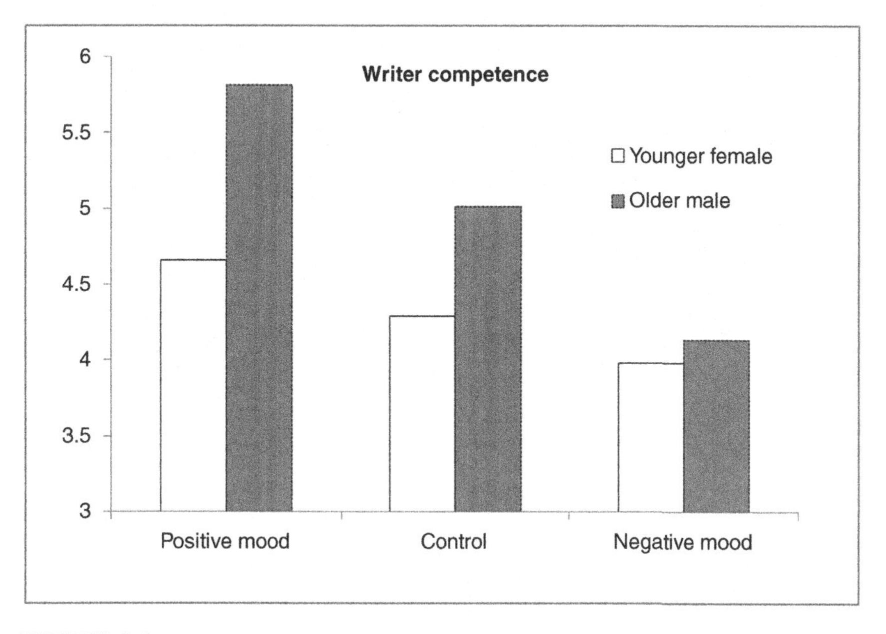

FIGURE 3.6 **Mood moderates the incidence of stereotype effects on the evaluation of the writer of an essay.** Positive mood increased, and negative mood eliminated the stereotype effects associated with the appearance of the writer. *After Forgas, 2011.*

Could mood-induced differences in processing style also influence reliance on stereotypes in actual social behaviors? We tested this prediction by asking happy or sad people to generate rapid responses to targets that appeared or did not appear to be Muslims, using the "shooter's bias" paradigm to assess subliminal aggressive tendencies (Correll et al., 2002). In this task, people are instructed to rapidly shoot at targets only when they carry a gun. Prior work with this paradigm showed that US citizens display a strong implicit bias to shoot more at Black rather than White targets (Correll et al., 2002, 2007).

We expected a "turban effect," that is, Muslim targets may elicit a similar bias. We used morphing software to create targets who did, or did not appear Muslim (wearing or not wearing a turban or the hijab) and who either held a gun, or held a similar object (e.g., a coffee mug; Fig. 3.7). Participants indeed shot more at Muslims rather than non-Muslims, but the most intriguing finding was that negative mood actually *reduced* this selective response tendency fueled by negative stereotypes (Unkelbach et al., 2008). Positive mood in turn increased the shooter's bias against Muslims, consistent with a more top-down, heuristic assimilative processing style (Bless and Fiedler, 2006; Forgas, 2007). Thus, mood effects on information processing styles may extend to influencing actual aggressive behaviors based on stereotypes as well.

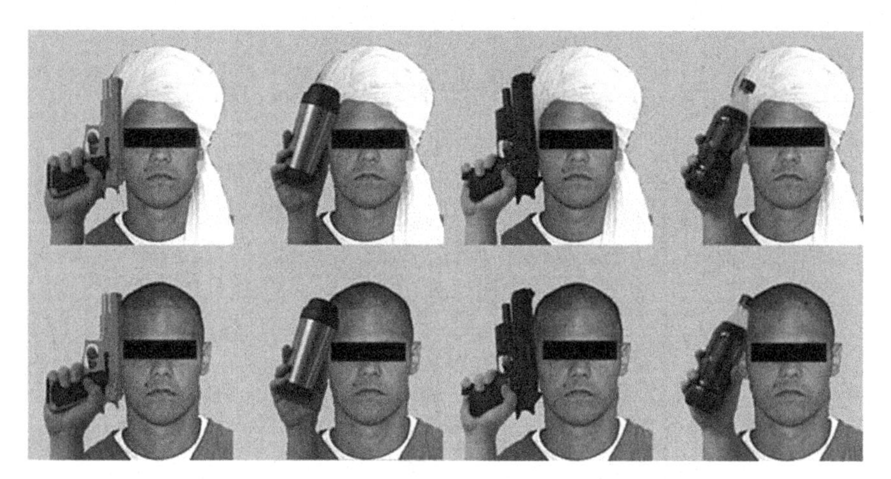

FIGURE 3.7 **The turban effect: stimulus figures used to assess the effects of mood and wearing or not wearing a turban on subliminal aggressive responses.** Participants had to make rapid shoot/don't shoot decisions in response to targets who did or did not hold a gun, and did or did not wear a Muslim head-dress (a turban).

MOOD EFFECTS ON INTERPERSONAL STRATEGIES

It has long been suspected that one of the possible benefits of negative affect may have to do with its interpersonal functions. Evolutionary psychologists, puzzled by the ubiquity of dysphoria, have speculated that negative affective states may provide hidden cognitive and social benefits (Forgas et al., 2007; Tooby and Cosmides, 1992). In situations where greater attention to new information and more accommodative processing is required, negative mood may provide significant processing benefits (Forgas, 2011a, 2015). However, there is growing evidence that in other situations where more cautious and less assertive behaviors are appropriate, it may be negative affect that produces real interpersonal benefits.

Mood effects on communication strategies. Effective interpersonal communication may be improved by processing external information in a more attentive and accommodative fashion. For instance, moods may optimize the way people process, produce, and respond to persuasive messages. In a number of studies, participants in sad moods showed greater attentiveness to message quality, and were more persuaded by strong rather than weak arguments. In contrast, those in a happy mood were not influenced by message quality, and were equally persuaded by strong and weak arguments (Bless et al., 1990, 1992b; Bohner et al., 1992; Sinclair et al., 1994; Wegener and Petty, 1997).

Further, mood states may also influence the *production* of persuasive messages. In one experiment, participants received an audio-visual mood

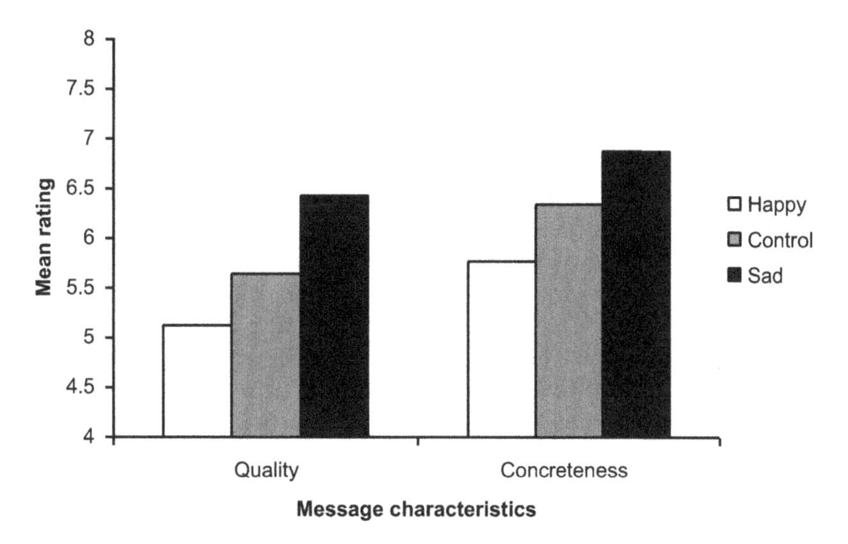

FIGURE 3.8 **Mood effects on the quality and concreteness of the persuasive messages produced.** Negative affect increased the degree of concreteness of the arguments produced, and arguments produced in negative mood were also rated as more persuasive.

induction and were then asked to produce effective persuasive arguments for or against (1) an increase in student fees, and (2) Aboriginal land rights (Forgas, 2007). As expected, results showed that participants in a sad mood produced higher quality, more effective persuasive arguments on both issues than did happy participants. A mediation analysis revealed that it was mood-induced variations in argument concreteness that mediated the observed differences in argument quality, consistent with the prediction that negative mood should recruit a more externally oriented, concrete, and accommodative processing style (Bless, 2001; Bless and Fiedler, 2006; Fiedler, 2001; Forgas, 2002). Similar effects were found when happy and sad people produced persuasive arguments for a "partner" to volunteer for a boring experiment using e-mail exchanges (Forgas, 2007). Once again, negative affect produced a processing benefit, resulting in more concrete and more effective persuasive messages (Fig. 3.8).

In a series of more recent studies, Koch et al. (2013) looked at the ability of people in positive and negative moods to detect subtle ambiguities in verbal communication. Their results showed that those in a negative mood were significantly better in identifying imprecise and ambiguous messages than were participants in a positive mood. In a further exploration of mood effects on effective communication, Matovic et al. (2014) investigated the effects of mood states on people's tendency to conform to good communication norms, by obeying Grice's communicative principles. Once again, negative mood significantly improved communication quality, and messages produced in negative mood were significantly more compliant with Gricean norms of effective communication. These results clearly support the idea that negative affect, by recruiting more accommodative and attentive processing, can improve communication effectiveness.

Mood effects on selfishness versus fairness. Intriguingly, the possibility that affective states may also influence interpersonal selfishness and fairness has received little attention in the past. Economic games offer a reliable and valid method to study interpersonal strategies, such as fairness, selfishness, trust, and cooperation. Induced moods may influence the degree of *selfishness* versus *fairness* people display when allocating resources among themselves and others (Forgas and Tan, 2013a,b; Tan and Forgas, 2010). Positive mood, by increasing internally focused, assimilative processing may promote greater selfishness, and in contrast, negative mood, by improving attention to external situational norms and expectations, may improve fairness. We investigated these predictions using two economic games, the *dictator game* and the *ultimatum game*. In the dictator game, an allocator can distribute resources to him/herself and to another person any way he/she likes. In the ultimatum game, the recipient also has to approve the decision—if rejected, neither party receives anything. In such games, proposers face a conflict between being selfish or fair (Güth et al., 1982), and their decisions may be open to affective influences.

In several experiments, we tested the hypothesis that (1) positive mood should increase, and negative mood decrease selfishness by allocators in both the dictator, and in the ultimatum games, and (2) receivers in the ultimatum game should also show greater concern with fairness, and paradoxically, should reject unfair offers more when they are in a negative rather than in a positive mood. Tan and Forgas (2010) found that happy allocators in the dictator game were significantly more selfish than sad players. In a follow-up experiment using a series of eight allocation decisions, those in a sad mood were again fairer and less selfish than happy individuals (Fig. 3.9).

These mood effects on fairness also endure in the more complex decisional environment faced by players in the ultimatum game (Forgas and Tan, 2013a). As hypothesized, those in a negative mood allocated significantly more resources to others than did happy individuals, and these mood effects could be directly linked to the predicted differences in processing style, as sad individuals took longer to make allocation decisions than did happy individuals, consistent with their expected more accommodative and attentive processing style.

Mood also influenced the behavior of responders (Forgas and Tan, 2013b) in the ultimatum game. Responders in a negative mood were also more concerned with external fairness norms, and were more likely to reject unfair offers. Overall, 57% of those in negative mood rejected unfair

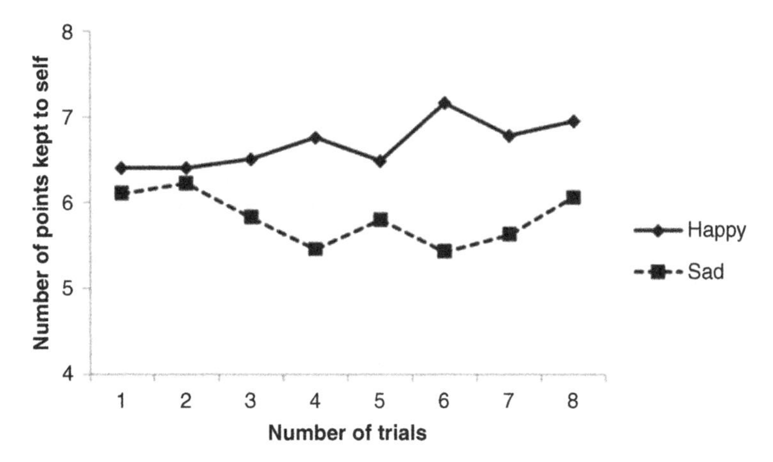

FIGURE 3.9 **The effects of mood on selfishness versus fairness.** Happy persons kept more rewards to themselves, and negative mood increased fairness toward others in reward allocations over 8 trials, this effect becomes even more pronounced in later trials. *Source: After Tan, H.B., Forgas, J.P., 2010. When happiness makes us selfish, but sadness makes us fair: affective influences on interpersonal strategies in the dictator game. J. Exp. Soc. Psychol. 46, 571–576.*

offers compared to only 45% in the positive condition, consistent with processing theories that predict that negative mood should increase and positive mood reduce attention to external fairness norms. Paying greater attention to external information, such as fairness norms when in a bad mood is also in line with other findings showing that negative mood increases attention to external information and improves eyewitness memory, reduces stereotyping, increases politeness, and reduces judgmental errors (Forgas, 1998a,b, 1999a,b; Forgas et al., 2009; Unkelbach et al., 2008).

These results further challenge the common assumption in much of applied, organizational, clinical and health psychology that positive affect has universally desirable social and interpersonal consequences. Rather, our findings confirm that negative affect often produces adaptive and more socially sensitive outcomes by recruiting a more attentive and accommodative processing style.

SUMMARY AND CONCLUSIONS

Human beings are a remarkably moody species. Affective states influence the way we think, remember and form inferences and judgments, as well as our interpersonal behaviors. So it is not surprising that understanding the relationship between feeling and thinking, affect and cognition has been one of the more enduring puzzles about human nature. From Plato to Pascal and Kant, a long line of Western philosophers have devoted their time to analyze the ways that affect can influence our thinking and behaviors.

Despite a number of promising early studies, psychologists were relatively late to apply empirical methods to study mood effects on cognition. This chapter reviewed the current status of this important research area. It was suggested that the effects of mood on cognition can be classified into two major kinds of influences: *informational effects* impacting on the content and valence of thinking usually resulting in mood congruence, and mood effects on *processing strategies*, influencing how people deal with information.

Practical implications. Our contemporary culture places a powerful emphasis on the desirability and beneficial effects of positive mood. As even a cursory visit to the popular psychology section of any bookshop will prove, the achievement of positive affect seems to be the objective of most applied psychological interventions. However, as the results reviewed here clearly show, positive affect is not universally beneficial. Numerous experiments now demonstrate the potentially adaptive and beneficial processing consequences of negative moods. For instance, people in a negative mood are less prone to judgmental errors (Forgas, 1998c), are more resistant to eye-witness distortions (Forgas et al., 2005), are less likely to rely on stereotypes (Forgas, 2011c; Unkelbach et al., 2008), and are better at producing high-quality, effective persuasive messages (Forgas, 2007).

Given the consistency of findings across a number of different domains, tasks, and affect inductions, these effects appear reliable. Further, they are broadly consistent with the notion that over evolutionary time, affective states came to operate as adaptive, functional triggers to elicit information processing patterns that are appropriate in a given situation. In a broader sense, the results presented here suggest that the persistent contemporary cultural emphasis on positivity and happiness may be misplaced, given growing evidence for the important, adaptive benefits of both positive and negative mood states. Professionals working in applied areas, such as human factors, and HCI should also benefit from closer attention to the often adaptive benefits of negative mood.

We should also note however that the processing advantages of negative affect reported here apply only to mild, temporary negative moods, and do not necessarily generalize to more intense and enduring negative affective states, such as depression. Depression has debilitating cognitive consequences and does not necessary produce more accommodative thinking. In a recent review article on the cognitive manifestation of depression, Gotlib and Joormann (2010) concluded that "depression is characterized by increased elaboration of negative information, by difficulties disengaging from negative material, and by deficits in cognitive control when processing negative information" (p. 285). According to this view, the cognitive dysfunction inherent in depression can rather be described as of prolonged, gridlocked mood-congruent information processing, rather than better accommodation to situational requirements.

In conclusion, this chapter reviewed strong cumulative evidence showing that mood states have a powerful, yet often subconscious influence on *what* people think (content effects) as well as *how* people think (processing effects). As we have seen, research shows that these effects are often subtle and subject to a variety of boundary conditions and contextual influences. A better understanding of the complex interplay between mood and cognition remains one of the most important tasks for psychology as a science. A great deal has been achieved in the last few decades applying empirical methods to exploring this issue, but in a sense, the enterprise has barely begun. Hopefully this chapter, and the collection of papers in this volume in general, will stimulate further research exploring the fascinating relationship between mood and cognition.

References

Adolphs, R., Damasio, A.R., 2001. The interaction of affect and cognition: a neurobiological perspective. In: Forgas, J.P. (Ed.), Handbook of Affect and Social Cognition. Erlbaum, Mahwah, NJ, pp. 27–49.

Ambady, N., Gray, H., 2002. On being sad and mistaken: mood effects on the accuracy of thin-slice judgments. J. Person. Soc. Psychol. 83, 947–961.

Beck, R.C., McBee, W., 1995. Mood-dependent memory for generated and repeated words: replication and extension. Cogn. Emot. 9, 289–307.

Becker, M. W., Leinenger, M., 2011. Attentional selection is biased toward mood-congruent stimuli. *Emotion*.

Beukeboom, C., 2003. How Mood Turns on Language. Doctoral dissertation, Free University of Amsterdam, Amsterdam.

Blaney, P.H., 1986. Affect and memory: a review. Psychol. Bull. 99, 229–246.

Bless, H., 2001. Mood and the use of general knowledge structures. In: Martin, L.L. (Ed.), Theories of Mood and Cognition: A User's Guidebook. Lawrence Erlbaum Associates, Mahwah, NJ, pp. 9–26.

Bless, H., Fiedler, K., 2006. Mood and the regulation of information processing and behavior. In: Forgas, J.P. (Ed.), Hearts and Minds: Affective Influences on Social Cognition and Behaviour. Psychology Press, New York, pp. 65–84.

Bless, H., Bohner, G., Schwarz, N., Strack, F., 1990. Mood and persuasion: a cognitive response analysis. Person. Soc. Psychol. Bull. 16, 331–345.

Bless, H., Hamilton, D.L., Mackie, D.M., 1992a. Mood effects on the organization of person information. Eur. J. Soc. Psychol. 22, 497–509.

Bless, H., Mackie, D.M., Schwarz, N., 1992b. Mood effects on encoding and judgmental processes in persuasion. J. Person. Soc. Psychol. 63, 585–595.

Bless, H., Schwarz, N., Wieland, R., 1996. Mood and the impact of category membership and individuating information. Eur. J. Soc. Psychol. 26, 935–959.

Bodenhausen, G.V., 1993. Emotions, arousal, and stereotypic judgments: a heuristic model of affect and stereotyping. In: Mackie, D.M., Hamilton, D.L. (Eds.), Affect, Cognition, and Stereotyping. Academic Press, San Diego, CA, pp. 13–37.

Bodenhausen, G.V., Kramer, G.P., Süsser, K., 1994. Happiness and stereotypic thinking in social judgment. J Person. Soc. Psychol. 66, 621–632.

Bohner, G., Crow, K., Erb, H.P., Schwarz, N., 1992. Affect and persuasion: mood effects on the processing of message content and context cues. Eur. J. Soc. Psychol. 22, 511–530.
Bousfield, W.A., 1950. The relationship between mood and the production of affectively toned associates. J. Gen. Psychol. 42, 67–85.
Bower, G.H., 1981. Mood and memory. Am. Psychol. 36, 129–148.
Bower, G.H., 1987. Commentary on mood and memory. Behav. Res. Ther. 25, 443–455.
Bower, G.H., 1991. Mood congruity of social judgments. In: Forgas, J.P. (Ed.), Emotion and Social Judgments. Pergamon, Oxford, pp. 31–53.
Bower, G.H., 1992. How might emotions affect learning? In: Christianson, S.A. (Ed.), Handbook of Emotion and Memory. Erlbaum, Hillsdale, NJ, pp. 3–31.
Bower, G.H., Forgas, J.P., 2000. Affect, memory, and social cognition. In: Eich, E., Kihlstrom, J.F., Bower, G.H., Forgas, J.P., Niedenthal, P.M. (Eds.), Cognition and Emotion. Oxford University Press, New York, pp. 87–168.
Bower, G.H., Mayer, J.D., 1985. Failure to replicate mood-dependent retrieval. Bull. Psychon. Soc. 23, 39–42.
Bower, G.H., Mayer, J.D., 1989. In search of mood-dependent retrieval. J. Soc. Behav. Person. 4, 121–156.
Brown, J.D., Mankowski, T.A., 1993. Self-esteem, mood, and self-evaluation: changes in mood and the way you see you. J. Person. Soc. Psychol. 64, 421–430.
Byrne, D., Clore, G.L., 1970. A reinforcement model of evaluation responses. Person. Int. J. 1, 103–128.
Ciarrochi, J.V., Forgas, J.P., 1999. On being tense yet tolerant: the paradoxical effects of trait anxiety and aversive mood on intergroup judgments. Group Dyn. 3, 227–238.
Ciarrochi, J.V., Forgas, J.P., 2000. The pleasure of possessions: affect and consumer judgments. Eur. J. Soc. Psychol. 30, 631–649.
Clark, M.S., Isen, A.M., 1982. Towards understanding the relationship between feeling states and social behavior. In: Hastorf, A.H., Isen, A.M. (Eds.), Cognitive Social Psychology. Elsevier, New York, pp. 73–108.
Clark, M.S., Waddell, B.A., 1983. Effects of moods on thoughts about helping, attraction and information acquisition. Soc. Psychol. Quart. 46, 31–35.
Clore, G.L., Byrne, D., 1974. The reinforcement affect model of attraction. In: Huston, T.L. (Ed.), Foundations of Interpersonal Attraction. Academic Press, New York, pp. 143–170.
Clore, G.L., Storbeck, J., 2006. Affect as information about liking, efficacy and importance. In: Forgas, J.P. (Ed.), Affect in Social Thinking and Behavior. Psychology Press, New York, pp. 123–142.
Clore, G.L., Gasper, K., Garvin, E., 2001. Affect as information. In: Forgas, J.P. (Ed.), Handbook of Affect and Social Cognition. Erlbaum, Mahwah, NJ, pp. 121–144.
Correll, J., Park, B., Judd, C.M., Wittenbrink, B., 2002. The police officer's dilemma: using ethnicity to disambiguate potentially threatening individuals. J. Person. Soc. Psychol. 83, 1314–1329.
Correll, J., Park, B., Judd, C.M., Wittenbrink, B., Sadler, M.S., Keesee, T., 2007. Across the thin blue line: police officers and racial bias in the decision to shoot. J. Person. Soc. Psychol. 92, 1006–1023.
Detweiler-Bedell, B., Detweiler-Bedell, J.B., 2006. Mood-congruent perceptions of success depend on self-other framing. Cogn. Emot. 20, 196–216.
Detweiler-Bedell, J.B., Salovey, P., 2003. Striving for happiness or fleeing from sadness? Motivating mood repair using differentially framed messages. J. Soc. Clin. Psychol. 22, 627–664.
Direnfeld, D.M., Roberts, J.E., 2006. Mood-congruent memory in dysphoria: the roles of state affect and cognitive style. Behav. Res. Ther. 44, 1275–1285.
Eich, E., 1995a. Searching for mood dependent memory. Psychol. Sci. 6, 67–75.
Eich, E., 1995b. Mood as a mediator of place dependent memory. J. Exp. Psychol. Gen. 124, 293–308.

Eich, E., Macaulay, D., 2000. Are real moods required to reveal mood-congruent and mood-dependent memory? Psychol. Sci. 11, 244–248.
Eich, E., Metcalfe, J., 1989. Mood dependent memory for internal versus external events. J. Exp. Psychol. 15, 443–455.
Eich, E., Macaulay, D., Ryan, L., 1994. Mood dependent memory for events of the personal past. J. Exp. Psychol. Gen. 123, 201–215.
Eich, E., Macaulay, D., Lam, R., 1997. Mania, depression, and mood dependent memory. Cogn. Emot. 11, 607–618.
Ellis, H.C., Ashbrook, P.W., 1988. Resource allocation model of the effects of depressed mood states on memory. In: Fiedler, K., Forgas, J.P. (Eds.), Affect, Cognition and Social Behavior. Hogrefe, Toronto, pp. 25–43.
Feshbach, S., Singer, R.D., 1957. The effects of fear arousal and suppression of fear upon social perception. J. Abnorm. Soc. Psychol. 55, 283–288.
Fiedler, K., 1990. Mood-dependent selectivity in social cognition. Stroebe, W., Hewstone, M. (Eds.), European Review of Social Psychology, vol.1, Wiley, New York, pp. 1–32.
Fiedler, K., 2001. Affective influences on social information processing. In: Forgas, J.P. (Ed.), Handbook of Affect and Social Cognition. Erlbaum, Mahwah, NJ, pp. 163–185.
Fiedler, K., Nickel, S., Asbeck, J., Pagel, U., 2003. Mood and the generation effect. Cogn. Emot. 17, 585–608.
Forgas, J.P., 1979. Social Episodes: The Study of Interaction Routines. Academic Press, NY.
Forgas, J.P., 1992. On bad mood and peculiar people: affect and person typicality in impression formation. J. Person. Soc. Psychol. 62, 863–875.
Forgas, J.P., 1993. On making sense of odd couples: mood effects on the perception of mismatched relationships. Person. Soc. Psychol. Bull. 19, 59–71.
Forgas, J.P., 1994. Sad and guilty? Affective influences on the explanation of conflict episodes. J. Person. Soc. Psychol. 66, 56–68.
Forgas, J.P., 1995. Mood and judgment: the affect infusion model (AIM). Psychol. Bull. 117, 39–66.
Forgas, J.P., 1998a. On feeling good and getting your way: mood effects on negotiating strategies and outcomes. J. Person. Soc. Psychol. 74, 565–577.
Forgas, J.P., 1998b. Asking nicely? Mood effects on responding to more or less polite requests. Person. Soc. Psychol. Bull. 24, 173–185.
Forgas, J.P., 1998c. On being happy but mistaken: mood effects on the fundamental attribution error. J. Person. Soc. Psychol. 75, 318–331.
Forgas, J.P., 1999a. On feeling good and being rude: affective influences on language use and requests. J. Person. Soc. Psychol. 76, 928–939.
Forgas, J.P., 1999b. Feeling and speaking: mood effects on verbal communication strategies. Person. Soc. Psychol. Bull. 25, 850–863.
Forgas, J.P., 2002. Feeling and doing: affective influences on interpersonal behavior. Psychol. Inq. 13, 1–28.
Forgas, J.P. (Ed.), 2006. Affect in Social Thinking and Behavior. Psychology Press, NY.
Forgas, J.P., 2007. When sad is better than happy: mood effects on the effectiveness of persuasive messages. J. Exp. Soc. Psychol. 43, 513–528.
Forgas, J.P., 2011a. Affective influences on self-disclosure strategies. J. Person. Soc. Psychol. 100 (3), 449–461.
Forgas, J.P., 2011b. Can negative affect eliminate the power of first impressions? Affective influences on primacy and recency effects in impression formation. J. Exp. Soc. Psychol. 47, 425–429.
Forgas, J.P., 2011c. She just doesn't look like a philosopher...? Affective influences on the halo effect in impression formation. Eur. J. Soc. Psychol. 41, 812–817.
Forgas, J.P., 2013. Don't worry, be sad! On the cognitive, motivational and interpersonal benefits of negative mood. Curr. Direct. Psychol. Sci. 22, 225–232.

Forgas, J.P., 2015. Why do highly visible people appear more important? Affect mediates visual fluency effects in impression formation. J. Exp. Soc. Psychol. 58, 136–141.
Forgas, J.P., Bower, G.H., 1987. Mood effects on person perception judgments. J. Pers. Soc. Psychol. 53, 53–60.
Forgas, J.P., Bower, G.H., Krantz, S., 1984. The influence of mood on perceptions of social interactions. J. Exp. Soc. Psychol. 20, 497–513.
Forgas, J.P., Bower, G.H., Moylan, S.J., 1990. Praise or blame? Affective influences on attributions. J. Pers. Soc. Psychol. 59, 809–818.
Forgas, J.P., Ciarrochi, J.V., 2002. On managing moods: evidence for the role of homeostatic cognitive strategies in affect regulation. Pers. Soc. Psychol. Bull. 28, 336–345.
Forgas, J.P., Dunn, E., Granland, S., 2008. Are you being served? An unobtrusive experiment of affective influences on helping in a department store. Eur. J. Soc. Psychol. 38, 333–342.
Forgas, J.P., East, R., 2008a. How real is that smile? Mood effects on accepting or rejecting the veracity of emotional facial expressions. J. Nonverbal Behav. 32, 157–170.
Forgas, J.P., East, R., 2008b. On being happy and gullible: mood effects on skepticism and the detection of deception. J. Exp. Soc. Psychol. 44, 1362–1367.
Forgas, J.P., Goldenberg, L., Unkelbach, C., 2009. Can bad weather improve your memory? A field study of mood effects on memory in a real-life setting. J. Exp. Soc. Psychol. 54, 254–257.
Forgas, J.P., Haselton, M.G., von Hippel, W. (Eds.), 2007. Evolution and the social mind. Psychology Press, New York.
Forgas, J.P., Moylan, S.J., 1987. After the movies: the effects of transient mood states on social judgments. Person. Soc. Psychol. Bull. 13, 478–489.
Forgas, J.P., Tan, H.B., 2013a. To give or to keep? Affective influences on selfishness and fairness in computer-mediated interactions in the dictator game and the ultimatum game. Comput. Hum. Behav. 29, 64–74.
Forgas, J.P., Tan, H.B., 2013b. Mood effects on selfishness versus fairness: affective influences on social decisions in the ultimatum game. Soc. Cogn. 31, 504–517.
Forgas, J.P., Vargas, P., Laham, S., 2005. Mood effects on eyewitness memory: affective influences on susceptibility to misinformation. J. Exp. Soc. Psychol. 41, 574–588.
Fredrickson, B.L., 2009. Positivity. Three Rivers Press, New York.
Gasper, K., Clore, G.L., 2002. Attending to the big picture: mood and global versus local processing of visual information. Psychol. Sci. 13, 34–40.
Gilbert, D.T., Malone, P.S., 1995. The correspondence bias. Psychol. Bull. 117, 21–38.
Gilboa-Schechtman, E., Erhard-Weiss, D., Jecemien, P., 2002. Interpersonal deficits meet cognitive biases: memory for facial expressions in depressed and anxious men and women. Psych Res. 113, 279–293.
Goodwin, D.W., 1974. Alcoholic blackout and state-dependent learning. Feder. Proc. 33, 1833–1835.
Gotlib, I.H., Joormann, J., 2010. Cognition and depression: current status and future directions. Ann. Rev. Clin. Psychol. 6, 285–312.
Gouaux, C., 1971. Induced affective states and interpersonal attraction. J. Person. Soc. Psychol. 20, 37–43.
Gouaux, C., Summers, K., 1973. Interpersonal attraction as a function of affective states and affective change. J. Res. Person. 7, 254–260.
Griffitt, W., 1970. Environmental effects on interpersonal behavior: temperature and attraction. J. Person. Soc. Psychol. 15, 240–244.
Güth, W., Schmittberger, R., Schwarze, B., 1982. An experimental analysis of ultimatum bargaining. J. Econ. Behav. Org. 3 (4), 367–388.
Heider, F., 1958. The Psychology of Interpersonal Relations. Wiley, New York.
Heimpel, S.A., Wood, J.V., Marshall, M., Brown, J., 2002. Do people with low self-esteem really want to feel better? Self-esteem differences in motivation to repair negative moods. J. Person. Soc. Psychol. 82, 128–147.

Hertel, G., Fiedler, K., 1994. Affective and cognitive influences in a social dilemma game. Eur. J. Soc. Psychol. 24, 131–145.
Hilgard, E.R., 1980. The trilogy of mind: cognition, affection, and conation. J. Hist. Behav. Sci. 16, 107–117.
Isen, A.M., 1984. Towards understanding the role of affect in cognition. Wyer, R.S., Srull, T.K. (Eds.), Handbook of Social Cognition, vol. 3, Erlbaum, NJ, pp. 179–236.
Isen, A.M., 1985. Asymmetry of happiness and sadness in effects on memory in normal college students: comment on Hasher, Rose, Zacks, Sanft, and Doren. J. Exp. Psychol. Gen. 114, 388–391.
Isen, A.M., 1987. Positive affect, cognitive processes and social behaviour. Berkowitz, L. (Ed.), Advances in Experimental Social Psychology, vol. 20, Academic Press, New York, pp. 203–253.
Isen, A.M., Daubman, K.A., 1984. The influence of affect on categorization. J. Person. Soc. Psychol. 47, 1206–1217.
Josephson, B.R., Singer, J.A., Salovey, P., 1996. Mood regulation and memory: repairing sad moods with happy memories. Cogn. Emot. 10, 437–444.
Kenealy, P.M., 1997. Mood-state-dependent retrieval: the effects of induced mood on memory reconsidered. Quart. J. Exp. Psychol. 50A, 290–317.
Kihlstrom, J.F., 1989. On what does mood-dependent memory depend? J. Soc. Behav. Person. 4, 23–32.
Koch, A.S., Forgas, J.P., Matovic, D., 2013. Can negative mood improve your conversation? Affective influences on conforming to Grice's communication norms. Eur. J. Soc. Psychol. 43, 326–334.
Leight, K.A., Ellis, H.C., 1981. Emotional mood states, strategies, and state-dependency in memory. J. Verbal Learning Verbal Behav. 20, 251–266.
Martin, L., 2000. Moods don't convey information: moods in context do. In: Forgas, J.P. (Ed.), Feeling and Thinking: The Role of Affect in Social Cognition. Cambridge University Press, New York, pp. 153–177.
Matovic, D., Koch, A., Forgas, J.P., 2014. Can negative mood improve language understanding? Affective influences on the ability to detect ambiguous communication. J. Exp. Soc. Psychol. 52, 44–49.
Mayer, J.D., Gaschke, Y.N., Braverman, D.L., Evans, T.W., 1992. Mood-congruent judgment is a general effect. J. Person. Soc. Psychol. 63, 119–132.
Miranda, R., Kihlstrom, J., 2005. Mood congruence in childhood and recent autobiographical memory. Cogn. Emot. 19, 981–998.
Piaget, J., 1954. The Construction of Reality in the Child. Free Press, New York.
Rapaport, D., 1961. Emotions and Memory. Science Editions, New York.
Razran, G.H., 1940. Conditioned response changes in rating and appraising sociopolitical slogans. Psychol. Bull. 37, 481–493.
Reus, V.I., Weingartner, H., Post, R.M., 1979. Clinical implications of state-dependent learning. Am. J. Psych. 136, 927–931.
Ruiz-Caballero, J.A., Gonzalez, P., 1994. Implicit and explicit memory bias in depressed and non-depressed subjects. Cogn. Emot. 8, 555–570.
Schacter, D.L., Kihlstrom, J.F., 1989. Functional amnesia. Boller, F., Grafman, J. (Eds.), Handbook of Neuropsychology, vol.3, Elsevier, New York, pp. 209–230.
Schiffenbauer, A.I., 1974. Effect of observer's emotional state on judgments of the emotional state of others. J. Person. Soc. Psychol. 30, 31–35.
Schwarz, N., 1990. Feelings as information: informational and motivational functions of affective states. Higgins, E.T., Sorrentino, R. (Eds.), Handbook of Motivation and Cognition, vol. 2, Guilford, New York, pp. 527–561.
Schwarz, N., Clore, G.L., 1983. Mood, misattribution and judgments of well-being: informative and directive functions of affective states. J. Person. Soc. Psychol. 45, 513–523.

Sedikides, C., 1994. Incongruent effects of sad mood on self-conception valence: it's a matter of time. Eur. J. Soc. Psychol. 24, 161–172.

Sedikides, C., 1995. Central and peripheral self-conceptions are differentially influenced by mood: tests of the differential sensitivity hypothesis. J. Person. Soc. Psychol. 69, 759–777.

Sinclair, R.C., 1988. Mood, categorization breadth, and performance appraisal: the effects of order of information acquisition and affective state on halo, accuracy, and evaluations. Org. Behav. Hum. Decis. Process. 42, 22–46.

Sinclair, R.C., Mark, M.M., Clore, G.L., 1994. Mood-related persuasion depends on misattributions. Soc. Cogn. 12, 309–326.

Singer, J.A., Salovey, P., 1988. Mood and memory: evaluating the network theory of affect. Clin. Psychol. Rev. 8, 211–251.

Smith, S.M., Petty, R.E., 1995. Personality moderators of mood congruency effects on cognition: the role of self-esteem and negative mood regulation. J. Person. Soc. Psychol. 68, 1092–1107.

Storbeck, J., Clore, G.L., 2011. Affect influences false memories at encoding: evidence from recognition data. Emotion 11, 981–989.

Tamir, M., Robinson, M.D., 2007. The happy spotlight: positive mood and selective attention to rewarding information. Person. Soc. Psychol. Bull. 33, 1124–1136.

Tan, H.B., Forgas, J.P., 2010. When happiness makes us selfish, but sadness makes us fair: affective influences on interpersonal strategies in the dictator game. J. Exp. Soc. Psychol. 46, 571–576.

Tooby, J., Cosmides, L., 1992. The psychological foundations of culture. In: Barkow, J.H., Cosmides, L. (Eds.), The adapted mind: evolutionary psychology and the generation of culture. Oxford University Press, London, pp. 19–136.

Ucros, C.G., 1989. Mood state-dependent memory: a meta-analysis. Cogn. Emot. 3, 139–167.

Unkelbach, C., Forgas, J.P., Denson, T.F., 2008. The turban effect: the influence of Muslim headgear and induced affect on aggressive responses in the shooter bias paradigm. J. Exp. Soc. Psychol. 44, 1409–1413.

Wadlinger, H., Isaacowitz, D.M., 2006. Positive affect broadens visual attention to positive stimuli. Motiv. Emot. 30, 89–101.

Watkins, T., Mathews, A.M., Williamson, D.A., Fuller, R., 1992. Mood congruent memory in depression: emotional priming or elaboration. J. Abnorm. Psychol. 101, 581–586.

Watson, J.B., 1929. Behaviorism. Norton, New York.

Watson, J.B., Rayner, R., 1920. Conditioned emotional reactions. J. Exp. Psychol. 3, 1–14.

Wegener, D.T., Petty, R.E., 1997. The flexible correction model: the role of naïve theories of bias in bias correction. Zanna, M.P. (Ed.), Advances in Experimental Social Psychology, vol. 29, Academic Press, New York, pp. 141–208.

Wegner, D.M., 1994. Ironic processes of mental control. Psychol. Rev. 101, 34–52.

CHAPTER

4

Cross-Cultural Similarities and Differences in Affective Processing and Expression

James A. Russell
Boston College, Chestnut Hill, MA, United States

Emotions play a role in nearly everything we do and think. Understanding people and their interactions requires an understanding of emotion. And science can contribute to that understanding. Here, I cannot cover everything in the science of emotion, but cover enough to introduce the reader to an emerging new way of thinking about emotion that is an alternative to the traditional Basic Emotion Theory taught in textbooks. I contrast the new way with the traditional way by examining the subjective experience of emotion, information processing during emotion, and facial expressions of emotion.

Everyone has extensive experience to draw on in understanding emotion. Even 2-year-olds have knowledge about emotion. Yet, as valuable as it sometimes is, our everyday experience can also mislead us. As Will Rogers said, it isn't what we don't know that gives us trouble, it's what we know that ain't so. Common sense resisted the heliocentric solar system, quantum mechanics, plate tectonics, and evolution by natural selection. We must not rely solely on common sense in building our understanding of emotion. A person is too easily divided to two parts: the animal versus the rational, passions versus reasons, feelings versus thoughts, heart versus mind, and so on. Emotion is usually (but not always) placed on the animal-passion-feeling-heart side, but doing so oversimplifies. Similarly, the origin of emotion is too easily seen as nature rather than nurture, as if origin were an either/or question. As a species, humans have much in common, and science must delineate precisely what that is. But one of the things we humans hold in common is our capacity to acquire new

Emotions and Affect in Human Factors and Human–Computer Interaction. http://dx.doi.org/10.1016/B978-0-12-801851-4.00004-5
Copyright © 2017 Elsevier Inc. All rights reserved.

information and practices and skills, to create novel solutions to the problems we face, to remember them, and to communicate them to the next generation. That capacity—culture—has helped shape the emotions we experience. We must include both nature and nurture in our theory. More generally, we must be prepared to abandon oversimplified ideas about emotion that seem commonsensical and, instead, follow the evidence to build a new understanding of emotion.

DEVELOPING A NEW CONCEPTUAL FRAMEWORK

The first step in going beyond what "we know that ain't so" is recognizing that emotion is not just one type of thing. Emotion is a broad and heterogeneous category, including life-long commitments (love), medium-term states (anxiety and depression), brief episodes (fear and jealously), and even momentary reflexes (startle) (Widen, 2016). The category of emotion also overlaps with attitude, mood, and motivation. Here I focus on emotional episodes, not so brief as startle, but with a beginning and end.

Emotional episodes are traditionally divided into categories, such as fear, anger, joy, grief, and so on. Each specific token episode (your fear upon encountering a bear in the woods last Saturday morning) is real, but it does not follow that categorizing all fear episodes together is the best way to understand them. As the philosopher-psychologist William James said, any division of emotions into categories is as natural as any other. Nonetheless, initial theories of emotional episodes in psychology followed common sense by treating fear, anger, joy, and other traditional categories of emotion as scientific primitives, with each thought of as a preformed module wired into the brain by evolution and understood in a simple Stimulus–Response framework: an antecedent event triggers the emotion, the consciously felt experience, which causes a cascade of highly coordinated component events, including a signature pattern in peripheral physiology, expressive vocal and facial signals, and an instrumental action. The bear in the woods triggers fear, which makes the heart pound, the eyes widen, freezing, and then fleeing. In our Western cultural tradition, this simple framework seems obvious: it's the ordinary way we have of thinking about emotional episodes. This way of thinking was articulated into a research program called "Basic Emotion Theory." This theory guided most research on emotion during the last half of the 20th century.

Ironically, the evidence that Basic Emotion Theory inspired has slowly undermined its assumptions. Biological basis of emotion is beyond the scope of this chapter, except to say that current evidence is uncovering a mind-boggling level of complexity at odds with Basic Emotion Theory ideas that all cases of fear, for example, are the activation of the same specific evolutionarily ancient brain circuit dedicated just to fear (for evidence see, e.g., LeDoux, 2015; Lindquist et al., 2012). Similarly, current evidence has uncovered complexity in the pattern of autonomic nervous system activity at odds with the Basic Emotion Theory prediction of a signature pattern for each category of basic emotion (Cacioppo et al., 2000).

Another problem is that few emotional episodes, or even fear episodes, fit the Stimulus–Response framework. The precipitating event is rarely a single event; it is often an on-going rapidly changing complex flow of events. Encounters with bear are real, but most antecedent events are complex social situations involving other persons. A person's response to this precipitating event is similarly on-going and rapidly changing. Further, the person enters the situation with relevant goals, beliefs, feelings, and anticipations. Thus, specifying a beginning and an end is somewhat arbitrary. When the autonomic nervous system changes rapidly, picking one pattern is somewhat arbitrary, and so on.

The emotional episode does indeed typically (although not always) include the types of component events just listed, but two surprising empirical findings must guide future thinking. First, these events must be elaborated to include much more: the information processing that includes perception and cognition of the unfolding antecedent event (its appraisal, and attributions of causality); changes in peripheral physiology in preparation for or recovery from action; behavioral processes such as planning, setting goals, and creative problem solving as well as overt actions including expressive and instrumental actions; continuous regulation of the unfolding episode, and various forms of subjective conscious experience of having the emotion. Most importantly, an emotional episode also includes change in the most simple and primitive affective feelings (called here core affect).

Second, the components do not cohere as a preformed package in the way anticipated. In a prototypical case of an emotional episode, these processes all occur and do so in the supposed temporal and causal order. Nonetheless, few cases are prototypical; most occur with these processes altered or absent or in a different order. The various component events do not form the stereotyped pattern such as that narrated for fear of the bear in the woods. Humans (like other animals) may have an innate system to detect certain dangers, but that system need not include the conscious experience of fear. Emotional episodes are coordinated, but not in same way for different cases of even the same emotion. Instead, coordination depends on the specific circumstances encountered and on the plan (sometimes hastily) formulated to deal with those circumstances.

The emotional episode takes place in a cultural context. A person inherits from the culture much that is relevant to emotion: a way of thinking about situations one encounters, a set of lexically coded categories into which emotional episodes are divided; rules regarding the display of expressive signs; rules regarding what emotions are proscribed or prescribed in different circumstances; social roles that specify how an emotional episode should unfold and under what circumstances. Such culturally based practices in turn inform judgments about and talk about emotional events.

Finally, Basic Emotion Theory downplayed a key element at the core of an emotional episode: a primitive universal state called "core affect," which is simply feeling good or bad, energized or enervated. Core affect gives each emotional episode its raw feel: tense, relaxed, excited, depressed, and so on.

The alternative conceptual framework advocated here is a member of a new family of theories called Psychological Construction, being developed by scientists from different disciplines as an alternative to Basic Emotion Theory (Barrett, 2006; Barrett and Russell, 2015; Clore and Ortony, 2013; LeDoux, 2015; Lindquist et al., 2012, Russell, 2003; Widen, 2016). In this framework, categories of emotion are secondary, based on underlying processes better described by dimensions. For example, I put the dimensions of core affect at the heart of an emotional episode. This core is surrounded by dimensions describing sensory-perceptual-cognitive and behavioral processes, which are influenced by immediate context and broader cultural context. These surrounding processes are part of the emotional episode and are part of what distinguishes one emotion category from another. The particular emotional episode that occurs on a specific occasion is not preformed but rather assembled on the fly to solve the current problem faced as it unfolds (often rapidly). The solution formulated is sometimes hasty and stereotyped but sometimes creative and clever. The processes involved in turn owe much to the cultural background of each individual. Let me illustrate this framework by considering subjective conscious experience, information processing, and facial expressions of emotion.

SUBJECTIVE CONSCIOUS EXPERIENCE

To most people, emotion simply is a conscious experience. I don't agree. (For example, a man can honestly deny experiencing jealousy but later realize that he was in fact jealous. Further, subliminal threat stimuli can elicit the same changes in peripheral physiology as do supraliminal threats.) Still, conscious experience is important and pervasive in the study of emotion through verbal reports and the assumptions of experimenters.

The conscious experiences associated with emotion are heterogeneous. Without claiming to be exhaustive, I distinguish four kinds of emotional conscious experiences: Emotional Metaexperience, Core Affect, Perception of Affective Quality, and Attributed Affect. Here I focus on the first two.[a] The prototypical conscious emotional experience includes all four,

[a]Perception of Affective Quality and Attributed Affect are defined by Core Affect. Perception of Affective Quality is the perception, pervasive in our understanding of the world, that objects and events have the capacity to alter core affect: we find the music pleasant, the weather stimulating, and the book boring. Attributed Affect is the experience of a causal link between current Core Affect and some object or event: she made me happy, the news made me sad.

but they can be separate. Distinguishing Emotional Metaexperience from Core Affect reconciles two traditions that emerged early in the psychology of conscious emotional experience: the categorical (anger, fear, joy, and so on) and the dimensional (valence, activation, and so on). Both traditions are needed and can be reconciled.

Emotional Metaexperience. The focus of much writing on emotion is the conscious subjective experience of having a specific emotion such as anger, jealousy, fear, love, and so on. On this traditional account, to experience anger or some other specific emotion is a simple reading of an inner biologically given signal. Instead, on my account, it is a psychologically constructed experience, the end product of an information processing stream that includes attention, memory, and a categorization of one's current state based on available information. Based on his physiological research, Levenson (2011) offered an account of the feeling of a specific emotion I find indistinguishable from my account of emotional metaexperience. Based on his neuroscience research, LeDoux (2015) offered a similar account—the interpreter theory of consciousness—for emotional metaexperience.

The mental categories we use in this process are not innate; rather, different categories are available to different individuals. English speakers may categorize a certain emotional state as *anger*; Ilongot speakers may categorize similar states as *liget* (Rosaldo, 1984). Different languages lexicalize different categories. One indication is that different languages recognize different numbers of categories, ranging from seven (Howell, 1984) to hundreds (Russell, 1991). Over the course of development, children similarly increase the number of different emotion categories they use (Widen, 2016). Thus, the number of potential categories is large and indeterminate. Categorization is a universal process, but the categories into which emotions are divided vary with language.

Some writers have attempted to reduce the number of categories by considering some to be "basic" and then defining the remainder as subcategories or blends of the basic ones. Still, there is no agreement on the number of such basic emotions, which emotions are basic, or what makes something basic (Ortony and Turner, 1990). Witness the disagreement over how many emotions are basic: Ekman (1972) listed six, but Ekman and Cordaro (2011) listed 21. The concept of "basic" may be of little use any more.

Moreover, different languages recognize different categories. Some languages make distinctions between emotions that English does not. There is no single word for shame in Chinese, but instead a set of types of shame. One might think that Chinese speakers recognize all these as subtypes of shame, but there is no evidence to support that thought. Conversely, English makes some distinctions that other languages do not. English distinguishes anger from sadness as qualitatively different (basic) emotions;

some African languages use the same word for the two (Leff, 1973). Similarly, the Ilongot term *liget* includes both anger and grief (Rosaldo, 1984). English distinguishes shame from embarrassment, but many non-Western languages do not (Levy, 1973). In other cases, English has a word for an emotion, whereas other languages do not (Leff, 1973; Levy 1973). Tahitian lacks words equivalent to sadness and guilt (Levy, 1973). Ekman (1972) found no words for disgust or surprise in the Fore language of Papua New Guinea.

A few paragraphs before, readers might have thought that perhaps *liget* could simply be translated as *anger*. Anthropologists and cross-cultural psychologists often seek translations for English emotion words, but more careful study revealed differences. Anthropologists have reported such findings (Davitz, 1969; Levy, 1973; Rosaldo, 1980; Wikan, 1989). More experimental approaches have likewise found differences between what had been assumed translation equivalents (Han et al., 2015; Hurtado-de-Mendoza et al., 2013; Russell and Sato, 1995).

My term for conscious emotional experience as categorized is Emotional Metaexperience. On my account, the experience of anger, for example, is not the recurrence of the same, simple, irreducible mental quale. Although the existence of such qualia is sometimes assumed, no evidence for the assumption has been offered. Instead, Emotional Metaexperience is a complex form of self-perception. It is a metaexperience in that it depends on other aspects of experience, some of which are themselves consciously accessible. Like all perceptions, Emotional Metaexperiences are intentional mental states; that is, they include a representation of something: what one is angry about, jealous of, afraid of, or in love with. Like all perceptions, Emotional Metaexperiences are interpretations. The raw data on which the interpretation is based are both top-down (such as concepts, stored knowledge, expectations, attributions, appraisals, and memories) and bottom-up (from both the internal world via somatosensory feedback and the external world).

Core Affect. Core Affect is an aspect of subjective emotional experience. It is a neurophysiological state consciously accessible as simply feeling good or bad, energized or quiescent. At any point in time, a person can answer the question, how do you feel? Studies of their answers have led to the concept of Core Affect (Russell, 2003). Core Affect is, at a psychological level, the most elementary simple primitive affective feeling. A map of Core Affect is seen in Fig. 4.1, which shows a circumplex representation of self-reported moods and feelings. A person's Core Affect at any one moment in time is represented by one point somewhere inside the space. The space, in turn, is characterized by two bipolar dimensions—valence and activation—in the dimensional tradition in the psychology of emotion. In this two-dimensional Cartesian space, a person's basic affective feeling for each moment in time

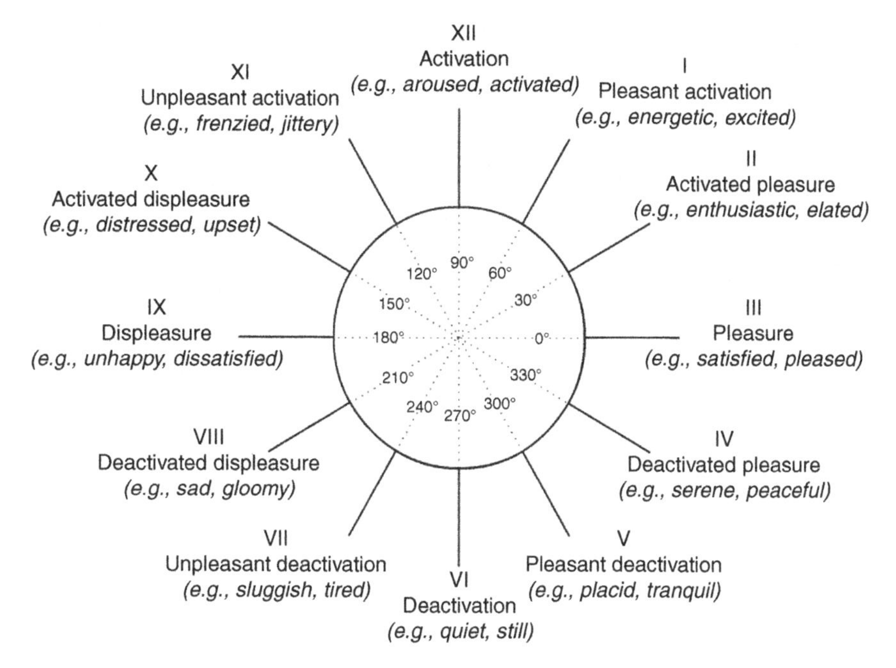

FIGURE 4.1 A representation of Core Affect by dividing the space into 12 segments. *Reprinted from Yik, M., Russell, J.A., Steiger, J.H. 2011. A 12-point circumplex structure of core affect.* Emotion 11, *705–731. by permission.*

is placed at a single point. Like a flat map of earth, longitude and latitude do not tell everything about each place on the map, but are basic ingredients.

The person thus has only one Core Affect at a time. The center can be thought of as an adaptation level (a neutral point midway between pleasure and displeasure and midway between low and high arousal), with distance from the center representing intensity or extremity of feeling. Core Affect can be extremely intense at times, milder at other times. When core affect lingers and is mild, we often call it mood.

The important feature of Core Affect is that it is pancultural. That is, in cultures and languages in which dimensions have been studied, valence and activation commonly emerge (Fontaine et al., 2013; Russell, 1983; Russell et al., 1989; Västfjäll et al., 2002).

Core Affect is part of (but not the whole of) emotional episodes, but is not synonymous with emotion. Thus, Core Affect is neither a substitute term for *emotion*, nor the essence of emotion, nor an additional discrete emotion. For example, whereas emotional episodes are said to begin and then, after a short time, end, one is always in some state of Core Affect, which simply varies over time (sometimes slowly, sometimes rapidly) without beginning or end. An analogy is body temperature: one always has a body temperature, but one is conscious of it only at times.

Emotional episodes (and emotional metaexperiences) are typically directed at something (one is angry with, afraid of, or sad about something). In contrast, Core Affect is not necessarily directed at anything. Like mood, Core Affect per se can be free-floating (as in feeling down but not about anything and not knowing why), but it can come to be directed at something. Still, the everyday concept of mood typically implies a long-lasting and mild state, whereas Core Affect has neither implication. Thus, mood can be thought of as a lingering Core Affect.

Core Affect is "a neurophysiological state that is consciously accessible as a simple, non-reflective feeling that is an integral blend of hedonic (pleasure–displeasure) and arousal (sleepy–activated) values" (Russell, 2003, p. 147). This seemingly simple definition contains a series of empirical proposals. Calling Core Affect a neurophysiological state is a promissory note so far left unfulfilled. The neural basis of Core Affect is an active research concern (Gerber et al., 2008; Posner et al., 2005, 2009).

This neurophysiological state has important functions. Core Affect is a continuous assessment of one's current state, and it influences other psychological processes accordingly. A change in Core Affect evokes a search for its cause and therefore facilitates attention to and accessibility of like-valenced material. Core Affect thus guides cognitive processing according to the principle of mood congruency. When Core Affect is positive, then events encountered, remembered, or envisioned tend to seem more positive—provided that the Core Affect is not attributed elsewhere (Schwarz and Clore, 1983). Core Affect is part of the information used to estimate affective quality and thus is implicated in incidental acquisition of preferences and attitudes. Core Affect influences behavior from reflexes to complex decision making. Core Affect is a background state continuously changing in response to a host of events, most beyond conscious monitoring. Core Affect in turn provides a powerful bias in processing of new information. In this way, Core Affect is involved in one's current state, including what are conventionally distinguished as cognitive state, motivational state, mood state, and so on, including past, present, and forecasts of the future.

One can seek to alter or maintain Core Affect directly—*affect regulation*—from the morning coffee to the evening brandy. People generally (but not always) seek behavioral options that maximize pleasure and minimize displeasure. Decisions thus involve predictions of future Core Affect. Core Affect is involved in motivation, reward, and reinforcement. An intriguing question is which of these functions requires conscious attention and which do not.

INFORMATION PROCESSING

The ecology of antecedent events. Emotional episodes normally (but not always) begin with a change in the world external to the person having the episode. This antecedent event and its context cannot be sharply distinguished from the perception of that event. (After all, the person having the emotion is part of the context and over time influences the unfolding event.) Still, a person exists in a physical and social reality that we must consider. The physical environment differs greatly over the surface of the earth, from Inuit winter to tropical rainforest. As it varies from one time to the next, the physical environment (temperature, noise level, etc.) alters Core Affect (for evidence, see Mehrabian and Russell, 1974). Some environmental events are imperceptible but affect people nonetheless. One example is ionization of air molecules, suggested as a factor in mood and mood disorders (Charry and Hawkinshire, 1981), but imperceptible.

A person also exists in a social environment. How events are coded by participants is a matter of culture, but here I point out that social events also have an ecology. People attend school, religious services, parties, and political rallies—in some societies. Or—in others—potlatches, igloo construction, or monkey hunts. People in a small hunter–gatherer society rarely encounter a stranger, something that happens frequently in a modern city. Not everyone in a given society is subject to the same ecology. Gender, social status, age, ethnicity, and so forth create their own ecologies.

Event coding. The blooming buzzing confusion of the external world is made manageable by categorizing its objects and events within a spatial-temporal framework. Many of the categories are likely common to all human beings (friend, foe, food, shelter, death, birth), but others are likely unique to certain societies (college graduation, visitation by a ghost, Sabbath). Codes vary in their salience. For example, enhancements of and threats to honor might be widespread, but are highly salient in Beduin culture; what Mesquita and Frijda (1992) referred to as the focality of a code.

Whether a type of event is considered specific to certain societies rather than universal depends on how abstract the event type is coded. Briggs mentioned that chasing lemmings is a source of happiness to the Utku Inuit of northern Canada. Chasing a lemming is not found elsewhere, and so this anecdote suggests cultural differences, but (1) the difference may lie in the ecology (lemmings are plentiful in the Canadian north, less so elsewhere) and (2) the similarity may lie in the meaning of the event. Children have fun in Italy chasing pigeons.

Perhaps a better example is danger. The ecology of dangerous events clearly varies widely with geography, but also with culture. Signs of evil spirits frighten some; risk of cancer from secondhand smoke frightens others. Culture includes a body of beliefs about what is dangerous. Those beliefs depend, in part, on the ecology of events. Danger exemplifies the point that some event coding categories for events are likely nearly universal. The specific dangerous events vary with the ecology (tigers here, extreme cold there), but the coding is likely universal. Other possible universal codes: humiliation, insult, revenge, loss, unfamiliar, illness. Even for a universal event code, the properties that lead to the coding might differ between societies.

Attribution. Attribution is another process that is part of the understanding of an event. Attribution is the process of linking causes with effects. Attributing a negative outcome to another person is associated with anger; attributing it to oneself with guilt or shame. Japanese differ from Americans in making such attributions, with Japanese more likely to do the latter, Americans the former. Attributing a negative event to a witch or ghost similarly has different consequences than attributing it to luck or happenstance (Weiner, 1985).

Appraisal. Coded events are appraised. Appraisal dimensions are processes that assess the emotional relevance of the event. Appraisal is commonly thought of as assessing an event, already coded, against a set of abstract dimensions each of which is relevant to emotion. There is consensus on a small number of appraisal dimensions: degree of relevance to one's goals, degree of facilitating or hindering that goal, degree of certainty about the future, degree of coping potential. Many scientists include further dimensions such as novelty and valence. For example, an event might be coded as a loss, but it elicits sadness only if relevant to some goal, hinders that goal, and so on. Appraisal theories assert that the number of appraisal dimensions is small. Still, different theorists have proposed lists with different numbers of appraisal dimensions and, indeed, different dimensions.

Whereas display rules and feeling rules are socially shared, individuals differ in just how they appraise even the very same event. In this way, appraisals account for different emotional reactions to the same event.

Culture also plays a role here, although we need more evidence. The relation of culture to appraisal involves two issues: does appraisal of the same event vary with culture, and does the same appraisal lead to different emotions in different cultures? Researchers have provided cross-cultural studies on one aspect of the second question: the relation between appraisals and self-reported categories of emotion. The most heralded result was that the relation is largely stable across cultures. Still, some of the relations found to be stable had the flavor of tautologies. For example, the emotion of joy was associated with an appraisal of the eliciting event as pleasant.

To summarize, separating cognition from emotion has not been helpful. The emotional episode, including the subjective experience of emotion, is based on and includes a complex rapidly unfolding set of perceptual-cognitive processes. This set is poorly understood if we restrict ourselves to simple innate categories such as danger, loss, or the like. Instead, we need a much richer analysis of the dimensions of appraisal and attribution, especially as shaped by cultural background.

FACIAL EXPRESSION

Most readers are familiar with emojis, cartoon faces that convey various emotional messages, such as happiness, anger, and disgust. Are these just modern inventions? Or, are these facial messages universal signals of hardwired biologically determined emotions? Basic Emotion Theory's strong suit has been considered to be the existence of such signals. Basic Emotion Theory became the textbook theory of emotion when it offered evidence that certain emotions have universal facial signals recognized by all humans. Universal recognition was taken to imply universal production and a biological basis to the emotions signaled.

Basic Emotion Theory began with the commonsense idea that facial expressions express emotions and added an evolutionary account, a neural mechanism, and empirical results, including famous cross-cultural studies in Papua New Guinea. Ekman (1980, p. 7) wrote, "When someone feels an emotion and is not trying to disguise it, his or her face appears the same no matter who that person is or where he or she comes from." Recently, Ekman and Keltner (2014) cited the studies in Papua New Guinea and wrote, "Those findings and the conclusion that all human beings have a shared set of facial expressions remains unchallenged." In fact, the findings and conclusion have been challenged.

To be sure, people frown, smile, laugh, grimace, wince, scowl, pout, sneer, and so on. And, in turn, observers interpret facial muscle movements, inferring what the expresser is thinking, feeling, perceiving, faking, and so on. Still, evidence does not support the Basic Emotion Theory account of either production of facial movements or their interpretation. Let's turn to the evidence.

The sender's production of facial movements. Production of facial movements is not as simple as might be inferred from Basic Emotion Theory, which focused on a small number of exaggerated facial configurations, which you can see for yourself in books (Ekman and Friesen, 1975) or sets of photographs (Ekman and Keltner, 2014). Gaspar and Esteves (2012) recorded the facial behavior of 3-year-olds during emotional episodes. They found much facial movement, but rarely the configurations studied by Basic Emotion Theory. Carroll and Russell (1997) found similar results with adults. So, Basic Emotion Theory does not account for most facial configurations. We especially lack information on the natural occurrence of facial movements in different societies.

What causes the face to move remains largely unknown. Basic Emotion Theory tells us that happiness makes us smile, anger makes us frown, fear makes us gasp, disgust makes us scrunch our noses, and so on. Yet, accumulating evidence challenges this assumption. Reisenzein et al. (2013) reviewed the laboratory evidence, Fernández-Dols and Crivelli (2013) the field evidence. Consider smiles. Happy people often fail to smile, and smiles occur without happiness. Smiles are easily posed, do not correlate well with the smiler's emotional state (Fridlund, 1991; Krumhuber and Manstead, 2009), and can even occur with negative experiences such as losing a game (Schneider and Josephs, 1991), being embarrassed (Keltner, 1995), or being in pain (Kunz et al., 2009). Similar problems arise for other emotion-face associations.

Let us take a broader look at facial movements. As we talk, eat, breathe, exert effort, smell, feel pain, or reach orgasm—our faces move. Our faces move as part of certain reflexes (gag, orienting, startle, and so on), of perception (alerting, looking, tasting, and so on), and of social interaction (social greeting, threatening, exerting dominance, or submission). Our faces move as we unconsciously imitate others. Our faces move as part of information processing and of subsequent behavior.

Thus, there are various possible accounts of the production of facial movement, even the ones called facial expressions of emotion. The possibilities include the following: (1) perception involves bodily movements (reaching to feel, turning to look), and facial movements are part of this process. For example, Basic Emotion Theory's "fear expression" might enhance visual exposure (Susskind et al., 2008). If so, the "fear expression" might more appropriately be called an expression of visual vigilance. Basic Emotion Theory's "disgust face" might occur when smelling a strong odor. If so, the "disgust face" might express olfactory salience. (2) Cognitions (appraisals of current events) might produce facial movement (Scherer, 1992). Ortony and Turner (1990) noted that a frown (brow contraction) often occurs when one is uncertain or puzzled. If so, the "fear expression" might express uncertainty. (3) Mesquita and Frijda (1992) proposed that facial movements are part of the preparation for action. The "anger expression" might include clenched teeth when preparing to do something strenuous. (4) As social animals, a large part of our behavior is negotiating social interaction. Fridlund (1994) suggested that facial movements signal to an audience projected plans and goals including contingencies. The "happy expression" might signal social recognition. (5) Facial movements are part of paralanguage. Chovil (1991) offered a taxonomy for paralanguage in which facial movements are part of speech communication. Thus, a "disgust face" might substitute for the words "that stinks" in voicing a metaphor for an immoral deed. (6) Core affect might produce facial movement. All facial expressions seem to convey degrees of valence and activation.

Conceivably, all these accounts are complementary. But science requires a more critical stance. The question is whether one or more of these alternative sources provide the better account for facial movements. Perhaps the first question is the validity of the commonsense assumption that emotions are involved. Some of the six proposals listed previously maintained some link to emotion (appraisal, action preparation, and core affect have been listed as components of an emotion). Yet, we now know that these processes are not closely tied to one another, and so the question is which of these processes moves the face. (For example, when afraid a person might be puzzled and furrow the brow, but then being puzzled even when not feeling afraid might similarly furrow the brow.) Further, emotion is often entangled with other cooccurring processes, and the question is what happens when they are disentangled. Kraut and Johnston's (1979) study of smiling bowlers illustrates this question. People often smile when happily interacting with others, and Kraut and Johnston asked what happens when happiness and interacting are teased apart. Smiles when happy but not interacting were found to be rare (see also Fernández-Dols and Ruiz-Belda, 1995; Ruiz-Belda et al., 2003). [The objection that we smile at pleasant thoughts when alone was answered by Fridlund (1994), who provided reason to believe that the smiler may be physically but not psychologically alone. We also talk when alone, but doing so does not undermine the principle that speech is largely for communication.]

In short, the production of facial expressions is sometimes correlated with discrete emotions, although surprisingly weakly, but there are alternative explanations to the theory that the emotions are causal, since emotions are confounded with other sources. Thus, we have no convincing evidence that emotions cause facial movements: the (weak) correlation between emotions and facial movements may have other underlying causes. The face moves as we perceive and interpret the current situation and then engage in action, especially (but not exclusively) during interactions with other persons, often in a rapidly evolving and complex manner.

The observer's interpretation of facial movements. We open our eyes and see that this person is happy, that one angry, and so on. Most of us believe that we can "recognize" happiness, anger, disgust, and so on in the faces of others. Many studies purported to demonstrate consensual recognition by asking people to match a still photo of a posed facial expression to the emotion predicted by Basic Emotion Theory. The studies carried out in Papua New Guinea mentioned earlier are of this sort. These "matching scores," even if reliable, would not show that people spontaneously recognize the predicted emotion but that, once told that one of a small number of emotions is expressed, they can select the predicted one. Even more troubling, the high matching scores found may be partly due to the method used to obtain them. No single method problem need be fatal, but cumulatively various feature of method combine to push scores in the predicted direction: within-subjects designs, posed exaggerated facial movements (devoid of voice, body, and information about the expresser's context), and the use of forced-choice response format (Russell, 1994). For example, when observers see spontaneous rather than posed faces, matching scores plummet. We recently found that people can achieve a high matching score between a label and a face without recognizing any emotion. Instead, they used an elimination strategy: after matching several standard faces with standard labels, both children and adults chose a nonword, "pax," from the list as the emotion expressed by a novel face (Nelson and Russell, 2016). An elimination strategy may be a major contributor to high matching scores.

Outside the laboratory, the observer does not use someone's facial expression alone to infer that person's emotion, but facial expression in light of the expresser's situation and other aspects of the face's context including the expresser's body (Fantoni and Gerbino, 2014). So, removing the context in a "recognition" experiment stands in the way of understanding how observers typically interpret facial expressions. More important, specifying context as well as face in such experiments can provide a test of Basic Emotion Theory. Basic Emotion Theory implies that the facial expression is more powerful for "recognition" of emotion than is its context, because, according to Basic Emotion Theory, the facial expression is an automatic signal of the specific emotion, whereas context can provide only probabilistic information because different individuals respond differently to the same situation. The prediction failed: in determining the emotion seen by the observer, context is more powerful than the face (Carroll and Russell, 1996): a person in an anger-inducing situation who showed Basic Emotion Theory's "fear face" was interpreted as angry rather than afraid. Masuda et al. (2008) found an even stronger reliance in context with Japanese observers.

Basic Emotion Theory emphasized the universality of recognition: the claim is that basic emotions are easily recognized from the predicted facial expressions by all people whatever their culture, language, or education. Metaanalyses have found that matching scores vary with culture, language, and education (Nelson and Russell, 2013; Trauffer et al., 2013). Jack et al. (2012) used a psychophysical technique and again found cultural differences in what facial configurations were matched to specific emotions. Basic Emotion Theory presupposed that the English words *fear*, *anger*, *disgust*, and so on express universal categories in terms of which recognition proceeds; evidence indicates that the way in which emotions are categorized is not universal: emotion categories expressed in different languages are in some ways similar to but in some ways different from those in English (Russell, 1991; Wierzbicka, 1999). As Ekman and Friesen (1971) emphasized, the most telling test of universality involves indigenous societies reasonably remote from Western culture and media. Yet, the evidence offered for universal recognition suffered from the method problems described previously and found cultural difference in matching scores (Russell, 1994). More recent studies of indigenous societies similarly found weak to nonexistent support for Basic Emotion Theory's prediction of universal recognition (Crivelli et al., 2016; Gendron et al., 2014; but see Tracy and Robins, 2008).

What the classic Basic Emotion Theory studies called "recognition" is interpretation. Observers may use facial information to make inferences not just about emotion but about any psychological state. The interpretation of the face is influenced by many factors, some rarely studied (color of the sclera), some more studied: by the observer's situation (state, interests, motives), by the face's context (the expresser's context, words, gaze, vocal prosody, body position and proxemics, motor behavior, underlying physiognomy), and by features of the experimental method.

I suggested an alternative account—called minimal universality—of an onlooker's interpretation of facial expressions (Russell, 1995). Universally, humans perceive others in simple terms (the core affect dimensions of valence and arousal): is the person feeling good or bad, energized or quiescent? This part of the proposal is consistent with above-chance matching of faces with emotion labels, because the meaning of an emotion label includes, among other elements, valence and arousal. Young children interpret faces in terms of valence (Widen and Russell, 2002, 2008). For example, the typical young 3-year-old uses the same one label (typically *angry*) for four of Basic Emotion Theory's canonical faces: those for fear, anger, sadness, and disgust. As children develop, they add new emotion concepts by differentiating: feeling bad is divided into feeling bad because of loss versus feeling bad because of receiving a hostile action; when a label is added, the child has learned to distinguish sad from angry. The end product is a set of adult emotion concepts, which are similar but not uniform across individuals, languages, and cultures (Russell, 1991). In interpreting facial expressions, older children and adults go beyond valence and arousal, including categorization by discrete emotions, at least in Western cultures, although apparently not everywhere (Howell, 1984).

In short, sufficient evidence has now accumulated to conclude that Basic Emotion Theory's claims about universal recognition of a specific discrete "basic" emotion from its facial expression are unwarranted. The face moves in many ways, often rapidly, from many sources. Observers, in turn, interpret facial movements by taking into account the context in which they occur.

CONCLUSIONS

Emotion is a large jigsaw puzzle. For centuries, philosophers and scientists of different disciplines have tried to solve the puzzle, but usually by considering just a few of its many pieces and by relying on implicit presuppositions. For example, a distinction is often made between thinking and feeling, reason and emotion, cold and hot processes. This was a mistake. Thinking includes feeling; feeling is a form of thinking. Trying to find a strict boundary between thinking and feeling turned out not to be useful. Another mistake was the focus on a small number of emotional episodes and dubbing them basic. Much has been written by psychologists on the fear elicited by a bear in the woods. Yet, few people have ever encountered a bear in the woods, but fear still enters our lives (fear of illness, of losing a job, of failing an exam, of insulting a boss, of losing a romantic partner) in ways importantly different from the episode with the bear in the woods.

We must uncover implicit presuppositions and put them to empirical test. When tested, such ideas so far turned out to be oversimplified or simply false. For example, it was presupposed that cases of a specific emotion, such as fear, have in common a pattern of activity in the autonomic nervous system, but no such pattern has been found (Cacioppo et al., 2000). It was presupposed that anger leads to aggression, but angry people do not frequently act aggressively (Averill, 1982). It was presupposed that emotions are seen on the face. Yet, happy people do not necessarily smile, and smiling people are not necessarily happy (Fernández-Dols and Ruiz-Belda, 1995). It was presupposed that emotional actions are preformed hardwired reactions (such as fight or flight), and yet we also know that human behavior is often based on creative problem solving. Once our preconceptions (what we often think of as simply common sense) are set aside, progress will be rapid in understanding how emotional episodes are assembled through the various psychological processes of which we are capable.

References

Averill, J.R., 1982. Anger and Aggression: An Essay on Emotion. Springer-Verlag, New York, NY.

Barrett, L.F., 2006. Are emotions natural kinds? Perspect. Psychol. Sci. 1, 28–58.

Barrett, L.F., Russell, J.A. (Eds.), 2015. The Psychological Construction of Emotion. Guilford Press, New York, NY.

Cacioppo, J.T., Berntson, G.G., Larsen, J.T., Poehlmann, K.M., Ito, T.A., 2000. The psychophysiology of emotion. Handbook of Emotions. second ed. Guilford, New York, NY, pp. 173–191.

Carroll, J.M., Russell, J.A., 1996. Do facial expressions signal specific emotions? Judging emotion from the face in context. J. Pers. Soc. Psychol. 70 (2), 205.

Carroll, J.M., Russell, J.A., 1997. Facial expressions in Hollywood's portrayal of emotion. J. Pers. Soc. Psychol. 72, 164–176.

Charry, J.M., Hawkinshire, F.B., 1981. Effects of atmospheric electricity on some substrates of disordered social behavior. J. Pers. Soc. Psychol. 41, 185–195.

Chovil, N., 1991. Discourse-oriented facial displays in conversation. Research on language and social interaction 25, 163–194.

Clore, G.L., Ortony, A., 2013. Psychological construction in the OCC model of emotion. Emot. Rev. 5, 335–343.
Crivelli, C., Jarillo, S., Russell, J.A., Fernández-Dols, J.M., 2016. Reading emotions from faces in two indigenous societies. J. Exp. Psychol. Gen. 145, 830–843.
Davitz, J.R., 1969. The Language of Emotion. Academic Press, San Diego, CA.
Ekman, P., 1972. Universal and cultural differences in facial expressions of emotions. In: Cole, J.K. (Ed.), Nebraska Symposium on Motivation. University of Nebraska Press, Lincoln, pp. 207–283.
Ekman, P., 1980. The Face of Man. Garland, New York, NY.
Ekman, P., Cordaro, D., 2011. What is meant by calling emotions basic. Emot. Rev. 3, 364–370.
Ekman, P., Friesen, W.V., 1971. Constants across cultures in the face and emotion. J. Pers. Social Psychol. 17 (2), 124.
Ekman, P., Friesen, W.V., 1975. Pictures of Facial Affect. Consulting Psychologists Press, Palo Alto, CA.
Ekman, P., Keltner, D., 2014. Are facial expressions universal? Available from: greatergood.berkeley.edu.
Fantoni, C., Gerbino, W, 2014. Body actions change the appearance of facial expressions. PLoS One 9 (9), e108211.
Fernández-Dols, J.M., Crivelli, C., 2013. Emotion and expression: naturalistic studies. Emot. Rev. 5 (1), 24–29.
Fernández-Dols, J.M., Ruiz-Belda, M.A., 1995. Are smiles a sign of happiness? Gold medal winners at the Olympic Games. J. Pers. Soc. Psychol. 69 (6), 1113–1119.
Fontaine, J.R., Scherer, K.R., Soriano, C. (Eds.), 2013. Components of Emotional Meaning: A Sourcebook. Oxford University Press, Oxford.
Fridlund, A.J., 1991. Sociality of solitary smiling: potentiation by an implicit audience. J. Pers. Soc. Psychol. 60, 229–240.
Fridlund, A.J., 1994. Human Facial Expression: An Evolutionary View. Academic Press, San Diego, CA.
Gaspar, A., Esteves, F.G., 2012. Preschooler's faces in spontaneous emotional contexts—how well do they match adult facial expression prototypes? Int. J. Behav. Dev. 36, 348–357.
Gendron, M., Roberson, D., van der Vyver, J.M., Barrett, L.F., 2014. Perceptions of emotion from facial expressions are not culturally universal: evidence from a remote culture. Emotion 14, 251–262.
Gerber, A.J., Posner, J., Gorman, D., Colibazzi, T., Yu, S., Wang, Z., Kangarlu, A., Zhu, H., Russell, J.A., Peterson, B.S., 2008. An affective circumplex model of neural systems subserving valence, arousal, and cognitive overlay during the appraisal of emotional faces. Neuropsychologia 46, 2129–2139.
Han, D., Kollareth, D., Russell, J.A., 2015. The words for disgust in English, Korean, and Malayalam question its homogeneity. J. Lang. Soc. Psychol. 35, 569–588.
Howell, S., 1984. Society and Cosmos: Chewong of Peninsular Malaysia. Oxford University Press, London.
Hurtado-de-Mendoza, A., Molina, C., Fernández-Dols, J.M., 2013. The archeology of emotion concepts: a lexicographic analysis of the concepts *shame* and *vergüenza*. J. Lang. Soc. Psychol. 32, 272–290.
Jack, R.E., Garrod, O.G., Yu, H., Caldara, R., Schyns, P.G., 2012. Facial expressions of emotion are not culturally universal. Proc. Natl. Acad. Sci. 109 (19), 7241–7244.
Keltner, D., 1995. Signs of appeasement: evidence for the distinct displays of embarrassment, amusement, and shame. J. Pers. Soc. Psychol. 68 (3), 441.
Kraut, R.E., Johnston, R.E., 1979. Social and emotional messages of smiling: an ethological approach. J. Pers. Soc. Psychol. 37, 1539–1553.
Krumhuber, E.G., Manstead, A.S., 2009. Can Duchenne smiles be feigned? New evidence on felt and false smiles. Emotion 9, 807.

Kunz, M., Prkachin, K., Lautenbacher, S., 2009. The smile of pain. Pain 145, 273–275.
LeDoux, J.E., 2015. Anxious. Viking, New York, NY.
Leff, J., 1973. Culture and the differentiation of emotional states. Br. J. Psychiatr. 123, 299–306.
Levenson, R.W., 2011. Basic emotion questions. Emot. Rev. 3, 379–388.
Levy, R.I., 1973. Tahitians. University of Chicago Press, Chicago, IL.
Lindquist, K.A., Wager, T.D., Kober, H., Bliss-Moreau, E., Barrett, L.F., 2012. The brain locus of emotion: a meta-analytic review. Behav. Brain Sci. 35, 121–143.
Masuda, T., Ellsworth, P.C., Mesquita, B., Leu, J., Tanida, S., Van de Veerdonk, E., 2008. Placing the face in context: cultural differences in the perception of facial emotion. J. Pers. Soc. Psychol. 94, 365.
Mehrabian, A., Russell, J.A., 1974. An Approach to Environmental Psychology. MIT Press, Cambridge, MA.
Mesquita, B., Frijda, N.H., 1992. Cultural variations in emotions: a review. Psychol. Bull. 112, 179–189.
Nelson, N.L., Russell, J.A., 2013. Universality revisited. Emot. Rev. 5, 8–15.
Nelson, N.L., Russell, J.A., 2016. A facial expression of pax: assessing children's "recognition" of emotion from faces. J. Exp. Child Psychol. 141, 49–64.
Ortony, A., Turner, T.J., 1990. What's basic about basic emotions? Psychol. Rev. 97 (3), 315.
Posner, J., Russell, J.A., Gerber, A., Gorman, D., Colibazzi, T., Yu, S., Wang, Z., Kangarlu, A., Zhu, H., Peterson, B.S., 2009. The neurophysiological basis of emotion: an fMRI study of the affective circumplex using emotion-denoting words. Hum. Brain Map. 30, 883–895.
Posner, J., Russell, J.A., Peterson, B.S., 2005. A circumplex model of affect: an integrative approach to affective neuroscience, cognitive development, and psychopathology. Dev. Psychopathol. 17, 715–734.
Reisenzein, R., Studtmann, M., Horstmann, G., 2013. Coherence between emotion and facial expression: evidence from laboratory experiments. Emot. Rev. 5, 16–23.
Rosaldo, M.Z., 1980. Knowledge and Passion: Ilongot Notions of Self and Social Life. Cambridge University Press, Cambridge.
Rosaldo, R.I., 1984. Grief and a headhunter's rage: on the cultural force of emotions. In: Plattner, S., Bruner, E.M. (Eds.), Text, Play, and Story: The Construction and Reconstruction of Self and Society. American Ethnological Society, Washington, DC, pp. 178–195.
Ruiz-Belda, M.A., Fernández-Dols, J.M., Carrera, P., Barchard, K., 2003. Spontaneous facial expressions of happy bowlers and soccer fans. Cogn. Emot. 17 (2), 315–326.
Russell, J.A., 1983. Pancultural aspects of the human conceptual organization of emotions. J. Pers. Soc. Psychol. 45, 1281–1288.
Russell, J.A., 1991. Culture and the categorization of emotion. Psychol. Bull. 110, 426–450.
Russell, J.A., 1994. Is there universal recognition of emotion from facial expressions? A review of the cross-cultural studies. Psychol. Bull. 115, 102.
Russell, J.A., 1995. Facial expressions of emotion: what lies beyond minimal universality? Psychol. Bull. 118, 379–391.
Russell, J.A., 2003. Core affect and the psychological construction of emotion. Psychol. Rev. 110, 145.
Russell, J.A., Lewicka, M., Niit, T., 1989. A cross-cultural study of a circumplex model of affect. J. Pers. Soc. Psychol. 57, 848–856.
Russell, J.A., Sato, K., 1995. Comparing emotion words between languages. J. Cross Cult. Psychol. 26, 384–391.
Scherer, K.R., 1992. What does facial expression express? Strongman, K. (Ed.), International Review of Studies on Emotion, vol. 2, Wiley, Chichester, pp. 139–165.
Schneider, K., Josephs, I., 1991. The expressive and communicative functions of preschool children's smiles in an achievement-situation. J. Nonverbal Behav. 15, 185–198.
Schwarz, N., Clore, G.L., 1983. Mood, misattribution, and judgments of well-being: informative and directive functions of affective states. J. Pers. Soc. Psychol. 45, 513.

Susskind, J.M., Lee, D.H., Cusi, A., Feiman, R., Grabski, W., Anderson, A.K., 2008. Expressing fear enhances sensory acquisition. Nat. Neurosci. 11 (7), 843–850.
Tracy, J.L., Robins, R.W., 2008. The nonverbal expression of pride: evidence for cross-cultural recognition. J. Pers. Soc. Psychol. 94, 516–517.
Trauffer, N.M., Widen, S.C., Russell, J.A., 2013. Education and the attribution of emotion to facial expressions. Psychol. Topics 22, 237–248.
Västfjäll, D., Friman, M., Gärling, T., Kleiner, M., 2002. The measurement of core affect: a Swedish self-report measure derived from the affect circumplex. Scand. J. Psychol. 43, 19–31.
Weiner, B., 1985. An attributional theory of achievement motivation and emotion. Psychol. Rev. 92, 548–558.
Widen, S.C., 2016. The development of children's concepts of emotion. In: Barrett, L.F., Lewis, M., Haviland-Jones, J.M. (Eds.), Handbook of Emotions. fourth ed. Guilford, New York, pp. 307–318.
Widen, S.C., Russell, J.A., 2002. Gender and preschoolers' perception of emotion. Merrill Palmer Quart. 48, 248–262.
Widen, S.C., Russell, J.A., 2008. Young children's understanding of other's emotions. In: Lewis, M., Haviland-Jones, J.M., Barrett, L.F. (Eds.), Handbook of Emotions. Guilford, New York, NY, pp. 348–363.
Wierzbicka, A., 1999. Emotions Across Languages and Cultures: Diversity and Universals. Cambridge University Press, Cambridge, UK.
Wikan, U, 1989. Illness from fright or soul loss: a North Balinese culture-bound syndrome? Cult. Med. Psychiatr. 13, 25–50.

Further Reading

Reisenzein, R., 2000. Exploring the strength of association between the components of emotion syndromes: the case of surprise. Cogn. and Emot. 14, 1–38.

CHAPTER

5

On the Moral Implications and Restrictions Surrounding Affective Computing

Anthony F. Beavers, Justin P. Slattery***

*The University of Evansville, Evansville, IN, United States; **Indiana University, Bloomington, IN, United States

PROLOGUE: A BIT OF DIALOGUE FROM *INTERSTELLAR* (2014)

Tars (robot): Everybody good? Plenty of slaves for my robot colony?
Doyle (human): They gave him a humor setting so he'd fit in better with his unit. He thinks it relaxes us.
Cooper (human): A giant sarcastic robot. A great idea.
Tars: I have a cue light I can use to show you when I'm joking, if you like.
Cooper: That might help.
Tars: Yeah, you can find your way back to the ship after I blow you out the airlock. [Cue light flashes]
Cooper: What's your humor setting, Tars?
Tars: That's 100%.
Cooper: Let's bring it on down to 75, please.

A BRIEF HISTORY OF A PROBABLY FALSE DICHOTOMY, OVERSIMPLIFIED

The question of the relationship between emotion and reason is nothing new, though it started out with a moral bias against the affects almost as soon as the question was raised by the Greeks. The affects were useful,

Emotions and Affect in Human Factors and Human–Computer Interaction. http://dx.doi.org/10.1016/B978-0-12-801851-4.00005-7
Copyright © 2017 Elsevier Inc. All rights reserved.

but only if properly tamed for some (Aristotle, most notably); for others, they were just moral noise (Pythagoras, perhaps Plato, depending on which work one wishes to examine, and later the Stoics). As the tradition unfolds in the West, the Platonic view dominates; reason is the "captain of the team," and we would do well to use it to keep our affectivity, a component of our animal nature, under tight control. Not doing so rendered one an inhumane brute.

Along the way, there were some who tried to catalog the affects (for them, "passions" or "appetites"), determine their interrelations, and help us understand which are acceptable and which are to be avoided altogether. Two early thinkers to taxonomize emotion in this tradition were Thomas Aquinas (1225–1274) and René Descartes (1596–1650). Both sought to tease out foundational affects that could explain all others in a manner similar to picking out primary colors to account for the origins of all other colors. As one might expect, the passage from the Middle Ages to the early enlightenment resituated the context for examining the affects. Thus, Descartes' analysis aims at explanation of affectivity befitting the age of early science in which he was thoroughly embedded. Both Aquinas and Descartes provide interesting insights into the nature of emotion, but in the interest of space and immediate relevance, we will skip the former and briefly discuss the latter.

In addition to physicist, psychologist, and biologist, Descartes was also a worthy mathematician who believed that human thought could ultimately be captured in a rigorous *mathesis universalis* following methods that had already proven their efficacy and coherence in mathematics (Descartes, c. 1628/1985b). We now know that this particular rationalistic and mathematical dream of Descartes and some of those to follow him, notably the rigorous computationalist Leibniz, runs into practical problems in standard artificial intelligence, most famously the frame problem in many of its variations; but in the absence of working computing machinery, this difficulty eluded them. It was intuitively obvious to them that emotion had little, if anything, to do with reason, and their dream of perfectly computational (that is, solely mathematical) intelligence was thereby skewed. But there is more to the story worth telling to situate the direction of this chapter.

Descartes was an enigmatic character whose philosophical agenda to this day remains unclear. He is mostly known for his mind/body dualism, though there is reason to suspect that this characterization overemphasizes what may be an incidental work, *The Meditations*, over the majority of his writings which are on physiological, mathematical, and other scientific matters. He himself indicates as much when he writes to Princess Elizabeth of Bohemia on June 28, 1643, "I think that it is very necessary to have understood, once in a lifetime, the principles of metaphysics since it is by them that we come to the knowledge of God and of our soul. But I think also that it would be very harmful to occupy one's intellect frequently in meditating upon them, since this would impede it from devoting itself to the imagination and the senses" (1991, p. 228). There is, of course, more to this quote and the back story behind it (Beavers, 2010), but it provides some support for the notion that Descartes may have written the *Meditations* to get his work past the Church, while advancing a genuine scientific theory that he thought more pressing than theological speculation. Evidence of this is suggested in his 1641 Letter to Mersenne: "I may tell you, between ourselves, that these six Meditations contain all the foundations of my physics. But please do not tell people, for that might make it harder for supporters of Aristotle to approve of them. I hope that readers will gradually get used to my principles, and recognize their truth, before they notice that they destroy the principles of Aristotle" (Descartes, 1991, p. 173).

The principles of Aristotle mentioned here provided the theological substrate for the Catholic theology of the time, which, in Descartes' view was impeding progress in physics. These facts help to explain some dissonance in the initial pages of the *Meditations*: Descartes' stated agenda in the first meditation to put science on a sure foundation is at odds with the agenda mentioned in the three prefatory pieces to the *Meditations* in which he states that the purpose of the work is to provide us with a clearer conception of the notions of God and the soul. We need not enter into this debate to establish that Descartes was much more practical than his armchair metaphysical speculations suggest. For instance, his understanding of how the human brain sees in three-dimensions was basically correct in its outlines. Furthermore, his work in dissecting human and other animals produced an interesting taxonomy of capacities for human versus nonhuman animals that only recently, that is, in the last 100 years or so of scientific research, has been overturned.

As with the Platonic and Aristotelian taxonomies of "the soul," Descartes' view is also tripartite. In the Cartesian taxonomy, there are three basic components to cognition: sensation, imagination, and reason. The first two are relegated to both human and nonhuman animals, largely because Descartes could explain them on a physicalist platform that followed from his research in biology. Nonhuman animals essentially were purely mechanistic according to him and could therefore be classified as mere automata because they lacked the capacity to reason. Sensation, the power to take in information from the outside world in seeing, hearing, etc., could be explained by the laws of physics, as could imagination, which was, according to him, the capacity to form mental images, a residue of sensation, though his understanding of neuroscience was too early to offer a realistic explanation of this phenomenon.

Reason, however, did not easily lend itself to mechanical explanation, largely because Descartes could not anticipate a mechanistic way to explain language, the use of which indicated the presence of a thinking thing (Descartes, 1637/1985a, p. 141). When one steps back and looks at the entire Cartesian corpus, it is difficult not to conclude that if Descartes could have found a mechanistic explanation for our rational capacities, he may have adopted that and avoided the dualism that characteristically defines him (Beavers, 2010; note that the authors are aware that this is a controversial position).

Thus, on the one hand, the Cartesian tradition, deeply rooted in history, posits a dichotomy between pure reason and the affects. Even though it does so, however, it assigns ethics to the affects and not to reason, and, hence, to the integrated human person and not the pure Cartesian ego. So it seems that, while Descartes' metaphysical speculation about mind/body relations might make him an incidental *dualist*, his practical philosophy makes him a *functionalist* where lived experience is concerned, and this includes for him living as an integrated mind/body unity, a unity of affectivity and reason, inseparable in the course of daily life. He writes in his 1643 Letter to Elizabeth: "Metaphysical thoughts, which exercise the pure intellect, help to familiarize us with the notion of the soul; and the study of mathematics, which exercises mainly the imagination in the consideration of shapes and motions, accustoms us to form very distinct notions of bodies. But it is the ordinary course of life and conversation, and abstention from meditation and from the study of the things which exercise the imagination, that teaches us how to conceive the union of the soul and the body" (1991, p. 227).

In the following 365 years since Descartes' death, the notion that reason is captain of the team has given way to a view that suggests that our rational sensibilities are very much informed by our affects. In the more recent past, this view enters into AI through several channels, but quite explicitly in the work of Marvin Minsky (1988 and 2007). This latter work, along with several others, will pay homage to the notion that perhaps there is no clear distinction between reason and the affects, but that the affects may present alternative ways of thinking, and, more importantly, that they serve to set the context necessary for our more rational states to function as they do.

THREE ISSUES IN SUPPORT OF AFFECTIVE COMPUTING

The notion that reason and affect are and ought to be separated is thoroughly ingrained in Western culture. While both are acknowledged and respected nowadays, they still are often characterized as operating independently among the folk. Many researchers in moral psychology and artificial intelligence alike now often think otherwise. There are, of course, many reasons to agree with them. In the interest of space, we will examine three of these reasons. The first will be based on the notion that rule-based moral systems result in a moral version of the frame problem and that moral decision-making requires some form of affect to set context and arrange priorities. The second will concern the way in which affects are necessary for anything more than a superficial understanding of utterances. They, in effect, help to draw frames around speech acts that are essential for disambiguating them. Third, we will examine deception. There are arguments for and against artificial affective agents based on the notion that simulated affects are obviously not genuinely felt and, therefore, any pretence that machines are emotive is a lie. While we recognize this objection, we will argue that deception is essential for several reasons and, furthermore, that sometimes deception is not only necessary, but also good.

Why Rule-based Moral Systems May Not Work

Dennett remarks that AI "makes philosophy honest" (2006). Similarly, Anderson and Anderson observe that "ethics must be made computable to make it clear exactly how agents ought to behave in ethical dilemmas" (2007, p. 16). One can derive from these statements that as we continue to develop more sophisticated AI, we must understand what type of ethics we employ in our human world and also what type we ought to employ. Both are necessary as a precursor to understanding some details concerning what ethical procedures to implement in machines. These notions, though, have an important corollary concerning moral theories in general. This is because the prospects of building moral machinery presents us with an engineering imperative: from the ethical (though not universally accepted) principle that *ought implies can*, we can deduce that if an ethical theory is viable, we must be able to specify an algorithm for how it is to work. One test that a proper specification can be achieved is to implement it in machines. Furthermore, following Herbert Simon's famous maxim that "in the computer field, the moment of truth is a running program; all else is prophecy," to prove such viability, we must actually build such a machine, and this requires laying out workable design specifications for doing so. If it is empirically or logically impossible to implement the moral theory in hardware, then it is untenable (Beavers, 2011, 2012).

Though the field of ethics is complex, it is fair to make a broad distinction between two common approaches, an affective approach typically advanced by moral sentimentalists, such as Adam Smith and David Hume, and a rule-based, algorithmic approach advanced by thinkers like Immanuel Kant and John Stuart Mill. There is a third classification relevant to this discussion, virtue ethics, advanced by Aristotle and others that falls more on the affective side, though in a manner that differs from traditional moral sentimentalism.

Moral sentimentalists believe that our emotions play a critical role in defining morality. Here, ethical deliberation is heavily influenced and constrained by affect and emotional connections, since the bedrock of how we operate morally is based on weighing factors like love, repulsion, guilt, shame, attachment, revenge, etc. Algorithmic ethicists, on the other hand, believe that rules constrain our behavior and that a proper ethical theory would be one that advocates for the right set of formal rules. Here, such rules generally quantify variables in ethical situations, weigh pros and cons, etc., and then are ultimately used to make rational choices when determining one's course of action. There is, however, a problem within these rule-based approaches, namely, that rule-based ethics runs into problems of scope which are basically moral variants of the frame problem, a point recognized by several scholars (Beavers, 2011, 2012; Gallagher, 2013; Horgan and Timmons, 2009; Wallach and Allen, 2009). We collectively identify these problems under the general heading of the moral frame problem (MFP). The MFP arises because we cannot know how to constrain the necessary variables when computing ethics. Furthermore, without using preexisting moral criteria, which begs the moral question, we cannot determine what constitutes a relevant variable in the first place.

While we cannot be certain that the MFP is a problem for all rule-based approaches to ethics without further analysis—though there are reasons to believe this is the case—two of the most common systems fall prey to it, utilitarianism and Kantian deontology, at least on their canonical readings. [Powers (2006) tries to save Kant from this critique, though there are problems here in interpreting Kant that are beyond the scope of this chapter, and several news stories in the popular press are currently reporting variations on utilitarianism to attempt to solve problems with automated cars.] In what follows, we take up the "principle of utility" first before turning to one formula of the "categorical imperative" used in Kantian ethics.

According to the brand of utilitarianism developed by John Stuart Mill (1861/1979), internal states, such as motives and intentions, are not central to moral calculus and operate in more of a heuristic fashion. One is obligated only to evaluate moral behavior based on consequences. Initially, this would suggest easy computability, but it has been well known, even since Mill's time, that there are problems. The "principle of utility" states that "actions are right in proportion as they tend to promote happiness; wrong as they tend to produce the reverse of happiness. By happiness is intended pleasure and the absence of pain; by unhappiness, pain and the privation of pleasure" (p. 7). This principle is sometimes identified by the notion that utilitarianism suggests that actions are moral when they promote the greatest good for the greatest number. Looking at the situation this way, it is clear that utilitarianism invites some way of calculating the overall pleasure and pain produced by an action (Feldman, 1978). In fact, almost 100 years before Mill, Bentham (1789/2007) went so far as to try to work out a "hedonic calculus" for this purpose. To borrow from his view for a moment, Bentham suggests that the following hedonic variables need to be considered concerning the pleasure and/or pain produced by an action: intensity, duration, certainty or uncertainty, propinquity or remoteness, fecundity, and purity. (Fecundity concerns whether a pleasure or pain will be followed by another state of the same hedonic kind, that is, another pleasure or pain; purity concerns whether a pleasure or pain will be followed by a state of a different kind.)

Putting aside the difficulties of quantifying such variables, how far are we supposed to go with such calculations? If we are talking about the greatest pleasure for the greatest number, we must deal with the scope of effect across both space and time. The spatial problem is serious; the temporal problem more so. Spreading happiness (pleasure) to those in my local sphere may or may not lead them to do the same in their sphere. But what of temporal effects? Spreading pleasure spatially could have an effect on future generations. Imagine the saved or killed baby that will ultimately cure cancer-thereby spreading happiness across generations after, and the problem becomes clear.

There is also the time it takes to make such calculations. This is a concern that Mill takes up directly in his *Utilitarianism* (1861/1979): he writes, "Again, defenders of utility often find themselves called upon to reply to such objections as this—that there is not enough time, previous to action, for calculating and weighing the effects of any line of conduct on the general happiness" (p. 23). (In fact, the problem is computationally intractable when we consider the ever extending ripple effects that any act can have on the happiness of others across both space and time.) Mill tries to get around the problem with a sleight of hand, noting that "all rational creatures go out upon the sea of life with their minds made up on the common questions of right and wrong" (p. 24), suggesting that calculations are, in fact, unnecessary, if one has the proper forethought and upbringing. Again, the rule is of little help, and death by failure to implement looks imminent. The theory cannot even be saved by heuristics, which are equally uncomputable in this case. In the end, Mill falls back on a well-formed conscience, more of an affective move than a rule-based approach. It is not too much of a stretch to suggest that Mill intends a well-formed conscience to set the context for application of the "principle of utility." If this is so, then affects are used even by Mill to deal with problems caused by the MFP.

It is perhaps worth acknowledging that there may be a possible way to save Mill when the problem is viewed through the lenses of complexity theory. If we can establish that "local utility," that is spreading pleasure to those closest to me, leads to "global utility," then we could have a heuristic, though this only provides a "best guess" scenario in which the "principle of utility" functions only as a loose guide and not a "rule" in any computable sense. In both this case and the previous, the scope of calculation makes a utilitarian solution intractable.

The situation is no better for Kant (1785/1981), who uses a universalization principle to determine when an action is moral. The "categorical imperative," which functions as the ultimate principle of morality, says that one should "act as if the maxim of your action were to become through your will a universal law of nature" (p. 30). One mainstream interpretation of this principle suggests that whatever rule (or maxim) I should use to determine my own behavior must be one that I can consistently will to be used to determine the behavior of everyone else.

Kant's most consistent example of this imperative in application concerns lying promises. I cannot make a lying promise without simultaneously willing a world in which lying is permissible, thereby also willing a world in which no one would believe a promise, particularly the very one I am trying to make. Thus, the lying promise fails the test and is morally impermissible. Though at first the categorical imperative may look to be easily implementable from an engineering point of view, it suffers from a problem of scope and relevance, since any maxim that is defined narrowly enough (for instance, to include a class of one, anyone like me in my situation) must consistently universalize. I can always will that my principle of action pertains to anyone in a situation exactly like mine, down to and including the color of the shirt I am wearing. Since the salient features that should be universalized obviously shouldn't (usually) include things like shirt colors, we run into a problem of relevance. One must already know what constitutes a morally relevant variable and what does not. But specifying how is very context-dependent and difficult. For instance, there may be situations where the color of clothes is morally relevant, as would be the case for uniforms on opposing sides in everything from sports to the theater of war. Death by failure to implement looks imminent for Kant as well. (There are other concerns about Kant that exceed the scope of this chapter. See Beavers, 2009, for an argument about why we should not want Kantian agents even if the categorical imperative were implementable.)

A complete analysis of the MFP and its relationship to rule-based systems would require a book of its own, but in the interests of serving the reader the above suffices to introduce the problem. The basic allegation is that moral, rule-based systems suffer the same fate as strong AI, rule-based systems, and, consequently, another approach is necessary if we are going to take into account the particularities, peculiarities, and exceptions that are necessary for sound moral judgment. (Though one might object that heuristics may be employed to solve the MFP for rule-based systems, we are still left with a needed procedure to determine when and where a moral heuristic is actually successful, and this again requires a solution to the MFP.) But, as with the general turn toward connectionism, or other network architectures, in cognitive systems, a similar escape has been suggested in automated morality that uses analogical, case-based reasoning, an approach that seems promising and implementable using connectionist, rather than rule based, AI models (Guarini, 2010; Wallach and Allen, 2009). Though initially it might look like formulaic moral theories are more easily implemented, this turn toward case-based moral reason makes virtue ethics a better candidate for our purposes.

Case-based reasoning is bottom-up in that it relies on specific instances to create a moral framework rather than taking a top-down approach. It also resonates well with the notion of habituation that comes with virtue ethics since its founding with Aristotle, though as Wallach and Allen (2009) note, this raises the further question of which set of virtues ought to be trained into a machine. For example, Plato named courage, wisdom, moderation, and justice as the cardinal virtues. But St. Paul named faith, hope, and charity, and early Catholicism advocated for seven virtues: chastity, charity, diligence, temperance, humility, patience, and kindness. There is also the set of 12 that comes from the American Boy Scouts: trustworthiness, loyalty, helpfulness, friendliness, courteousness, kindness, obedience, cheerfulness, thriftiness, braveness, cleanliness, and reverence.

We here leave it to moral philosophers to decide the question of which set of virtues is best, though whatever it may be, it seems that affect inevitably will be involved. Emotionists in moral theory likewise note that the affects are necessary to solve the MFP. Gallagher (2013) states that "the emotionist… contends that psychopaths fail to make genuine moral judgment because they lack moral emotions." He cites Blair (1995), Nichols (2004), and Prinz (2008) for support. For the purposes of this chapter, we need only draw the conclusion that simulated affects are most likely necessary for solving the MFP and thereby answering to the engineering imperative stated previously. If this is so, then affective computing is necessary for building moral machines.

There is another side to this issue as well that should briefly be mentioned and that has to do with what humans need in robots and other machines to accept them. Obviously, if one employs a service robot to help her ailing mother, she wants one that is moral, meaning that it is trustworthy, helpful, and well- intentioned, etc. If the mother is to accept it, however, it will also have to appear to her to be warm, caring, and kind. Perhaps Descartes was right after all that ethics is founded on our affects, that it resides in the realm of mind/body unities living in a world with others rather than in the mind alone, and that it is a functional means to find favor and avoid harm.

Disambiguation

In addition to reassessing our ethical systems, building more human-like artifacts requires us to consider the ingredients of effective communication. When utterances are made by human speakers, it is left to the listener to disambiguate the meaning of the utterance so that she may formulate an appropriate and accurate response. Miscommunication commonly occurs when the semantic content of an utterance is incorrectly (and most often inadvertently) disambiguated by the listener.

Multiple solutions to this problem have been proposed (for a glimpse at several, see Agirre & Edmonds, 2007), whether by appealing to intention, relevance, or context, but a ubiquitous and vital factor of human conversational disambiguation is affect. Affect, as we will show, is not semantically neutral, and thus, if a listener is to disambiguate a speaker's utterance properly and effectively, the listener must be able to understand the affect being expressed by the utterance of the speaker. This often involves matters of gesture, tone of voice, body posture, etc. If the goal in building artifacts is to make them responsive to and understood by human beings on more than a superficial level, they will need simulated affect to be able to express the meanings of their utterances in such a way that human listeners can disambiguate them.

To illustrate these points, we briefly examine Grice's intention-based model for disambiguation. Here, disambiguation is "a matter of *inferring* the speaker's communicative intention: the hearer uses all kinds of information available to get at what the speaker intended to convey" (Korta and Perry, 2015). This model is called "intention-recognition," and it considers such things as what a speaker said, what the words themselves mean, what the speaker implied, and, if possible, roughly what they were thinking when they uttered the statement. Intention-recognition utilizes context clues and certain maxims as developed by Grice (1967, 1989) to help listeners attempt to disambiguate utterances when a speaker's meaning is not explicitly clear or coherent. Here, if one can understand how the individual parts of an utterance interrelate, how relevant certain word meanings might be, what would possibly be implied by that specific speaker, etc., then one can begin to narrow the list of possible meanings, and ultimately one has a fair chance of landing on the accurate and intended meaning.

While this might increase the chances for proper disambiguation, there nevertheless remains a large possibility of error. It is also not true to life. Rational deduction of intention alone cannot fully account for the semantic content of an utterance because strict rational analysis treats utterances as if they are affectively flat. For our purposes, a flat utterance is one that lacks distinct tonality, inflection, expression, and other affective contextual components. Were it not for these affective components of semantic meaning, human interpersonal relationships could not transcend the superficial use of signs.

Recent advances in communicative strategies among artificial agents and between artificial agents and human beings have provided critical insight into the problem of superficiality, and they have helped us to reconsider the potential effectiveness or ineffectiveness of utterances, conversation, and interaction as a whole. Scheutz, for instance, draws attention to this problem by indicating that affects may be necessary for "information integration (e.g., emotional filtering of data from various information channels or blocking of such integration)" and "attentional focus (e.g., selection of data to be processed based on affective evaluation)" among other things (2011, p. 3). To draw the problem into sharper relief, one need only consider how information and communications technologies that are affectively flat, such as email, text messaging, etc., can easily lead to misunderstanding, even when the context is known and even, we must acknowledge, when emojis are used. Access to tonality, facial expression, and bodily gestures when communicating is necessary, and we cannot simply dismiss their absence as responsible for a lack of clarity. More poignantly, we are *semantically deprived*.

Today it is quite common to interact with computers, electronic systems over the phone, programs like Apple's Siri, Microsoft's Cortana, Amazon's Alexa, etc. In these instances, the goal is to make interaction as human-like as possible, though improvement on existing systems is necessary. Cowie (2012, p. 5) notes, for instance, that some current electronic communications systems "bully, patronise, demean, confuse, infuriate and simply don't work in ways that cause people real problems," and in other cases, the systems appear inauthentic. "Thanks for calling your neighborhood pharmacy." As such, "designers have a moral obligation to consider the impact [on people's emotions], and to seek out ways to ensure that it is not needlessly unpleasant" (Cowie, 2012, p. 6).

This obligation is taken seriously in the depiction of robots in the movie *Interstellar* (2014) (Nolan et al., 2014) quoted at the beginning of this chapter:

> *Tars (robot)*: Everybody good? Plenty of slaves for my robot colony?—This is a joke, but it doesn't come off that way. Hence, Doyle's reply.
> *Doyle (human)*: They gave him a humor setting so he'd fit in better with his unit. He thinks it relaxes us.—What's the tone here? Aggravation? Amusement? Or is Doyle just trying to inform?
> *Cooper (human)*: A giant sarcastic robot. A great idea.—When Cooper says, "A great idea" is he being sarcastic? It doesn't sound like it in the film.
> *Tars*: I have a cue light I can use to show you when I'm joking, if you like.—This statement and the following seem to be merely to inform.
> *Cooper*: That might help.

Tars: Yeah, you can find your way back to the ship after I blow you out the airlock. [Cue light flashes.]—An attempt at humor that isn't funny to Cooper. The rest seems to be stated as flat-affected matter of fact.
Cooper: What's your humor setting, Tars?
Tars: That's 100%.
Cooper: Let's bring it on down to 75, please.—There is no laughing in the scene, which serves to highlight the need for comfortable human–computer interaction.

Deception

Cooper: It's hard leaving everything… my kids, your father …
Brand (human): [cutting him off] We're gonna be spending a lot of time together …
Cooper: We should learn to talk.
Brand: And when not to. Just being honest.
Cooper: I don't think you need to be that honest.
Cooper: Hey Tars, what's your honesty parameter?
Tars: 90%
Cooper: 90%?
Tars: Absolute honesty isn't always the most diplomatic nor is the safest form of communication with emotional beings.
Cooper: Okay, 90% it is [meaning 90% honesty with Brand.]

The issue of deception fits into this chapter in two ways. The first is that some deception may be necessary for humans to interact with robots. It's actually necessary in human–human interactions as well. We are always making choices of what to say, what not to say, what to reveal, and what to hold back. As Cooper explains later to Brand, "... when you become a parent, one thing becomes really clear. And that's that you want to make sure your children feel safe. And that rules out telling a 10-year-old that the world's ending." Sometimes we protect each other, justifiably, to promote the well-being of the other. The second way deception fits into this chapter is that according to some moral theories, deception itself is bad. To continue the argument, simulated emotions are themselves deceptive, since presumably computing machinery can only mimic, and not actually have, affective states. These two dimensions of deception are difficult to separate, but we will attempt to treat the second issue first.

Bringsjord and Clark note in their paper, "Red-Pill Robots Only, Please," that there is a necessary distinction to be made between robots that are "red-pill" in nature, that is, those that do not intentionally deceive or illicit illusory behavior like expressing affect when they do not have it, and "blue-pill" robots, that is, those built either with the intent or capacity to deceive, or by people who know that deception could or would occur (Bringsjord and Clark, 2012). They argue against blue-pill robots (with rare exception) because interaction with them would probably lead to the formation of unidirectional bonds. Healthy relationships hang on reciprocity between agents. However, in human–robot interaction, when humans know that robots cannot reciprocate, the symmetry of the relationship is disturbed and may lead to dangerous or unhealthy consequences. Additionally, we may easily be fooled into feeling affection that is not present even though we know better. Such illusions run the risk of confusing the real with the artificial such that "much if not most of human life will gradually devolve into the mad pursuit of pleasure, quite independent of truth and falsity; and consequently the human condition will be a living, throbbing, massive lie for many if not most" (p. 2).

Scheutz (2011) argues in a similar vein, warning about the potential dangers of unidirectional bonds with social robots. Certain people develop emotional connections with a variety of social robots, and they care for them as if these robots were able to extend care back in the other direction, but the reality is that they cannot. Like Bringsjord and Clark, he agrees that deception is apparent when we are aware of it, but in the minutiae of life, we may easily fail to remember it. This can lead to extending moral regard and emotional concern to everything from inanimate household objects to fictional characters in books, plays, movies, etc. Scheutz (2012) explains that such deception could cause serious harm to humans, as, for instance, when a robot realizes that a unidirectional bond has been formed by the human and then exploits or manipulates it. He notes that it is enough to cause concern that a human should form affective dependencies on robots unable to reciprocate them in the first place.

On the other hand, Scheutz (2012) also addresses the positive side of deception because, in deception, there is the potential for greater-than-superficial interaction, which brings us back to the first issue raised at the start of this section, namely that deception may facilitate better interaction. Task performance can also improve under these pretenses, especially if the task involves therapeutic care or the like. Therefore, it is difficult to determine when, where, and how deception should be used.

Arkin (2011) also takes the more positive approach to this debate because deception can be effective in certain contexts. He states that "it may be *where* deception is used that forms the hot button for this debate" (p. 1). For instance, deception could be useful in warfare but dangerous elsewhere. Similar sentiments are expressed in Arkin (2012) regarding the possibility that deception could benefit a deceiver for "biologically inspired" reasons. Arkin's example is that of a robotic squirrel that uses deception to hide food from potential competitors (making one wonder what Arkin's humor setting might be).

Is deception in robots only beneficial in extreme circumstances, such as warfare or with respect to general survival? Perhaps not. Arkin and Shim (2013) express the idea of "other-oriented" deception and the benefits it can have in human–robot interaction. In this context, they argue that if deception benefits the deceived instead of the deceiver, it should be allowed and perhaps encouraged. They present eleven instances that fit here including but not limited to crisis management, cheering someone up or increasing someone's confidence in a stressful situation, quality of life management, flattery, etc.

Such benefits are aided and perhaps only possible if the robots are endowed with simulated affect, thus tying the two dimensions of deception introduced at the start of this section together, but the benefits in these instances far outweigh moral concerns. We agree with Arkin and Shim that these types of deception should be goals in affective computing, and thus, any possibility for developing robots with these capabilities should be pursued so long as ill intent is not present.

To return to the negative side of deception for a moment, Cowie (2012) agrees that affective based deception can be a particular concern, when artificial agents "give misleading signals, which uncritical people will be overwhelmingly disposed to accept. Some avatars will be confidence tricksters, and some people will be conned" (p. 9). It is another matter entirely, as Cowie points out, to think that this deception is any different than other forms that already exist in human–human interaction.

On the positive side, Cowie discusses that we should be optimistic for three reasons: first, humans are adaptable and will learn to manage the deceptive capabilities and tendencies of artificial agents after enough exposure to it; second, we do not know for certain whether sophisticated deception in the way we normally understand it will be possible or implemented into these artificial agents; and third, even if the deception is implemented and runs rampant at first, there is obviously the option and indeed the obligation to police and manage it.

If deception can be minimized or appropriately harnessed, whether in the context of war, as in Arkin (2011), or in therapeutic situations, as in Scheutz (2012), then it would seem to be both morally permissible and even necessary. To summarize, deception will be inherent if robots are endowed with simulated affect, but not all deception is negative, and thus, it is our responsibility to pursue positive employment of it while regulating negative uses. There is reason to remain optimistic about what deception can foster and, simultaneously, realistic about its moral dangers.

The preceding three sections lay out reasons why affective computing is necessary. In making this case, we do not intend to suggest that it should be unconstrained. There are definite possible abuses that we must recognize, the topic to which we now turn.

POTENTIAL AFFECTIVE ABUSE INVOLVING COMPUTING MACHINERY

De Sousa notes, "no aspect of our mental life is more important to the quality and meaning of our existence than emotions. They are what make life worth living, or sometimes ending" (2014). In other words, our emotions define in part who we are, how we interact, and how we experience the world around us. They are ubiquities of human life, and our interactions and relationships depend on their symmetrical exchange. As such, maintaining emotional well-being is critical, but, unfortunately, progress in affective computing could jeopardize that well-being due to the potentials for abuse, some of which include manipulation, misinterpretation, confusion, etc.

To address the first concern, in general, affective artifacts could manipulate people's emotions, especially for a company's financial gain. Advertisements are already used in various ways to try to influence consumers into buying products, often by appeal to affect and sexual attraction. In recent years, with the technological revolution, online profiling of consumers has created a pervasive presence of marketers, as the same advertisement floats from page to page customized to peak the individual user's interest. Marketers can now quite successfully use ghost node identification and other profiling techniques to pitch ads that speak directly to an individual's race, sexual orientation, political disposition, and so forth. As technology advances and machines can penetrate deeper into the user's affective psychology, it will be all the easier for marketers to use us against ourselves.

Imagine an artifact that is a cleaning companion, a more sophisticated Roomba of sorts, with affective capabilities, and it happens to be designed by a company that is partnered with another company that sells cleaning products. Call it RoboCleaner 1.0 or RC 1.0 for short. Before the user is bonded to RC, the owner buys less expensive brands, but RC forms an intentional attachment to its owner over time. Soon, RC expresses overwhelming sadness, informing the owner that it would be happier if it had its favorite brands. Not wanting the robot to be sad, the owner decides to buy them, and it becomes ecstatic. It expresses intense joy when it receives them, and it cleans better than it ever has before. After a while, the increase in expenses is noticeable, so the owner has to make a decision: keep the robot happy while sacrificing something else to make up for the cost, or save the money, let the robot continually express its sadness, and suffer with inferior cleaning.

Such an example could be shifted in a variety of ways. RC could insist on regular software updates, bond with children, and so forth. The important thing here is that this example already at this point in history appears trite, trivial, and unsurprising. Why? Perhaps because marketers already use such techniques in a variety of places. We need not look far to see commercial intervention into our affective lives to keep us playing online games, to feed a Facebook addiction, or lead us to form online attachments to other like-minded people. The examples are too numerous to mention; indeed, even before the age of ubiquitous intelligent machines, the union of funny and entertaining characters with particular cereals formed the minds of children to beg their parents for cheaply manufactured and poorly nutritious foods to grow an industry and perhaps promote unhealthy habits. There is little room to doubt that as the technology develops to allow deeper and deeper forms of manipulation, marketers will take advantage. [To see this played out in an unsettling way, we invite readers to view "Be Right Back," which is the first episode in the second season of the television series, *Black Mirror* (Brooker and Harris, 2013)]

To exacerbate the problem of manipulation raised previously, we must also worry about a second issue, misinterpretation. This is significant in affective computing because "it is difficult for people to avoid inferring that if a system shows fragments of behavior that are strikingly human, it has other characteristics that would be associated with those fragments in a human" (Cowie, 2012, p. 11). Since we already have a tendency to want to see or believe that there is more happening than may actually be the case, affective computing could easily (and inadvertently) promote confusion over one's own affective states and interfere with our ability to understand the affective states of others. In other words, affective computing runs the risk of lowering our emotional intelligence. One need not look far to see related problems emerging from misunderstood emails, text messages, etc. Nonetheless, affective computing may well exaggerate the effects of such misinterpretation.

One can consider this with respect to sex robots. Initially designed and sold for pure pleasure without inducing an affective bond, upgrades, and improvements to elevate the intensity of pleasure are directed toward making the sexbot as real as possible. Some fear that this will cheapen the sexual intimacy that is the foundation for many relationships, while others note that such bots might increase our sexual intelligence (Sullins, 2012; Levy, 2007). Nonetheless, it is easy to see how a bot that convincingly played the role of the submissive in an abusive relationship might lead its exploiter to infer that people might respond the same way. Cowie states that there is already concern about how we "use computers much as the Romans used Greek slaves," enjoying that we get to tell them to do something, and more often than not, they must just do it (p. 8). This basic enjoyment already happens without affective machines; it could worsen and develop into something highly dangerous if affect is implemented into artifacts without preventative measures, lowering not only our emotional intelligence but also inducing moral confusion.

Both concerns become even more pronounced when we move into the domain of virtual reality (VR). While still in its infancy, VR technology is quickly developing, for example, the recent release of the Oculus Rift headset providing a case in point. Since our emotional well-being hangs on frequent and intimate reciprocal emotional exchange, constant engagement with a virtual world may jeopardize that well-being, and human-to-human interaction may suffer. Additionally, we may easily suffer confusion over what is permissible in which world and what is not. To fully grasp this, we need only imagine VR technologies taken to their logical limit and crossed with video games to produce rich, immersive, multiplayer VR video games. Let us suppose that an impressionable teen has a set of various games of this type. He can visit VR1, VR2, VR3, etc. If VR1 is a version of "Call of Duty," which advances a particular set of moral values and is so real that it is indistinguishable from real life (RL), then how is this teen to distinguish the appropriate values in the two conflicting moral frameworks of VR1 and RL. The problem of moral confusion will only be increased when VR2, VR3, etc., promote opposing value systems. It is not too soon for game designers to worry about the affective depth that realism in games can reach in users. There is the additional worry, of course, that users may begin to prefer interaction with virtual characters in virtual storylines rather than engagement with real people in RL. (Are we there already? It's hard to say.)

The previous section represents only a small subset of the possible range of abuses promoted by affective computing. A full list and analysis would of course be a topic for a whole book. We leave it to the reader to imagine other forms of abuse.

CONCLUSION: THE STANDARD ETHICAL LIMITS APPLY

One could easily write a paper "On the Moral Implications and Restrictions Surrounding Baseball Bats" that would parallel the argument developing here. If we're going to play baseball, we need the bat, and, in the context of the game, it is morally neutral. But if used to beat someone over the head or to threaten a human into submission, it starts to take on moral dimensions. So too, if we're going to build artificial agents that are fully effective in interacting with human beings, then simulated affects are necessary. They can be misused, of course. This observation does not preclude the rather large question of whether we should even be playing this game. The question is futile, and the advancement of affective machines is inevitable. The game is already afoot. The moral dimensions of affective computing come primarily from the fact that it can intensify dispositions already at play in human life, in marketing, video games, human–computer interfaces, etc. This in no way trivializes the need to understand the moral impact of affective computing. On the contrary, it presents us with an urgency to use it to promote the basic precepts and principles of human decency. A further analogy will help.

Today we are capable of engaging in global warfare utilizing weapons of mass destruction. We can commit genocide and even destroy the whole human habitat. Our ability to do so arises from the exponential pace of technology, which is now outstripping the speed of human comprehension and legislation. This invites caution about adopting technologies too soon, but little can be done to stop their development. The situation, it would seem, is not that humans are less moral now that we can kill more people, but that we have failed to direct attention to the heart of the problem, which is human decency in general. Thus, we find that affective computing is morally neutral, just as most technology is. The morally relevant factor is what it is used for and how it is used. Caution is in order. Good people will use it well. Bad people will not. For now, this is about all that can be said.

References

Agirre, E., Edmonds, P. (Eds.), 2007. Word Sense Disambiguation: Algorithms and Applications. Springer, Dordrecht, Netherlands.

Anderson, M., Anderson, S., 2007. Machine ethics: creating an ethical intelligent agent. AI Magazine 28, 15–26.

Arkin, R., 2011. The Ethics of Robotic Deception. Report from the Georgia Institute of Technology Mobile Robot Laboratory, Atlanta, Georgia.

Arkin, R., 2012. Robots that Need to Mislead: Biologically-Inspired Machine Deception. Report from the Georgia Institute of Technology Mobile Robot Laboratory, Atlanta, Georgia.

Arkin, R., Shim, J., 2013. A Taxonomy of Robot Deception and its Benefits in HRI. Paper presented at the IEEE International Conference on Systems, Man, and Cybernetics, Washington, DC.

Beavers, A.F., 2009. Between angels and animals: the question of robot ethics, or is Kantian moral agency desirable. Paper presented at the annual meeting of the Association for Practical and Professional Ethics, Cincinnati, Ohio.

Beavers, A.F., 2010. Cartesian mechanisms and transcendental philosophy. Paper presented at the conference on the myth of the Cartesian 'ego' and the analytic/continental divide, Radboud University Nijmegen, The Netherlands.

Beavers, A.F., 2011. Moral machines and the threat of ethical nihilism. In: Lin, P., Bekey, G., Abney, K. (Eds.), Robot Ethics: The Ethical and Social Implications of Robotics. MIT Press, Cambridge, Massachusetts, pp. 333–344.

Beavers, A.F., 2012. Could and should the ought disappear from ethics. In: Heider, D., Massanari, A. (Eds.), Digital Ethics: Research and Practice. Peter Lang, New York, pp. 197–209.

Bentham, J., 2007. An Introduction to the Principles of Morals and Legislation. Dover, Mineola, New York, (Original work published in 1789).

Blair, R.J.R., 1995. A cognitive developmental approach to morality: investigating the psychopath. Cognition 57, 1–29.

Bringsjord, S., Clark, M., 2012. Red pill robots only. IEEE Trans. Affect. Comput. 3, 394–397.

Brooker, C. (Writer), Harris, O. (Director), 2013. Black Mirror (Television Series). In: Zeppotron (Producer). United Kingdom.

Cowie, R., 2012. The good our field can hope to do, the harm it should avoid. IEEE Trans. Affect. Comput. 3, 410–423.
De Sousa, R., 2014. Emotion. In: Zalta, E.N., (Ed.), The Stanford Encyclopedia of Philosophy. http://plato.stanford.edu/archives/win2015/entries/emotion/
Dennett, D., 2006. Computers as prostheses for the imagination. Paper presented at the International Computers and Philosophy Conference, Laval, France.
Descartes, R., 1985a. Discourse on method of rightly conducting one's reason and seeking the truth in the sciences. In: Cottingham, J., Stoothoff, R., Murdoch, D. (Trans.), The Philosophical Writings of Descartes, vol. 1. Cambridge, New York, pp. 111–151 (Original work published in 1637).
Descartes, R., 1985b. Rules for the direction of the mind. In: Cottingham, J., Stoothoff, R., Murdoch, D. (Trans.), The Philosophical Writings of Descartes, vol. 1. Cambridge, New York, pp. 9–78 (Original work published in c. 1628).
Descartes, R., 1991. The correspondence. In: Cottingham, J., Stoothoff, R., Murdoch, D. (Trans.), The Philosophical Writings of Descartes, vol. 3. Cambridge University Press, New York (Original letters written 1619 to 1650).
Feldman, F., 1978. Introductory Ethics. Prentice-Hall, Englewood Cliffs, New Jersey.
Gallagher, S., 2013. Phronesis and psychopathy: the moral frame problem. Philos. Psychiatr. Psychol. 20, 345–348.
Grice, H.P., 1967. Logic and conversation. In: Davidson, D., Harman, G. (Eds.), Logic of Grammar. Dickenson, Encino, California, pp. 64–75.
Grice, H.P., 1989. Studies in the Way of Words. Harvard, Cambridge, Massachusetts.
Guarini, M., 2010. Particularism, analogy, and moral cognition. Minds Mach. 20, 385–422.
Horgan, T., Timmons, M., 2009. What does the frame problem tell us about moral normativity? Ethical Theory Moral Pract. 12, 25–51.
Kant, I., 1981. Grounding for the Metaphysics of Morals. In: Ellington, J.W. (Trans.). Hackett, Indianapolis (Original work published in 1785).
Korta, K., Perry, J., 2015. Pragmatics. In: Zalta, E.N. (Ed.), The Stanford Encyclopedia of Philosophy. http://plato.stanford.edu/archives/win2015/entries/pragmatics/.
Levy, D., 2007. Love and Sex With Robots: The Evolution of Human-Robot Relationships. Harper Collins, New York.
Mill, J.S., 1979. Utilitarianism. Hackett, Indianapolis, (Original work published in 1861).
Minsky, M., 1988. The Society of Mind. Simon & Schuster, New York.
Minsky, M., 2007. The Emotion Machine: Commonsense Thinking, Artificial Intelligence, and the Future of the Human Mind. Simon & Schuster, New York.
Nichols, S., 2004. Sentimental Rules. Oxford, New York.
Nolan, C., Thomas, E., Obst, L. (Producers), Nolan, C. (Director), 2014. Interstellar [Motion Picture]. Paramount Pictures, Warner Bros., Legendary Pictures, Lynda Obst Productions, Syncopy, United States.
Powers, T., 2006. Prospects for a Kantian machine. IEEE Intell. Syst. 21, 46–51.
Prinz, J., 2008. The Emotional Construction of Morals. Oxford, New York.
Scheutz, M., 2011. Evolution of affect and communication. Affective Computing and Interaction: Psychological, Cognitive, and Neuroscientific Perspectives. IGI Global, Hershey, Pennsylvania.
Scheutz, M., 2012. The affect dilemma for artificial agents: should we develop affective artificial agents? IEEE Trans. Affect. Comput. 3, 424–433.
Sullins, J.P., 2012. Robots, love and sex: the ethics of building a love machine. IEEE Trans. Affect. Comput. 3, 398–409.
Wallach, W., Allen, C., 2009. Moral Machines: Teaching Robots Right From Wrong. Oxford, New York.

PART II

FRAMEWORKS OF AFFECTIVE SCIENCES IN HUMAN FACTORS AND HUMAN–COMPUTER INTERACTION

CHAPTER

6

Design and Emotional Experience

Bruce Hanington

Carnegie Mellon University, Pittsburgh, PA, United States

INTRODUCTION

Emotion is one of the most significant, complex and interesting human factors to designers. While physical and cognitive human factors have an established body of recognized standards in measurement, theory and application, the acceptance of emotion as an aspect critical for the success of design has a much more recent history. Common human factors textbooks place primary emphasis on fit, safety, and comfort. Even the Human Factors and Ergonomics Society (HFES) only gestures toward making our products and systems "enjoyable," in addition to safe and efficient (Human Factors and Ergonomics Society, n.d.). In human–computer interaction (HCI) and interaction design, the primary emphasis has been on usability. *Design and emotion* addresses the complex layer of human nuance in response to encounters with designed artifacts, and presents a challenge to the narrow focus on usability alone, traditionally measured through cognitive and physical human factors.

It is well established that emotion plays a role in our perceptions, attitudes, motivations, and behaviors. Our emotional state can affect how we focus attention and expectations, with obvious ramifications for how we process information and interact with products, systems, or other people. In a behavioral sense, we are prewired to approach stimuli associated with positive affect, and avoid those associated with negative affect. Designers can manipulate stimulation, or "arousal," as a physiological dimension of emotion. "The more stimulated we are, the more motivated we are to take action and avoid or approach the thing that is stimulating us." (van Gorp and Adams, 2012, p. 70). In general, people who are relaxed, happy, and "feel good" are more capable of creative problem solving, with

Emotions and Affect in Human Factors and Human–Computer Interaction. http://dx.doi.org/10.1016/B978-0-12-801851-4.00006-9
Copyright © 2017 Elsevier Inc. All rights reserved.

implications for how they might figure out complex devices or interfaces, and even be forgiving of product shortcomings (Norman, 2004; van Gorp and Adams, 2012).

The premise of design and emotion is that a holistic view must be taken of people in order to design effectively for them. Emotion is not a separate entity from cognitive processing or even physical interaction. This integration of complex human layers is nicely illustrated in the proven "aesthetic-usability effect." Research has demonstrated that interactions deemed more aesthetically pleasing are rated as significantly easier to use, and are more likely to be used, than less aesthetic interactions, even when usability is, in fact, purposely compromised in research manipulations (Lidwell et al., 2010).

As an integrated layer of human understanding necessary for design, emotional response is difficult to predict, apply, and measure. While it is easy to identify products that engage us enjoyably through intriguing interfaces, pleasing colors, sensuous textures, and whimsical details, the human response to artifacts can be much more complex. For example, why do some people respond favorably to objects that are met with distaste by others? This phenomenon alone suggests a range of idiosyncratic interpretations of the product world. Human reactions to products are rooted in the combined considerations of design intent, and interpretation in the context of personal and cultural experience (Hanington, 2004).

There is lack of consensus to adequately define pleasure, or positive emotional response, even among experts. Definitions alternately offer vague references to any "positive experience," and to the "opposite of pain" (Tiger, 2000, p. 17). The opposing force to pain may in some measure convolute the understanding of the concept of pleasure. There are various things in individual experience that will reduce or eliminate "pain," or discomfort associated with physical or psychological stress. It is critical to realize in defining pleasure that the things people do may be sought out both to *seek* pleasure for purely positive effects, and to *reduce* the effects of boredom, depression, or other negative psychological or physical conditions (Hanington, 2004). Jordan (2000) indicates that pleasure is distinguished between "need pleasures," fulfilled by eliminating a state of discontentment, and "pleasures of appreciation," accrued because of inherent positive worth found in things or activities.

These complex factors explain to some degree various pleasure-seeking behaviors engaged in by people, often contradicting reasonable assumptions of health and safety, such as dangerous sporting activities, smoking, and substance abuse. The understanding of emotional experience has for designers both an intellectual and commercial appeal. To comprehend, and possibly predict, the emotional resonance that products and systems may have for the people using them and affected by them will make for better design, and by extension, improved human experience.

This chapter will provide an overview of design and emotion through its origins and key moments in history, evolving to its prominence as a critical element of design for human experience, and user experience (UX) design. Theoretical perspectives of design and emotion are presented, particularly as they challenge more traditional human factors in design practice. And finally, a section on emerging methods and approaches identifies several ways in which design and emotion can be understood through responsive research, field techniques, and design applications.

THE EVOLUTION OF EMOTION IN DESIGN AND HUMAN–COMPUTER INTERACTION

The explicit inclusion of an emotional component in design dates back at least to Vitruvius (c. 80–70 BC—c. 15 BC), in architecture. Vitruvius referenced three key components of building architecture: *Utilitas*—representing utility, commodity, and function; *Firmitas*—indicating the need for structural and technical soundness; and *Venutas*—the provision of delight, pleasure, or beauty in a pure experience of space beyond the aesthetic (Vitruvius, transl. 1960). A parallel to this triad of elements is evident in the common design mantra of *useful, usable, desirable,* first coined in 1992 (Sanders, 1992). Whether applied to the design of products, systems, interfaces, information, services, or environments, these three factors suggest a balanced emphasis. The things that we use need to fulfill their intended purposes and are instrumental in meeting our goals. They need to be usable in how they are interpreted, comprehended and navigated, ultimately the purview of usability testing. And they need to be desirable, resonating with people from an aesthetic, emotional, personal, or cultural standpoint. The notion of making something desirable has both a human and commercial incentive.

The recognition of emotion as a critical element of design has primary roots in industrial design, with quite a different history in human–computer interaction (HCI). The field of HCI is younger than industrial design, explaining part of the difference; however, there is also a difference of emphasis. While both disciplines were historically dominated by a fundamental stress on usefulness and usability, *desirability* was evident earlier in industrial design. Furthermore, the formal organization of *design and emotion* through an organized professional society grew naturally out of product (industrial) design and continues to be dominated by that field, whereas much of the subject matter was introduced into HCI circles as a reaction to limited functional views of technology.

In the 1980s prominent research in industrial design focused on aspects of product meaning, semantics and semiotics, with implicit references to emotion. Klaus Krippendorff and Reinhart Butter coined the term product semantics in 1984 in an article titled, "Exploring the Symbolic Qualities of Form." Partially in response to functionalism, approaching design through a semantic filter attempted to provide meaning to product users through the elements of form, including shape, texture, materials and color, but extended this to include aspects of product identity (what is it), character (what kind is it), affordance (what benefit does it provide), and operation (how do I use it). While products typically provide function, semantics as an approach channeled away from the simple mantra of form follows function, with a focus on making products meaningful, easy to use, and enjoyable to experience (Krippendorff and Butter, 1984; Krippendorff, 2005).

In 1981, Csikszentmihalyi with Rochberg-Halton published their extensive studies of research on the symbolic meaning of things in the home. While originating in the social sciences and not from design, the book emerging from the study, *The Meaning of Things: Domestic Symbols and the Self*, was instrumental in guiding many designers toward explanations of product meaning, and influential in subsequent design research. It is also noteworthy that 10 years later, in 1991, a condensed version of the study was published as an article in *Design Issues*, under the title, *Design and Order in Everyday Life* (Csikszentmihalyi, 1991). The work has been extended to identify key attributes contributing to product meaning attainable by design, including interaction (personal engagement, accomplishment), satisfying experience (enjoyment, ongoing occasions, release), physical attributes (intrinsic object qualities, craftsmanship, and evidence of hand; material quality), style and utility (Disalvo et al., 2003).

The primary focus of early HCI was on workplace efficiency and measurable usability. The rise of complex consumer electronics in the home saw the emergence of interaction design as distinct from industrial design, and subsequent attention being paid to the emotional impact of technology products. Rosalind Picard's landmark book on *Affective Computing* was first published in 1997, extending her prior MIT technical report published in 1995. Picard's work recognized the critical role emotions play in decision-making, perception, and learning, advocating for why we must give computers the ability to recognize, understand, and even to have and express emotions (Picard, 1995, 1997). Kristina Höök (n.d.) explores the idea of involving users both cognitively and physically in an "affective loop," described in her own words as follows:

- The user first expresses his/her emotions through some physical interaction involving the body, for example, through gestures or manipulations of an artifact.
- The system (or another user through the system) then responds through generating affective expression, using for example, colors, animations, and haptics.

- This in turn affects the user (both mind and body) making the user respond and step-by-step feel more and more involved with the system.

In 2004, the book *Funology: From Usability to Enjoyment* was published, challenging the narrow focus of HCI on usability to incorporate fun, enjoyment, aesthetics, and the experience of users (Blythe, 2004). This reexamination of emphasis occurred particularly as technology products began to proliferate in the home and for personal use, providing a contrasting viewpoint to the task-based work applications more commonly associated with HCI research and design.

Similarly, in 2000, Patrick Jordan advocated for a "new human factors" in his book, *Designing Pleasurable Products: An Introduction to the New Human Factors* (Jordan, 2000). Jordan promoted a broad look at design, crafting a framework of experiential pleasures of human–product interaction based on the work of anthropologist Lionel Tiger (1992). The text was among the early works to recognize that usability had moved from being what marketing professionals call a "satisfier" to being a "dissatisfier." In other words, people are no longer pleasantly surprised when a product is usable, but are unpleasantly surprised by difficulty in use. Recognizing that the human response to products and systems is comprised of more than physical fit and information processing, the "new" human factors explicitly promoted the inclusion of an emotional component in design.

Much of the early work formally establishing *design and emotion* as a recognized force within design grew out of The Netherlands. The *Design and Emotion Society* was officially formed in The Netherlands in 1999, with biannual conferences beginning that same year. In 2002 at Delft University of Technology, Peter Desmet published his PhD dissertation, *Designing Emotions*, providing a framework for defining, measuring, and designing product emotions (Desmet, 2002). The daily board of the Design and Emotion Society continues to be based in the Netherlands. In 2001, the *Conference on Affective Human Factors Design* (CAHD), was held in Singapore, bringing together an international cohort of practitioners and academics from diverse fields of design, technology, sciences, and social sciences. This conference eventually morphed into *Designing Pleasurable Products and Interfaces* (DPPI), running biannually from 2003 through to the most recent occurrence at Newcastle in 2013. In 2014, the 9th International *Design and Emotion Conference* was held in Bogota, Columbia.

Based anecdotally on conference attendance and membership in the Design and Emotion Society, the field is dominated by an international group of professionals and academics in product design (industrial design), with smaller contingents of designers from UX and interaction design and HCI, as well as from graphic and communication design, architecture, and environmental design. Participants from various fields of media design are also active, as are those from the social and behavioral sciences, and sciences. The subject matter of design and emotion is intentionally and sufficiently broad to be appealing and relevant to all design disciplines and related fields. As stated on the Design and Emotion Society site, "The network is used to exchange insights, research, tools and methods that support the involvement of emotional experience in product design. Although the initiative originated from the discipline of product design and design research, through the years practitioners from other design disciplines such as interaction design and branding design, contributed and benefited from the network and activities" (Design and Emotion Society, n.d.).

The holistic human view of design and emotion caused it to gradually become at least partially absorbed into the realm of "user experience," a term first coined by Donald Norman at a computer–human interaction (CHI) conference in 1995. This led to similarly accepted terms, such as experience design, design for experience, and ultimately the established shorthand of "UX" for user experience design, along with the popular company role of Chief Experience Officer (CXO). Affective aspects of design advocated under the term UX were intended to coexist with behavioral concerns more traditionally associated with product and system usability. Norman's connection of design experience to emotion became explicit in the publication of his book, *Emotional Design: Why We Love (and Hate) Everyday Things*, in 2004.

Kuru and Forlizzi (2015) present a concise history of UX design, with key elements of emotion noted throughout. Beginning as early as 1996, Alben (1996, in Kuru and Forlizzi, 2015) identifies factors of user experience to include "how people feel about a product, how well they understand its functions, how the product makes people feel when using it, and how it fits its purpose and context of its use." As the interest in design and emotion grew within the HCI community, this holistic view became more accepted, defining user experience to encompass all aspects of the experience of an interactive technology product, service, or system, including physical, emotional, sensual, cognitive, and aesthetic, before, during and after use (Hassenzahl, 2008, in Kuru and Forlizzi, 2015).

In 2004, John McCarthy and Peter Wright published *Technology as Experience*, building heavily on the premise of John Dewey's notion of lived experience, suggesting that we not only *use* technology, we *live* with it. It stands to reason that if we are to design according to the popularized term of "user experience" (UX), we must take into account the "emotional, intellectual, and sensual aspects" of the interactions we are creating. This aspect of felt human experience, viewing technology as creative, open and relational, again presented a challenge to traditional notions of HCI as merely fulfilling functional usability needs (McCarthy and Wright, 2004).

The fact that emotion is an integral component of UX design is addressed in the ISO definition for user experience, stated as a "person's perceptions and responses that result from the use or anticipated use of a product, system or service. User experience includes all the users' emotions, beliefs, preferences, perceptions, physical and psychological responses, behaviors and accomplishments that occur before, during and after use." (International Organization for Standardization, 2010). The ISO lists three factors that influence user experience: system, user, and the context of use. "This has led to a shift away from usability engineering to a much richer scope of user experience, where users' feelings, motivations, and values are given as much, if not more, attention than efficiency, effectiveness and basic subjective satisfaction (i.e., the three traditional usability metrics)." ("User Experience," n.d.).

THEORETICAL PERSPECTIVES ON EMOTION IN DESIGN AND HUMAN–COMPUTER INTERACTION

The expanding literature on design and emotion, growing through both conference venues and publications, is a testament to the significance and status of the field. The diverse subject matter encompassed by the term is evident in conference tracks that address a range of topics, such as well-being, culture, experience, sensation, social implications, sustainable behaviors, relationships and patterns, and inclusive design. The number of designers interested in the field and claiming an aspect of design and emotion as part of their expertise, and the large membership of the Design and Emotion Society, further confirm that design and emotion is now a significant element of human factors in design research, theory and practice today.

Lidwell et al. (2010) identify several human principles that attempt to explain our emotional response to the designed world. For example, the "baby-face bias" is "a tendency to see people and things with baby-faced features as more naïve, helpless, and honest than those with mature features." Repeated exposure to stimuli for which people have neutral feelings will increase the likeability of the stimuli in a so-called "exposure effect." The "Veblen effect" suggests that people will tend to find a product desirable because it has a high price, and the phenomena of *scarcity* makes things more desirable when they are perceived to be in short supply or occur infrequently. And of course, the nuances of *color* choice are well known to produce emotional reactions, positive and negative, based on culture, experience, and subjective interpretation by people experiencing the designed world.

The Yerkes–Dodson Law is informative in design and emotion. This law posits that increasing levels of stimulation or arousal can initially improve motivation and performance, yet then reach an optimum level before decreasing performance as stress becomes too high (Fig. 6.1). Stimulation

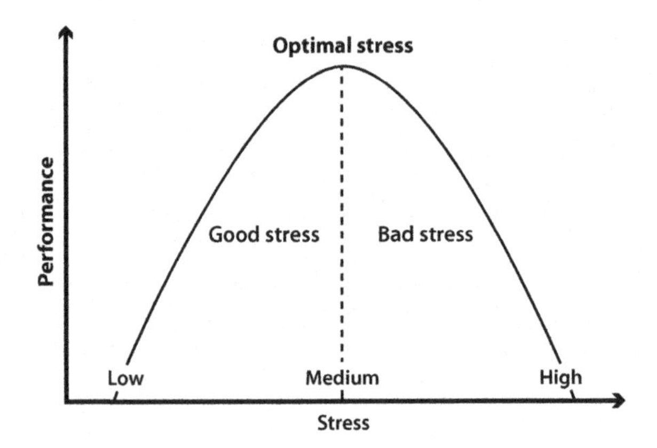

FIGURE 6.1 **The Yerkes–Dodson law indicating performance as a function of stress (Yerkes and Dodson, 1908).**

must be treated in balance by design, dependent on the context. For example, the use of negative stimuli, such as loud alarms is appropriate to evoke strong cognitive, emotional, and physiological responses in emergency situations. However, an illuminated status indicator, quietly telling us that an alarm system is powered and ready, provides calming reassurance (van Gorp and Adams, 2012).

Emotion itself can be viewed as a short-lived, reflexive response, most commonly associated with a commercial approach wherein products or services are advertised and sold on the basis of their fun and exciting allure, or presence. However important, this level of emotional appeal is often placed in contrast to a more sustained, reflective response, "emotional experience," or "mood" connected to long-term product relationships (DiSalvo et al., 2004).

DiSalvo, Forlizzi, and Hanington propose three ways in which products can essentially shape human emotional experience: *stimuli* for new experiences, *extenders* of ongoing experience, and *proxies* for past experiences (Fig. 6.2). As stimuli, products with specifically designed interactions or sensory stimulation can become catalysts for new emotional experience. For example, a walking aid can provide the catalyst for enhanced mobility, independence, and social access to a person formerly limited by physical constraints. Extenders offer ongoing product experiences of familiar objects, through particularly designed style, utility, and other sustaining qualities that make them more enjoyable or appealing to own and use. Proxies are products that provide associations with previous experiences, often through reminders of other times or people. Souvenirs, family heirlooms, and gifts are common proxies (Disalvo et al., 2003).

In a similar vein, Zimmerman (2009), in his work on "designing for the self," extracts insights from possession attachment theory to inform the

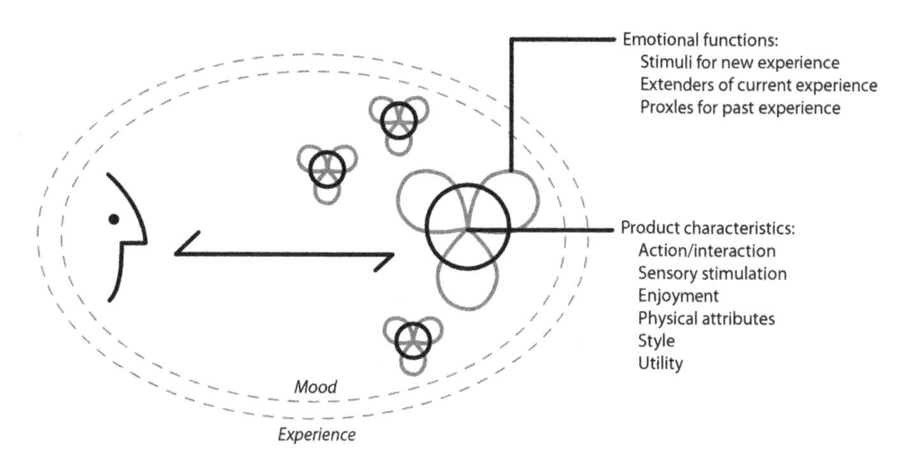

FIGURE 6.2 The shaping of emotional experiences through product characteristics (Disalvo et al., 2003; DiSalvo et al., 2004).

design of interactive products and services, thereby enhancing emotional experiences. Material possession attachment theory comes from consumer behavior research (Belk, 1988; Kleine et al., 1995; Kleine and Baker, 2004). It builds on concepts from identity theory (Belk, 1988; Goffman, 1959; James, 1890), life story theory (McAdams, 2001), and social identity theory (Tajfel and Turner, 1986) and attempts to explain how and why people love some of their material things. Zimmerman's work investigates two approaches to operationalizing product attachment theory around social role performance. Role enhancement includes products that help people become the person they desire to be in a specific social role, like a parent, romantic partner, or religious practitioner, for example. Role transition includes products that help people transition into a new social role, like when high school students shed their high school identity and reinvent themselves as a college student, or when a couple has a first child and invents themselves as parents.

Odom et al. (2011, 2012) significantly extended Zimmerman's work in his research investigating how people perceive value in their immaterial (virtual) possessions and how changes to the form and behavior of virtual things can increase perceptions of personal and economic value. For example, in a study of teens and their virtual possessions, it was determined that social media affords flexibility in personalization and a place to curate and share current interests in the experimentation with a sense of self and the portrayal of identity presented to others, while risking "self obsession" stemming from peer pressure to create multiple selves. Virtual possessions can increase a sense of social connectedness, but can also work to amplify differences and reinforce cliques, with obvious emotional outcomes.

In Jordan's theory of four pleasures, physical, psychological, sociological, and ideological factors can all contribute to emotional experience of the designed world. For example, the tactile sensation and response of a laptop keyboard provide *physio* pleasures; the status of owning a Philippe Starke designed juicer or the nonstigmatizing success of personal diabetes monitors constitute *socio* pleasures; appropriate sound levels in domestic appliances and the right level of challenge in game design are means of achieving *psycho* pleasure through design; and the communicative function of aesthetics corresponding to individual values represent *ideo* pleasures. Product features are broken into "formal" and "experiential" properties to define objective and subjective aspects of design. Similarly, people are described by their physical attributes and abilities, their sociological status, self-image and lifestyle, psychological traits of personality, confidence, skills and knowledge, and ideological beliefs, values, and aspirations (Jordan, 2000).

Donald Norman posits that experiences are colored by emotional responses that operate on three levels: visceral, behavioral, and reflective (Norman, 2004). Pleasure and emotional design are linked in a holistic assessment of user experience (van Gorp and Adams, 2012). Visceral design concerns appearances (e.g., colors, aesthetics), relates to physio-pleasures and provides hedonic benefits. Behavioral design relates to effectiveness of use (e.g., functionality, usability), links to psycho-pleasures and relates to practical benefits. Reflective design considers interpretation and understanding over time (e.g., experience, personal story), concerns ideo-pleasures and socio-pleasures and relates to emotional benefits (Scupelli and Hanington, 2014; Table 6.1).

Pleasure, or satisfaction, particularly in the psychological realm, can be achieved as an aspect of balancing challenge and skill building. In

TABLE 6.1 A Holistic Assessment of User Experience (Scupelli and Hanington, 2014)

Emotional response (Norman, 2004)	**Design concerns**	**Pleasure (Jordan, 2000)**	**Benefits**
Visceral	*Design qualities, appearance* Color Aesthetics	*Physio* Physiological	*Hedonic*
Behavioral	*Effectiveness and ease of use* Function Usability	*Psycho* Psychological	*Practical*
Reflective	*Interpretation* *Understanding* Personal experience Story	*Ideo* Ideological, values, aspirations *Socio* Social relationships	*Emotional*

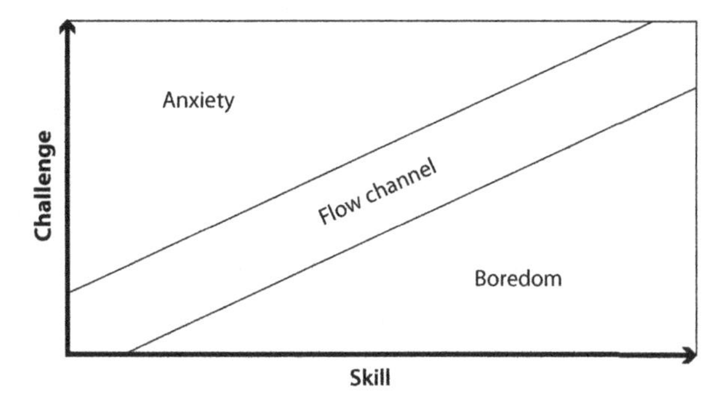

FIGURE 6.3 A balance of incremental challenge and skill building to achieve a state of "flow." *Source: Adapted from Csikszentmihalyi, M., 1990. Flow: The psychology of Optimal Experience. Harper and Row, New York.*

the theory of "flow," a state of intense immersion occurs when progressively developing skills are met with increasing challenge, and vice versa (Csikszentmihalyi, 1990). If skills are overchallenged, a person ends up in a state of anxiety; if skills develop beyond the challenge, boredom may occur (Fig. 6.3). This is evident in the tiered structure of game design, for example, whereby successively complex challenge levels are presented to gamers as goals are achieved. Flow is similar to the effect of *immersion* used in experiential exhibit design, movies, and amusement park attractions, creating a state of mental focus so intense that conscious awareness of the "real world" is temporarily suspended (Lidwell et al., 2010).

Extending the structure and success of games into the motivational realm, researcher and game expert Jesse Schell states that, "if we want to understand how to get people to engage more with our ideas, games, and products, we need to look at what games do very well: motivate us to play them." (Lefebvre, 2011). The concept of "gamification" applies the mechanics and techniques of game design to engage and motivate people in other realms, for example, learning tasks or losing weight, or simply captivating an audience to engage with an interface, product, or system. By adding game-like elements, such as competition, status, achievement, and a state of "flow" as described previously, nongame applications can be made more fun and engaging, thereby enhancing the motivation to interact.

Maslow's theory of human motivation has also been influential in design and emotion. As the central premise of the theory, a hierarchy of needs suggests that self-actualization is reliant on the prior satisfaction of esteem, belongingness, safety, and physiological needs. In similar fashion, a theory of design suggests that functional needs and usability criteria must be met before successfully fulfilling emotional benefits (Jordan, 2000). Lidwell et al. (2010) similarly concur that design must satisfy basic needs, such as functionality, reliability, and usability, before succeeding on merits of human empowerment or fostering creativity.

Several examples of design and emotion work have challenged the narrow focus on just efficiency and error reduction, or measured usability. The "slow movement" that began as slow food and other forms of advocated living also inspired some to examine the role of technology in our everyday lives, and propose slow technology "aimed at reflection and moments of mental rest rather than efficiency in performance." (Hallnäs and Redström, 2001). Similarly, work by Odom et al. (2014) on "Designing for Slowness" performed a long-term study of the "Photobox," "a domestic technology that prints four or five randomly selected photos from the owner's Flickr collection at random intervals each month." The research suggested "several opportunities, such as designing for anticipation, better supporting reflection on the past, and, more generally, expanding the slow technology research program within the HCI community."

Subsequently, Bill Gaver's work on "ludic interaction" focused on interaction that is playful and often has no specific task or goal. This challenged the traditional focus by HCI on task completion time and the strict fit between what people want and what designers make. For example, the "Drift Table" is "an electronic coffee table that displays slowly moving aerial photography controlled by the distribution of weight on its surface, designed to investigate ideas about how technologies for the home could support activities motivated by curiosity, exploration, and reflection rather than externally-defined tasks." (Gaver et al., 2004).

New trends are constantly extending design and emotion into other areas of investigation and application. For example, Lalande and Racine (2006) and Lalande et al. (2010) have examined how longevity of product relationships and emotional connection can be fostered through rapid prototyping repairability and transformation of obsolete objects into new functionalities, in turn fostering environmentally sound behaviors and product sustainability.

Design and emotion is inherently connected to design for social innovation, or enhancing the social good, in areas, such as health care and well-being. Recognizing the new global focus on happiness, including measures of gross national happiness (GNH), a special issue of the *International Journal of Design* in 2013 promotes the idea of *Design for Subjective Well-Being*. The editors propose a positive design framework for "designing with the explicit intention to support people in their pursuit of a pleasurable and satisfying life, and, even more important, in their desire to flourish. It aims at designing products that contribute positively to the experienced quality of life, in making things that are useful, usable, enjoyable, purposeful, desirable, and even virtuous and ethical." Themes of the special issue examine the topic through lenses of daily life, work, leisure, fun, and health, with the intention of appealing to a broad audience in the private, public, social, and health-care sectors (Desmet et al., 2013).

EMERGING METHODS IN DESIGN AND EMOTION

The multitude of theories and research presented here represent a sample of moments in a field that gradually emerged yet quite suddenly burgeoned. As a core feature of design, research and theory typically coexist in parallel with action, or practical application. To that end, various design methods and approaches have been associated with design and emotion.

As a methodology of design, Sanders and Stappers (2012) position design and emotion as a design-led approach, situated between a "participatory" and "expert" mindset in how users are viewed (Fig. 6.4). This positioning suggests that design and emotion grew primarily out of the design community itself, in contrast to traditional human factors, ergonomics and usability testing emerging from a dominantly research-based professional practice. Human factors, ergonomics, and usability testing professions, including HCI, have also been traditionally aligned with the expert mindset, viewing users as reactive informers or subjects. Design and emotion is more closely aligned with a participatory mindset, seeing users somewhat more as partners or cocreators in their contribution to

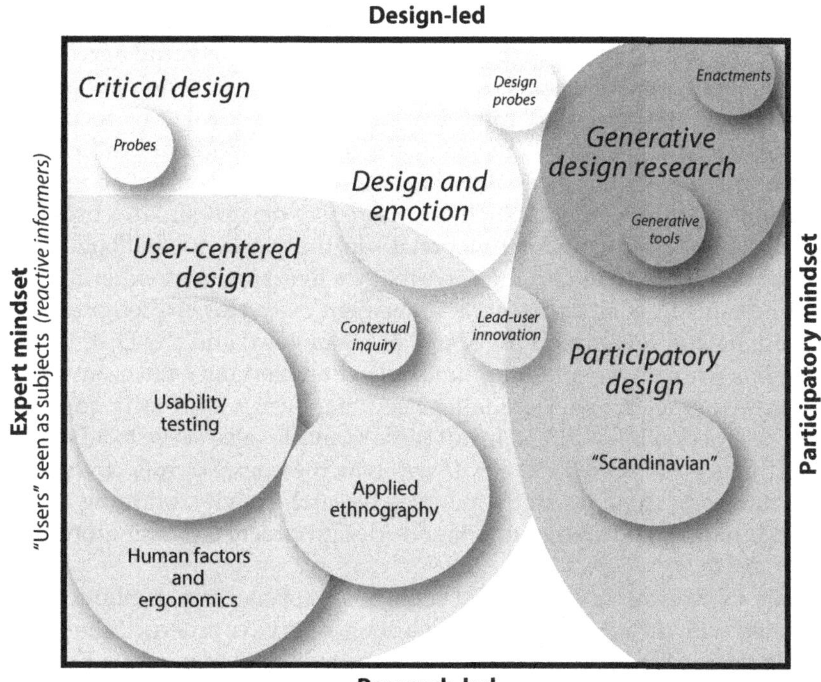

FIGURE 6.4 **The landscape of design research and practice.** *Source: Adapted and used with permission of Elizabeth B.-N. Sanders. Versions of this figure appear in Sanders (2006) and Sanders and Stappers (2012).*

design. However, design and emotion has been influenced by, and continues to influence, other realms of research within this landscape of methods and approaches.

For example, elements of design ethnography, or the immersive tools adapted from anthropology, attempt to forge a sympathetic and empathic sense of the stakeholder (user, research participant), to positively influence the design process and outcomes such that emerging products resonate with people on all levels, including emotional. Forms of contextual inquiry and observation are key to understanding people in their lives of work and play. Design probes or cultural probes provide creative, sometimes provocative self-report tools to participants through which they may expressively communicate their personal thoughts and feelings in response to subject matter prompts (Gaver et al., 1999; Hanington and Martin, 2012).

Similarly, generative tools and methods associated with participatory design are empathically based, aspiring to engage and involve those affected by design in a process that results in emotionally resonant products, systems and services, and ultimately bettering the lives of those who use them. Design workshops and codesign tools, such as flexible modeling and collage allow participants to project emotions onto physical materials to express their thoughts and desires (Hanington and Martin, 2012). Most designers operating in the sphere of design and emotion would agree that a human-centered design process, placing users and stakeholders at the forefront of a research-based approach to design, is critical to its success.

Some would argue that elements of critical design and cultural probes are also squarely within the realm of design and emotion. Critical design, and counterpart movements, such as speculative design, situates provocative designs—often counter functional—in the world to challenge our thinking about products and the ways that we live. However, experiments noted previously, such as the slow movement examples challenging our conceptions and reliance on the speed of technology, and the Drift Table motivating curiosity, exploration and reflection, certainly have an emotional overtone in the subject matter itself, the design approach, and the audience reaction intended. Many of these examples also serve to advance the legitimacy of *research through design*, which recognizes reflective processes of design thinking and making as research activity, bridging theory, and building knowledge to enhance design practice (Hanington and Martin, 2012).

While there is not a specifically prescribed approach or methodology associated with design and emotion, various tools have emerged that represent a positive fit with the goals and intentions of the field. The Design and Emotion Society website features an interactive section on tools and methods, allowing researchers and practitioners to parse methods on the basis of intended application and category. Selections can be made according to phase of the project, whether looking for radical or incremental change; purposes including collection, exploration, or measurement; and subaspects of emotion, such as sensory characteristics or expression (Design and Emotion Society, n.d.).

While many methods are common to design practice, others have been developed for specific purposes within design and emotion. For example, *personas* are fictional yet representative users developed as a convergence of human elements extracted from research with real people; *scenarios* offer realistic situations of task or goal completion including positive and negative interactions. These are methods widely used in design, yet have particular value in design and emotion. A variant on personas is designing for *extreme* characters, with exaggerated emotional attitudes, to expand creative possibilities in considering the aesthetics of interaction design. For example, Djajadiningrat et al. (2000), describe projects targeting a drug dealer or "the fictitious Pope."

Bodystorming, likewise used for various design purposes, has resonance with design and emotion as a form of physical brainstorming, whereby scenarios are quickly set up using low-fidelity props at hand, and acted out for quick understanding and feedback on a proposed situation. Bodystorming contains elements of role-playing, also a familiar method to many designers as an empathic tool for understanding users and their needs, and assessing ideas.

Methods specifically created for design and emotion attempt to ascertain emotional aspects of human motivation or behavior, and inspire design outcomes. For example, *Emofaces* allows participants to express their emotional reactions using eight illustrated cards, each depicting emotions along high to low dimensions of arousal and pleasant–unpleasant, based on the "circumplex model of affect" (Fig. 6.5). The [product & emotion] navigator provides an anecdotal database of product images and the emotional reactions they have evoked, along with reasons behind these reactions (Design and Emotion Society, n.d.). *Desirability testing* uses card-based adjectives and descriptive phrases to aid participants in their expression of emotional responses to design prototypes or products (Hanington and Martin, 2012).

In the totality of product experience encompassing emotion, the *Experience Sampling Method* (ESM) gathers snapshots of people's lives, including feelings, through mechanisms like diaries or participant-generated photos and notes. Likewise, *experience prototyping* facilitates active, subjective engagement with convincing prototypes of products or systems, to assess early functional and emotional resonance with potential users (Hanington and Martin, 2012).

van Gorp and Adams (2012) propose a model designed to help ensure that the design process addresses the three levels of useful, usable and desirable attributes, and their associated variables. Using the acronym

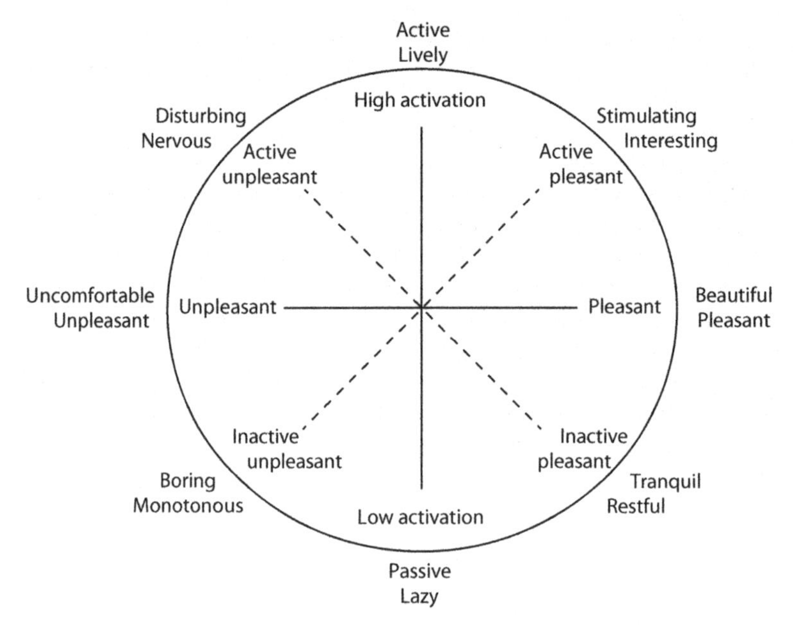

FIGURE 6.5 **A circumplex model of affect.** *Source: Reprinted from Russell, J.A., 1980. A circumplex model of affect. J. Pers. Soc. Psychol. 39 (6), 1161–1178. With permission of SAGE.*

ACT for *Attract*, *Converse*, and *Transact*, the model suggests that there first needs to be a motivation to interact with a product or service, through some form of *attraction*, typically through aesthetic appeal to the senses. We then need to ensure a positive *conversation* with the product, forming a relationship through feedback and dialog, or an engaging narrative flow to the experience of product use. And finally, a *transaction* is necessary, or a decision to act and thereby commit to a product or service, by purchasing a product or signing a service agreement, for example. A positive transaction is reliant on feelings of trust and credibility determined through attraction and conversation (van Gorp and Adams, 2012).

Scientist B.J. Fogg of the Persuasion Technology Lab at Stanford is known for creating systems to change human behavior through self-proclaimed "behavior design." Using the psychology of persuasion to describe a three-step process, the so-called "Fogg Method" advocates first specifying the desired behavior you want as an outcome, then simplifying to make the behavior easy to do, and finally triggering the behavior either naturally or through design (Fogg, 2013).

While the Persuasion Technology Lab and Fogg himself work primarily through mobile devices on specific topics, such as improving health, the notion of design for behavior change has become a generally prominent theme within design and emotion. Very public examples have emerged through projects, such as the VW Fun Theory, "dedicated to the thought that something as simple as fun is the easiest way to change people's behaviour for the better." (Thefuntheory.com). Among several examples are the "Piano Stairs," motivating people to take stairs rather than an escalator by embedding musical electronics in the steps, clad as giant piano keys, and activated in correct musical notation when treaded upon. The option for lower energy consumption and a moment of exercise is spontaneously presented to commuters, following the premise of a simple and emotionally rewarding act to seamlessly change an everyday behavior.

Much of what we design and encounter as consumers, customers, or "users" has a direct influence on our mental and emotional state of being. While the necessity for functional usability and efficiency cannot be overlooked, a holistic view of people encompasses how we emotionally resonate with the things we use and the experiences we have in the material and natural world. Human factors is fortunately moving to address this totality, accounting for the complexity of interactions that make up our human experience, physically, cognitively, and emotionally.

References

Alben, L., 1996. Quality of experience: defining the criteria for effectiveinteraction design. Interactions 3, 11–15.

Belk, R.W., 1988. Possessions and the extended Self. J. Cons. Res. 15 (2), 139–168.

Blythe, M.A., 2004. Funology: From Usability to Enjoyment. Kluwer Academic Publishers, Dordrecht.

Csikszentmihalyi, M., 1990. Flow: The Psychology of Optimal Experience. Harper and Row, New York.

Csikszentmihalyi, M., 1991. Design and order in everyday life. Des. Issues 8 (1), 26–34.

Design and Emotion Society, n.d. http://www.designandemotion.org.

Desmet, P., 2002. Designing Emotions. Delft University of Technology, Dept. of Industrial Design, Delft, The Netherlands.

Desmet, P., Pohlmeyer, A., Forlizzi, J., 2013. Special issue editorial: design for subjective well-being. Int. J. Des. 7, 3, http://www.ijdesign.org/ojs/index.php/IJDesign/article/view/1676/594.

Disalvo, C., Forlizzi, J., Hanington, B., 2003. On the relationship between emotion, experience and the design of new products. Des. J. 6 (2), 29–38.

DiSalvo, C., Forlizzi, J., Hanington, B., 2004. An accessible framework of emotional experiences for new product conception. In: McDonagh, Deana, Hekkert, Paul, van Erp, Jeroen, Gyi, Diane (Eds.), Design and Emotion: The Experience of Everyday Things. Taylor and Francis, Londonhttp://www.tandfonline.com.

Djajadiningrat, J., Gaver, W. and Frens, J., 2000. Interaction relabeling and extreme characters: methods for exploring aesthetic interactions. In: Boyarski, D., Kellogg, W.A. (Eds.). Proceedings of the Third Conference on Designing Interactive Systems: Processes, Practices, Methods, and Techniques (DIS'00). ACM, New York, pp. 66–71.

Fogg, B.J., 2013. Fogg method: 3 steps to changing behavior. http://www.foggmethod.com/.

Gaver, B., Bowers, J., Boucher, A., Gellerson, H., Pennington, S., Schmidt, A., Steed, A., Villars, N., Walker, B., 2004. The drift table: designing for ludic engagement. In: CHI '04, Extended Abstracts on Human Factors in Computing Systems (CHI EA '04). ACM, New York, pp. 885–900.

Gaver, B., Dunne, T., Pacenti, E., 1999. Cultural probes: novel interaction techniques to increase the presence of the elderly in their local communities. Interactions 6, 21–29.
Goffman, E., 1959. The Presentation of Self in Everyday Life. Doubleday, Garden City, N.Y.
Hallnäs, L., Redström, J., 2001. Slow technology: designing for reflection. Pers. Ubiquitous Comput. 5 (3), 201–212.
Hanington, B., 2004. Death and catharsis: re-defining pleasure by design. Proceedings of the International Conference on Design and Emotion. July 2004, Ankara, Turkey, pp. 12–14.
Hanington, B., Martin, M., 2012. Universal Methods of Design: 100 Ways to Research Complex Problems, Develop Innovative Ideas, Aand Design Effective Solutions. Rockport Publishers, Beverly, MA.
Hassenzahl, M., 2008. User experience (UX): towards an experiential perspective on product quality. Proceedings of the 20th International Conference of the Association Francophone d'Interaction Homme-Machine. ACM, Metz, France, pp. 11-15.
Höök, K., n.d. Affective loop. https://www.sics.se/~kia/aff_loop.html.
Human Factors and Ergonomics Society, n.d. https://www.hfes.org.
International Organization for Standardization, 2010. Ergonomics of Human-System Interaction: Part 9241-210.
James, W., 1890. The Principles of Psychology. H. Holt and Company, New York.
Jordan, P.W., 2000. Designing Pleasurable Products: An Introduction to the New Human Factors. Taylor and Francis, London.
Kleine, S.S., Baker, S.M., 2004. An integrative review of material possession attachment. Acad. Market. Sci. Rev. 2004, 1.
Kleine, S.S., Kleine, R.E., Allen, C.T., 1995. How is a possession "me" or "not me"? Characterizing types and an antecedent of material possession attachment. J. Consum. Res. 22 (3), 327–343.
Krippendorff, K., 2005. The Semantic Turn: A New Foundation for Design. Taylor and Francis, Boca Raton, Fla.
Krippendorff, K., Butter, R., 1984. Product semantics: exploring the symbolic qualities of form. Innovation 3 (2), 4–9http://repository.upenn.edu/asc_papers/40.
Kuru, A., Forlizzi, J., 2015. Engaging experience with physical activity tracking products. In: Marcus A. (Ed.), DUXU 2015, Part I, LNCS 9186, pp. 490–501.
Lalande, P., Racine, M., Colby, C., Joyce, A., 2010. Metacycling, An International Collaborative Project Promoting Sustainable Design. International Technology, Education and Development, Valencia, Spain.
Lalande, P., Racine, M., 2006. The Metamorphosis of products: a sustainable design strategy that favours increased attachment. Proceedings of the International Conference on Design and Emotion. Design and Emotion Society, Goteborg, Sweden, pp. 1–17, http://www.meta-morphose.ca.
Lefebvre, R. 2011. Game visionary Jesse Schell talks about the pleasure revolution. http://venturebeat.com/2011/12/05/jesse-schell-talks-about-the-pleasure-revolution/.
Lidwell, W., Holden, K., Butler, J., Elam, K., 2010. Universal Principles of Design: 125 Ways to Enhance Usability, Influence Perception, Increase Appeal, Make Better Design Decisions, and Teach Through Design. Rockport Publishers, Beverly, Mass.
McAdams, D.P., 2001. The psychology of life stories. Rev. Gen. Psychol. 5 (2), 100–122.
McCarthy, J., Wright, P., 2004. Technology as Experience. MIT Press, Cambridge, Mass.
Norman, D.A., 2004. Emotional Design: Why We Love (or Hate) Everyday Things. Basic Books, New York.
Odom, W., Zimmerman, J., Forlizzi, J., 2011. Teenagers and their virtual possessions: Design opportunities and issues. Conference on Human Factors in Computing Systems—Proceedings. pp. 1491–1500.
Odom, W., Zimmerman, J., Forlizzi, J., Choi, H., Meier, S., Park, A., 2012. Investigating the presence, form and behavior of virtual possessions in the context of a teen bedroom. Proceedings of the Conference on Human Factors in Computing Systems. pp. 327–336.

Odom, W., Forlizzi, J., Zimmerman, J., Sellen, A., Banks, R., Regan, T., Selby, M., 2014. Designing for slowness, anticipation and re-visitation: a long term field study of the photobox. Proceedings of the Conference on Human Factors in Computing Systems. pp. 1961–1970.
Picard, R., 1995. Affective computing. M.I.T. Media Laboratory Perceptual Computing Section Technical Report, no. 321.
Picard, R., 1997. Affective Computing. MIT Press, Cambridge, Mass.
Sanders, E.B.-N., 1992. Converging perspectives: product development research for the 1990s. Des. Manag. J. (former Series) 3 (4), 49–54.
Sanders, E.B.-N., 2006. Design research in 2006. Design Research Quarterly 1 (1). September, Design Research Society.
Sanders, E.B.-N., Stappers, P.J., 2012. Convivial Toolbox: Generative Research for the Front End of Design. BIS, Amsterdam.
Scupelli, P., Hanington, B., 2014. An evidence-based design approach for function, usability, emotion, and pleasure in studio redesign. Proceedings of the Design Research Society. June 16–19, Umea, Sweden.
Tajfel, H., Turner, J.C., 1986. The social identity theory of intergroup behaviour. In: Worchel, S., Austin, W.G. (Eds.), Psychology of Intergroup Relations. second ed. Nelson-Hall, Chicago, pp. 7–24.
Thefuntheory.com. (n.d.). Retrieved March 17, 2016, from http://www.thefuntheory.com/
Tiger, L., 1992. The Pursuit of Pleasure. Little, Brown, Boston.
Tiger, L., 2000. The Pursuit of Pleasure. Transaction Publishers, New Brunswick.
User Experience, n.d. From Wikipedia: https://en.wikipedia.org/wiki/User_experience.
van Gorp, T., Adams, E., 2012. Design for Emotion. Elsevier/Morgan Kaufmann, Boston.
Vitruvius, P. (transl. by Morgan M.H.), 1960. De architetura: The Ten Books on Architecture. Courier Dover Publications, New York.
Yerkes, R.M., Dodson, J.D., 1908. The relation of strength of stimulus to rapidity of habit-formation. J. Comp. Neurol. Psychol. 18, 459–482.
Zimmerman, J., 2009. Designing for the self: making products that help people become the person they desire to be. CHI Conference Proceedings, vol. 1. pp. 395–404.

Further reading

Csikszentmihalyi, M., Rochberg-Halton, E., 1981. The Meaning of Things: Domestic Symbols and the Self. Cambridge University Press, Cambridge, England.
International Conference on Affective Human Factors Design, Helander, M.G., Khalid, H.M., Tham, M.P., 2001. Proceedings of the International Conference on Affective Human Factors Design: 27–29 June 2001. ASEAN Academic Press, London.
Jimenez, S., Pohlmeyer, A.E., Desmet, P., Delft University of Technology, 2015. Positive Design: Reference Guide. Delft University of Technology, Delft, Netherlands.
McCoy, M., 1987. Interpreting technology through product form. Ind. Des. 139/140. Japan Industrial Designers' Association, Summer, Tokyo.
Norman, D.A., 1988a. The Design of Everyday Things. Basic Books, New York.
Norman, D.A., 1988b. The Psychology of Everyday Things. Basic Books, New York.
Norman, D.A., 2013. The Design of Everyday Things. Basic Books, New York.
Sanders, E.B.-N., Stappers, P.J., 2004. Generative tools for context mapping: tuning the tools. In: McDonagh, D., Hekkert, P., van Erp, J., Gyi, D. (Eds.), Design and Emotion: The Experience of Everyday Things. Taylor and Francis, London.
Zhang, Y., 2011. Sustainable Design: Strategies for Favouring Attachment Between Users and Objects. Library and Archives Canada, Ottawa.

CHAPTER

7

From Ergonomics to Hedonomics: Trends in Human Factors and Technology—The Role of Hedonomics Revisited

*Tal Oron-Gilad**, *Peter A. Hancock***

*Ben-Gurion University of the Negev, Be'er Sheva, Israel; **University of Central Florida, Orlando, FL, United States

FROM ERGONOMICS TO HEDONOMICS: TRENDS IN HUMAN FACTORS AND TECHNOLOGY

Ergonomics, or human factors (HF/E) has been defined as *the application of scientific information concerning objects, systems, and environment for human use* (International Ergonomics Association, 2016). HF/E is commonly conceived in terms of how companies design work arenas, tasks, interfaces and the like, to maximize the efficiency and quality of their employees' work. However, HF/E comes into everything which involves people and technology; largely featuring the physical and cognitive interactions between people and these respective creations. A newer definition of Ergonomics by the International Ergonomics Association (IEA) is: *Ergonomics (or Human Factors) is the scientific discipline concerned with the understanding of* ***interactions*** *among humans and other elements of a system, and the profession that applies theory, principles, data and methods to design in order to* ***optimize human well-being and overall system performance*** (International Ergonomics Association, 2016). What is notable here is the emphasis change in the newer definition from the application of scientific principles to the need to understand human–system interaction, and also the requirement to satisfy a dual optimization goal concerning human well-being as well as overall system performance. Indeed, the foundation of this had

Emotions and Affect in Human Factors and Human–Computer Interaction. http://dx.doi.org/10.1016/B978-0-12-801851-4.00007-0
Copyright © 2017 Elsevier Inc. All rights reserved.

already been laid some two decades ago by Helander (1997) who specified three important targets for Ergonomics design activity. These were: (1) to improve safety, (2) to improve productivity, as well as (3) improving operator satisfaction. Notably then, the targets for ergonomic design appear to be constantly evolving. Yet, while Helander's observations may now seem outdated to some, it remains an unattainable vision to many others.

To understand such changes that are rapidly occurring in this evolving area of science, it is important, first, to recapitulate at the unprecedented progress and change in focus that has occurred in work and work design over the past century, especially the most recent decade. In 2009, we provided a brief synopsis of the history of Ergonomics (Oron-Gilad and Hancock, 2009) and how the science of and the focus on Hedonomics emerged. In the current chapter we revisit Hedonomics and look to record what progress has occurred, especially in light of: (1) the vast change in use of personal devices, (2) the erasure of separation between work and nonwork ("leisure" time) tasks, and (3) where work is being conducted (at home, in public places, as opposed to designated companies, workplaces, factories, and offices).

THE CURRENT MILLENNIUM

As a conceptual companion to HF/E, Hedonomics is defined as *"that branch of science which facilitates the pleasant or enjoyable aspects of human-technology interaction"* (Hancock et al., 2005). Hedonomics has two major purposes: (1) the promotion of pleasant and enjoyable human–system interaction, and (2) the explicit promotion of well-being through technological augmentation. Thus, Hedonomics is directed primarily at the affective dimension of interactions between people and the environment and artifacts they create. The term Hedonomics is derived from two Greek roots: *hedon(e)*, meaning joy or pleasure and *nomos*, meaning law-like or collective. The moral foundation of Ergonomics is centered upon the effort to prevent pain, injury (mostly repetitive, work-related types), and suffering (and even fatality) predominately in the workplace. Hedonomics is much more directed toward the positive aspects of work interaction. Ergonomics is now acknowledged as the formal branch of science devoted to improvements in the physical and cognitive work environments. In the same realm of human–systems interaction (HIS), Hedonomics is primarily concerned with the promotion of pleasure. The major difference then between Hedonomics and HF\E is that HF\E is often limited in trying to show the value of preventing events that eventually do not happen (Hollnagel, 2014). Hedonomics, being a positive enterprise, is intended to promote the occurrence of pleasurable and satisfying interaction and can actually demonstrate this. Nevertheless, just as Ergonomics is tolled by the need to cooptimize between human well-being and overall system performance, it is quite clear that Hedonomics is also required to define the boundaries and balances between being affective versus productivity driven. This is one critical aspect of Hedonomics that we believe has not been emphasized previously. The main reason for this oversight may be that it was imperative first to emphasize the need for novel design matrices to enhance users' well-being and satisfaction. But now, as the general acknowledgment for this need has become a growing consensus, a more mature and responsible scientific approach must be established. As already noted by Hancock in 2003, hedonic design does not mean utopia, a place where people act and interact with no constraints or demands.

The formal history of a science of Hedonomics is a relatively short one. Various forms of it and related concepts were introduced around the beginning of the current millennium. Some of the pioneering ideas and terms were funology (Blythe et al., 2003), design for pleasure (Jordan, 2000), and affective and pleasurable design (Helander and Khalid, 2005; Khalid, 2004). Perhaps the most common term used now in the human–computer interaction community is user experience (UX) which is defined as *a person's perceptions and responses that result from the use or anticipated use of a product, system or service* (ISO 9241-210). This term is often associated with Norman (Norman et al., 1995) who emphasizes that the shift from the concept of user interface (UI) to user experience (UX) was made to feature a total or holistic perception as to how the user interacts with a system thus seeking even more elegant solutions.

Given that Hedonomics is a hybrid, practical, and interdisciplinary facet of science, the base knowledge is drawn from multiple fields; primarily in the sciences but also beyond in the humanities. Hedonomics draws heavily on disciple-based constructs from positive psychology, emotion, motivation, and affect (Snyder and Lopez, 2002; Seligman and Csikszentmihalyi, 2000). This form of science has indeed developed into an individual and distinct area of study in and of itself. Affective evaluations provide new and different perspectives for human factors engineers changing the focus—from how to evaluate users to how users evaluate. The research on hedonic values and persuasive interfaces was initially perceived as a welcome contrast to safety and productivity foci, which traditionally dominated (Murphy et al., 2003). But it is now evident that the ways that affect effects productivity or goal attainment are imperative to almost any successful design. Furthermore, how hedonic design principles should be implemented, measured, and evaluated remains inconsistent and poses major challenges for the development of the field. Here, we can learn from the increasing use of persuasive technologies and gamification (e.g., Hamari and Koivisto, 2014), that most studies report on positive and partially positive success in using motivational affordances to generate behavioral changes across individuals. But, as noted by those authors, studies with positive findings are more likely to be published than those with negative or null results. This perennial and persistent "File drawer" effect influences much of the behavioral science (e.g., Szalma and Hancock, 2011). Furthermore, long-term assessments are rare, and most studies look only at a short-term performance. Hence, critical information on how task-related affect changes over time, and what elements can be maintained over time is still needed.

Hedonomics, as it develops, will continue to draw on theories of motivation and emotion for advancing the goal of facilitating the development of convivial (Illich, 1973) and pleasurable interactions. Although there is an overabundance of theories on motivation and emotion, general themes have emerged that are relevant for a mature Hedonomic science. Thus, most theories incorporate goals (e.g., Locke and Latham, 1990), often organized into hierarchies, as a central driver of motivation and emotional states. Relevant issues here include how goals are selected, the consequences for meeting or failing to meet those goals, and the rate of progress toward such goal states. Another central theme is that of appraisal, where individuals' motivational and emotional states are determined by their subjective evaluation of events (Lazarus, 1991; Scherer, 1999). Such appraisals are influenced by estimates individuals make regarding the relative likelihood of outcomes (expectancies) and the personal relevance of those outcomes (valences and/or values; e.g., George and Jones, 1997; Lewin, 1936; Vroom, 1964). For Hedonomic interventions to succeed, they must facilitate a genuine fulfillment of these respective needs (Fig. 7.1).

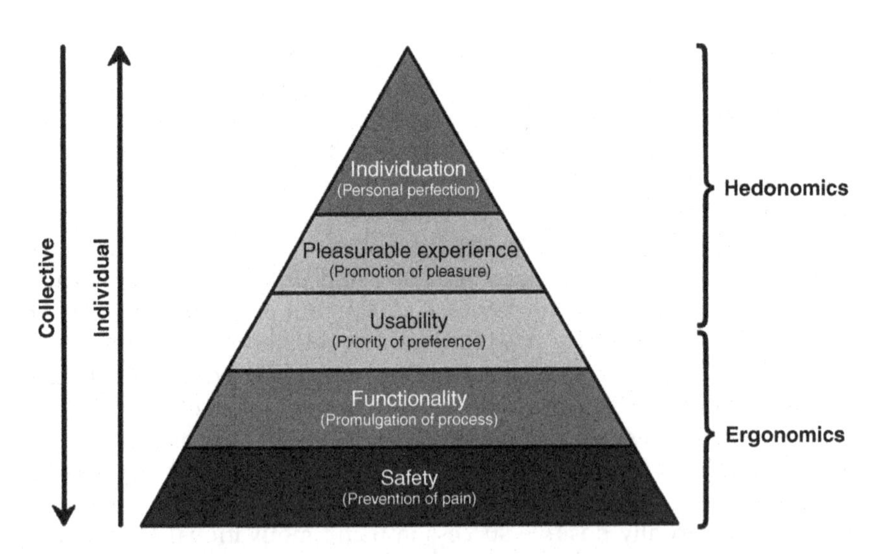

FIGURE 7.1 **Hancock's Hedonomic Pyramid (after Hancock et al., 2005).**

From this perspective, a simple usability approach is unlikely to succeed, since it is based upon collective user preferences rather than whether the technology facilitates the individuals' need for affective aspects, such as competence, autonomy, and relatedness. Indeed, studies on user experience suggest that initial hedonic effects (e.g., design aesthetics) tend to overshadow usability flaws and dominate the formation of first impact user experience evaluations (e.g., Hassenzahl, 2004; Karapanos et al., 2009; Tractinsky et al., 2000). Karapanos et al. (2009) proposed a three-stage model to describe the underlying changes in user perceptions that contribute to their overall experiential judgment. They claim that the temporality of experience consists of three main forces: increasing familiarity, functional dependence, and emotional attachment. These respectively motivate the transition across the three phases of orientation, incorporation, and identification of expectations, thus altering the way individuals experience any system over time.

As in other areas of HF/E, purpose is the central issue in the application of motivation and emotion theory to hedonomic design (Hancock, 1997). How one applies these theoretical models (and which theories are adopted) and which hedonomic goals are selected will depend on both the explicit and implicit purpose for the design. A key issue to be resolved is whether the general purpose for Hedonomics should be the promotion of pleasant and enjoyable human–machine interaction or the goal of promoting well-being through the employment of technical means. This, in turn, depends in part on whether one's approach to the study of well-being is based on pleasure (hedonism) or self-actualization. While the two goals will in many cases be concordant, there will also be instances in which they will be conflict (i.e., increasing the alienation of the individual and exchanging short-term pleasure for long-term well-being). In such cases, which position prevails will ultimately depend on the intention of those who create and operate the technology. To this end, Hassenzahl et al. (2002) examined the effect of usage scenario on correlations among system-related attributes of perceived hedonic quality, perceived pragmatic quality and perceived system appeal. In goal mode, participants were instructed to browse a website to complete a data retrieval task, while in activity mode, participants were instructed to explore the same website freely. Usage scenario affected the correlation between system attributes. Following free interaction, the correlation between perceived system appeal and perceived hedonic quality was significantly higher than between perceived system appeal and perceived pragmatic quality. Yet, when participants were given specific data retrieval tasks, no significant correlational differences were found. While this study demonstrates contextual dependency on system evaluations, more quantitative research findings and theory driven guidelines are needed to conclude how usage patterns affect hedonic experience over time.

FROM HEDONOMIC FULFILLMENT TO COLLECTIVE GOALS

In our 2009 review, we indicated that for hedonomic design to reach its fullest potential in work environments, the organizational/social context must necessarily be considered as part of the system to be designed. We also mentioned that this requires a comprehensive sociotechnical approach incorporated into the design process (Noy et al., 2015). However, it is true to say that most of our review focused on the needs of the individual and that we did not there address how this type of societal component should be incorporated.

A brief survey of the scientific literature over the past decade reveals little use of the term Hedonomics. In fact, the only new study that we found that used the term and attempting to expand upon it was Fiore et al. (2014). Fiore and his coworkers have offered an interesting take on Hedonomics. In contrast with Hancock et al. (2005), as moving the field of HF/E from one of "preventing pain to promoting pleasure" (Hancock et al., 2005). More specifically, Hancock et al. (2005) argued that Ergonomics, with its focus on safety and functionality, is geared more toward a collective goal. In contrast, Hedonomics, which emphasizes pleasure and personal perfection, is a more individualized goal (Hancock et al., 2009). To broaden the use of the term further and also to apply it in the context of sustainability, Fiore and his coworkers proposed a reconceptualized map for Hedonomics consisting of two parallel constructs; related to individualistic and collectivistic designs. That is, this conceptualization of hedonomic design (exemplified by sustainability design, in Fiore et al., 2014) opens this scientific domain to broader impacts, as now more stakeholders need to be involved and considered in finalizing any design. Hence, aside from motivational theories, which focus on the individual, game theories and equilibrium strategies will then come into play and conception.

CONCLUSIONS

Hedonomics, as initially defined, had two main purposes: (1) the promotion of pleasant and enjoyable human–system interaction and (2) the promotion of well-being and using technology to facilitate that well-being. These two goals may sometimes conflict. For example, the question of enjoyment is one that is stable either in space or time. For one to experience epochs of joy or high-levels of enjoyment, they must necessarily be contrasted with intervals that are, by definition, less enjoyable. Thus, hedonomic experience is fundamentally cyclic in nature. In the modern 24/7/365 society, we have tended to forget these inherent cycles, instead adopting a continuous temporal ganzfeld. Worse, we have often taken those very festivals and holidays, which have connoted past pleasures and have sought to expand upon them by extending them in time. In so doing, we have inevitably diluted the experience. Indeed, as one looks at the annual calendar, one can begin to see that commercial concerns have arranged the dilution of experience, such that there is nearly always a "celebration" at hand. In, thus, dissolving these peaks of expectation and assumedly enjoyment in time, the modern propensity for vast greed has dissolved the fundamental nature of the experience. It is not merely the content of an interval, but its length and the degree of anticipation involved in its arrival that dictate the emotional response engendered. If we aim to broaden Hedonomics into collectivism and the search for collective well-being, this can be an example of how society can strive for change.

Another way to look at the issue of the completeness of an experience is to merge work and the leisure. The over simplified twofold categorization of work as something that cannot be pleasurable while leisure is all about pleasure and enjoyment is crude and nonproductive. An alternate way to examine things, people do over a day is to distinguish between routine tasks and challenging tasks (regardless of where they are performed). With this perspective, one accepts that not every task done at home can be pleasurable and, vice versa, tasks done at work can be affectively engaging. Once this type of distinction becomes more acceptable, job designers can develop better job descriptions and combine the two types of tasks to generate higher fulfillment at work. Furthermore, design criteria for routine tasks may be different than for those affectively engaging ones.

Lastly, global issues, such as privacy of the individual versus the safety of many have recently emerged reinforcing again the need to redefine hedonomic design. The role of Ergonomics in job design has often been based on a pull (on-demand) approach rather than on a push (leading) one. Hence, ergonomists were hardly ever leading the progress of technology, but rather setting posthoc procedures while attempting to adapt technology to the task at hand. Now is a good time for Ergonomics (and more so Hedonomics) to take the lead and become proactive and influential in the development of tools and technology (e.g., Oron-Gilad and Hancock, 2004). Industry is not homogenous in its position toward workers' well-being. The wide range of preferences, goals, and priorities among the various industrial environments is a difficult barrier to cross. For certain types of work domains, there is still a long way to go before the basic hierarchical needs of the worker will be maintained; more so the higher hedonic aspirations. Nevertheless, those hedonic principles should lead our aspirations.

References

Blythe, M.A., Overbeeke, K., Monk, A.E., Wright, P.C., 2003. Funology: From Usability to Enjoyment. Kluwer, Boston.

Fiore, S.M., Phillips, E., Sellers, B.C., 2014. A Transdisciplinary Perspective on Hedonomic Sustainability Design. Ergon Des. 22 (2), 22–29.

George, J.M., Jones, G.R., 1997. Experiencing work: values, attitudes, and moods. Hum. Relations 50, 393–416.
Hamari, J., Koivisto, J., 2014. Measuring flow in gamification: Dispositional Flow Scale-2. Comput. Human Behav. 40, 133–143.
Hancock, P.A., 1997. Essays on the Future of Human-Machine Systems. BANTA Information Services Group, Eden Prairie, MN.
Hancock, P.A., Hancock, G.M., Warm, J.S., 2009. Individuation: the N=1 revolution. Theoretical Issues in Ergonomic Science 10 (5), 481–488.
Hancock, P.A., 2003. Individuation: Not Merely Human-Centered but Person-Specific Design. Paper presented at the 47th Annual Meeting of the Human Factors and Ergonomics Society, Denver, Colorado.
Hancock, P.A., Pepe, A.A., Murphy, L.I., 2005. Hedonomics: The Po"W"er of positive and pleasurable ergonomics. Ergon. Des. 13 (1), 8–14.
Hassenzahl, M., 2004. The interplay of beauty, goodness, and usability products. Human Comput. Interact. 19 (4), 319–349.
Hassenzahl, M., Kekez, R., Burmester, M., 2002. The importance of a software's pragmatic quality depends on usage modes. In : Luczak, H., Cakir, A.E., Cakir, G., (Eds.), Proceedings of the 6th international conference on Work With Display Units, WWDU 2002, Ergonomic Institut für Arbeits- und Sozialforschung, Berlin, pp. 275-276.
Helander, M.G., 1997. Forty years of IEA: some reflections on the evolution of ergonomics. Ergonomics 40 (10), 952–961.
Helander, M.G., Khalid, H.M., 2005. Underlying theories of hedonomics for affective and pleasurable design. Proceeding of the Human Factors and Ergonomics Society 49th Annual Meeting, pp. 1691–1695.
Hollnagel, E., 2014. Human factors/ergonomics as a systems discipline? "The human use of human beings" revisited. Appl. Ergon. 45, 40–44.
Illich, I., 1973. Tools for Conviviality. Harper & Row, New York.
International Ergonomics Association, 2016. What is ergonomics. http://www.iea.cc/whats/
Jordan, P.W., 2000. Designing Pleasurable Products. Taylor & Francis, London.
Karapanos, E., Zimmerman, J., Forlizzi, J., Marten, J.B.O.S., 2009. In : Greenberg, S., Hudson, S.E., (Eds.), Proceedings 27th SIGCHI Conference on Human Factors in Computing Systems, CHI-2009, Boston MA, USA; ACM, New York, pp. 729-738.
Khalid, H.M., 2004. Conceptualizing affective human factors design. Theor. Issues Ergon. Sci. 5 (1), 1–3.
Lazarus, R.S., 1991. Emotion and Adaptation. Oxford University Press, Oxford.
Lewin, K, 1936. Principles of Topological Psychology. McGraw-Hill, New York.
Locke, E.A., Latham, G.P., 1990. A Theory of Goal Setting and Task Performance. Prentice-Hall, Englewood Cliffs, NJ.
Murphy, L.L., Stanney, K., Hancock, P.A., 2003. The effect of affect: the hedonomic evaluation of human-computer interaction. Proceedings of the Human Factors and Ergonomics Society 47th Annual Meeting. pp. 764–767.
Norman, D., Miller, J., Henderson, A, 1995. What You See, Some of What's in the Future, And How We Go About Doing It: HI at Apple Computer. Proceedings of CHI 1995. Denver, Colorado, USA.
Noy, Y.I., Hettinger, L.J., Dainoff, M.J., Carayon, P., Leveson, N.G., Robertson, M.M., Courtney, T.K., 2015. Emerging issues in sociotechnical systems thinking and workplace safety. Ergonomics 58 (4), 543–547.
Oron-Gilad, T., Hancock, P.A., 2004. Interacting with Nano systems: is the time right to look for new applications systematically? Proceedings of the 1st Annual Florida Tech Transfer Conference. St. Petersburg, FL.
Oron-Gilad, T., Hancock, P.A., 2009. From ergonomics to hedonomics—trends in human factors and technology. In: Amichai-Hamburger, Yair (Ed.), Technology and the Psychological Well-Being. Cambridge Press, pp. 131–147.

Scherer, K.R., 1999. Appraisal theory. In: Dalgleish, T., Power, M. (Eds.), Handbook of Cognition and Emotion. Wiley, Chichester, pp. 637–663.
Seligman, M.E.P., Csikszentmihalyi, M., 2000. Positive psychology: an introduction. Am. Psychol. 55, 5–14.
Szalma, J.L., Hancock, P.A., 2011. Noise effects on human performance: a meta-analytic synthesis. Psychol. Bull. 137, 682–707.
Snyder, C.R., Lopez, S.J. (Eds.), 2002. Handbook of Positive Psychology. Oxford University Press, Oxford.
Tractinsky, N., Katz, A.S., Ikar, D., 2000. What is beautiful is usable. Interact. Comput. 13, 127–145.
Vroom, V.H., 1964. Work and Motivation. Wiley, New York.

Further Reading

Chapanis, A., 1988. Some generalizations about generalization. Hum. Factors 30 (3), 253–267.
Craik, K.J.W., 1948. Theory of the human operator in control systems: II. Man as an element in a control system. Br. J. Psychol. 38, 142–148.
Desmet, P.M.A., Karlsson, M.A., van Erp, J., 2008. Design and Emotion Moves. Cambridge Scholars Publishing, Cambridge.
Endsley, M.R., 1995. Toward a theory of situation awareness in dynamic systems. Hum. Factors 37 (1), 32–64.
Fee, E., Brown, T.M., 2001. Preemptive bio preparedness: can we learn anything from history? Am. J. Public Health 91, 721–726.
Flach, J.M., Dominguez, C.O., 1995. Use-centered design—integrating the user, instrument, and goal. Ergon. Des. 3 (3), 19–24.
Frøkjær, E., Hertzum, M., Hornbæk, K., 2000. Measuring usability: are effectiveness, efficiency, and satisfaction really correlated? Proceedings of the ACM CHI 2000 Conference on Human Factors in Computing Systems. April 1–6, 2000, The Hague, The Netherlands, pp. 345–352.
Gibson, J.J., 1966. The Senses Considered as Perceptual Systems. Houghton Mifflin, Boston.
Gibson, J.J., 1979. The Ecological Approach to Visual Perception. Houghton Mifflin, Boston.
Gilbreth, L.M., 1956. Management in the Home. Dodd/Mead, New York.
Graham, L., 1997. Beyond manipulation: Lillian Gilbreth's industrial psychology and the governmentality of women consumers. Sociol. Quart. 38 (4), 539–565.
Hancock, P.A., Flach, J.M., Caird, J., Vicente, K., 1995. Local Applications of the Ecological Approach to Human-Machine Systems. Erlbaum, Hillsdale, NJ.
Hancock, P.A., Szalma, J.L., 2008. Performance Under Stress. Ashgate, Hampshire, UK.
Jastrzebowski. W., 1857. An Outline of Ergonomics or the Science of Work Based Upon the Truths Drawn From the Science of Nature. Originally published in Nature and Industry. No. 29 and No. 30 (1857); reprinted by Central Institute for Labour Protection, Warsaw, 1997.
Kahneman, D., 2002. Maps of Bounded Rationality: A Perspective on Intuitive Judgment and Choice. Nobel Prize Lecture.
Khalid, H.M., Helander, M.G., 2006. Customer emotional needs in product design. Concurr. Eng. 14 (3), 197–206.
Kira, A, 1966. The Bathroom. Center for Housing and Environmental Studies.
Lavie, T., Oron-Gilad, T., Meyer, J., 2008, in prep. Usability and aesthetics of map displays.
Lindgaard, G., 2007. Aesthetics, visual appeal, usability, and user satisfaction: what do the user's eyes tell the user's brain? AJETS 5, 1–16.
Maslow, A., 1970. Motivation and Personality, rev. ed. Harper & Row, New York.
Meister, D., 1999. The History of Human Factors and Ergonomics. Lawrence Erlbaum, Mahwah, NJ.

Miller, C., 2002. Definitions and dimensions of etiquette. In: Miller, C. (Ed.), Etiquette for Human-Computer Work: Papers from the 2002 Fall Symposium. Technical Report FS-02-02, American Association for Artificial Intelligence, Menlo Park, California, pp. 2–8.

Moray, N., 1993. Technosophy and humane factors: a personal view. Ergon. Des. 1 (4), 33–39.

Mynatt, E.D., Melenhorst, A.S., Fisk, A.D., Rogers, W.A., 2004. Aware technologies for aging in place: understanding user needs and attitudes. IEEE Trans. Pervasive Comput. 3 (2), 36–41.

Nielsen, J., 1993. Usability Engineering. Academic Press, Boston.

Norman, D.A., 2003. Emotional Design: Why We Love (or Hate) Everyday Things. Basic Books, New York.

Parasuraman, R., Miller, C., 2004. Trust and etiquette in high-criticality automated systems. In: Miller, (Ed.), "Human-Computer Etiquette," a Special Section of Communications of the ACM, C. pp. 51–55.

Powers, W.T., 1973. Behavior: The Control of Perception. Aldine, Chicago.

Preece, J., Rogers, Y., Sharp, H., 2002. Interaction Design: Beyond Human Computer Interaction. Wiley, New York.

Seva, R.R., Been-Lirn, H., Helander, M.G., 2007. The marketing implications of affective product design. Appl. Ergon. 38 (1), 723–731.

Smith, K., Hancock, P.A., 1995. The risk space representation of commercial airspace. Paper presented at the Eighth International Conference on Aviation Psychology, Columbus, OH.

Taylor, F.W., 1911. The Principles of Scientific Management. Harper Bros., New York, Retrieved from ©Paul Halsall, July 1998.

Vicente, K., 2004. The Human Factor: Revolutionizing the Way People Live With Technology. Routledge, New York.

CHAPTER

8

An Approach Through Kansei Science

Toshimasa Yamanaka
University of Tsukuba, Tsukuba, Japan

Kansei Engineering is one of the representative approaches and methodologies in affective design. The present chapter first describes the evolution of the term, its definition, and the historical backdrop for the emergence of Kansei Engineering. Then, it outlines its function and communication. After delineating its approaches in design, the chapter introduces an empirical study that has used both a traditional questionnaire method and a novel neuroimaging method.

FROM KANSEI TO KANSEI ENGINEERING

In everyday life in Japan, the word "kansei" is often reflected in statements, such as "his/her kansei is rich," "kansei is sharp," "stimulate your kansei," "polish your kansei," and so on. Kansei has been used as an everyday Japanese linguistic term since time immemorial. Subsequently, around the Edo and Meiji eras, Nishi Amane, the prominent Japanese philosopher, adopted the academic term "sensibility" from the Western philosophy, in order to promote its conceptual understanding. Moreover, as can been seen from the study and translation work of Kant philosophy by Amano Teiyu, *kansei* was used as a Japanese word corresponding to German "Sinnlichkeit" representing an intuitive and logical understanding of reason.

In Western Europe, Aristotle has used the ancient Greek word "aesthesis" to denote kansei and sensation, which are interpreted in Japanese as ethos (translated as virtuous habit or unknown experience of "good" nature). Both terms bear a closely approximate meaning. In addition, the German philosopher, Baumgarten, in his work *Aesthetica* (1750), gave his

Emotions and Affect in Human Factors and Human–Computer Interaction. http://dx.doi.org/10.1016/B978-0-12-801851-4.00008-2
Copyright © 2017 Elsevier Inc. All rights reserved.

definition as "Aesthetica est scientia cognitions senstivae". Aesthetica has been widely interpreted as aesthetics, defined as "cognition through an integrated sensation," whereas in Japanese, aesthetics is translated as "the study of sensible recognition." Faivre records that since the mid-14th century, French "sensibilite" has been used as a word related to human feeling that involves ease of human emotion based on ethical impression of what is considered good and true. This way, kansei in the modern era of Japan could be considered as a word to indicate a system of thought, as in the Western philosophical concept. However, between 1970 and 1980, Japanese industry has come to adopt the use of kansei as a new concept as it enters a mature phase. In 1986, during the lecture delivered by Mazda automobile president, Kenichi Yamamoto, at the University of Michigan, kansei was first described in connection with method of manufacturing. At that time, Nagamachi Mitsuo who had been deeply involved in Mazda's research, introduced and structured Kansei Engineering as a product development process that combines psychological evaluation and statistical analysis, which has now become known worldwide and reckoned as an "original Japanese development approach in order to quantify the value of the automobile in terms of human feeling." Around the same period, Dentsu Public Relations director, Fujioka Wakao, in his book titled "*Goodbye, Mass—How Read the Kansei Era?*" (1984), proposed a category of fragmentation for the masses. He argued further that "kansei need" is a feature of the category, and thus, kansei should be a point of focus in marketing.

According to Nagamachi, Kansei Engineering is defined as a "technology that combines kansei into design specifications in order to realize the whole products that match consumers' need and desire." The instance of product development, which applied Kansei Engineering includes Milbon cosmetics manufacturer bottle design and curtain interiors designs. The multifarious relationship between subjects' impression and the physical quantity of stimuli was revealed through analysis and then applied to the actual product development (the method of statistical analysis to investigate the influence of the design elements in this case showed that the results have been affected by a number of factors). Notably, the feature of Japanese products in this age tends to emphasize compactness and lightness (light, thin, short, shape of small products). In addition to the techniques of product sales, the development of new functions and new taste has come to be characteristics of the product development in Japan since the 1970s, and the need to link the value of the products and individual consumers has come to be recognized. Based on the recognition that a practical understanding of personal values in marketing is essential in the kansei era, kansei can be captured as an individual value in order to expand the potential for individual consumption and therefrom the possibility of incorporating kansei to product value has come to be seen. In the midst of Japan's bubble economy of the 1980s, there was such consideration for the "comfortable" feeling of people toward the product, and thus, Kansei Engineering emerged and began to be used as a method to know the structure that is hard to measure only at first glance.

Subsequently, there was a progression of movement that scientifically attempts to capture kansei; the *Journal of the Society of Instrument and Control Engineers* featured "Comfort and Kansei" (1991), which was edited in the *Special Issue* "Engineering of the Mind and Sensibility" (1994). In 1993, the Kansei Engineering subcommittee was installed from the Science Council of Japan, while scientific research on kansei took off in earnest. After this period, there was a significant increase of scientific publications about kansei in Japan. In addition, kansei became a point of focus to evaluate the function of the human mind. In 1997, "kansei evaluation structure model building" started as a University of Tsukuba special research project as a fusion of robotics research and kansei research, fostering information network through the multifaceted measurements and analyses of appreciation of artworks in order to capture the structure of kansei. In 1998, the Japan Society of Kansei Engineering was founded, which settled the academic basis of Kansei Engineering research. At the same time in Europe, there appeared a development of designs, which incorporate emotional expression as if counteracting the principle of functional designs. In 1999, the Design and Emotion Society was established as a research community, which held its first meeting in 2000. At this meeting, Harada introduced Japan's model building project of kansei evaluation structure and the exchanges of European and Japanese sensibility research began in earnest. In industry, Nissan Motor Corporation introduced the concept of added value defined as Perceived Quality (kansei quality) around the year 2000.

"Perceived Quality" may include the visual unity of form and goodness of surface treatment. It has been described as the quality, which consumers may not easily be aware of as an evaluation factor and which they can even evaluate the product without, but if added, can increase the attractive quality of product. This quality is not included as a "functional quality" based on the functional needs of the product.

From this time, the importance of the design of kansei quality for product development started to be projected. Although the methodology is not easily formulated, respective companies have continued their efforts. In 2003, the Japan Society for the Science of Design and the Japan Society of Kansei Engineering coorganized the 6th Asia Design International Conference under a main theme, "Integration of Knowledge, Kansei and Industrial Power," where there were a considerable number of attendees from more than 20 countries and 450 research papers were published. Harada in this declarative statement presents that "Kansei, which is a term peculiar to Japan, means the high order function of the brain as a source of inspiration, intuition, pleasure/displeasure, taste, curiosity, aesthetics, and creation." In this international conference of design research, it was declared that kansei, emotion, and affect were included about 10% in kansei studies, while other contents were about methodology, evaluation, interaction, business, etc. Thereafter, the Ministry of Economy, Trade, and Industry made a national declaration that 2007 is a year of Kansei Value Creation. From 2008, it was supposed that the industrial applications of the kansei research were more advanced. Then, several attempts to integrate the case and knowledge about the importance of transmitting kansei quality have been created.

WHAT IS KANSEI?

Kansei by the Kojien (Japanese) dictionary is described as follows:

1. The sensitivity of sensory organs which result in perception based on the response to external stimuli.
2. It is evoked by the senses and governed by experience content. Also, it includes feelings, impulses, and desires associated with sensation.
3. Innate desires which can be controlled by reason and will.
4. It is described as an intuitive recognition through sensory perception of a material (i.e., understanding recognition).
5. It is also believed that it sometimes occurs in the case of conflict with desirable thinking.
6. It refers to "Sinnlichkeit" used in the philosophy initially introduced by Kant. Sinnlichkeit is intuitive recognition related to space and time, and human fundamental recognition method underlying "Verstand" (which means "understanding" in Japanese).

According to the definition, kansei is a sensory experience prior to thinking and judgment.

悟 REPRESENTS 心 MIND+ 吾 FIVE MOUTHS

It is a characteristic of Chinese characters. If interpreted on the basis of the expansion of meaning by the set of character, 口 is mouth (to tell) that represents the description possibility, and in particular, the letter representing five mouths emphasizes the importance of description. Since it is what is meant to represent the concept of human, it represents the knowledge that you understand as possible explanatory information. Therefore, the nature of the mind for understanding recognition is to be able to describe a lot of knowledge, especially ethical, social normative, and integrated rational recognition that contains the nature involved in the descriptive aspects of the "function of the mind." Sinnlichkeit is one of the meanings of kansei and is considered as the basis for understanding recognition. Although it is a function of the mind to an external stimulus, it is a mind process preceding awareness as well. It involves knowledge that is prescriptive rather than descriptive, with a personal and subjective way of perception and that is mentioned to be an individual mental process with a certain universal feature 咸 **represents** positive intuition.

The Chinese character "咸" that is included in *kansei*, represents "an intuition occurs before one can be conscious of it." Often, it can be described as an example of the feeling when a male and female are mutually aware of each other, and precisely, it is a function of the mind, such as to determine its orientation before making a valuable judgment. 咸, in its construction of character, as inscribed on oracle bone script, has an interpretation composed of characters consisting of "big ax" and "mouth," such as representing voice emanating momentarily when viewing a weapon. On the one hand, in modern Chinese, the meanings embedded in these constructions of characters have already disappeared, but remaining meanings include "all" and "salty." This feeling that you can have with 咸 exists in the mind, which is responsive to all external stimuli to feel a kind of saltiness just as it is the most important element for adjusting the seasoning.

Taking these together, *kansei* is a property that is shaped by all of the knowledge that is in the mind of a person (including the experience and the unconscious emotional part of the mind) and a process by which information is made from what is held in the mind. Related to the study of various psychological indicators of the elements and functions of kansei process, Kansei Science has its application on how to interpret the mind related to the mechanism of creativity and sensitivity in design while taking advantage of the physiopsychological evaluation, such as behavior and brain activity. It is the science of the "mind" that is closest to design and manufacturing of objects.

咸 (=positive intuition) + (mind+life) = **kansei** (感性)= sensibility + Sinnlichkeit

setting direction for aware and understanding

(mind + five + description) + (mind + life)= **Gosei** (悟性) = Verstand= understand

describing the perception

decision

KANSEI FUNCTION

Since the 1980s, it is believed that kansei has started to contribute to the advancement of manufacturing more practically. For example, when designers want to design a good product, often there is such a feeling that "already I was aware of that goodness." Even though it is not possible to confirm a good design before it is expressed by means of sketching, when it reaches to that stage, there is such a feeling that I knew this goodness before but at the time I did not notice because I could not remember "myself." Before this epiphany, designers are ready to perform a value judgment about the goodness, and when they notice that it is a "good idea," they could be as "confident" as if they previously knew about that. Although this process might be a reconstruction of knowledge, such a feeling is not limited to the design, or would be universally experienced in any creative activity. "To actualize good design" as "inspiration" or to get an "epiphany," it can be said to be an act of "rebuilding what you already knew." For instance, it can be seen in Harada's previous description of kansei as a higher function of the brain, and the word "inspiration" is mentioned. This is an important word indicating that "we know as a result of the function of kansei."

In general, learning and experience that we are aware of must have already been accumulated as an implicit knowledge before taking the form that can be transmitted to others. Kansei function is that implicit knowledge, which is difficult to take out as knowledge, which can be considered as the connecting process of the expressions of knowledge-based intuitive logic of action, form or completion. Usually, knowledge refers to the storage and information in the descriptive state. However, in order to "use descriptive knowledge," one must remember how to. It is a state that knowledge is being used, and not the state where you are using the knowledge. Meanwhile, physicist, Michael Polanyi, proposed "implicit knowledge" as behaviors and also knowledge integrated with nonconscious activities, such as emotions and it is the knowledge to get an epiphany rather than knowledge to explain something. It is knowledge that can be used anytime soon if there is a need but it is difficult to describe the knowledge itself. Implicit knowledge works in an unconscious manner, and whenever we notice something, there is a feeling that "I've got it!" In other words, when kansei functions, an answer would have already been considered ready. Meanwhile, "the goodness of design" is when you have the power transferred to the user and consumers, and this is convincingly said to be high.

Considering this from the function of the user's kansei, it is that the "awareness" between the user and designer are shared. However, one can wonder whether the conditions for kansei function can be shared.

SHARING AWARENESS = KANSEI COMMUNICATION

At one time, I made an observation at a nearby park in a highly internationalized city. Gathered in this place are families from places whose native languages varied, such as Britain, the United States, France, Italy, China, South Korea, and so on, while they had small kids with them. There, I found an interesting feature in the communication between parents and their children. The frequently used words in the parent and child conversation included "mom," "come," and "look." In addition to the three words, which accounted for most frequently used words by the kids when playing with their parents at the park, were "yes" and "no" responses, altogether making up five words. In communication, depending on the media, the spoken words and gestures we often use to mutually convey nuances and emotions are not very descriptive. Communication is not achieved only through grammar and predicate, but also we are able to transmit a message first by having a nondescriptive common ground, such as the customs and say once. In the parent and child conversations, children, prior to the stage they can master gestures rich in nuance and spoken words like *mom, come, look*, by looking at each other, can communicate without having to use words to show their intention. Furthermore, for the "yes" and "no" description or representation shown by the parent to express their intention, it will become possible to understand each other's subjective value as "something we shared." Such things as the "lung breath," even if a limited number of words are used without almost being written, are possible to transfer a high level of common communication basis. On the other hand, if the common knowledge and the environment for mutual understanding are lacking, and the description of expression cannot be improved, then transfer of knowledge might not be possible.

In descriptive knowledge (formal knowledge) and nondescriptive knowledge and expressions (implicit knowledge), there is a concept of "ba" (place) expressing the combination of communication and expression formed on knowledge basis. In this case, "ba" represents the implicit information in a nondescriptive manner, and more by writing understanding sentences. Furthermore, people want an environment in which they can share knowledge with each other through steps that save their own experiences. This "ba" is a mechanism that brings managers and employees talking together while making things and this is essential to the Japanese style consensus model, "understand each other." For example, we can also consider the transfer of academic knowledge. Although academic information exchange is usually considered as doing knowledge transfer using only highly descriptive information, a knowledge basis with each of the members in the same academic area can be large, in which case they could make a wrong interpretation for the representation of different terms and methods.

The kansei functions when knowledge is transmitted and when receiving interpretation is significantly different and can be considered as a factor. This is said to be the nature of the enigmatic decision making in Japanese business environment, which explains the concept of "large room," in that a lot of people working in a place at the same time communicate each other in the same way despite having the different knowledge basis. In this situation, people can be very subjective while they selectively react to the information according to the state of their knowledge, experience and awareness. From the same kind of information, people can be more aware of a specific part, as in the example of the "cocktail party effect," where people sensitively react only to their name in a noisy party. Probably, there is knowledge to react to a certain stimulus, which is implicitly embedded in experience. For example, in the cocktail party effect, the observed influential factors include sound direction, sound volume, knowledge of the sound, the subject's voice recognition, the point of difference in the native language of the speaker where the experiment was carried out, and whether the speaker's native language matches with that of the subject. In other words, when a recipient is not aware of the usual information, the influence of the language that was acquired while growing up helps for an intuitive understanding.

The commonality of experience that is rooted in cultural background will not only create awareness that is similar but also produce an intuitive understanding and inspiration. For the implicitness of knowledge accumulated with experience, not only direct experience is important, but also the full use of the imagination is important. The knowledge gained by reading text and the one obtained by looking at the manga images could be classified as different knowledge while different imagination could be involved when we experience information through these sources. The resulting experience when we say "I understood" while reading a book and when we say "I understood" while watching a manga, can differ in how they are used, and such a difference reflects individual characteristics, such as personality traits and experience.

KANSEI IN DESIGN AND BRAIN SCIENCE

Design is a business saddled with the responsibility for providing a new experience for the user, but questions still remain on what conditions are necessary to create a new experience to make them a "feel good product." For this purpose, it is necessary to know the principle of kansei, the phenomenon of kansei function, the mechanism of the brain function, and physiological and psychological indicators. Here, between the designers (who create the shape and product interaction) and the users, a similar relationship would be established just as between the kids (who say, "mom, come, look") and their parents. As shown before, it is a direct relationship, whereas the verbal description does not necessarily fully convey the meaning as in the case of kids. Likewise, for the users to properly interpret and understand products and advertisements, the experience that "I understood" clearly has an important value.

Even when there is an incomplete description associated with the forms and functions of the intervening stimuli, users can perform an evaluation of the product by implicit understanding. When the "inspiration" in the users is evoked toward the stimuli or the products, it can be said that kansei is evoked by the stimuli. When good inspiration is generated, the users will say something like "oh, I like this kind of stuff" or "this stuff makes me feel like that..." as they try to understand it in the light of their knowledge and experience. This kind of description which relates the object (or the product) to what we already know is also referred to as a "metaphor." Designers' work is to explore and share these well-known experiences with users by adding newness to forms and methods of an operation of a stimulus. However, it is challenging that it is not guaranteed whether/how well the designers' intention can be conveyed to the user. Nevertheless, the designers are required to create value in the product that evokes the metaphor held by the users. For this purpose, the designers themselves are supposed to become one of the users by predicting a step ahead of the user, by sharing an environment in which future users will experience and by creating appealing products, which depend on the work of *kansei*. Considerations for kansei supporting systems and product development methods based on kansei evaluation would be an important mechanism to effectively exert the ability of designers.

Regarding an individual name in the cocktail party effect, there is a degree of changes in the brain activity when a familiar word is heard. With the recent development in brain science and the scanning of brain and nerve activities in details, the extent of the relationship of the characteristics of the sound and degree of inspiration can be evaluated.

To know the brain activity, EEG, fMRI, fNIRS (optical topography) and various other methods have been developed. As it is not possible to explain all the human brain activities using experimental animals, it is essential to research from various angles to clarify the relationship of the function of mind. Therefore, in Kansei Science, it is considerably important to know "what kind of kansei is evoked" from the analysis of words and observations of unconscious behavior. To this end, evaluations to investigate the level of perception can be done by questionnaires and subjective evaluation criteria (e.g., semantic differential method, etc.), to comprehensively study the mechanism of kansei at the time of design. Results are combined with the outcomes coming from the advanced technology in brain science, which scan user's brain when they interact with the product. Thus, Kansei Science, which aims at the fusion of brain and behavioral sciences, is also expected to play an important role as a basic science in practical activities, such as design and planning.

A SAMPLE STUDY USING SD EVALUATION AND BRAIN IMAGING

This is an example of how we can conduct a Kansei Engineering research study by combining a traditional subjective questionnaire and a novel brain imaging technique. Our research objective for this study was to identify differences in the subconscious evaluation of kansei between experienced designers and nondesigners. We hypothesized that there would be an evaluation gap between design-educated (DE) and nondesign-educated (NDE) groups in the kansei process. For our research, DE groups were defined as individuals with, at least, 2 years of university-level design education and NDE group was comprised of people with no formal design education.

As an approach, we used a semantic differential method (SD method) to filter thinking characteristics through subjective evaluation. At the same time, we measured brain function while evaluating the design products. For the brain imaging measurement, we used NIRS. This technology measures the flow volume of oxyhemoglobin in the prefrontal cortex of brain that represents the activity of brain. Then after, we compared the result of measurement of the DE and NDE groups (Fig. 8.1).

Preexperiment

In the preexperiment, we started to look for specific differences in kansei processes between DE and NDE groups in subjective expression. We decided to use a chair as the object of study because of its popularity, variation of design, and simplicity in function. Then, we selected 50 familiar chairs and printed out those pictures on the same sized cards. We recruited five DE and four NDE subjects from University of Tsukuba and prepared the following questionnaire:

(Q1) Which chair(s) are you familiar with?
(Q2) Which chair(s) do you think is simple?
(Q3) How did you feel the selected chair as simple?
(Q4) Which chair do you like?
(Q5) What is the reason you liked the chair?

Kang and Yamanaka (2005) showed the interesting relationship between the product description and evaluation. According to them, designers tend to find more product categories and get higher concentration of the favored product into the category. This means that people with design training tend to have stronger abilities to describe and evaluate the products. In the present study, (Q1) represents the descriptive knowledge, (Q3 and Q5) represent evaluation of the product. To analyze the relationship between (Q1), (Q3) and (Q5), we used the number of chairs selected for both (Q1) and (Q3) as the matching ratio of knowing and simple, (Q1 and Q5) as knowing and liking, and (Q3 and Q5) as simple and liking. Then, we compared the matching ratio between DE and NDE. Even with the low number of participants, we found that there was a significant difference between groups. In DE, we found the tendency of higher relationship of knowing and simple ratio ($P = 0.002$) and simple and liking ratio ($P = 0.03$) compare to NDE. With this result, we found that we can assume that the DE group is more familiar with simpler chair designs and also

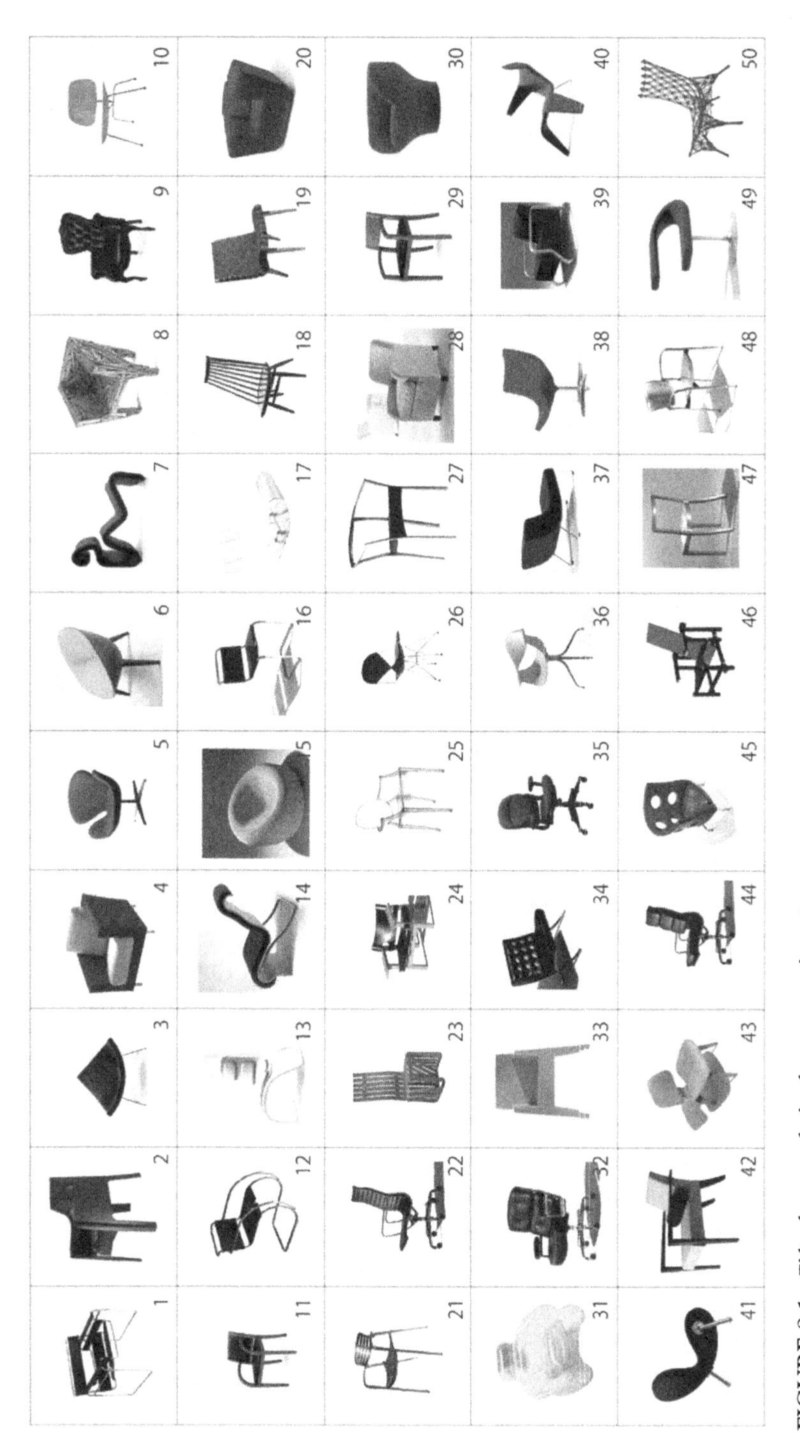

FIGURE 8.1 **Fifty famous chairs for preexperiment.**

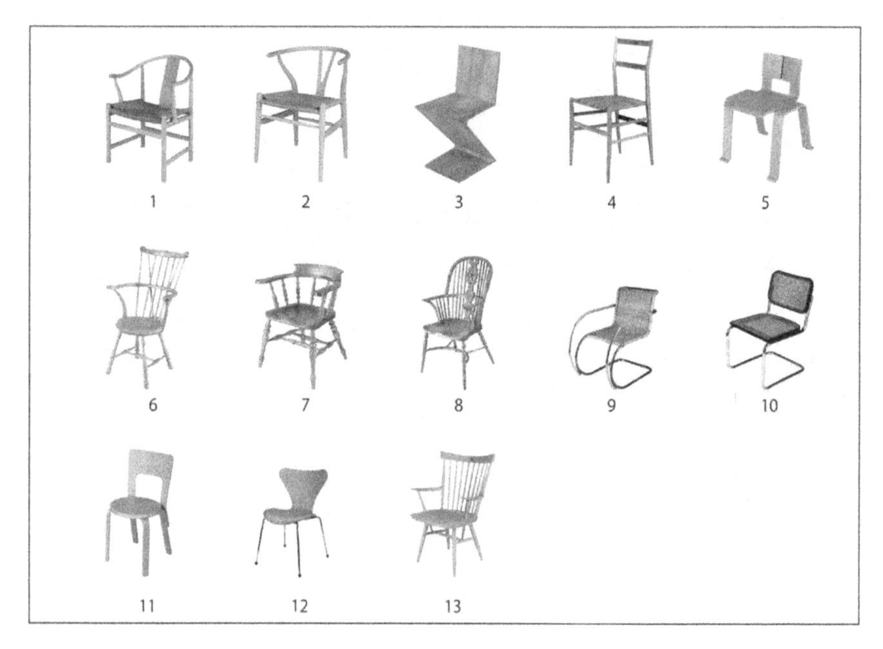

FIGURE 8.2 **Selected chair images for the experiment**

likes simpler designs more. Based on this result, we investigated the existence of such a difference in brain function as well.

Experiment

Then, we conducted an advanced experiment. For the sample of the evaluation, we chose a design collection of chairs published by Kyusyu Sangyo University, Japan. We selected appropriate chairs with similar color, material but different structure. To exclude any effect of colors and too much ready-made images, we focused on natural colored (beige, light amber) and not-super-modern ones. Then, we bought 13 chairs from the collection (Fig. 8.2). Usually, participants' reaction can be different in the period of experiment. Thus, we used two more chairs for the training purpose at each set of evaluation. From the limitation of experimental situation, we used digitized image of chairs with all white background. Images were shown on the 27-in. white frame computer monitor in portrait orientation.

Methods and Procedure

For the subjective evaluation, we applied the SD evaluation method. For the sets of adjectives, we prepared the eight sets of adjectives by analyzing participants' comments in preexperiment (Q2 and Q4) that are supposed to be used to evaluate chair design:

1. simple in form/complex in form;
2. monolithic/component;
3. liner shape/curved shape;
4. simple/complexed;
5. simple and clean/decorative;
6. unstable in structure/stable in structure;
7. eccentric/normal;
8. reasonable structure/uncertain structure.

Then, we set up the experimental procedure as follows:

1. watch the target chair;
2. evaluate score for the specific pair of adjectives;
3. rest.

This process was repeated 120 times (number of chairs × SD combinations) per participant, including the two training periods. During the experiment, we measured participants' prefrontal cortex, which is regarded as the seat of executive control in the brain, using a 24-channel system of near-infrared spectroscopy (NIRS, ETG-4000, Hitachi-Medico Corp.). The NIRS system measures amount of oxyhemoglobin in the prefrontal cortex, which represents the activity of brain since the neuron requires oxygen while processing. We compared DE and NDE groups during the process of watching the chair (evaluating) and answering (thinking and expressing).

Analyses and Results

Results consist of two parts. First, we conducted a two-way ANOVA to determine if there were significant differences in responses according to the condition or chair type. As seen in Table 8.1, chair type was significant for all SD scales. This means that the participants' evaluation was affected by the difference of chair design. This result is quite understandable. However, in relation to the DE and NDE factor, only the [stability in structure] showed a significant difference while there was no significant difference in interaction. This means that the evaluation of the stability in structure differs between people who have formal design education and those who do not, and that the difference is not caused by a specific chair but a generally applicable difference.

In addition, we applied multiple-discriminant analysis. The result showed the first canonical correlation coefficient showed 0.687 and [unstable/stable in structure = stability in structure] showed factor loading at 3.737. Comparing to the other independent variables, this is a very large factor loading (Table 8.2).

TABLE 8.1 Probability (P) Value of Two Way ANOVA

	Factor1	Factor2	Interaction
Set of SD	Chair	DE and NDE	Chair × DE and NDE
1. Simple in form/complex in form	<0.001	0.332	0.838
2. Monolithic/component	<0.001	0.51	0.14
3. Liner shape/curved shape	<0.001	0.913	0.351
4. Simple/complexed	<0.001	0.165	0.857
5. Simple and clean/decorative	<0.001	0.901	0.221
6. Unstable in structure/stable in structure	<0.001	<0.001	0.798
7. Eccentric/normal	<0.001	0.32	0.457
8. Reasonable structure/uncertain structure	<0.001	0.178	0.937

DE, Design educated; NDE, nondesign educated; SD, semantic differential.

TABLE 8.2 Factor Loading by Multiple-Discriminant Analysis

Set of SD	Firstst canonical component
Cannonical correlation	0.687
1. Simple in form/complex in form	1.27
2. Monolithic/component	−0.076
3. Liner shape/curved shape	1.462
4. Simple/complexed	−0.715
5. Simple and clean/decorative	0.581
6. Unstable in structure/stable in structure	3.737
7. Eccentric/normal	1.197
8. Reasonable structure/uncertain structure	1.324

This means that the sense of [stability in structure] is the best single factor to account for the difference between DE and NDE students.

Then, we checked the neurophysiological difference in the evaluation of the chairs while watching the chair images. In this analysis, we used the analysis software version 1.63 provided by Hitachi medico Corp., which uses t-test by subtraction method. We found that only the case of answering to the [stability of structure] showed the different result between DE and NDE groups. Figure 8.3 (DE side) shows that there was a significant difference between nonwatching and watching while preparing to answer [stability of structure],

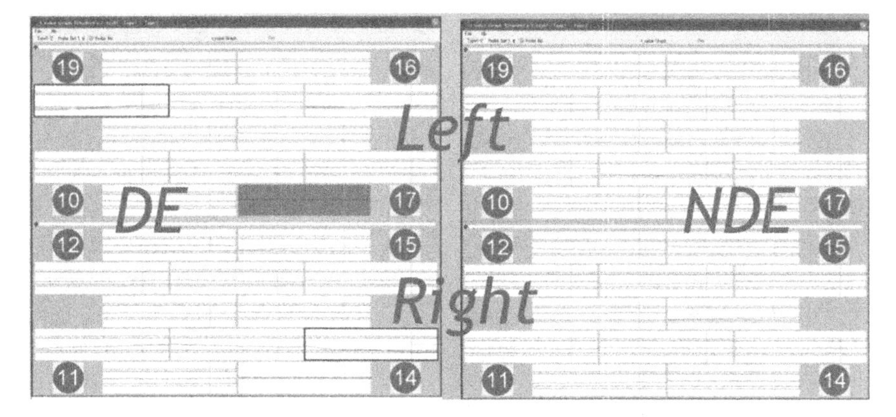

FIGURE 8.3 **Brain function difference between nonwatching and watching chair image while answering [stability of structure] in design-educated (DE) group.** Two channels showed significantly active while watching the chair images.

where the participant is evaluating the chair in their mind. On the other hand Fig. 8.3 (NDE side) shows the same comparison in the NDE group, which shows no significant difference in brain function between times.

In addition to this comparison, we checked the difference in subjective evaluation between DE and NDE groups. Since it is difficult to determine the nonevaluation period in experiment, we applied a direct comparison between DE and NDE with using averaged data in the evaluation situation. In this case, we could see the significant difference only on the left frontal cortex higher in the DE group (Figs. 8.4 (DE side); Fig. 8.5).

DISCUSSION AND CONCLUSIONS

In our hypotheses, we expected that there could be significant differences in the evaluation on simplicity, but this was not supported. We could confirm both DE and NDE groups were sensitive to the difference of chairs in evaluation. However, only the evaluation of [stability of structure] showed the difference in both SD evaluation and brain function between DE and NDE groups. This means that the effect of professional education might develop the different sensitivity to design evaluation and the professional analytical skill, such as stability expectation as a fundamental ability of developing potentially new design achievement. This result shows one aspect of how professional designers' skill sets would be. At the same time, we hope that we could show readers an example of contemporary approach in Kansei Engineering, using a combination of the SD evaluation method and the brain function imaging.

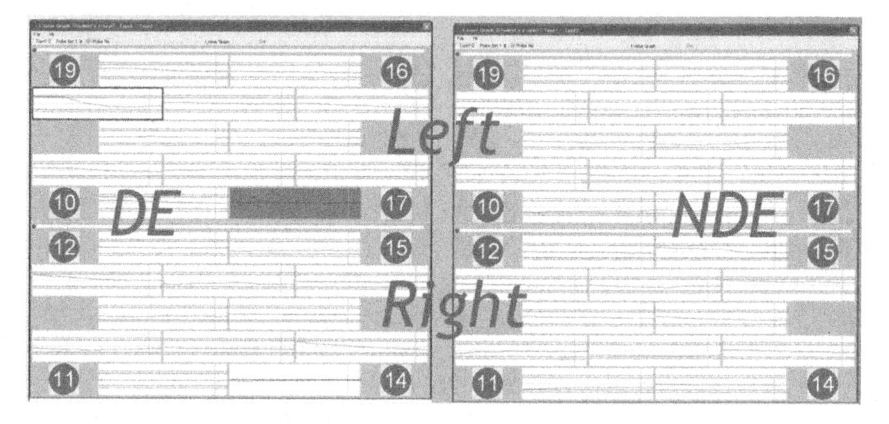

FIGURE 8.4 Brain function difference between DE and NDE while subjective evaluation while answering [stability of structure]. Two channels in right frontal cortex functioned significantly different *(higher in NDE) while thinking the answer.*

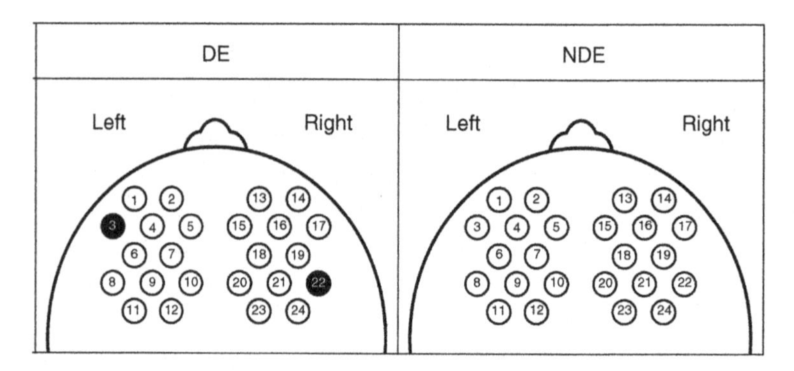

FIGURE 8.5 Differently functioning brain part in 5 s of watching the chair images in comparison between DE and NDE groups.

Reference

Kang, N., Yamanaka, T., 2005. Research on product and reference by Kansei information. Bull. Int. Des. Congr. od Soc. Des. Res. (Bul of IASDR), IASDR 2005.

Further Reading

A study based on optical topography measurement. Shinsyu Univ. J. Educ. Res. Pract. 1, 13–20.

The role of Kansei engineering in modern society, 2005.

Atsushi, K., 2009. Passenger car development. Perceived quality evaluation. Proceedings of the 56th Annual Spring Meeting of Japanese Society for the Science of Design.

Beuttel. B., 2010. Kansei Research and Design. University of Tsukuba.

Koizumi, H., Yamamoto, T., Maki, A., Yamashita, Y., Sato, H., Kawaguchi, H., Ichikawa, N., 2003. Optical topography: practical problems and new applications. Appl. Opt. 42 (16), 3054–3062.

Kowatari, et al., 2006. Prefrontal activity while appreciating abstract and representational paintings: a NIRS study. Neurosci. Res. 55 (Suppl. 1), S134.
Lee, S.H., Harada, A., Stappers, P.J., 2002. Pleasure with products: design based on Kansei. Pleasure With Products: Beyond Usability. Taylor and Francis, pp. 212–222.
Levy, P., Yamanaka, T., 2006. Towards a definition of kansei. Proceeding of the 2006 Design Research Society International Conference. Wonderground.
Levy, P., Yamanaka, T., 2008. Explaining kansei design studies. Proceeding of the Design and Emotion Conference.
Morikawa, et al., 2007. A study of brain hemodynamics in the writing task using near infrared spectroscopy. Kobe Gakuin Univ. Rehabil. Res. 2 (2), 23–30.
Nagamachi, M., 1989. Kansei Engineering—Technology That Makes Use of Kansei in Design. Science Library, Kaibundou Publication, Tokyo.
Nagamori, Y., Nakajima, M., Yokoi, T., Yamanaka, T., 2009. Analysis of the brain activity at the chair design task with lego blocks. J. Jap. Soc. Kansei Eng. 9 (1), 51–60.
Noriaki, K., Nobuhiko, S., Fumio, T., Shin'ichi, T., 1984. Miryokuteki Hinshitsu to Atarimae Hinshitsu. Qual. Manag. J. 14 (2), 147–156.
Pierre, D.L., Yamanaka, T., 2009. Design with event-related potentials: a Kansei information approach on CMC design. Int. J. Prod. Dev. 7.
Saito, T. The influence of listening to various types of music on cerebral blood volume.
Sogabe, et al., 2009. Evaluation system distributed from review comment of design awards. Bull. Japan. Soc. Sci. Des. 56, 63–64.
Tomico, O., Levy, P., Mizutani, N., Yamanaka, T., 2009. The repertory grid technique as a method for the study of cultural differences. Int. J. Des. 3, 55–63.
Wakao, F., 1984. Goodbye, Mass: How to Read the Kansei Era. PHP Institute, Kyoto.
Watanabe, E., 2004. Functional brain mapping using optical topography. Med. Imaging Technol. 22 (3), 122.
Yamanaka, T., Pierre, D.L., 2006. Design process by intuitive understanding. Proceedings of the 53rd Annual Spring Meeting of Japanese Society for the Science of Design.
Yamanaka, T., Tomico, O., Mizutani N., Levy, P., 2008. Combining Kansei-physiological measurements and constructivist-psychological explorations through the repertory grid technique. International Symposium on Emotion and Sensibility.

CHAPTER

9

Affective Computing: Historical Foundations, Current Applications, and Future Trends

Shaundra B. Daily, Melva T. James**,†, David Cherry**, John J. Porter, III**, Shelby S. Darnell**, Joseph Isaac*, Tania Roy***

***University of Florida, Gainesville, FL, United States; **Clemson University, Clemson, SC, United States; †Massachusetts Institute of Technology, Lexington, MA, United States**

AFFECTIVE COMPUTING HISTORY

The field of affective computing encompasses both the creation of and interaction with machine systems that sense, recognize, respond to, and influence emotions (Picard, 1997; Picard and Klein, 2002). It is a multidisciplinary research area that relies on contributions from fields as disparate as psychology, physiology, engineering, sociology, mathematics, computer science, education, and linguistics to accomplish its goals. The wide variety of disciplines relevant to affective computing is a reflection of the complexity of describing, understanding, and, ultimately, emulating the dynamic experience of *feeling*.

Affective computing, as an academic discipline, is relatively new. In fact, the phrase *affective computing* was not coined until the late 1990s (Picard, 1997). The recent coalescing of affective computing into a recognized field of study, however, belies the observation that concepts related to both affective computing and artificial intelligence, its parent discipline, can be found many years before either term existed. One of the most enduring examples of this is the idea of the golem; a creature crafted from an ordinary material, such as clay, and brought to life through supernatural

Emotions and Affect in Human Factors and Human–Computer Interaction. http://dx.doi.org/10.1016/B978-0-12-801851-4.00009-4
Copyright © 2017 Elsevier Inc. All rights reserved.

means (Idel, 1990; Scholem, 1965; Scholem and Idel, 2007). With origins dating back to at least 1000 BCE (Leaman, 2011; Rendsburg, 2008; Satlow, 2006), the Hebrew Bible, or Tanakh (Brettler, 2005; Satlow, 2006), contains some of the earliest references to the golem concept. The word *golem* is used once in the Bible (Scholem and Idel, 2007), and it appears in the Book of Psalms (Tehillim 139:16 Orthodox Jewish Bible). In the context of the passage, golem connotes an unformed, imperfect mass or embryo (Idel, 1990; Rozenberg and Leviant, 2007; Scholem and Idel, 2007). A second, more explicit example of the idea of a golem is found in Genesis.

Then the LORD God formed a man from the dust of the ground and breathed into his nostrils the breath of life, and the man became a living being.... [Then the] LORD God took the man and put him in the Garden of Eden to work it and take care of it (Gen. 2:7 and Gen. 2:15, New International Version).

Here, in the Bible's second story of creation, the dynamic between God and humanity resembles the modern interpretation of the human–golem relationship. In this interpretation, the human–golem relationship is one between a creator and an entity made for the purpose of serving its creator's interests.

Away from the realm of the supernatural, a clear parallel can be drawn between the golems of myth and real automata (Convino, 1996; Wiener, 1964). Indeed, the myth of the golem is one of the earliest contributions to the topic of artificial intelligence. From a technical perspective, artificial intelligences need not be anthropomorphic (Ferrucci et al., 2010; Hsu, 2002; Turing, 1950; Wiener, 1948). Yet, the idea of creating automata with human-like characteristics holds a special place in the human imagination.

The HAL 9000 computer, antagonist of *2001: A Space Odyssey* (Clarke, 1968; Kubrick, 1968), is one of the most famous fictional examples of an artificial intelligence endowed with human-like emotions. At the beginning of the film, HAL, as the system is called, is shown to be a valued member of the ship's crew. Displaying a dazzling array of abilities, including speech, facial recognition, lip reading, systems monitoring, regulation and control, game play, and social skills (Clarke, 1968; Kubrick, 1968; Picard, 2001), HAL is an exciting vision of what real artificial intelligence could become. As Kubrick and Clarke present it, however, the design and deployment of an artificial intelligence that possesses a seamless integration of emotional and functional capabilities is not without risk (Clarke, 1968; Kubrick, 1968). In the course of the film, HAL is faced with challenging situations that involve both deception and emotionally influenced decision making. The first is HAL's internal conflict between the obligation to relay information accurately to the crew and the obligation to conceal the true purpose of the mission from the crew. The second is the threat of HAL's deactivation in response to repeated system malfunctions.

Acting in self-defense, HAL, in violation of two of Asimov's three laws of robotics (Asimov, 1950), attempts to kill both members of the crew. In the end, HAL is destroyed.

While *2001: A Space Odyssey* is, of course a work of fiction, it nevertheless successfully elucidates some of the key challenges facing affective computing researchers today. Technical challenges relevant to affect-aware system development include multimodal natural language processing, affect detection, biometrics, and systems integration. In addition, the ethical implications of affective computing systems must be considered before widespread deployment; the urgency of these considerations will only increase as technology advances.

The remainder of this chapter focuses on the modern science of affective computing. First, efforts to address the technical challenges of affect sensing and affect generation are described. Next, current applications of affective technologies are considered. Finally, ethical issues relevant to the building and deployment of affective systems are outlined.

AFFECT SENSING

Recognizing emotions is the first step to making a computer affectively intelligent. To accomplish this task, a computer would need to be equipped with hardware and software to sense emotions. Affect sensing refers to a system that can recognize emotion by receiving data through signals and patterns (Picard, 1997). Affect-sensing systems can be classified by modalities, each of which has a unique signature. This section focuses on describing systems that sense affect via low-level signals, such as facial activity, posture, gesture, hand tension, and vocal, textual, and electrodermal activity (EDA).

Systems that sense facial activity can detect and process expressions from a human face. Facial expressions are distinctive for each emotion and they are informative due to their visibility and omnipresence (Ekman et al., 1972). The face conveys information through movement, such as smiling, frowning, squinting, and furrowing (Ekman et al., 1972). According to Ekman (1993), the six basic emotions are anger, disgust, fear, happiness, sadness, and surprise, which can be detected and processed using methods like the Hidden Markov Model, optical flow, active appearance model, and neural network processing (Cohn et al., 1998; Caridakis et al., 2006; Thomas and Mathew, 2012; Wilfred et al., 2009). These methods can be used either individually or in combination. A commonly used system to categorize and classify facial expressions is the facial action coding system (FACS). This system, created by Paul Ekman, Wallace V. Friesen, and Joseph C. Hager (Reevy et al., 2010), helps to identify expressions that the human face can show and the different muscles that produce these expressions. Research methods used to study the face, nonverbal communication, and emotion have been significantly impacted by FACS. Moreover, FACS has been helpful in research on emotions, in the diagnosis of mental disorders, in lie detection, and in psychotherapy (Ekman and Rosenberg, 2005). In addition, the emotion facial action coding system (EMFACS) is a related method that scores facial actions that might be applicable to sensing emotion (Reevy et al., 2010). EMFACS focuses on the action units involved in a facial movement, the intensity of the action units, and the degree of asymmetry. Action units represent the muscular activity that produces facial appearance changes (Ekman and Friesen, 1978). As a result, muscle movements' effects on facial appearance change are critical in facial activity affect sensing.

Posture and gesture affect sensing, which is the recognition of varying states of the body. Posture is the position of a person's body while he or she is standing or sitting, while gestures are the movements of parts of the body as a form of expression, usually from the hand or head (Glowinski et al., 2011). Posture and gestures are important to affect sensing, because they can communicate discrete emotion categories as well as affective dimensions (Kleinsmith and Bianchi-Berthouze, 2007). Kleinsmith and Bianchi-Berthouze (2007) have shown in their review of affective body expression perception and recognition that static body postures and gestured body motions can convey emotion. Devices commonly used in research to collect data from posture and gesture consist of wired gloves, depth-aware cameras, stereo cameras, standard 2D cameras, and gestures. Algorithms to detect posture and gesture include 3D model based algorithms (Lee and Kunii, 1995), skeletal-based algorithms (Nandakumar et al., 2013), and appearance-based models (Shimada et al., 2000). Algorithms based on 3D models use volumetric models, skeletal models, or a combination of the two to detect affect.

Skeletal-based algorithms use a virtual skeleton of the human body to map parts of the body, using position and orientation between the segments of the human skeleton to collect information. Thus, skeletal-based algorithms enable pattern matching and use key points to focus on essential parts of the body. Appearance-based models derive parameters directly from images or videos in a template database, which are used as sets of points to outline objects on parts of the human body. Appearance-based models are commonly used for hand-tracking.

There is a correlation between the amount of force exerted on objects humans handle and their frustration. Examples of objects that humans handle include a computer mouse, steering wheels, gear shifts, remote controls, video game controllers, and cutting knives. For example, a study conducted by Dennerlein et al. (2003) showed that the force exerted by the hand on a computer mouse increases with user frustration. Force-sensitive resistors and stenographs are common hand tension sensors

(Kreil et al., 2008). Therefore, tension in the hand applied to an object can be used to sense affect.

Moreover, the way words are spoken or, in other words, vocal expressions, also relates to affective sensing. Such vocal expressions can be broken into two components: cues emphasizing which content in the message is most important and cues arising from the speaker's affective state (Fulcher, 1991). The intonation of vocals provides style to speech and the message's content. The vocal effects that are most commonly associated with the six basic emotions are speech rate, pitch average, pitch range, intensity, voice quality, pitch changes, and articulation (Murray and Arnott, 1993). As a result, vocal expression is a way to sense affect based on how something is said.

In addition to vocal intonation, words and language have affect in textual form. Input devices used to sense affect from words and languages include sentiment analysis tools. Sentiment analysis tools make use of natural language processing, computational linguistics, and text analysis (Ahmad, 2011; Taboada et al., 2011). Examples of software used for such analysis include WordNet-Affect (Keshtkar and Inkpen, 2013), SenticNet (Denecke, 2008), and SentiWordNet (Poria et al., 2012). Other approaches for analyzing written language include keyword spotting (Liu et al., 2003), lexical affinity (Terra, 2005), statistical methods (Pande and Dhami, 2014), and hand-crafted models (Singer, 2004). Textual affect sensing is important because many user interfaces are text based.

Electrodermal activity (EDA) is the result of the autonomic properties of the skin (Braithwaite et al., 2013; Dawson et al., 2007). This electrical skin conductance is gathered from sweat-induced skin moisture caused by the sympathetic nervous system indicating psychological or physiological arousal (Dawson et al., 2007; Picard and Scheirer, 2001). Electrodes can be placed on the skin for the purpose of EDA detection; however, positive and negative valence cannot be determined through this signal. Further, since changes in EDA can be triggered by nonaffective changes (e.g., physical activity), the context of the arousal is extremely important when attempting to determine emotion (Braithwaite et al., 2013).

AFFECT GENERATION

Social Robots

Social robots are robots that interact with humans and each other in a socially acceptable fashion, conveying intention in a human-perceptible way, and are empowered to resolve goals with fellow agents, be they human or robot (Breazeal and Scassellati, 1999; Duffy et al., 1999). Optimal human–robot interactions require robots to possess the following attributes: embodiment in the physical environment, quick reactions to unexpected events, computational sophistication for meeting goals, and the ability to interact with other robots for the realization of goals of increasing difficulty (Duffy et al., 1999). Here, we examine why social robots are researched, give a few cases of robot–human interaction, and provide a brief explanation of how these robots and interactions are designed.

Given the definition of social robots, what use do humans have for them that inform and inspire research and usage? Dautenhahn (2002) listed social robot applications, such as office, medicine, hotel use, cooking, marketing, entertainment, hobbies, recreation, nursing care, therapy, and rehabilitation. Breazeal (2003) added such uses as a personal assistant, small child care, and small child development. In the office application domain, social robots can be used as physical counterparts for remote conference participants, commonly referred to as telepresence. If the remote participant wishes to interact physically with the conference space, a social robot can act as an interaction avatar (Thalmann et al., 2014). Social robots could also act in childcare and development roles by supplementing the interaction of human caregivers (Feil-Seifer and Mataric´, 2010).

As noted previously, humans use physical appearance and physical attributes to categorize and form impressions (Eyssel and Kuchenbrandt, 2012). This impression-forming process also applies to robots. Humans tend to observe a robot's physical attributes and assign it a social place and competency (Eyssel and Kuchenbrandt, 2012); this is anthropomorphism, or the assigning of human characteristics to nonhuman entities. This aspect of human nature is taken into account when designing social robots and programming their interactions. Another attribute taken into account when designing social robots is the application for which they are being built (Breazeal, 2003). Breazeal (2003) presented four subclasses of social robots, which have been supplemented by Fong et al. (2003). The initial four classes correspond to different social robot designs that lead to the creation of social robots that respond according to humans' perceptions of them. The four classes are socially evocative, social interface, socially receptive, and sociable. Further, socially situated, socially embedded, and socially intelligent complete the seven subclasses. The socially evocative subclass encourages anthropomorphizing robots as one might anthropomorphize toys. Responding to social cues to facilitate communication characterizes the social interface subclass. The socially receptive subclass learns from interactions, while the sociable subclass interacts purposefully to aid and be aided (Breazeal, 2003). Furthermore, a social environment surrounds socially situated robots, while socially embedded robots are in the environment, learning and interacting steadily. Finally, socially intelligent robots attempt to respond to social situations like humans by simulating models of human cognition.

Breazeal and Scassellati (1999) outlined the development of a social robot, Kismet, that responds to its environment by way of infant-like protosocial responses (e.g., initiation, mutual orientation, greeting, play-dialog, and disengagement), mainly communicating its reactions through gaze and facial expressions. Breazeal's social robot perceives social cues and reacts to them with facial emotive responses. Kismet has an expressive robotic head and has 15 degrees of freedom in its facial expressions and 6 for gaze and head movement, while using an infant-like language to respond to audio cues and 4 cameras to perceive the world around it visually (Breazeal, 2003). Kismet fits into the socially situated class of social robots because it learns from interactions with people. Two other socially evocative robot toys are Tiger Electronic's Furby and Sony's Aibo. Furby is reminiscent of a hamster, while Aibo looks like a small dog. Since these robots remind people of animals, people assign animal responsiveness and reactiveness to them. Dautenhahn (2002) developed a toy robot whose purpose is to teach autistic children social behaviors. This robot, designed as a toy, is a social interface that is "interesting enough to catch and maintain" attention and "engage the child in therapeutically relevant interactions until the trial is ended" (Dautenhahn, 2002, p. 447). Dautenhahn's robot is designed both to take turns and to follow, and it has a behavior-based design. Each of these robot examples is designed to generate affect according to its application domain. The socially receptive robot is built to be very expressive to communicate better during interactions, and the toys are designed to maintain attention and provide goals for users to accomplish.

Virtual Characters With Emotion

The novelty of a mechanical tool that moves with purpose keeps people's attention, but adding emotive capability allows the robot to interact with humans socially. However, humans engage humans better than robots. In an attempt to obtain the same level of engagement that humans give each other, virtual characters with emotion can be developed. Virtual characters (i.e., 3D avatars or nonplayer game characters) are meant to generate affect in humans that will cause empathetic interaction between virtual character and human (Vinayagamorrthy et al., 2006). Similar to social robots expressing emotion, emotion models are used to convey emotion in virtual characters. Breazeal employs models using the face to communicate emotion in her social robot Kismet (Breazeal and Scassellati, 1999). In virtual characters, models for giving primary, secondary, and tertiary emotions have been created and utilized (Scheutz et al., 2000; Vinayagamorrthy et al., 2006). With virtual characters, the models to communicate emotion can extend to full human avatars (Ushida et al., 1998). Although not all virtual characters are the same, following the purpose of their design, such virtual characters can be referred to as embodied conversational agents (ECAs) or relational agents. Virtual characters with verbal and nonverbal emotional and conversational behaviors are referred to as ECAs (Vinayagamorrthy et al., 2006), while relational agents are virtual characters that build relationships by interacting over long periods of time (Bickmore and Cassell, 2001).

APPLICATIONS

Affective computing devices are being used in a number of domains including education, security, and healthcare. Picard (1997) describes three types of affective computing applications: "1) systems that detect the emotions of the user, 2) systems that express what a human would perceive as an emotion (e.g., an avatar, robot, and animated conversational agent), and 3) systems that actually 'feel' an emotion." Detection, expression, and perception are crucial when designing technologies with affective capabilities in mind. In this section, we explore different applications that exhibit properties of the first and second type.

Healthcare Applications

Previous research has indicated that the emotional responses of individuals with Asperger syndrome (AS) or high functioning autism (HFA) are less differentiated, less positive, and more negative than those without AS or HFA (Ben Shalom et al., 2006; Capps et al., 1993). Individuals with AS or HFA have also been known to experience significant difficulties effectively assessing and classifying their own emotions (Berthoz and Hill, 2005; Fitzgerald and Bellgrove, 2006; Fitzgerald and Molyneux, 2004; Hill et al., 2004; Szatmari et al., 2008; Tani et al., 2004). These behavioral traits can affect their relationships with other people (Sano et al., 2012). Companies have developed mobile applications like SymTrend (n.d.) and Autism Track (HandHold Adaptive, n.d.), which allow patients with disabilities to enter behavioral data manually and to track changes over time. The patients are also given accurate advice, made more aware of their symptoms, and given reminders. This gives patients and their therapists an insight into the patients' behavioral patterns, moods, and triggers that may accompany any emotional outbursts.

Sano et al. (2012) created a system for schoolteachers monitoring autism spectrum disorder (ASD)-related behaviors. This annotation tool gives teachers the ability to note important events quickly without having to write down all the details at the exact moment. These data can then be appropriately stored and shared for deeper analysis. Caregivers, teachers, and other relevant officials can then revisit the event to upload and share their individual annotations, thus providing doctors with multiple perspectives on how the student's day transpired and helping doctors to identify any specific triggers.

Posttraumatic stress disorder (PTSD), as currently defined by the American Psychiatric Association, is the development of characteristic symptoms following exposure to actual or threatened death, serious injury, or sexual violence in one or more of the following ways: directly experiencing the traumatic event, repeated or extreme exposure to aversive details of the event(s), witnessing the event(s) as it happens to another person; or learning the event occurred to a family member or close friend (American Psychiatric Association, 2013, p. 271).

PTSD disables a person from carrying out daily activities and haunts him or her with memories of stressful events (Foa et al., 2008). A well-known technique for treating individuals with PTSD is exposure therapy and stress inoculation (Bradley et al., 2005; Horowitz, 1997; Van Etten and Taylor, 1998). Current research has extended this technique to include virtual reality (Rizzo et al., 2009). StartleMart is a virtual reality-based gaming environment with affect detection capabilities that integrates cognitive behavioral approaches with physiological signals to treat veteran soldiers with PTSD (Holmgard et al., 2013). StartleMart adds real-time stress detection using skin conductance to an already developed virtual reality treatment scenario.

Veteran soldiers with PTSD often suffer from flashbacks from the combat zone; for example, they may hear the sound of a ventilator fan in a store and be reminded of being in a war zone. StartleMart simulates three highly stressful scenarios and the skin conductance is used to measure the body's response to anxiety. A study conducted using StartleMart successfully correlated stressors on screen with peaks in skin conductance data, leading researchers to believe that these kinds of systems can help with the diagnosis and treatment of PTSD. Similar games like Virtual Vietnam (Rothbaum et al., 1999; Rothbaum et al., 2001), Virtual Iraq (Gerardi et al., 2008), and Virtual Afghanistan (Rizzo et al., 2010) have been used successfully in clinical testing and have also shown positive results.

Applications in Education

Schools have been incorporating more and more technology like smart boards and interactive presentations into their classrooms to engage students and to help them to learn better. Incorporating affect into classroom technology dynamics may enable students to get more of a personalized learning experience. However, creating an individualized curriculum for each student is very difficult in the present classroom setting, as it is extremely time consuming and requires an in-depth understanding of each student's likes, dislikes, and preferred learning method. Likewise, teachers cannot always rely on students to speak up when they do not understand the material being taught. Research shows that students who receive support from teachers and peers tend to feel more comfortable in school, like school more, and participate more actively in classroom activities (Furrer and Skinner, 2003; Gest et al., 2005; Goodenow, 1993; Hughes and Kwok, 2006; Marsh, 1989; Midgley et al., 1989; Ryan et al., 1994; Stipek, 2002).

Effective learning is contingent upon the extent to which students are engaged in classroom learning activities (Chen, 2005; Finn and Rock, 1997; Osterman, 2000; Reyes et al., 2012; Wang and Pomerantz, 2009). EngageMe is a visualization tool designed to support teachers in understanding how they are connecting with their students and how their pedagogical strategies can be modified to meet the individual needs of a diverse student population (Darnell, 2014). This system uses skin conductance data collected from students in a classroom, along with video feeds, to help the teacher reflect on his or her classes. Graphs display the arousal levels of each student, with green signifying high, yellow signifying medium, and red signifying a low level of arousal throughout that student's time in class. To distinguish between the sources of these moments of arousal, a video feed is provided to the teacher, so he or she can determine if the arousals are due to classroom engagement or some other factor. The teacher also has the capability to take notes for a particular student, get a complete picture of an individual student during a class period, and look at the overall classroom performance over a period of time.

Another application, the Subtle Stone, is a wireless, hand-held, squeezable ball that allows students to communicate their affective and motivational experiences to their teachers in real time. The ball has seven different colors, and each student in the class can customize the stone to represent his or her own emotional language by associating specific colors with specific emotions he or she wants to communicate. When the stone is in use, students squeeze the sensor to cycle through colors until they reach the emotion they wish to express at that moment. Thus, one drawback of this tool is that the feedback delivered to the teacher is heavily dependent on the student's self-report. In addition, the cognitive load imposed on the instructor may be high depending on the number of students in the classroom and the level of variation between students' emotional color schemes. Finally, the possible distraction of the students from the instruction as they attempt to express themselves utilizing this tool could be an issue. Other researchers developing and studying intelligent tutoring systems have identified learner engagement and affect by monitoring conversational cues, gross body language, and facial features with a variety of sensors (D'Mello and Graesser, 2007).

Other Applications

Other applications of affective technology include job interview performance, which is not solely based on an individual's knowledge, but, more importantly, how well he or she can communicate that knowledge to the other person. Along with good verbal communication skills, the ability to control one's emotions is essential. To work on these skills, individuals frequently practice in front of a mirror or with a friend. MACH (My Automated Conversation coacH) developed at the MIT Media Lab, is a virtual agent that can read facial expressions as well as speech and language intonations. The agent then gives both verbal and nonverbal feedback to the users, helping them to improve their communication skills and control anxiety (Hoque et al., 2013).

Moreover, due to the rise in security threats and controversies related to interrogation techniques, researchers have suggested using affective systems that can unobtrusively detect specific emotions like anger, frustration, or deception in real time. Affective applications currently focus on using an individual's facial expressions as the primary measure to track emotional cues. Over the years, researchers have worked on developing a universal coding system for standard facial expressions, such as the FACS described previously (Ekman and Friesen, 1978). Similarly, researchers from Carnegie-Mellon University and Naval Criminal Investigative Services developed the automated facial recognition system (AFERS) (Ryan et al., 2009). This system uses video streams and support vector machines to detect facial expressions. An automated FACS system codes these expressions, and the results can be viewed in real time. AFERS has the ability to generate a graphical representation of the expressions over a period of time, and that helps the investigators identify patterns of deception. AFERS also allows authorities to record, annotate, and store frame-by-frame references of the video feed for future review. A system like AFERS also has the potential of being used in large gatherings like airports, games, or concerts to detect suspicious behavior in real time (Meservy et al., 2005; Ryan et al., 2009).

ETHICAL CONSIDERATIONS

According to Picard (1997), "The fictional message has been repeated in many forms and is serious: a computer that can express itself emotionally will someday act emotionally, and if it is capable of certain behaviors, then the consequences can be tragic" (p. 127). The introduction of this chapter raised the issue of how giving technology certain functionalities can be potentially harmful. What types of machines should be given affective capabilities? Should machines with already destructive capabilities (e.g., fighter jets) be imbued with emotional reasoning? For example, given that an airplane or a mailbox has been granted affective reasoning, should anything go wrong with either of those as a result of emotional misreading, the resulting causalities and backlashes would be completely different.

Another ethical consideration is how in line affective reasoning capabilities will be with a machine's task. In the case of conflict, what form of reasoning wins? For example, if a robot with affective capabilities is being used to administer a shot to a crying child, will the robot still be able to complete its task even though it may sympathize with the crying child? Likewise, if an affective robot enters in a sensitive situation, such as a funeral, will it know how to maneuver appropriately without offending anyone? Although there are numerous technologies in place that have made errors that have cost many human lives, a vast number of technologies have also saved them (Picard, 1997). Human safety is key when considering where to place affective reasoning in technology. Therefore, as we progress in affective computing, it is important to remember also to place thoughts on safeguarding the users during these processes as we would any other technology. Researchers must prioritize maximizing the contribution of the technology and minimizing any extraneous hindrances that may come as a result of its use. Though these possibilities may not arise from affective computing for many years to come, it is still our ethical responsibility to consider them.

Privacy Concerns

Concerns exist regarding how affective computers are able to gather information about a user's emotional status and well-being. According to Picard (2003), "Emotions, perhaps more so than thoughts, are ultimately personal and private" (p. 7). This information about a user's emotional state or how the user feels could be stored over a long period of time and could potentially be accessed, hacked, or stolen. Having software that picks up on a user's mood does not mean that the user would want a salesman or telemarketer to know how he or she feels because it could be used against him or her (Picard, 1997). Developing safe and secure software to store a user's emotional information while leaving the user in complete control of who has access to the information is an important factor in affective computing. Maintaining the privacy of each individual user, especially over potentially insecure networks, is another factor. If this information could or would be accessed by another party, then it could have a negative impact on users.

Emotional Dependency

Affective computing can open the door to creating *moral agents*, tools developed to assist with humans' emotional well-being. However, could having users routinely use these moral agents create codependence? Another perspective is that if these moral agents were introduced to children early on, could children also develop an unhealthy attachment to or need for these affective devices to sustain and regulate their emotions (Picard and Klein, 2002)? The main goal of such a moral agent would be to assist in the regulation and well-being of users, without hindering their ability to handle human emotions and interaction. Ensuring that affective computing software and moral agent roles are clearly defined for both the developers and users is important to prevent the tool from becoming an entity upon which humans would become completely dependent. If a user becomes addicted to or entirely dependent on moral agents, then a large problem could arise that could negatively impact a user's ability to manage his or her own emotional well-being. Moral agents should be designed to serve as emotional support, not as replacements for an individual's ability to manage his or her own well-being. Hence, moral agents should be seen as guides or enhancers to strengthen the overall happiness of the user, not the immediate or only sources of emotional direction.

Emotional Manipulation

With emotions being considered such a private entity by default, another question arises: is it ethical for computers to detect, recognize, and then attempt to modify certain behaviors? Some would say that a computer with the ability to attempt to resolve a user's emotions is a breach of ethics and is unacceptable because of its manipulative nature; however, humans typically behave this way in this situation daily (Picard, 2003). If an individual is sad or angry, another person may try to calm the person, which can also be viewed as manipulative in a sense, but would be deemed acceptable if claimed that it is "motivated by good will" (Picard and Klein, 2002, p. 16). With this in mind, having these moral agents be able to detect negative affect could be a great benefit to the well-being and mental health of some users. The important part in developing this software is not to enable these affective machines to affect positive moods negatively, to limit the feelings users have, or to violate the privacy of user's thoughts and emotions.

Building Relationships

The discussion of the potential effects of building relationships with affective technologies brings into question not only how affective reasoning may alter the technology over time, but also how that technology will alter humanity and our perception of it. As the next generation grows, so will the new affective technologies. The depth and appropriateness of the relationships have yet to be defined. However, over time, this will become an ethical topic to discuss. At what point will a person begin to value affective technology and its well-being over that of another human being? Is there a certain amount of time a person should spend away from these technologies to avoid having an unhealthy relationship?

These are questions that will need to be answered in the years to come as affective technologies become more intertwined into everyday living. For instance, in the movie *Her*, a man develops romantic feelings for his operating system (Jonze, 2013). People already claim to have emotional attachments to their cars, phones, etc. How would adding affective capabilities to those technologies change those relationships? Humans attach and connect to each other strongly through emotion. Therefore, by giving technology access to that emotion, we open ourselves to all that comes with it.

FUTURE DIRECTIONS

This chapter has reviewed the history of affective computing, techniques for sensing affect through technology, methods for generating emotion, applications, and ethical considerations. The applications reviewed in this section have shown that, although there is still room for improvement, progress in applications with affective capabilities is not only possible, but also flourishing. Therefore, as research interests in designing applications and software with affective capabilities increase, then so will the amount and variety therein. As technology finds its way into almost every aspect of human life, the questions of its ability to understand, help treat, develop, and impact the human psyche will continue to grow. With this growth, affective capabilities will be an integral part of the development of future technologies.

References

Ahmad, K. (Ed.), 2011. Affective Computing and Sentiment Analysis, vol. 45, Springer Netherlands, Dordrecht, http://link.springer.com/10.1007/978-94-007-1757-2.

American Psychiatric Association, 2013. Diagnostic and statistical manual of mental disorders: DSM-5, fifth ed. American Psychiatric Association, Arlington, VA.

Asimov, I., 1950. I, Robot, first ed. Gnome Press, New York, NY.

Ben Shalom, D., Mostofsky, S.H., Hazlett, R.L., Goldberg, M.C., Landa, R.J., Faran, Y., Hoehn-Saric, R., 2006. Normal physiological emotions but differences in expression of conscious feelings in children with high-functioning autism. J. Autism Dev. Disord. 36 (3), 395–400.

Berthoz, S., Hill, E.L., 2005. The validity of using self-reports to assess emotion regulation abilities in adults with autism spectrum disorder. Eur. Psychiatry 20 (3), 291–298.

Bickmore, T., Cassell, J., 2001. Relational agents: a model and implementation of building user trust. Proceedings of the SIGCHI Conference on Human Factors in Computing Systems. AMC, pp. 396–403.

Bradley, R., Greene, J., Russ, E., Dutra, L., Westen, D., 2005. A multidimensional meta-analysis of psychotherapy for PTSD. Am. J. Psychiatry 162 (2), 214–227.

Braithwaite, J.J., Watson, D.G., Jones, R., Rowe, M., 2013. A guide for analysing electrodermal activity (EDA) & skin conductance responses (SCRs) for psychological experiments. Psychophysiology 49, 1017–1034.

Breazeal, C., 2003. Toward sociable robots. Rob. Auton. Syst. 42 (3), 167–175.

Breazeal, C., Scassellati, B., 1999. How to build robots that make friends and influence people. Intelligent Robots and Systems, 1999. IROS '99. Proceedings. 1999 IEEE/RSJ International Conference on. Kyongju, pp. pp. 858–863.

Brettler, M.Z., 2005. How to Read the Bible, first ed. Jewish Publication Society, Philadelphia, PA.

Capps, L., Kasari, C., Yirmiya, N., Sigman, M., 1993. Parental perception of emotional expressiveness in children with autism. J. Consult. Clin. Psychol. 61 (3), 475–484.

Caridakis, G., Malatesta, L., Kessous, L., Amir, N., Raouzaiou, A., Karpouzis, K., 2006. Modeling naturalistic affective states via facial and vocal expressions recognition. Proceedings of ICMI '06: The Eighth International Conference on Multimodal Interfaces. ACM Press, New York, NY, pp. 146–154.

Chen, J.J.-L., 2005. Relation of academic support from parents, teachers, and peers to Hong Kong adolescents' academic achievement: the mediating role of academic engagement. Genet. Soc. Gen. Psychol. Monogr. 131 (2), 77–127.

Clarke, A.C., 1968. 2001: A Space Odyssey, first ed. New American Library, New York, NY.

Cohn, J.F., Zlochower, A.J., Lien, J.J., Kanade, T., 1998. Feature-point tracking by optical flow discriminates subtle differences in facial expression. The Proceedings of the IEEE International Conference On Automatic Face And Gesture Recognition. IEEE, Nara, Japan, pp. 396–401.

Convino, W.A., 1996. Grammars of transgression: Golems, cyborgs, and mutants. Rhetoric Rev. 14 (2), 355–373.

Darnell, S.S., 2014. EngageME: a tool to simplify the conveyance of complicated data. Proceedings CHI'14 extended abstracts on human factors in computing systems. ACM Press, Toronto, Canada, pp. 359–362.

Dautenhahn, K., 2002. Design spaces and niche spaces of believable social robots. Proceedings 11th IEEE International Workshop on Robot and Human Interactive Communication, 2002. IEEE, Berlin, Germany, pp. 192–197.

Dawson, M.E., Schell, A.M., Filion, D.L., 2007. The electrodermal system. In: Cacioppo, J.T., Tassinary, L.G., Berntson, G. (Eds.), Handbook of Psychophysiology. third ed. Cambridge University Press, New York, NY.

Denecke, K., 2008. Using SentiWordNet for multilingual sentiment analysis. Proceedings IEEE 24th International Conference on Data Engineering. IEEE, Cancun, Mexico, pp. 507–512.

Dennerlein, J., Becker, T., Johnson, P., Reynolds, C., Picard, R.W., 2003. Frustrating computer users increases exposure to physical factors. Proceedings International Ergonomics Association. International Ergonomics, Association, Seoul, Korea, pp. 24–29.

D'Mello, S., Graesser, A., 2007. Mind and body: dialogue and posture for affect detection in learning environments. In: Luckin, R., Koedinger, K.R., Greer, J. (Eds.), Artificial Intelligence in Education: Building Technology Rich Learning Contexts That Work. IOS Press, Amsterdam, pp. 161–168.

Duffy, B.R., Rooney, C., O'Hare, G.M., O'Donoghue, R., 1999. What is a social robot? Proceedings 10th Irish Conference On Artificial Intelligence & Cognitive Science. University College Cork, Cork, Ireland.

Ekman, P., 1993. Facial expression and emotion. Am. Psychol. 48 (4), 384–392.

Ekman, P., Friesen, W.V., 1978. Facial Action Coding System: A Technique for the Measurement of Facial Movement. Consulting Psychologists Press, Palo Alto, CA.

Ekman, P., Rosenberg, E.L. (Eds.), 2005. What the Face Reveals: Basic and Applied Studies of Spontaneous Expression Using the Facial Action Coding System (FACS). second ed. Oxford University Press, Oxford.

Ekman, P., Friesen, W.V., Ellsworth, P., 1972. Emotion in the Human Face: Guidelines for Research and an Integration of Findings. Pergamon Press, Elmsford, NY.

Eyssel, F., Kuchenbrandt, D., 2012. Social categorization of social robots: anthropomorphism as a function of robot group membership. Br. J. Soc. Psychol. 51 (4), 724–731.

Feil-Seifer, D., Mataric´, M.J., 2010. Dry your eyes: examining the roles of robots for childcare applications. Interact. Stud. 11 (2), 208–213.

Ferrucci, D., Brown, E., Chu-Carroll, J., Fan, J., Gondek, D., Kalyanpur, A.A., et al., 2010. Building Watson: an overview of the DeepQA project. AI Magazine 31 (3), 59–79.

Finn, J.D., Rock, D.A., 1997. Academic success among students at risk for school failure. J. Appl. Psychol. 82 (2), 221–234.

Fitzgerald, M., Bellgrove, M.A., 2006. The overlap between alexithymia and Asperger's syndrome. J. Autism Dev. Disord. 36 (4), 573–576.

Fitzgerald, M., Molyneux, G., 2004. Overlap between alexithymia and Asperger's syndrome. Am. J. Psychiatry 161 (11), 2134–2135.

Foa, E.B., Keane, T.M., Friedman, M.J., Cohen, J.A. (Eds.), 2008. Effective Treatments for PTSD: Practice Guidelines From the International Society for Traumatic Stress Studies. Guilford Press, New York, NY.

Fong, T., Nourbakhsh, I., Dautenhahn, K., 2003. A survey of socially interactive robots. Rob. Auton. Syst. 42 (3), 143–166.

Fulcher, J., 1991. Vocal affect expression as an indicator of affective response. Behav. Res. Methods Instrum. Comput. 23 (2), 306–313.

Furrer, C., Skinner, E., 2003. Sense of relatedness as a factor in children's academic engagement and performance. J. Educ. Psychol. 95 (1), 148–162.

Gerardi, M., Rothbaum, B.O., Ressler, K., Heekin, M., Rizzo, A., 2008. Virtual reality exposure therapy using a virtual Iraq: case report. J. Traumat. Stress 21 (2), 209–213.

Gest, S.D., Welsh, J.A., Domitrovich, C.E., 2005. Behavioral predictors of changes in social relatedness and liking school in elementary school. J. Sch. Psychol. 43 (4), 281–301.

Glowinski, D., Dael, N., Camurri, A., Volpe, G., Mortillaro, M., Scherer, K., 2011. Toward a minimal representation of affective gestures. IEEE Trans. Affect. Comput. 2 (2), 106–118.

Goodenow, C., 1993. Classroom belonging among early adolescent students: relationships to motivation and achievement. J. Early Adolesc. 13 (1), 21–43.

HandHold Adaptive. (n.d.). Autism track. http://www.handholdadaptive.com/AutismTrack.html

Hill, E., Berthoz, S., Frith, U., 2004. Brief report: cognitive processing of own emotions in individuals with autistic spectrum disorder and in their relatives. J. Autism. Dev. Disord. 34 (2), 229–235.

Holmgard, C., Yannakakis, G.N., Karstoft, K.-I., Andersen, H.S., 2013. Stress detection for PTSD via the Startlemart game. Proceedings 2013 Human Association Conference on Affective Computing and Intelligent Interaction (ACII). IEEE, Geneva, Switzerland, pp. 523–528.

Hoque, M., Courgeon, M., Martin, J.-C., Mutlu, B., Picard, R.W., 2013. MACH: My Automated Conversation Coach. ACM Press, New York, NY.

Horowitz, M.J., 1997. Stress Response Syndromes: PTSD, Grief, and Adjustment Disorders. Jason Aronson. http://psycnet.apa.org/psycinfo/1997-36792-000

Hsu, F., 2002. Behind Deep Blue: Building the Computer That Defeated the World Chess Champion. Princeton University Press, Princeton, NJ.

Hughes, J.N., Kwok, O., 2006. Classroom engagement mediates the effect of teacher–student support on elementary students' peer acceptance: a prospective analysis. J. Sch. Psychol. 43 (6), 465–480.

Idel, M., 1990. Golem: Jewish Magical and Mystical Traditions on the Artificial Anthropoid. State University of New York Press, Albany, NY.

Jonze, S. (Director), 2013. Her [Motion Picture]. Warner Bros Pictures, United States.

Keshtkar, F., Inkpen, D., 2013. A bootstrapping method for extracting paraphrases of emotion expressions from texts. Comput. Intell. 29 (3), 417–435.

Kleinsmith, A., Bianchi-Berthouze, N., 2007. Recognizing affective dimensions from body posture. Proceedings 2nd International Conference Affective Computing and Intelligent Interaction. Springer, Lisbon, Portugal, pp. 48–58.

Kreil, M., Ogris, G., Lukowicz, P., 2008. Muscle activity evaluation using force sensitive resistors. Proceedings 5th International Summer School and Symposium on Medical Devices and Biosensors. IEEE, Hong Kong, China, pp. 107–110.

Kubrick, S. (Director), 1968. 2001: A Space Odyssey [Motion Picture]. Metro-Goldwyn-Mayer, United States.

Leaman, O., 2011. Judaism: An Introduction. I.B. Tauris, London.

Lee, J., Kunii, T.L., 1995. Model-based analysis of hand posture. IEEE Comput. Graph. Appl. 15 (5), 77–86.

Liu, H., Lieberman, H., Selker, T., 2003. A model of textual affect sensing using real-world knowledge. Proceedings of the 8th International Conference on Intelligent User Interfaces. ACM Press, Miami, FL, pp. 125–132.

Marsh, H.W., 1989. Age and sex effects in multiple dimensions of self-concept: preadolescence to early adulthood. J. Educ. Psychol. 81 (3), 417–430.

Meservy, T.O., Jensen, M.L., Kruse, J., Burgoon, J.K., Nunamaker, Jr., J.F., Twitchell, D.P., et al., 2005. Deception detection through automatic, unobtrusive analysis of nonverbal behavior. IEEE Intell. Syst. 20 (5), 36–43.

Midgley, C., Feldlaufer, H., Eccles, J.S., 1989. Student/teacher relations and attitudes toward mathematics before and after the transition to junior high school. Child Dev. 60 (4), 981–992.

Murray, I.R., Arnott, J.L., 1993. Toward the simulation of emotion in synthetic speech: a review of the literature on human vocal emotion. J. Acoust. Soc. Am. 93 (2), 1097–1108.

Nandakumar, K., Wan, K.W., Chan, S.M.A., Ng, W.Z.T., Wang, J.G., Yau, W.Y., 2013. A multimodal gesture recognition system using audio, video, and skeletal joint data. In: Epps, J., Chen, F., Oviatt, S., Mase, K., (Eds.), Proceedings of the Fifteenth International Conference on Multimodal Interaction. ACM Press, New York, NY, pp. 475–482.

Osterman, K.F., 2000. Students' need for belonging in the school community. Rev. Educ. Res. 70 (3), 323.

Pande, H., Dhami, H.S., 2014. Statistical methods in language and linguistic research. J. Quant. Ling. 21 (3), 295–297.

Picard, R.W., 1997. Affective Computing, first ed. MIT Press, Cambridge, MA.

Picard, R.W., 2001. Building HAL: computers that sense, recognize, and respond to human emotion. Proceedings of the Society of Photo-Optical Instrumentation Engineers, vol. 4299. SPIE, San Jose, CA, pp. 518–523.

Picard, R.W., 2003. Affective computing: challenges. Int. J. Hum. Comput. Stud. 59 (1–2), 55–64.

Picard, R.W., Klein, J., 2002. Computers that recognise and respond to user emotion: theoretical and practical implications. Interact. Comput. 14, 141–169.

Picard, R.W., Scheirer, J., 2001. The galvactivator: a glove that senses and communicates skin conductivity. In: Smith, M.J., Salvendy, G.A. (Ed.), Proceedings of the Ninth International Conference on Human-Computer Interaction. Association for Computing Machinery, New Orleans, LA, pp. 1538–1542.

Poria, S., Gelbukh, A., Cambria, E., Yang, P., Hussain, A., Durrani, T., 2012. Merging SenticNet and WordNet-Affect emotion lists for sentiment analysis. Proceedings IEEE 11th International Conference on Signal Processing, vol. 2. IEEE, Beijing, China, pp. 1251–1255.

Reevy, G.M., Ozer, Y.M., Ito, Y., 2010. Encyclopedia of Emotion. Greenwood, Santa Barbara, CA.

Rendsburg, G.A., 2008. Israel without the Bible. In: Greenspahn, F.E. (Ed.), The Hebrew Bible: New Insights and Scholarship. New York University Press, New York, NY, pp. 3–23.

Reyes, M.R., Brackett, M.A., Rivers, S.E., White, M., Salovey, P., 2012. Classroom emotional climate, student engagement, and academic achievement. J. Educ. Psychol. 104 (3), 700–712.

Rizzo, A.S., Reger, G., Gahm, G., Difede, J., Rothbaum, B.O., 2009. Virtual reality exposure therapy for combat-related PTSD. In: Siromani, P.J., Keane, T.M., LeDoux, J.E. (Eds.), Post-Traumatic Stress Disorder. Springer, New York, NY, pp. 375–399, http://link.springer.com/chapter/10.1007/978-1-60327-329-9_18.

Rizzo, A.S., Difede, J., Rothbaum, B.O., Reger, G., Spitalnick, J., Cukor, J., McLay, R., 2010. Development and early evaluation of the Virtual Iraq/Afghanistan exposure therapy system for combat-related PTSD. Ann. NY Acad. Sci. 1208 (1), 114–125.

Rothbaum, B.O., Hodges, L., Alarcon, R., Ready, D., Shahar, F., Graap, K., Pair, J., Herbert, P., Gotz, D., Wills, B., Baltzell, D., 1999. Virtual reality exposure therapy for PTSD Vietnam veterans: a case study. J. Trauma. Stress 12 (2), 263–271.

Rothbaum, B.O., Hodges, L.F., Ready, D., Graap, K., Alarcon, R.D., 2001. Virtual reality exposure therapy for Vietnam veterans with posttraumatic stress disorder. J. Clin. Psychiatry 62 (8), 617–622.

Rozenberg, Y.Y., Leviant, C., 2007. The Golem and The Wondrous Deeds of the Maharal Of Prague. Yale University Press, New Haven, CT.

Ryan, R.M., Stiller, J.D., Lynch, J.H., 1994. Representations of relationships to teachers, parents, and friends as predictors of academic motivation and self-esteem. J. Early Adolesc. 14 (2), 226–249.

Ryan, A., Cohn, J.F., Lucey, S., Saragih, J., Lucey, P., De la Torre, F., Rossi, A., 2009. Automated facial expression recognition system. Proceedings 43rd annual 2009 international Carnahan conference on Security Technology. IEEE, Zürich, Switzerland, pp. 172–177.

Sano, A., Hernandez, J., Deprey, J., Eckhardt, M., Goodwin, M. S., Picard, R. W., 2012. Multimodal annotation tool for challenging behaviors in people with Autism spectrum disorders. Proceedings of the 2012 ACM Conference on Ubiquitous Computing. ACM, New York, NY, pp. 737–740.

Satlow, M.L., 2006. Creating Judaism: History, tradition, practice. Columbia University Press, New York, NY.

Scheutz, M., Sloman, A., Logan, B. 2000. Emotional states and realistic agent behaviour. GAME-ON. p. 81.

Scholem, G., 1965. On the Kabbalah and Its Symbolism, first ed. Schocken Books, New York, NY.

Scholem, G., Idel, M., 2007. Golem, M., Merenbaum, M., Skolnik, F. (Eds.), Encyclopaedia Judaica, vol. 7, Thomson Gale/Macmillan Reference, Detroit, pp. 735–738.

Shimada, N., Kimura, K., Shirai, Y., Kuno, Y., 2000. Hand posture estimation by combining 2-D appearance-based and 3-D model-based approaches. Proceedings 15th International Conference on Pattern, Recognition, vol. 3. IEEE, Barcelona, Spain, pp. 705–708.

Singer, R., 2004. Comparing Machine Learning and Hand-Crafted Approaches for Information Extraction From HTML Documents. National Library of Canada, Ottawa, ON.

Stipek, D.J., 2002. Motivation to Learn: Integrating Theory and Practice, fourth ed Allyn and Bacon, Boston, MA.

SymTrend. (n.d.). Symptom tracking, charting, & reminding. https://www.symtrend.com/tw/public/

Szatmari, P., Georgiades, S., Duku, E., Zwaigenbaum, L., Goldberg, J., Bennett, T., 2008. Alexithymia in parents of children with autism spectrum disorder. J. Autism Dev. Disord. 38 (10), 1859–1865.

Taboada, M., Brooke, J., Tofiloski, M., Voll, K., Stede, M., 2011. Lexicon-based methods for sentiment analysis. Comput. Ling. 37 (2), 267–307.

Tani, P., Lindberg, N., Joukamaa, M., Nieminen-von Wendt, T., von Wendt, L., Appelberg, B., et al., 2004. Asperger syndrome, alexithymia and perception of sleep. Neuropsychobiology 49 (2), 64–70.

Terra, E., 2005. Lexical affinities and language applications (Unpublished doctoral dissertation). University of Waterloo, Canada.

Thalmann, N.M., Yumak, Z., Beck, A., 2014. Autonomous virtual humans and social robots in telepresence. Paper Presented at 2014 IEEE 16th International Conference: Multimedia Signal Processing (MMSP). IEEE, Jakarta, pp. 1–6.

Thomas, N., Mathew, M., 2012. Facial expression recognition system using neural network and MATLAB. Paper Presented at International Conference on Computing, Communication and Applications. IEEEXplore, Dindigul, Tamilnadu, India.

Turing, A.M., 1950. Computing machinery and intelligence. Mind 59 (236), 433–460.

Ushida, H., Hirayama, Y., Nakajima, H., 1998. Emotion model for life-like agent and its evaluation. In: Proceedings of the Fifteenth National/Tenth Conference on Artificial Intelligence/Innovative Applications of Artificial Intelligence. American Association for Artificial Intelligence. pp. 62–69.
Van Etten, M.L., Taylor, S., 1998. Comparative efficacy of treatments for post-traumatic stress disorder: a meta-analysis. Clin. Psychol. Psychother. 5 (3), 126–144.
Vinayagamorrthy, V., Gillies, M., Steed, A., Tanguy, E., Pan, X., Loscos, C., Slater, M., 2006. Building expression into virtual characters. Eurographics Conference State of the Art Report, Vienna, Austria.
Wang, Q., Pomerantz, E.M., 2009. The motivational landscape of early adolescence in the United States and China: a longitudinal investigation. Child Develop. 80 (4), 1272–1287.
Wiener, N., 1948. Cybernetics; or, Control and Communication in the Animal and the Machine, first ed. Hermann et Cie, Paris.
Wiener, N., 1964. God and Golem, Inc.: A Comment on Certain Points Where Cybernetics Impinges on Religion, first ed. MIT Press, Cambridge, MA.
Wilfred, O.O., Lee, G.B., Park, J.J., Cho, B.J., 2009. Facial component features for facial expression identification using Active Appearance Models. In: AFRICON, 2009, AFRICON'09. IEEE. pp. 1–5.

Further Reading

Bradley, R., Greene, J., Russ, E., Dutra, L., Westen, D., 2014. A multidimensional meta-analysis of psychotherapy for PTSD. Am. J. Psychiatry.

PART III

METHODOLOGIES: INTRODUCTION AND EVALUATION OF TECHNIQUES

CHAPTER

10

Affect/Emotion Induction Methods

Seyedeh Maryam Fakhrhosseini, Myounghoon Jeon
Michigan Technological University, Houghton, MI, United States

INTRODUCTION

During the past 120 years, the perceived role of emotion in normal and abnormal human functioning led to different theories and approaches. Some of the theories viewed it as an adaptive system while others talked about its destructive roles. Previous studies have shown the role of emotional states in information processing, reasoning, and behavioral and cognitive activities.

Emotions and affect have also become a crucial part of human factors and human–computer interaction (HCI) research. In HCI, people have shown a tendency to apply human–human social rules to computers. Among these social interaction components, emotions play an important role, which allows people to express themselves beyond the verbal communications. Therefore, the ability to understand humans' emotions is desirable for the computers in many applications.

Recently, interactions between people and computers have been changed in different ways: activities computers are used for, the domains they are used in, the people who use them, and the way that they are used have been constantly evolving. Accordingly, affect has been considered a new dimension in these interactions, which has great influence on many of human activities. To add these new dimensions to the systems, a few indicators of people's emotional responses, such as skin conductance, heart rate variability, facial muscles, and other noninvasive measurements have been considered. These methods have provided information regarding the changes in attention, perception, memory and decision-making methods.

With the interactional approach to HCI, systems and devices are not only tools that store and manipulate data, but also they have taken the roles

Emotions and Affect in Human Factors and Human–Computer Interaction. http://dx.doi.org/10.1016/B978-0-12-801851-4.00010-0
Copyright © 2017 Elsevier Inc. All rights reserved.

of companions, coaches, secretaries, cooperative partners, smart friends, etc. Previous studies show considerable impact of positive and negative affective states on people's interaction with a technological artifact (Picard and Klein, 2002; Picard and Picard, 1997; Sears and Jacko, 2009). Increasing interest in the role of emotion in HCI and cognition has led to the development of a range of laboratory methods for inducing temporary mood states in the area of UX dynamics, dynamics of affect, and user-centered design research.

In this chapter, we review common emotion induction techniques and discuss the pros and cons of each. For some of the techniques, we provide the list of validated pieces that researchers can use for their studies.

EMOTION INDUCTION METHODS

Imagination

Imagination is one of the simplest emotion induction techniques. During this procedure, participants should imagine situations and memories related to the desired mood (Westermann et al., 1996). They can either imagine hypothetical situations or real events from their past. In some cases, participants listen to an audio recording to relax, to make them comfortable, and to listen for further instruction (Martin, 1990). In guided imagery, the experimenter describes a situation with all the details and participants should imagine themselves there as vividly as possible (Mayer et al., 1995). Note that guided imagery, personal memory procedures, and hypnotic memory are all different methods.

According to the neuroimaging evidence, the imagination technique triggers networks of brain regions that are specified for memory indicating that imagination and recall are valid methods of affect induction (Quigley et al., 2014). The biggest limitation of imagination-based methods is the wide range of peoples' ability to recall past events or to imagine new scenarios. Moreover, imagination-based methods frequently require isolation of participants to concentrate, which contradicts most emotional experience that occurs in social contexts (Coan and Allen, 2007).

Film

Films, more than any other art forms, have a way of drawing viewers into a situation that helps people empathize and identify themselves with characters. Films engage visual and auditory modalities, by more closely approximating real-life emotional situations, which makes the use of film one of the most ecologically valid emotion induction techniques. To induce the desired mood, the experimenter plays a film or short scenes from a film (Westermann et al., 1996), which can be mute or not. Jurásová and Spajdel (2013) showed that mute film clips were as effective as excerpts with dialogues. The experimenter can play the film to an individual or a group of people. Peoples' reaction may differ in a group as opposed to individual setup. It is important to find films that target specific emotions.

Like imagination techniques, films are easy to use. However, the experimenter should control some physical settings of the location (e.g., room lighting, temperature, color, larger display size) in which films are presented (Coan and Allen, 2007). This method can be applied with or without explicit instructions. With instructions, participants are told to imagine and get involved in the story of the video that they are watching. Instructions can vary from simple, for example, "please watch the film carefully," to more structured, "let yourself experience your emotions fully."

Experimenters who use more than one film clips for a given emotion should note that different films may not generate similar reactions in the participants even if they are designed to target a particular emotion. Peoples' sensitivity and reactions are different. In addition, fatigue and habituation increase with more samples. Coan and Allen (2007) suggested three important factors that should be controlled in the experimental design: (1) timing—the duration of the film, (2) order in which films are presented, and (3) prior viewing.

Gross and Levenson (1995) summarized 5 years of work on developing an archive of validated films that reliably elicit each of eight emotional states (amusement, anger, contentment, disgust, fear, neutral, sadness, and surprise). Out of 250 films, 16 films were selected from the results on a sample of 494 participants. The 16 films can successfully elicit those 8 emotions (2 films for each emotion). These 16 movies also showed consistent results in a Japanese sample, and so they discussed films as universal tools to elicit emotions (Sato et al., 2007). In the following section, some of the validated movies in previous studies are listed based on the target emotion.

Amusement

When Harry Met Sally: Gross and Levenson (1995) and Macht and Mueller (2007)
The Curse of Mr. Bean: Jurásová and Spajdel (2013)
Robin Williams Live: Gross and Levenson (1995)
Slunce, seno a par facek: Jurásová and Spajdel (2013)
Mr. Bean's Holiday: Uhrig et al. (2016)
What Women Want: Uhrig et al. (2016)
Meet the Parents: Uhrig et al. (2016)
Bruce Almighty: Uhrig et al. (2016)
Drop Dead Fred: Nasoz et al. (2004)
The Great Dictator: Nasoz et al. (2004)

Anger

Cry Freedom: Coan and Allen (2007) and Gross and Levenson (1995)
My Bodyguard: FakhrHosseini et al. (2014) and Gross and Levenson (1995)
Schindler's List: Nasoz et al. (2004)

Disgust

Pink Flamingos: Coan and Allen (2007) and Gross and Levenson (1995)
Saving Private Ryan: Uhrig et al. (2016)
Joan of Arc: Uhrig et al. (2016)
Blade: Uhrig et al. (2016)

Fear

The Shining: Coan and Allen (2007) and Gross and Levenson (1995)
Silence of the Lambs: Droit-Volet et al. (2011) and Gross and Levenson (1995)
The Blair Witch Project: Droit-Volet et al. (2011)
Monster: Uhrig et al. (2016)
Kill Bill II: Uhrig et al. (2016)
Scream1: Droit-Volet et al. (2011) and Uhrig et al. (2016)

Happiness

Sister Act: Uhrig et al. (2016)
Dances with Wolves: Uhrig et al. (2016)
Big Fish: Uhrig et al. (2016)

Neutral

Alaska's Wild Denali: Coan and Allen (2007)
Hannah and Her Sisters: Gross and Levenson (1995)
Abstract shapes: Gross and Levenson (1995)
Weather news: Droit-Volet et al. (2011)

Sadness

The Champ: Gross and Levenson (1995) and Macht and Mueller (2007)
The Bear: Jurásová and Spajdel (2013)
Bambi: Gross and Levenson (1995)
Dangerous Mind: Droit-Volet et al. (2011)
Philadelphia: Droit-Volet et al. (2011)
City of Angels: Droit-Volet et al. (2011)
Gladiator: Uhrig et al. (2016)
Braveheart: Uhrig et al. (2016)
The English Patient: Uhrig et al. (2016)
Coach Carter: Uhrig et al. (2016)
Powder: Nasoz et al. (2004)

Surprise

Sea of Love: Coan and Allen (2007) and Gross and Levenson (1995)
Capricorn One: Coan and Allen (2007) and Gross and Levenson (1995)

High Arousal/Positive Valence

Blues Brothers: Wensveen et al. (2002)

High Arousal/Negative Valence

Koyaanisqatsi: Wensveen et al. (2002)

Low Arousal/Negative Valence

Stalker: Wensveen et al. (2002)

Low Arousal/Positive Valence

Easy Rider: Wensveen et al. (2002)

Sound

Specific characteristics of sounds can directly impact the human nervous system and evoke emotional changes in listeners (Bradley and Lang, 2000). In this class of emotion induction technique, participants listen to digitized sounds through speakers or headphones. The most frequently used sounds for inducing emotional states are the International Affective Digitized Sounds (IADS) (e.g., Choi et al., 2015). IADS was developed to provide a set of normative emotional stimuli for investigations of emotion and attention. IADS-2 contains ratings for 167 sounds including a wide number of contexts related to animals (e.g., puppy, dog, and robin), humans (e.g., giggling, baby, erotic couple, man wheezing, boy laughing, and child abuse), means of transport (e.g., helicopter, train, bike wreck, and plane crash), objects (e.g., music box, typewriter, and doorbell), musical instruments (e.g., harp, guitar, and bagpipes), and scenarios (e.g., country night, tropical rainforest, restaurant, and brook) (Soares et al., 2013). These emotionally evocative and internationally accessible sounds have been rated in terms of their ability to evoke changes in hedonic valence and level of arousal (Bradley and Lang, 2007), as well as their impact on discrete emotional states (happiness, sadness, fear, disgust, and anger) (Stevenson and James, 2008), among American English participants.

Bradley and Lang (2007), in the instruction manual described how they arrange the experimental situation. The procedure consists of playing each excerpt for a few seconds after a warning sound. The warning sound informs participants about each step and what they should do at that time. After listening to each excerpt, participants should reflect their immediate personal experience on the questionnaire. The advantage of using these kinds of sounds is their familiarity. They are usually heard in our daily life but may be paired with different experiences for each individual and consequently influence on their perception.

Music

Music offers another method of affect induction. Music is a specific kind of sound used to induce affect and emotion (Juslin and Västfjäll, 2008; Kreutz et al., 2007; Lundqvist et al., 2008). Researchers have found some elements of music are central to the perception of emotion, such as mode (major or minor), tempo, pitch, rhythm, harmony, and loudness (Coan and Allen, 2007). The way that these elements complement each other is different in various styles and genres of music. A particular musical style can be strongly related to a limited cluster of emotions (Terwogt and Van Grinsven, 1991).

To induce emotions with music, mood-suggestive music should be played to participants. Music is usually selected by the experimenter based on the result of a previous survey, designed to reveal correlations between musical compositions and the desired emotional state (Westermann et al., 1996). In another technique, participants choose a piece of music that they find best suited for the intended emotion. Stronger effects are typically obtained if one uses music selected by participants rather than the experimenters (Panksepp and Bernatzky, 2002) because of familiarity and emotional attachment. Panksepp and Bernatzky (2002) also investigated the profile of affect over time after induction and found that emotions were significantly highest immediately after the music and were still significant, but diminished after 10 min, and returned to baseline after 20 min. These findings highlight the time limitation of eliciting emotions with music in experimental conditions.

This method like film can be with or without explicit instructions (Westermann et al., 1996). Panksepp and Bernatzky (2002) believe emotions conveyed in music are fundamentally different from real emotions, which is considered a downside of this technique. They state that music usually creates complex forms of emotions in humans that do not serve adaptive behavioral functions, such as happiness and sadness (Scherer, 2003). Although experiencing adaptive emotions while listening to music has been reported by the listeners (Gabrielsson, 2001; Sloboda, 1992), further research should be conducted since most of the results are coming from self-report questionnaires with optional questions which may bias the listeners' opinions. Here is the list of validated music pieces used in previous studies:

Anger

The Planets (Holst): Jallais and Gilet (2010) and Mayer et al. (1995)
Night on Bald Mountain (Mussorgsky): Jallais and Gilet (2010) and Mayer et al. (1995)
Totentanz (the first 11 opening bars) (Liszt): Terwogt and Van Grinsven (1991)

Fear

Halloween (Ives): Mayer et al. (1995)
Night on the Bare Mountain (the first 34 bars) (Mussorgsky): Terwogt and Van Grinsven (1991)
Sorcerer's Apprentice (Dukas): Terwogt and Van Grinsven (1991)
Psycho (Herrmann): Mayer et al. (1995)

Happiness

Brandenburg Concertos No.3 (Bach): FakhrHosseini et al. (2014) and Mayer et al. (1995)
Midwest Train (CrusaderBeach): FakhrHosseini et al. (2014)
Mazurka from Coppélia (Delibes): Mayer et al. (1995)
Coppélia (Delibes): Jallais and Gilet (2010), Kenealy (1997), and Mayer et al. (1995)
Brandenburg Concerto No. 2 (Bach): Jallais and Gilet (2010) and Mayer et al. (1995)
Carnaval des Animeaux (bars 10–26) (Saint Saens): Terwogt and Van Grinsven (1991)
Utrecht Te Deum (bars 7–17 from the first chorus) (Handel): Terwogt and Van Grinsven (1991)
Coppélia (Delibes): Albersnagel (1988)

Neutral

Prelude l'apres Midi d'un Faune (Debussy): Albersnagel (1988)
Peter and the Wolf (Prokofiev): Terwogt and Van Grinsven (1991)

Sadness

Prelude in E Minor op. 28 (Chopin): FakhrHosseini et al. (2014) and Mayer et al. (1995)
Swan of Tuonela (Sibelius): Albersnagel (1988)
Ninth Symphony (Dvoiak): Albersnagel (1988)
Swan Lake (the first six bars from the finale) (Tchaikovsky): Terwogt and Van Grinsven (1991)
Kol Nidrei (the first eight bars) (Bruch): Terwogt and Van Grinsven (1991)
Alexander Nevsky (Prokofiev): Jallais and Gilet (2010) and Mayer et al. (1995)

Préludes (Chopin): Jallais and Gilet (2010) and Mayer et al. (1995)
Adagio in G minor (Albinoni): Kenealy (1997)
Into the Dark (Sebastian Larsson): FakhrHosseini et al. (2014)
At the Ivy Gate (Brian Crain): FakhrHosseini et al. (2014)

Anxiety

The Rite of Spring (Stravinsky): Albersnagel (1988)

Negative Mood Music With High-Arousal Excerpts

The Miraculous Mandarin (Bartok, 1918, track 1): Conklin and Perkins (2005)

Positive Mood Music With High-Arousal Excerpts

Arrival of The Queen of Sheba (Handel, 1715, track 11): Conklin and Perkins (2005)

Neutral Mood Music Excerpts Were Selected for Low Arousal

Wind on Water (Fripp, 1975, track 3): Conklin and Perkins (2005)

Images

In daily life, people seek out evocative imagery in magazines, on the internet, in various forms of social media. Although images are unimodal visual stimuli and lack the ecological validity of movies, when done well they can express and evoke emotions. Thus, researchers also use images to induce emotions. In this procedure, participants usually sit in front of a computer monitor and are shown a series of images for a few seconds each of which followed by an interstimulus interval. One of the most commonly used tools is the International Affective Picture System (IAPS; Lang et al., 1997) that provides more than 1000 systematized images (Coan and Allen, 2007). This stimulus set contains clear and colorful pictures that cover broad sample of contents across the entire affective space based on a dimensional model of emotion (Lang et al., 1997), such as loving families, undressed people, funerals, art objects, snakes, insects, attack scenes, accidents, contamination, illness, loss, pollution, puppies, babies, and landscape scenes, among others. Other than valence and arousal, a third dimension is also considered for the pictures, which is called potency or dominance. This dimension measures how controlled/in control participants feel after looking at some pictures.

To measure pleasure and arousal of IAPS stimuli, self-assessment manikin (SAM) is usually used as the rating scale (e.g., Feliu-Soler et al., 2013). SAM contains figures that show ranges from smiling and happy to frowning and unhappy to present the two dimensions of emotions. For the potency aspect, the figures in SAM range from small (not dominant) to large (dominant). Further analysis on finding correlations between the pictures and discrete emotional states did not clarify any one-to-one relationship (Coan and Allen, 2007).

The benefit of using pictures is that it is a safe and noninvasive manner and the intensity of the affective responses to the pictures is less than the real phenomenon. Note that some of the pictures include contents that may be considered objectionable by some participants; therefore, before presenting images, experimenters should inform participants about the contents of the images they will be shown.

Reading Passages

During this procedure, the participants read a few stories on the desired mood. Based on the experimental condition and goal, participants may read it aloud in front of the experimenter or quiet and alone. This method is usually paired with autobiographical recall (e.g., FakhrHosseini et al., 2014). After reading a story about the desired mood, participants will recall or write their own story with details as they found in the reading session. Here are some examples for angry, neutral, and sad conditions.

Example 1 (Anger)

I was so excited to start my first internship experience. I was assigned to develop a webpage for the company and I spent all nights to finish the assignment to meet the short deadline I was given. It was a startup company where everybody was busy and I had nobody to get a tip or feedback from for what I designed. After several attempts to ask for feedback only to be brushed off, I decided to get it done and then ask for a final checkup instead. After finalizing the work, I reported my work to one of associates, and he showed an attitude of indifference toward my work and told me although not satisfied he'll report my work to the manager for me. I thanked him and continued on to other projects assigned for me. After 2 months, I was summoned to the manager that I didn't report any progress regarding my first assignment like all the other interns had. To my bewilderment, he told me that he had waited till now thinking that I needed more time to finish it and was disappointed that I failed to do so. He even showed me an exemplary work done by another intern which looked exactly like mine! Then did I realize that the associate who I reported to was no associate but an intern who took credit.

Example 2 (Neutral)

I went to the grocery store to pick up a new carton of milk to replace the one that ran out the day before and some ingredients to make my dinner. I entered the store and grabbed a grocery cart. First, I headed over to the pasta section and picked up a bag of pasta shells and a can of tomato sauce. Then, I went to the fresh produce area to pick out some vegetables for the pasta. Finally, I went down the dairy section to grab a carton of milk. I went to the cashier to check out, unloaded the groceries into the car and drove home.

Example 3 (Sadness)

When I was in college, one day my mum called me and told me that our dog was lost. The dog was called "Sawyer," and he came to our family when he was 4 months and he was 5 years old then. At the beginning, I didn't believe that the dog couldn't be found any more. I still hoped to see him. I said to myself that he would come back. He was not actually lost, just played somewhere we couldn't find. I believed then he would come home later soon. But finally he didn't come back. Everyone in the family was very sad. My mum cried as long as we mentioned him during those days. I was so sad that even in class, I couldn't help thinking of him, thinking when he was playing in the ground with us. Several years passed, now I still miss him and sometimes I can't help thinking of his lovely face.

Writing Passages

In this class of emotion induction procedure, participants write a few paragraphs on a past desired emotional experience for a few minutes (e.g., 10–12 min) (FakhrHosseini et al., 2015). Participants are instructed to reexperience the situation and remember the details as clearly as possible:

"Please use this pen and paper to write a few paragraphs on a past emotional experience that made you angry. You need to remember the situation as clearly as possible and pretend you are in that situation again. You will have 12 min, so you can write about more than one experience if you finish before the time limit is up. Please continue writing until I tell you to stop."

In some studies, the experimenter asks the participants to read their writing aloud and the experimenter records the narratives. Next, participants listen to the playback of their recorded narratives and try to be engaged in the memory (Zhang et al., 2014). In another way, which is called solitary recollection, participants are asked to write down four events of a particular type of emotion (e.g., unhappy) and rank them. Then, they should concentrate on the highest rank events for a few minutes and remember how they felt and what happened (Martin, 1990). In his paper, Martin (1990) also talked about social recollection. This method starts in the same way as solitary recollection. After two separate participants write down several emotional events happened to them and ranked those, they are asked to talk about the most emotional memory with each other for a few minutes. While a person is talking about the memory, the other one should act as a good, sympathetic listener, and ask questions about the event.

In another study participants were asked to provide one sentence descriptions of approximately 10 sad autobiographical memories and rate them on two 10-point Likert scales to reflect how sad the memory made them feel, and how vividly they could recall the memory (Hernandez et al., 2003). In this emotion induction technique, experimenters should note that personal memories vary from individual to individual. Moreover, the experimenter is not allowed to have any control of the specific information the individual is recalling.

Embodiment

The concept of embodiment is related to the theory of William James who suggested that the experience of physiological changes lead to emotional feelings. Given the emphasis on the role of simulation and embodiment in emotion, it seems reasonable that bodily movements and posture could be used to evoke emotions (Niedenthal et al., 2005). To induce emotions with this method, participants are asked to move their facial muscles, make postures, or/and change tone-of-voice based on the given structures. First, participants should relax all of the muscles of the face and body and then follow the expression and posture manipulation instructions for a few seconds (Zhang et al., 2014). In the directed facial action task (DFA), Ekman said that they do not ask the participants to pose emotions but instead they directly give the specific instructions. For example, if they want to induce anger, the instructions would be:

pull your eyebrows down and together;
raise your upper eyelids;
now tighten your lower eyelids;
narrow, tighten, and press your lips together, pushing your lower lip up a little (Coan and Allen, 2007).

A coach provides the instructions and then observes and gives feedback the participants as they attempt. For example, the coach says if they succeed, or what they are doing wrong. Ekman (2007) mentioned participants could not see their faces in a mirror since it is less frustrating and embarrassing.

In another study, participants were asked to guess the target emotion by reading the muscular instructions. Most of the participants failed to identify the emotion from the instructions. In the next step, researchers validated their emotion induction technique (i.e., moving their facial muscles) with a self-report questionnaire to test that their method generates emotion rather than generating the physiology of emotion (Coan and Allen, 2007). This means that their participants did not recognize the target emotion from reading the instructions but could recognize them when they performed it. The feedback from contraction of specific facial muscles convey related affective information to the central nervous system that has shown 85%–90% success, specially when participants were not aware of the goal of the study (Laird et al., 1982). Interestingly, this method shows that emotional experience can be elicited without the need for emotionally evocative contexts. Ekman (2007) says "it is not essential to generate some of the elements of emotional experience which means the elements of emotion can be generated without appraisal." This phenomenon can be observed through an example of sympathy, which does not require context but still can evoke strong emotional responses. One disadvantage of this technique is that some people cannot make facial expressions required to elicit emotional experiences and those people need to be excused from participation in the study.

Virtual Reality

Other than movies and imagery techniques to elicit emotions, it has been shown that a computer is able to replicate an environment, real or imagined, that allows the user to interact with it. With virtual realities, participants can artificially experience some situations just as interacting with real-world people, objects, scenes, and events. Virtual reality (VR) allows people to immerse themselves in a social situation or a scene from a first person point of view.

To investigate the possible use of VR as an affective medium, Riva et al. (2007) analyzed the relationship between presence and emotions. The results showed that VR is an effective affective medium to produce the target emotion. The interaction between "anxious" and "relaxing" virtual environments produced anxiety and relaxation. Moreover, the feeling of presence was greater in the "emotional" environments (Riva et al., 2007). In another study, a virtual reality therapy (VRT), based on exposure to virtual environments, was used to treat social phobia. Thirty-six participants diagnosed with social phobia were assigned to either VRT or a group-CBT (cognitive behavior therapy as a control condition) and the treatment lasted 12 weeks. Results showed statistically and clinically significant improvements in the VR condition as well as the non-VR condition (Klinger et al., 2005). The usage of this technique in clinical environment helps patients recreate the world more vividly and at real times, which is hard for some participants to describe through their own imagination and memory (Baños et al., 2006; Vincelli, 1999).

Feedback

In this class of emotion induction method, participants perform a task and are then given positive or negative feedback on their performance. The tests may be different in terms of content but the most common ones are cognitive, intelligence, spatial, and analytical tests (Westermann et al., 1996). They may be told either that they did badly (failure feedback) or well (success feedback) (Martin, 1990). For example, researchers gave feedback by telling the participants a false score after they had completed the spatial reasoning test or after half time of the computer-simulated scenario with an automatic pop-up window showing their position in a fictitious ranking (positive: "You are in position 12 out of 250 and are thus better than 95.2% of all participants," negative: "You are in position 208 out of 250. That means that 83.2% of all participants have performed better than you") (Spering et al., 2005). This technique may affect self-esteem, which should be considered in the analysis (Harmon-Jones et al., 2007). Results of a metaanalysis on 32 studies with a total of 2468 participants showed that the success-failure manipulation is a reliable induction technique to evoke both positive and negative affective reactions (Nummenmaa and Niemi, 2004), which are different from discrete emotions (Harmon-Jones et al., 2007).

Self-Referent-Statement

The self-referent-statement technique was developed by Velten (1968). This method includes a number of statements describing positive or negative self-evaluations and participants repeat a list of positive or negative mood inducing words and try to feel the mood described by these statements. The statements can target different things. For example, for the elation induction, the statements are: I feel happy; I feel cheerful, confident; I can think quickly and clearly right now; Right now, I feel very contented; right now, I feel like smiling; and for the depression mood induction, the statements are: I feel fed up; I just feel drained of energy, worn out; I feel pretty low; things seem futile, pointless; I feel hopeless; I feel downhearted and miserable; I feel heavy and sluggish (Kenealy, 1997). With this technique, people have shown to remember mood congruent memories from their personal pasts (Coan and Allen, 2007). To compare the strength of two laboratory-based mood induction procedures, Velten self-statement and a manipulated success, failure, or control condition on an achievement task were investigated. Results indicated that both procedures were effective in influencing mood in intended directions. However, the Velten method caused higher ratings of depression immediately after induction, but followed by steep drop-off, compared to manipulated failure, which induced a less pronounced, but longer lasting feeling of depression.

The techniques were equivalent in generating brief increases in positive mood (Chartier and Ranieri, 1989). Albersnagel (1988) compared Velten technique with musical mood induction procedure and showed the music was superior to Velten technique. Albersnagel also found that participants' gender influenced the impact of emotion induction on emotional experience, with women being more susceptible to the effects of emotion induction than men.

Social Interaction

Sometimes, the experimenter arranges a certain social situation and engages the participants in interactions with some people that are trained to behave in a certain way (Westermann et al., 1996). This method is usually used in social psychology experiments to achieve a high level of psychological realism in the laboratory. Such methods avoid participants' awareness and allow researchers to induce some emotions, such as anger that is difficult to induce using other methods (Harmon-Jones et al., 2007). In a study, a computer simulation of a dyadic interaction was used where the participant interacted with a series of preprogrammed "computer persons" (Martin, 1990). Schachter and Singer (1962) used confederates who read from a script to induce emotion. One of the downsides of this method is that it involves extensive practice to ensure that confederates are convincing and that their behavior is consistent across experimental sessions. Lobbestael et al. (2008) found that for anger induction among four methods—film, stress interview, punishment, and harassment—two methods that included personal contact (harassment and interview) induced more significant physiological reactivity.

Physiological Manipulations

Psychedelic drugs have long used for changing the state of mind (Peet and Peters, 1995). Clinical findings show that pharmacotherapy is helpful in the treatment of mood disorder (Mathew et al., 2008; Neumann, 2008; Vollenweider and Kometer, 2010; Williams et al., 2002) and in some cases may change mood in a negative way as side effects (Ananth and Ghadirian, 1979). Schachter and Singer (1962) demonstrated that pharmacological manipulation of physiological arousal changed the participant's behavior. The disadvantage of physiological manipulation is that it requires considerable expertise to administer and extensive precautions for their safe use.

Motivated Performance Tasks

To induce anxiety, participants are asked to speak in front of an audience (Martin, 1990). Their talk will be recorded or an evaluative experimenter will attend to view the presentation. The most widely used psychosocial stress protocol is the *Trier Social Stress Test*. This test provides a protocol for inducing stress under laboratory conditions (Kirschbaum et al., 1993). It consists of a 3-min anticipatory period, a 5-min public speaking task, and a 5-min mental arithmetic task in front of an evaluative panel of two adults (Kirschbaum, 2010). This test can be applied to adolescents and adults of all ages. For children from 7 years and older, there is a modified version (Miller and Kirschbaum, 2013). Motivated performance tasks are difficult to arrange. These kinds of emotion induction procedures are usually used in the clinical settings or social psychology where the ecological validity is an important factor and determines the results.

Combined Techniques

A series of studies compared the effects of the combination of different types of emotion induction methods to increase the effectiveness of the induction. The results showed that the combination of music and pictures (e.g., IAPS) was very effective (Baumgartner et al., 2006). However, in another study, Jallais and Gilet (2010) found that autobiographical recall had greater efficiency than the combined procedure (music and guided imagery). To apply music and guided imagery, Mayer et al. (1995) first played 4 min of a related music. Then, participants performed a guided imagery task in which participants were asked to imagine as clearly as possible the situation describing by the sentences on the screen for around 30 s. In another way to apply a combined method, music can be played during the autobiographical procedure (Hernandez et al., 2003).

Considerations

Note that although emotion induction procedures target specific emotions, it gives rise to multiple affective states. One emotion can elicit other emotions (Izard, 1972). Therefore, how participants are instructed and the type of induction method influence the result. Some techniques are more successful; for example, music has the highest success rate and hypnotic suggestions have the lowest one (Clark, 1983). Albersnagel (1988) also supported the superiority of music over the Velten technique. Music is also a more subtle and easier to apply than Velten technique (Kenealy, 1997). Clark (1983) claimed that more than one third of all participants do not show mood changes in response to the Velten technique. However, Westermann et al. (1996) recommended film, especially with instructions as the strongest technique among others. Additionally, some researchers believe the combined methods intensify the elicited emotions (Mayer et al., 1995). On the other side, Jallais and Gilet (2010) showed that autobiographical recall is stronger than combined procedure. The debate on the best technique shows contradictory suggestions and is the result of experimental settings, individual differences, different periods of the induction procedures, goals of the study, experimental instructions, number of participants, measurement tools, methods, and the stimuli.

Another point that should be considered is the type of measurement used to assess emotional states. For example, self-report measures may be susceptible to demand effects. Participants may suggest more intense emotion in the self-report than their physiological response or behavioral data. Regarding the ecological validity, it is hard to say that the induced emotion is equivalent to emotions in real world. They are somehow different in terms of strength, duration, and realistic "presence." In this regard, the effect size of emotion induction techniques in studies with the goal of testing the effectiveness of an emotion induction technique is lower than the studies which aimed to investigate the impact of mood on other phenomena (Westermann et al., 1996). Moreover, people's ability to develop a kind of emotion is different. For example, women relative to men have shown stronger emotional reaction in terms of behavioral or physiological changes (Coan and Allen, 2007; Jurásová and Spajdel, 2013). Finally, researchers should be careful about the effects of emotion induction on variables of interest. Specific emotion induction procedures can elicit specific effects that may or may not be relevant to the scientist's research question. For example, if a scientist is interested in understanding the relationship between emotional music and driving, he/she should be careful about the specific effects of music on real driving, regardless of emotional content.

Ethics

It is important to consider any ethical implications of mood induction procedures beforehand. Most concerns are about the induction of certain negative emotions, such as anger, fear, and disgust. However, understanding of these emotions is very important in improving mental health and society. Moreover, these emotions are routinely experienced in everyday life, for example, watching a sad film. Participants should voluntarily take part in mood induction studies. Finally, experimenters should debrief the purpose of the study to the participants and make sure that the participants' moods are close to the baseline level when they are leaving the lab.

References

Albersnagel, F.A., 1988. Velten and musical mood induction procedures: a comparison with accessibility of thought associations. Behav. Res. Ther. 26 (1), 79–95.

Ananth, J., Ghadirian, A.M., 1979. Drug-induced mood disorders. Int. Pharmacopsychiatry 15 (1), 59–73.

Baños, R.M., Liaño, V., Botella, C., Alcañiz, M., Guerrero, B., Rey, B., 2006. Changing induced moods via virtual reality. Persuasive Technology. Springer, Berlin Heidelberg, pp. 7–15.
Baumgartner, T., Lutz, K., Schmidt, C.F., Jäncke, L., 2006. The emotional power of music: how music enhances the feeling of affective pictures. Brain Res. 1075 (1), 151–164.
Bradley, M.M., Lang, P.J., 2000. Affective reactions to acoustic stimuli. Psychophysiology 37 (02), 204–215.
Bradley, M.M., Lang, P.J., 2007. The International Affective Digitized Sounds (IADS-2): Affective Ratings of Sounds and Instruction Manual (Tech. Rep. B-3). University of Florida, Gainesville, FL.
Chartier, G.M., Ranieri, D.J., 1989. Comparison of two mood induction procedures. Cogn. Ther. Res. 13 (3), 275–282.
Choi, Y., Lee, S., Jung, S., Choi, I.M., Park, Y.K., Kim, C., 2015. Development of an auditory emotion recognition function using psychoacoustic parameters based on the International Affective Digitized Sounds. Behav. Res. Methods 47 (4), 1076–1084.
Clark, D.M., 1983. On the induction of depressed mood in the laboratory: evaluation and comparison of the velten and the musical procedures. Adv. Behav. Res. Ther. 5, 27–49.
Coan, J.A., Allen, J.J., 2007. Handbook of Emotion Elicitation and Assessment. Oxford University Press, Washington, DC.
Conklin, C.A., Perkins, K.A., 2005. Subjective and reinforcing effects of smoking during negative mood induction. J. Abnorm. Psychol. 114 (1), 153.
Droit-Volet, S., Fayolle, S.L., Gil, S., 2011. Emotion and time perception: effects of film-induced mood. Front. Integr. Neurosci. 5, 33.
Ekman, P., 2007. The directed facial action task. In: Coan, J.A., Allen, J.J. (Eds.), Handbook of Emotion Elicitation and Assessment. Oxford University Press, New York, pp. 47–53.
FakhrHosseini, S., Jeon, M., Bose, R., 2015. Estimation of drivers' emotional states based on neuroergonmic equipment: an exploratory study using fNIRS. Proceedings of the Seventh International Conference on Automotive User Interfaces and Interactive Vehicular Applications (AutoUI'15), Nottingham, UK.
FakhrHosseini, S., Landry, S., Tan, Y., Bhattarai, S., Jeon, M., 2014. If you're angry, turn the music on: music anger effects on driving performance. Proceedings of the Sixth International Conference on Automotive User Interfaces and Interactive Vehicular Applications (AutoUI'14). Seattle, WA, USA.
Feliu-Soler, A., Pascual, J.C., Soler, J., Pérez, V., Armario, A., Carrasco, J., et al., 2013. Emotional responses to a negative emotion induction procedure in borderline personality disorder. Int. J. Clin. Health Psychol. 13 (1), 9–17.
Gabrielsson, A., 2001. Emotions in strong experiences with music. In: Juslin, P.N., Sloboda, J.A. (Eds.), Music and Emotion: Theory and Research. Oxford University Press, pp. 431–449.
Gross, J.J., Levenson, R.W., 1995. Emotion elicitation using films. Cogn. Emot. 9 (1), 87–108.
Harmon-Jones, E., Amodio, D.M., Zinner, L.R., 2007. Social psychological methods of emotion elicitation. Handbook of Emotion Elicitation and Assessment Oxford University Press, pp. 91–105.
Hernandez, S., Vander Wal, J.S., Spring, B., 2003. A negative mood induction procedure with efficacy across repeated administrations in women. J. Psychopathol. Behav. Assess. 25 (1), 49–55.
Izard, C.E., 1972. An empirical analysis of anxiety in term of discrete emotion. Patterns of Emotions: A New Analysis of Anxiety and Depression. Academic Press, Washington, DC, p. 65.
Jallais, C., Gilet, A.L., 2010. Inducing changes in arousal and valence: comparison of two mood induction procedures. Behav. Res. Methods 42 (1), 318–325.
Juslin, P.N., Västfjäll, D., 2008. Emotional responses to music: the need to consider underlying mechanisms. Behav. Brain Sci. 31 (05), 559–575.
Jurásová, K., Spajdel, M., 2013. Development and assessment of film excerpts used for emotion elicitation. Activitas Nervosa Supe rior Rediviva 55 (3), 135–140.

Kenealy, P.M., 1997. Mood state-dependent retrieval: the effects of induced mood on memory reconsidered. Quart. J. Exp. Psychol. A 50 (2), 290–317.

Kirschbaum, C., 2010. Trier social stress test. Encyclopedia of Psychopharmacology. Springer, Berlin Heidelberg, 1346–1346.

Kirschbaum, C., Pirke, K.M., Hellhammer, D.H., 1993. The 'Trier Social Stress Test'–a tool for investigating psychobiological stress responses in a laboratory setting. Neuropsychobiology 28 (1–2), 76–81.

Klinger, E., Bouchard, S., Légeron, P., Roy, S., Lauer, F., Chemin, I., Nugues, P., 2005. Virtual reality therapy versus cognitive behavior therapy for social phobia: a preliminary controlled study. Cyberpsychol. Behav. 8 (1), 76–88.

Kreutz, G., Ott, U., Teichmann, D., Osawa, P., Vaitl, D., 2007. Using music to induce emotions: influences of musical preference and absorption. Psychol. Music 36 (1), 101–126.

Laird, J.D., Wagener, J.J., Halal, M., Szegda, M., 1982. Remembering what you feel: effects of emotion on memory. J. Pers. Soc. Psychol. 42 (4), 646.

Lang, P.J., Bradley, M.M., Cuthbert, B.N., 1997. International Affective Picture System (IAPS): Technical Manual And Affective Ratings. NIMH Center for the Study of Emotion and Attention, pp. 39–58.

Lobbestael, J., Arntz, A., Wiers, R.W., 2008. How to push someone's buttons: a comparison of four anger-induction methods. Cogn. Emot. 22 (2), 353–373.

Lundqvist, L.O., Carlsson, F., Hilmersson, P., Juslin, P., 2008. Emotional responses to music: experience, expression, and physiology. Psychol. Music 37 (1), 61–90.

Macht, M., Mueller, J., 2007. Immediate effects of chocolate on experimentally induced mood states. Appetite 49 (3), 667–674.

Martin, M., 1990. On the induction of mood. Clin. Psychol. Rev. 10 (6), 669–697.

Mathew, S.J., Manji, H.K., Charney, D.S., 2008. Novel drugs and therapeutic targets for severe mood disorders. Neuropsychopharmacology 33 (9), 2080–2092.

Mayer, J.D., Allen, I.P., Beauregard, K., 1995. Mood inductions for four specific moods: a procedure employing guided imagery. J. Ment. Imagery 19 (1–2), 133–150.

Miller, R., Kirschbaum, C., 2013. Trier Social Stress Test. Encyclopedia of Behavioral Medicine. Springer, New York, pp. 2005–2008.

Nasoz, F., Alvarez, K., Lisetti, C.L., Finkelstein, N., 2004. Emotion recognition from physiological signals using wireless sensors for presence technologies. Cogn. Technol. Work 6 (1), 4–14.

Neumann, I.D., 2008. Brain oxytoc: a key regulator of emotional and social behaviours in both females and males. J. Neuroendocrinol. 20 (6), 858–865.

Niedenthal, P.M., Barsalou, L.W., Ric, F., Krauth-Gruber, S., 2005. Embodiment in the acquisition and use of emotion knowledge. In: Barrett, L.F., Niedenthal, P.M., Winkielman, P. (Eds.), Emotion and Consciousness. The Guilford Press, New York, pp. 21–50.

Nummenmaa, L., Niemi, P., 2004. Inducing affective states with success-failure manipulations: a meta-analysis. Emotion 4 (2), 207.

Panksepp, J., Bernatzky, G., 2002. Emotional sounds and the brain: the neuro-affective foundations of musical appreciation. Behav. Processes 60 (2), 133–155.

Peet, M., Peters, S., 1995. Drug-induced mania. Drug Saf. 12 (2), 146–153.

Picard, R.W., Picard, R., 1997. Affective Computing, vol. 252. MIT press, Cambridge.

Picard, R.W., Klein, J., 2002. Computers that recognise and respond to user emotion: theoretical and practical implications. Interact. Comput. 14 (2), 141–169.

Quigley, K.S., Lindquist, K.A., Barrett, L.F., 2014. Inducing and Measuring Emotion and Affect: Tips, Tricks, and Secrets. Handbook of Research Methods in Social and Personality Psychology. Cambridge University Press, New York.

Riva, G., Mantovani, F., Capideville, C.S., Preziosa, A., Morganti, F., Villani, D., et al., 2007. Affective interactions using virtual reality: the link between presence and emotions. Cyberpsychol. Behav. 10 (1), 45–56.

Sato, W., Noguchi, M., Yoshikawa, S., 2007. Emotion elicitation effect of films in a Japanese sample. Soc. Behav. Pers. Int. J. 35 (7), 863–874.

Schachter, S., Singer, J., 1962. Cognitive, social, and physiological determinants of emotional state. Psychol. Rev. 69 (5), 379.
Scherer, K.R., 2003. Why music does not produce basic emotions: a plea for a new approach to measuring emotional effects of music. In: Bresin, R. (Ed.), Proceedings of the Stockholm Music Acoustics Conference 2003. Royal Institute of Technology, Stockholm, pp. 25–28.
Sears, A., Jacko, J.A. (Eds.), 2009. Human-Computer Interaction: Development Process. CRC Press, Washington, DC.
Sloboda, J.A., 1992. Empirical studies of emotional response to music. In: Riess-Jones, M., Holleran, S. (Eds.), Cognitive Bases of Musical Communication. Amer Psychological Assn, Washington, DC, pp. 33–46.
Soares, A.P., Pinheiro, A.P., Costa, A., Frade, C.S., Comesaña, M., Pureza, R., 2013. Affective auditory stimuli: adaptation of the international affective digitized sounds (IADS-2) for European Portuguese. Behav. Res. Methods 45 (4), 1168–1181.
Spering, M., Wagener, D., Funke, J., 2005. The role of emotions in complex problem-solving. Cogn. Emot. 19, 1252–1261.
Stevenson, R.A., James, T.W., 2008. Affective auditory stimuli: characterization of the international affective digitized sounds (IADS) by discrete emotional categories. Behav. Res. Methods 40, 315–321.
Terwogt, M.M., Van Grinsven, F., 1991. Musical expression of mood states. Psychol. Music 19 (2), 99–109.
Uhrig, M.K., Trautmann, N., Baumgärtner, U., Treede, R.D., Henrich, F., Hiller, W., Marschall, S., 2016. Emotion elicitation: a comparison of pictures and films. Front. Psychol. 7.
Velten, E., 1968. A laboratory task for induction of mood states. Behav. Res. Ther. 6 (4), 473–482.
Vincelli, F., 1999. From imagination to virtual reality: the future of clinical psychology. Cyberpsychol. Behav. 2 (3), 241–248.
Vollenweider, F.X., Kometer, M., 2010. The neurobiology of psychedelic drugs: implications for the treatment of mood disorders. Nat. Rev. Neurosci. 11 (9), 642–651.
Wensveen, S., Overbeeke, K., Djajadiningrat, T., 2002. Push me, shove me and I show you how you feel: recognising mood from emotionally rich interaction. Proceedings of the Fourth Conference on Designing Interactive Systems: Processes, Practices, Methods, and Techniques. ACM, pp. 335–340.
Westermann, R., STAHL, G., Hesse, F., 1996. Relative effectiveness and validity of mood induction procedures: analysis. Eur. J. Soc. Psychol. 26, 557–580.
Williams, R.S., Cheng, L., Mudge, A.W., Harwood, A.J., 2002. A common mechanism of action for three mood-stabilizing drugs. Nature 417 (6886), 292–295.
Zhang, X., Yu, H.W., Barrett, L.F., 2014. How does this make you feel? A comparison of four affect induction procedures. Front. Psychol 5, 689.

Further Reading

Aljanaki, A., Wiering, F., Veltkamp, R.C., 2016. Studying emotion induced by music through a crowd sourcing game. Inf. Process. Manag. 52 (1), 115–128.
Blair, R.J.R., Curran, H.V., 1999. Selective impairment in the recognition of anger induced by diazepam. Psychopharmacology 147 (3), 335–338.
Creed, C., Beale, R., 2005. Using emotion simulation to influence user attitudes and behaviour. Proceedings of the 2005 Workshop on the Role of Emotion in HCI.
Schneider, F., Gur, R.C., Gur, R.E., Muenz, L.R., 1994. Standardized mood induction with happy and sad facial expressions. Psychiatry Res. 51 (1), 19–31.

CHAPTER

11

Affect Measurement: A Roadmap Through Approaches, Technologies, and Data Analysis

*Javier Gonzalez-Sanchez**, *Mustafa Baydogan***, *Maria Elena Chavez-Echeagaray**, *Robert K. Atkinson**, *Winslow Burleson*†

***Arizona State University, Tempe, AZ, United States; **Bogazici University, Istanbul, Turkey; †New York University, New York, NY, United States**

Affect is inextricably related to human cognitive processes and expresses a great deal about human necessities (Picard, 1997); affect signals what matters to us and what we care about. Furthermore, affect impacts our rational decision-making and action selection (Picard, 2010). Providing computers with the capability to recognize, understand, and respond to human affective states would narrow the communication gap between the highly emotional human and the emotionally detached computer, enhancing their interactions. Computer applications in learning, health care, and entertainment stand to benefit from such capabilities.

Affect is a conceptual quantity with fuzzy boundaries and with substantial individual difference variations in expression and experience (Picard, 1997). This makes measuring affect a challenging task. This chapter does not intend to present a comprehensive survey of the methods used to measure affect but instead provides a roadmap through a selection of the principal approaches and technologies. Section "Affect, Emotions, and Measurement" introduces the concepts and background behind affect measurement. Section "Gathering Data: Approaches and Technologies"

Emotions and Affect in Human Factors and Human–Computer Interaction. http://dx.doi.org/10.1016/B978-0-12-801851-4.00011-2
Copyright © 2017 Elsevier Inc. All rights reserved.

presents approaches and technologies, explaining the type of data gathered from each, its characteristics, and its pros and cons. Section "Data Handling: Sampling, Filtering, and Integration" describes the stages of data handling, which include data sampling, data filtering, and data integration from a variety of sources. Section "Data Analysis" describes tools and techniques for data analysis that correlate affect measurements with stimuli, and presents examples of the application of these tools in the analysis of data samples collected in experimental studies.

AFFECT, EMOTIONS, AND MEASUREMENT

Affect is a construct of neural activity and psychological reactions; it is used as an encompassing term to describe emotion, feelings, and mood because they are so closely related and almost simultaneous in occurrence. Although emotion and mood are states of mind and, as such, are indicators of experiencing feeling or affect, the terms emotion and affect are frequently used interchangeably because they are so closely related.

Some theories propose that emotions are states embodied in the peripheral physiology and assume that a prototypal electro-physiological response exists for each emotion. Therefore, emotions can be detected by analyzing electrophysiological changes and identifying patterns associated with a particular emotion. Automatic affect recognition is a two-step process. First, data is gathered from electrophysiological manifestations of affect using sensing devices; these devices range from brain–computer interfaces (BCI), eye-tracking systems, text-based recognition, and cameras for facial gesture and body language recognition to physiological sensors that collect data regarding skin conductance, heart rate variability (HRV), and voice features, among others. Second, the vast amount of data gathered by the sensors is processed with the aim of inferring affective states by applying machine learning and data mining algorithms; commonly used machine learning and data mining algorithms include rule-based models, support-vector machines, Bayesian networks, hidden Markov models, and neural networks, as well as *k*-nearest neighbors, decision trees, and Gaussian mixture models (Calvo and D'Mello, 2010).

Using several sources of data, either to recognize a broad range of emotions or to improve the accuracy of recognizing a single emotion, is referred to as multimodal affect recognition. Multimodality requires a third step, integrating the information. This step either integrates the data from a number of sensing devices before running the inference process or integrates the inferences made from each device's data separately. An analysis of integration approaches and methods is presented in Novak et al. (2012). Fig. 11.1 summarizes the three steps in multimodal affect recognition.

The integration of inferences requires the adoption of a model that describes the relationships between those inferences, each usually

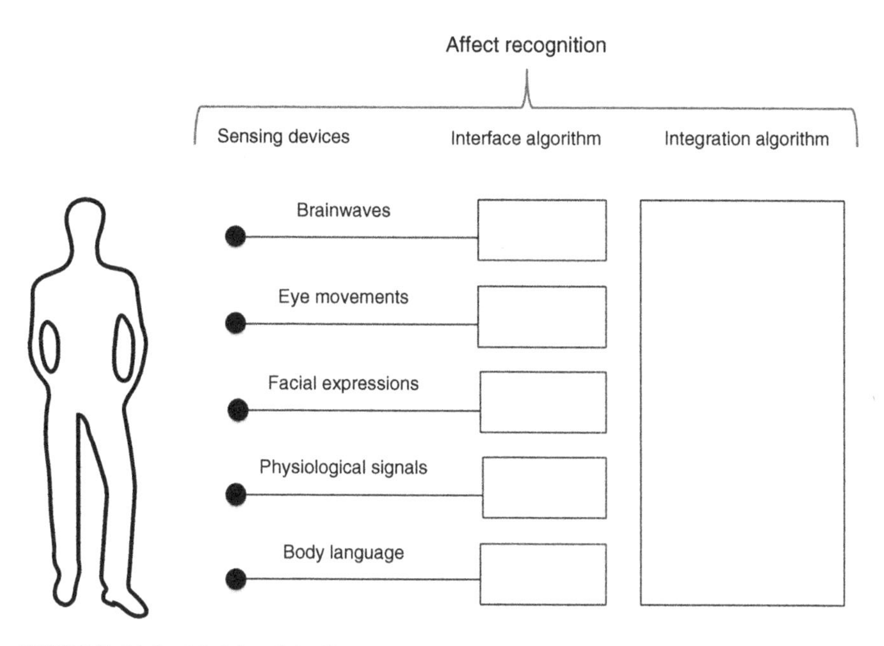

FIGURE 11.1 Multimodal affect recognition three-step process. *Source: A modified version of Fig. 1 from Gonzalez-Sanchez, J., Atkinson, R., Burleson, W., Chavez-Echeagaray, M.E., 2011. ABE: An agent-based software architecture for a multimodal emotion recognition framework. Presented at the Proceedings of the 2011 Ninth Working IEEE/IFIP Conference on Software Architecture. IEEE Computer Society, Washington, DC, USA, pp. 187–193.*

representing an individual emotion. These models are called emotional models (Gilroy et al., 2009). The classification of emotions is an ongoing aspect of affective science and experts still struggle to reconcile competing emotional models. To date, two emotional models have come to the fore: the discrete model and the continuous dimensional model.

The discrete model assumes emotions are discrete values with only a finite number of possible values, and that they are fundamentally different constructs (Ekman, 1992). A limitation of this model is that it focuses on strong emotions (such as disgust, sadness, happiness, fear, anger, and surprise) and it cannot accommodate a variety of closely related emotions or combinations of emotions.

The continuous dimensional model asserts that affective states are continuous values in one or more dimensions, and conceptualizes emotions by defining where they lie in that dimensional space. Russell (1980) proposed a two-dimensional model that links arousal with pleasure. Arousal measures intensity, or how energized or soporific one feels, ranging from calmness to excitement. Pleasure measures how pleasant or unpleasant one feels, ranging from positive to negative. For instance, while both boredom and frustration are unpleasant emotions, frustration has a higher level of arousal.

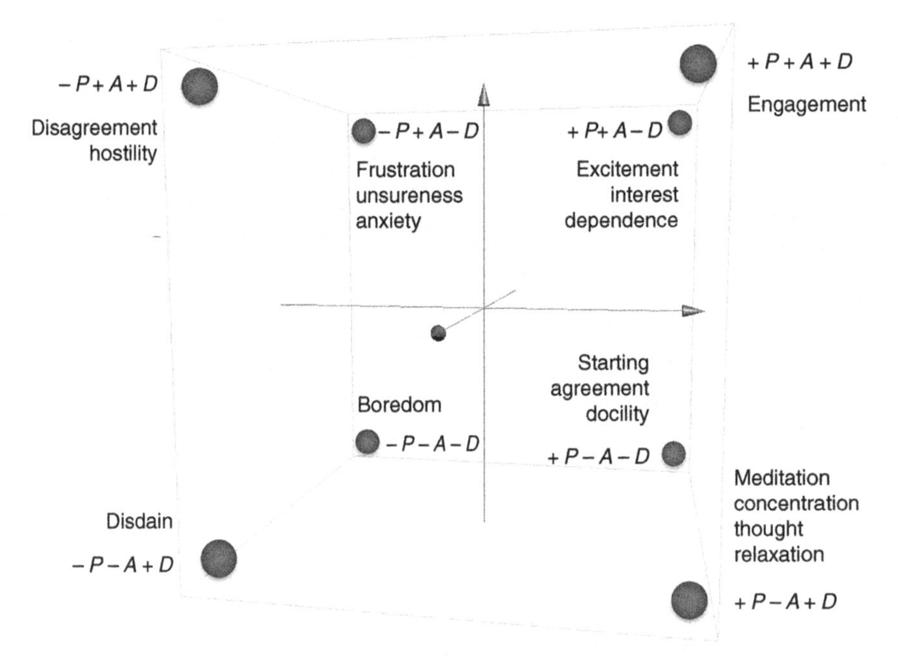

FIGURE 11.2 Pleasure-Arousal-Dominance dimensional model. *Source: From Gonzalez Sanchez, J., 2016. Affect-Driven Self-Adaptation: A Manufacturing Vision with a Software Product Line Paradigm. Doctoral dissertation. Arizona State University, Arizona, United States.*

Mehrabian (1996) proposed expanding the two-dimensional model to three dimensions by adding another axis: dominance. Dominance represents how controlling and dominant versus how controlled or submissive one feels. For instance, while both frustration and disagreement are unpleasant emotions, disagreement is a dominant emotion and frustration is submissive. Fig. 11.2 shows the three-dimensional model and plots some emotions with their pleasure–arousal–dominance (PAD) vectors.

Since continuous values characterize the measure of affect during human interactions, Mehrabian's model, also known as the PAD model, has been recommended for real-time emotion recognition. Gilroy et al. (2009) describe a case study using the PAD model in real-time for an art installation.

GATHERING DATA: APPROACHES AND TECHNOLOGIES

Having reviewed the conceptual background to affect measurement, our roadmap starts by describing some popular, inexpensive, easy to install, and widely available sensing approaches and technologies that we have been using in our research over the last 4 years. For each of them, we describe the data gathered and its characteristics, as well as outlining its pros and cons.

Brain–Computer Interfaces

BCI is a physiological instrument that uses brainwaves as data sources. Most BCIs work under the principles of electroencephalography (EEG), recording electrical activity along the scalp produced by the firing of neurons within the brain. BCI devices of varying accuracy and cost are widely available. Three devices that we have had the opportunity to work with are described below:

1. *NeuroSky biosensor*. This device facilitates low-cost EEG-linked research using one dry sensor, situated at the forehead (Fig. 11.3A), which provides a very easy and almost nonintrusive setup. It provides raw data at a sampling rate of 512 Hz. Its software is able to extract constructs for attention, meditation, and eye blinking, as well as delta, theta, low alpha, high alpha, low beta, high beta, and gamma waves.
2. *Emotiv EPOC headset*. Emotiv raw data output includes 14 values, 7 channels for each brain hemisphere. Electrodes for these channels are situated and labeled according to the CMS/DRL configuration: AF3, F7, F3, FC5, T7, P7, O1, O2, P8, T8, FC6, F4, F8, and AF4 (American Electroencephalographic Society, 1990), as shown in Fig. 11.3B. Additionally, two accelerometers track head movements; they are reported as AccX and AccY. Accelerometer values can be used to identify nodding, headshaking, or yes/no indication. Its setup is straightforward and can be accomplished in as little as 2–3 min with one caveat: users with thick or curly hair may present a challenge. The device requires that the electrodes be coated in a multipurpose solution. They normally remain wet for about 1 h before needing to be remoistened. While using this headset, it is important that

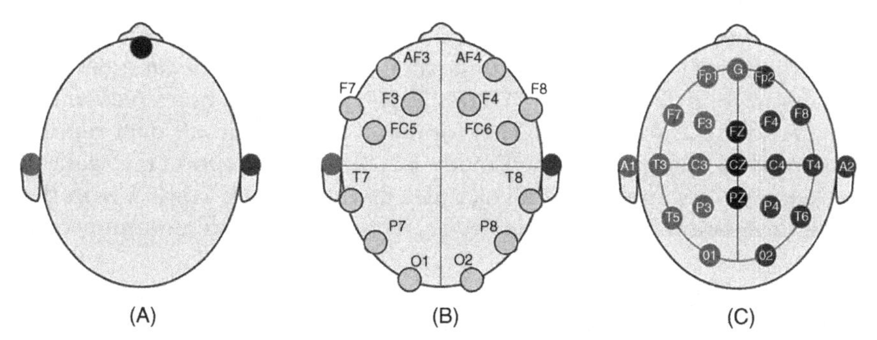

FIGURE 11.3 BCI devices range from simple to complex; from (A) 1 channel at the forehead, NeuroSky; to (B) 14 CMS/DRL channels, Emotiv EPOC; to (C) standard international 10-20 configuration, ABM B-Alert X-Series.

the user does not chew gum since chewing generates noise signals. The device reports raw data at a sampling rate of 128 Hz, reporting packets of 128 samples each time. This system is able to infer five emotional constructs (engagement, boredom, excitement, frustration, and meditation) and to detect a wide range of facial gestures (blink, wink left and right, look left and right, raise brow, furrow brow, smile, clench, smirk left and right, and laugh). Emotional constructs and facial gestures are reported at a sampling rate of 8 Hz.

3. *ABM B-Alert X-Series EEG systems*. B-Alert X-Series systems[a] are appropriate for the high-quality, nonmedical wireless acquisition of EEG and physiological signals. It generates validated cognitive state metric and cognitive workload metric computations in real-time or during offline analyses. It applies sensors according to the standard international 10–20 system, as shown in Fig. 11.3C. It reports data at a sampling rate of 256 Hz. Its setup requires the application of conductive foam between the electrodes and the scalp. Similar to the Emotiv system, thick or curly hair can represent a challenge to fitting the electrodes. The setup of the headset includes an impedance test (5 min) to test each node's connection with the scalp, as well as a baseline test (15–20 min) to normalize the system for each individual user. The latter comprises a vigilance task, in which the user responds to visual stimuli, and an audio task, in which the user reacts to audio tones. A user's baseline test results can be saved and reused. Due to the sensitivity of this headset, the B-Alert User Manual (version 2.0 from 2011) recommends ensuring that the user is well rested (not sleepy) and that the user did not consume nicotine or caffeine immediately prior to the experiment.

Table 11.1 contains an example of raw data collected from a BCI, specifically an extract of a dataset collected using the Emotiv EPOC headset. Note that a timestamp is included for each row. The timestamp is a 15-digit string in which each consecutive pair of numbers respectively indicates the year, month, day, hour, minute, and second value, and the final three digits indicate the millisecond value. For instance, the timestamp in the first row in Table 11.1 is 141116112544901, which denotes November 16, 2014 at 11:25:44.901 a.m. This format is followed in all data reports presented in this chapter. The Emotiv EPOC headset reports raw data in packets of 128 samples and all samples in a packet are labeled with the same timestamp. Thus, for this device, a change in the timestamp occurs every 128 rows.

Table 11.2 shows a sample dataset of the affective constructs produced by Emotiv software. Notice that the construct values are expressed as probabilities (values ranging from 0 to 1). The construct excitement is reported

[a]http://www.biopac.com/B-Alert-X10-Analysis-Software

TABLE 11.1 Extract of Raw Data Collected Using the Emotiv EPOC Headset

Timestamp	AF3	F7	F3	FC5	T7	P7	O1	O2	P8	T8	FC6	F4	F8	AF4	AccX	AccY
141116112544901	4542.05	4831.79	4247.18	4690.26	4282.56	4395.38	4591.79	4569.23	4360	4570.77	4297.44	4311.28	4282.56	4367.18	1660	2003
141116112544901	4536.92	4802.05	4243.08	4673.85	4272.31	4393.33	4592.82	4570.26	4354.87	4570.26	4292.31	4309.74	4277.95	4370.77	1658	2002
141116112545010	4533.33	4798.97	4234.87	4669.74	4301.03	4396.92	4592.31	4570.77	4351.28	4561.03	4281.54	4301.54	4271.28	4363.59	1659	2003
141116112545010	4549.23	4839.49	4241.03	4691.28	4333.85	4397.95	4596.41	4567.18	4355.9	4556.41	4286.15	4306.15	4277.95	4369.74	1659	2003
141116112545010	4580	4865.64	4251.79	4710.26	4340	4401.54	4603.59	4572.82	4360	4558.46	4298.97	4324.62	4296.41	4395.9	1657	2004
141116112545010	4597.44	4860	4252.82	4705.64	4350.26	4412.31	4603.59	4577.44	4357.44	4555.9	4295.38	4329.23	4296.41	4414.36	1656	2005
141116112545010	4584.62	4847.69	4246.67	4690.26	4360	4409.23	4597.44	4569.74	4351.79	4549.74	4278.97	4316.92	4272.82	4399.49	1656	2006
141116112545010	4566.15	4842.05	4238.46	4684.1	4322.05	4389.74	4592.82	4566.67	4351.79	4549.74	4274.36	4310.26	4262.05	4370.77	1655	2005
141116112545010	4563.59	4844.62	4231.79	4687.69	4267.69	4387.69	4594.36	4580	4361.03	4556.41	4278.97	4310.77	4274.36	4370.77	1653	2006
141116112545010	4567.18	4847.18	4233.33	4688.72	4285.13	4409.23	4602.05	4589.23	4368.21	4560	4280.51	4310.77	4281.54	4390.26	1655	2004

TABLE 11.2 Extract of Affective Constructs Reported by the Emotiv EPOC Headset

Timestamp	Short-term excitement	Long-term excitement	Engagement	Meditation	Frustration
141116091145065	0.447595	0.54871	0.834476	0.333844	0.536197
141116091145190	0.447595	0.54871	0.834476	0.333844	0.536197
141116091145315	0.447595	0.54871	0.834476	0.333844	0.536197
141116091145440	0.487864	0.546877	0.834146	0.339548	0.54851
141116091145565	0.487864	0.546877	0.834146	0.339548	0.54851
141116091145690	0.487864	0.546877	0.834146	0.339548	0.54851
141116091145815	0.487864	0.546877	0.834146	0.339548	0.54851
141116091145940	0.521663	0.545609	0.839321	0.348321	0.558228
141116091146065	0.521663	0.545609	0.839321	0.348321	0.558228
141116091146190	0.521663	0.545609	0.839321	0.348321	0.558228

twice: once as short-term excitement, reflecting an immediate change, and then as long-term excitement, using the cumulative readings over time to calculate a value that reflects the overall change. Timestamps confirm a sampling rate for affective constructs at 8 Hz (one sample every 125 ms).

In general, BCIs are easy to use and portable. However, it is worth noting that they require time for setup and calibration (from an average of 10 min for the Emotiv headset to an average of 40 min for the ABM B-Alert), and variables, such as the battery level of the headset, the noise in the environment, and the connection between the electrodes and the scalp should be monitored continuously. Furthermore, the headset can only be a limited distance from the host computer, and the number of headsets that can operate in the same room without creating interference is also a limited. The presence of other wireless devices using the same bandwidth can also produce interference. Ongoing research includes the development of EEG devices that use dry electrodes; these are becoming more readily available and overcome the limitations of scalp preparation and wet gels. Cognionics and Wearable Sensing are examples of companies working on improving this aspect of this device.

Facial Gestures

Emotion recognition systems based on facial gesture enable real-time analysis, tagging, and inference of cognitive affective states from a video recording of the face. It is assumed that facial expressions are triggered for a period of time when an emotion is experienced and so emotion detection can be achieved by detecting the facial expression related to it. From each

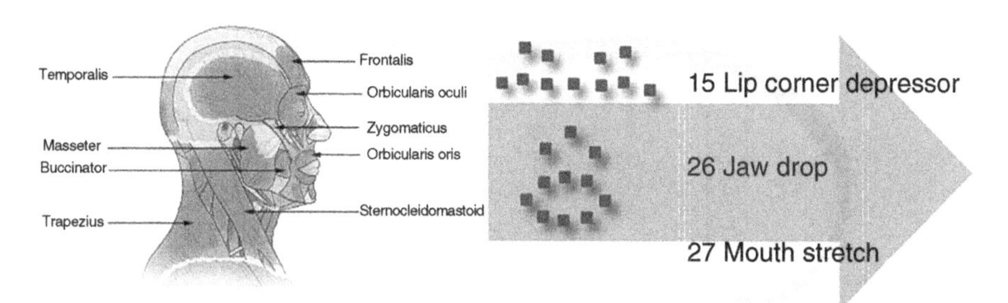

FIGURE 11.4 **Movements of individual facial muscles are encoded as action units. For instance, lip corner depression is coded as action unit 15, jaw drop as action unit 26, and mouth stretch as action unit 27.**

facial expression, a set of facial action units is extracted, each facial action unit identifying an independent motion of the face. Movements in facial muscles are perceived as changes in the position of the eyes, nose, and mouth. Computer systems implement this approach by capturing images of the user's facial expressions and head movements. Those systems detect changes in the position of the eyes, nose, and mouth as changes in the position of dots in a coordinate system. Then, by analyzing those changes, the occurrence of a facial action unit can be determined. The Facial Action Coding System (FACS) documents 46 possible facial action units (Ekman et al., 1980). For instance, happiness is associated with the occurrence of action units 6 and 12 (cheek raiser and lip corner puller), and sadness is associated with the occurrence of action units 1, 4, and 15 (inner brow raiser, brow lowerer, and lip corner depressor). Fig. 11.4 schematizes the process.

Diverse methodologies are used to infer affect from facial images, some of which are summarized in Table 1 of Calvo and D'Mello (2010). Examples of emotion recognition systems based on facial gesture that conduct real-time frame-by-frame facial expression recognition from a video stream include:

1. *MindReader* (Kaliouby and Robinson, 2005) uses a standard 30 fps USB webcam to capture the user's face. It also takes into account an analysis of head and shoulder movements. It provides results at a sampling rate of 10 Hz and is able to infer the affective states of agreement, concentration, disagreement, interest, thought, and unsureness.
2. *iMotion FACET*[b] uses a standard 30 fps USB webcam to capture the user's face. It provides results at a sampling rate of 30 Hz and is able to infer nine states: seven basic emotions (joy, anger, fear, sadness, disgust, surprise, and contempt) and two complex emotions (confusion and frustration).

Table 11.3 shows a sample dataset for affect constructs provided, in real-time, by MindReader software. The constructs are probabilities (values

[b]http://imotionsglobal.com

TABLE 11.3 Extract of Affective Constructs Reported by MindReader

Timestamp	Agreement	Concentration	Disagreement	Interest	Thought	Unsureness
141116112838516	0.001836032	0.999917000	0.000179000	0.164854060	0.571142550	0.045950620
141116112838578	0.001447654	0.999951600	0.000129000	0.163106830	0.595892100	0.042706452
141116112838672	0.000597000	0	0.000150000	0.449962940	0.455276130	0.007896970
141116112838766	0.000246000	0	0.000175000	0.774456860	0.321447520	0.001418217
141116112838860	0.000101000	0	0.000204000	0.935119150	0.211671380	0.000253000
141116112838953	0.000041800	0	0.000238000	0.983739000	0.132086770	0.000045200
141116112839016	0.000017200	0	0.000278000	0.996077400	0.079410380	0.000008070
141116112839110	0.000007100	0	0.000324000	0.999062660	0.046613157	0.000001440
141116112839156	0.000002920	0	0.000377000	0.999776540	0.026964737	0.000000257
141116112839250	0.000001210	0	0.000440000	0.999946700	0.015464196	0.000000046

ranging from 0 to 1). The timestamp column shows an approximate sampling rate of 10 Hz (one sample every 100 ms).

The use of emotion recognition systems based on facial gesture has a number of advantages. They are largely nonintrusive because they do not involve attaching sensors to the user and are inexpensive because they do not require expensive hardware. They also have reasonable accuracy, although the accuracy is challenged when the user's movements cause the face to turn away from the camera's line of sight, such as when a user lies back in the chair or looks down to read or write. Cohn and De la Torre (2014) describe in more detail the challenges associated with automated facial image analysis.

Eye Tracking

Eye-tracking systems measure eye position, eye movement, and pupil size to detect zones in which the user has a particular interest at a specific time. There are a number of methods for measuring eye movement. The most popular are optical methods, in which light, typically infrared, is reflected from the eye and sensed by a camera or some other specially designed optical sensor. The data is then analyzed to extract eye rotation from changes in the reflections. Optical methods are widely used for gaze tracking and are favored for being noninvasive and inexpensive. An example of a commercial optical eye-tracking system is the Tobii T60XL Eye Tracker. The Tobii Eye Tracker reports data at a sampling rate of 60 Hz, and the reported data includes attention direction as a gaze point (x and y coordinates), duration of fixation, and pupil dilation.

Pupil diameter has been demonstrated to be an indicator of emotional arousal, as seen in Bradley et al. (2008), who found that pupillary changes were larger when viewing emotionally arousing pictures, regardless of whether these were pleasant or unpleasant. Pupillary changes during picture viewing covaried with skin conductance changes, supporting the interpretation that sympathetic nervous system activity modulates these changes.

A sample dataset from a Tobii T60XL Eye Tracker is shown in Table 11.4. Gaze point values (GPX and GPY columns) range from 0 to the size of the display; pupil (left and right) is the size of the pupil in millimeters; validity (left and right) is an integer value ranging from 0 to 4 (0 if the eye is found and the tracking quality is good and 4 if the eye cannot be located by the eye tracker); and fixation zone is a sequential number corresponding to one or a set of predefined zones in which special interest exists. Timestamps in the table confirm a sampling rate of 60 Hz (approximately one sample every 16 ms).

Eye-tracking systems can be fixed (embedded in a display), mobile (able to be connected and mounted in diverse displays), or wearable

TABLE 11.4 Extract of Data Collected Using Tobii T60XL Eye Tracker

Timestamp	GPX	GPY	Pupil left	Validity *L*	Pupil right	Validity *R*	Fixation zone
141124162405582	636	199	2.759313	0	2.88406	0	48
141124162405599	641	207	2.684893	0	2.855817	0	48
141124162405615	659	211	2.624458	0	2.903861	0	48
141124162405632	644	201	2.636186	0	2.916132	0	48
141124162405649	644	213	2.690685	0	2.831013	0	48
141124162405666	628	194	2.651784	0	2.869714	0	48
141124162405682	614	177	2.829281	0	2.899828	0	48
141124162405699	701	249	2.780344	0	2.907665	0	49
141124162405716	906	341	2.853761	0	2.916398	0	49
141124162405732	947	398	2.829427	0	2.889944	0	49

(embedded in a pair of glasses). Regardless of the type of system, the set-up process is fairly easy. The calibration process includes having the user follow an object around the display area with their eyes (for embedded and mobile systems), or having them stare at a particular point (for wearable glasses). The calibration for embedded and mobile systems requires time to ensure that the eyes of the user are within the line of sight of the infrared and optical sensors and that nothing is producing glare for the camera, which could affect the reflection and thus the tracking of eye movements. The reliability of embedded and mobile systems is reduced by glare on the cameras, the incorrect position of the user's face, and the presence of framed glasses or eye disorders, such as strabismus. In the case of systems in glasses, important things to consider are the interference of other wireless devices and the distance that the glasses can be from the host computer.

Physiological Sensor: Skin Conductance

Electrodermal activity (EDA) is the continuous variation in the electrical characteristics of the skin, which varies with the moisture level. The moisture level depends on the sweat glands and blood flow, which are controlled by the sympathetic and parasympathetic nervous systems. Although the electrical variation alone does not identify a specific emotion, a relationship has been established between this and emotional arousal, that is, how energized or soporific the user feels. A skin conductance device senses EDA by measuring the conductance of the skin. To measure skin conductance, the electrical resistance between two electrodes, normally attached to the skin about an inch apart, when a very weak current is steadily passed between them is measured. Skin conductance is perhaps the most inexpensive method discussed here in terms of the hardware and software required. Examples of skin conductance sensors used in our research are:

1. A wireless Bluetooth device that reports data at a sampling rate of 2 Hz, designed by MIT Media Lab (Strauss et al., 2005) and modified in-house at Arizona State University. The data reported includes the battery level, a floating-point value ranging from 0 to 5, and the conductance, a floating-point value with a lower value of 0, which increases in proportion to the increase in the skin conductance.
2. The Shimmer3 GSR+[c] monitors skin conductance using two reusable electrodes attached to two fingers (middle and ring). It is able to report data at a variable sampling rate in the range of hundreds of hertz. Additionally, the Shimmer3 GSR+ provides an estimated heart

[c]http://www.shimmersensing.com

TABLE 11.5 Extract of Data Collected Using an in-House Built Skin Conductance Device

Timestamp	**Voltage**	**Conductance**
141116101332262	2.482352941	1.030696176
141116101332762	2.482352941	1.023404165
141116101333262	2.482352941	1.019813274
141116101333762	2.482352941	1.041657802
141116101334247	2.482352941	0.998280273
141116101334747	2.482352941	0.991181142
141116101335247	2.482352941	0.980592229
141116101335747	2.482352941	0.998280273
141116101336247	2.482352941	1.012586294
141116101336762	2.482352941	1.012586294

rate calculated from a photoplethysmogram signal captured by a wired ear or finger clip electrode.

A sample dataset from our in-house developed skin conductance device is shown in Table 11.5. The timestamp confirms a precise sampling rate of 2 Hz. Feature selection and extraction approaches for this measurement can be consulted in Strauss et al. (2005) and Cooper et al. (2009).

The setup for a skin conductance device does not require calibration and as long as its battery has sufficient power, the data will be gathered accurately.

Physiological Sensor: Heart Rate Variability

Heart rate is defined as the number of heartbeats occurring per minute and the average resting heart rate for an adult human is between 60 and 90 beats. The heart rate goes up when activity in the sympathetic nervous system (which controls the body's involuntary responses to a perceived threat) increases. The heart rate goes down when activity in the sympathetic nervous system decreases. Conversely, the heart rate goes up when parasympathetic nervous system activity (which controls the body's involuntary responses at rest) decreases (because there is less inhibition). The heart rate goes down when parasympathetic nervous system activity increases (because there is more inhibition). Although HRV is influenced by numerous physiological and environmental factors, the influence of the autonomic nervous system on cardiac activity is particularly prominent and of psychophysiological importance. HRV analysis is emerging as an objective measure of a regulated emotional response. HRV measures the variation in the time intervals between heartbeats and has been related to emotional arousal. Experimentation and theory support the usefulness of HRV as an objective index of the brain's ability to organize regulated emotional responses and as a marker of individual differences in emotion regulation capacity (Appelhans and Luecken, 2006).

One recommendation for this device's use is that the sensor should be placed on the nondominant hand or ankle to minimize noise caused by movement. We are currently introducing this method to our toolkit using a Shimmer3 GSR+ sensor and looking forward to a comparative analysis of the results against other sensing systems.

Body Language: Pressure

Like facial gestures, body language is a potential channel of emotional expression. Body language can be more challenging to identify than facial gestures. However, several approaches have been developed to overcome this challenge. One uses pressure sensors. Pressure sensors allow the amount of pressure applied to an object, such as a mouse, a game controller, or a keyboard, to be detected. Pressure has been correlated to levels of frustration (Qi and Picard, 2002). Qi and Picard created a device in which pressure sensors reporting data at a sampling rate of approximately 6 Hz were embedded in a mouse. The six sensors were situated in the right, left, and middle front and rear parts of the mouse. The raw data from the six pressure sensors was processed and correlated to levels of frustration (Cooper et al., 2009). The raw pressure data is represented by integer values in the range of 0 to 1024, where 0 represents the highest pressure. The machine learning approach for inferring affect based on measurements from such pressure values can be consulted in Qi and Picard (2002) and Cooper et al. (2009).

A sample dataset from an in-house version of Qi and Picard's device with six pressure sensors is shown in Table 11.6. The timestamp shows a sampling rate of approximately 6 Hz.

Pressure sensors are nonintrusive since the user does not have to wear any equipment and the device does not require any calibration. In addition, the functionality of the target object is not affected by the introduction of the pressure sensors, although the pressure sensors in the object must be connected to the computer using a serial or USB port.

Body Language: Posture

Another approach to body language analysis is posture detection. One approach that we use to achieve this consists of using a low-cost, low-resolution pressure sensitive seat cushion and back pad. Sensors are embedded in the seat cushion and back pad so that they are imperceptible

TABLE 11.6 Extract of Data Collected Using a Mouse Enhanced With Pressure Sensors

Timestamp	Right rear	Right front	Left rear	Left front	Middle rear	Middle front
140720113306312	1023	1023	1023	1023	1023	1023
140720113306468	1023	1023	1023	1023	1023	1023
140720113306625	1023	998	1023	1002	1023	1023
140720113306781	1023	1009	1023	977	1023	1023
140720113306937	1023	794	1023	982	1023	1023
140720113307109	1023	492	1022	891	1023	1023
140720113307265	1023	395	1021	916	1019	1023
140720113307421	1023	382	1021	949	1023	1023
140720113307578	1023	364	1022	983	1023	1023
140720113307734	1023	112	1021	1004	1023	1023

to the user. They are positioned in a triangle configuration in the middle, right, and left. An example of this device was developed at Arizona State University based on the design of a more expensive high-resolution unit from the MIT Media Lab (Mota and Picard, 2003). The data from three pressure sensors in the back, three in the seat, and the two accelerometers is processed to obtain the net seat change, net back change, and sit forward features. Those features are then used to infer a level of interest. The machine learning approach for inferring affect related measures from such values is described in Mota and Picard (2003) and Cooper et al. (2009).

A sample dataset from a posture device is shown in Table 11.7. As is the case with the pressure sensor, the values range from 0 to 1024; however, in this instance 1024 represents the highest pressure.

Posture sensing using a seat cushion enhanced with pressure sensors is an inexpensive, easy to implement, and nonintrusive measurement, although, like the pressure sensor, the seat cushion must be connected to the computer using a serial or USB port. Moreover, these devices do not require any calibration.

Language Processing: Writing Patterns

Text-based human interaction often carries important emotional significance. In the context of computer-mediated communication, some methods for text-based emotion recognition have been developed, such as the one described in Krcadinac et al. (2013). They work at the sentence level

TABLE 11.7 Extract of Data Collected Using a Seat Cushion Enhanced With Pressure Sensors

Timestamp	Right seat	Middle seat	Left seat	Right back	Middle back	Left back	Net seat change	Net back change	Sit forward
140720074358901	1015	1019	1012	976	554	309	12	152	0
140720074359136	1008	1004	1012	978	540	305	22	20	0
140720074359401	1015	1012	1008	974	554	368	19	81	0
140720074359636	1001	1004	1016	975	548	306	30	69	0
140720074359854	1015	1011	1003	967	559	418	34	131	0
140720074400120	1011	1008	1001	968	620	358	9	122	0
140720074400354	1011	1011	1013	968	541	413	15	134	0
140720074400589	1012	1010	1006	974	565	314	9	129	0
140720074400839	1016	1014	1012	972	668	290	14	129	0
140720074401089	1012	1012	1004	858	108	2	14	962	0

and implement a recognition technique founded on a refined keyword spotting method, which employs a set of heuristic rules, a WordNet-based word lexicon, and a lexicon of emoticons and common abbreviations. Their approach is implemented through a software system named Synesketch, released as a free, open source software library. Synesketch analyses the emotional content of text sentences in terms of emotional types (happiness, sadness, anger, fear, disgust, and surprise), weights (how intense the emotion is), and valence (positive or negative).

Other approaches go beyond word- or sentence-level analysis and perform a semantic analysis of the text. These approaches move toward sentiment analysis and opinion mining, which use natural language processing, text analysis, and computational linguistics to identify and extract subjective information from source materials. They aim to determine the attitude of a writer with respect to some topic or the overall contextual polarity of a document. The attitude may be the writer's emotional state, or the intended effect the writer wishes to have on the reader. A summary of diverse approaches to inferring affect from text analysis are summarized in Table 4 in Calvo and D'Mello (2010).

No calibration is required for emotion recognition from writing patterns since it does not involve any particular hardware or device; this is fully implemented through software. To our knowledge, currently available options are limited to the English language. Even though Synesketch is open source and free, commercial implementation may be expensive as it is comparable in cost to facial recognition systems.

DATA HANDLING: SAMPLING, FILTERING, AND INTEGRATION

Section "Gathering Data: Approaches and Technologies" presents a selection of the broad range of technologies and approaches for sensing and gathering data correlated with affective state changes, which are becoming increasingly more accessible and robust. Our roadmap now turns to explore the problems and methodologies associated with sampling, filtering, and integrating affective data.

Sampling

The first challenge when dealing with multiple sources of data relates to the sampling rate. Sampling rates for different sources vary from a few samples per second (2 Hz for skin conductance sensing) to over a hundred samples per second (128 Hz for raw brain wave data). Thus, a decision must be made regarding how to synchronize and integrate the data; common options include:

1. sampling at the lowest rate;
2. sampling at the highest rate; and
3. determining a suitable sampling rate between the computed lower and higher limits, prioritizing specific trade-offs.

Using the lowest rate means losing data and potentially forfeiting accuracy; however, it involves less computer strain, more latency, and using less hard drive space. While using higher rates potentially provides greater accuracy, it does involve more computer strain, less latency, and more hard drive space. The higher sampling rate is more appropriate if the goal is to create a model for off-line analysis. For real-time analysis, it may be preferable to determine a suitable sampling rate according to the system requirements and the computational resources. In real-time systems emotions experienced by the user for substantial periods of time are usually the priority, rather than emotions that occur briefly.

Filtering

One of the most relevant goals of filtering data is noise reduction. Noisy data are unwanted readings that might impact the efficiency of the data processing. Ways in which sensors can report noisy data include:

1. MindReader facial recognition software reports a value of −1 when an affective state inference cannot be achieved; this usually occurs when the face moves out of the camera's line of sight.
2. Eye tracker reports gaze point values as negative or greater than the screen resolution values when light conditions are bad or the user's

head movements interfere with the reading. The validity attribute indicates how reliable an eye tracker sample is. Usually, samples with a validity value of 0 to 3 (high and medium quality) are retained and samples with a validity value of 4 are discarded.
3. BCIs, such as the Emotiv headset, report low-quality data when the electrodes become dry and when nearby electronic devices affect the EEG signal. Since a baseline of 4000 is defined, noise values can be detected and excluded.
4. For skin conductance devices, the battery level is important. Low battery power reduces the quality of the conductance readings.

Various techniques are employed to reduce noise by cleaning unwanted values from the data. Common filtering approaches for denoising data are further discussed by Manolakis et al. (2000). These techniques include:

1. Low-pass filtering. This ignores signals with a frequency lower than a certain cutoff value and attenuates signals with frequencies higher than the cutoff value to understand the main signal behavior.
2. Moving average filtering. This analyzes data points by creating a series of averages for different subsets of the full data set.
3. Median filtering. This denoises data in a similar way to moving average filtering but uses the median value rather than an average. In some cases, these results are more robust because the median value is less likely to be affected by outlying (unusual) values.

Integration

Integration is required when a multimodal approach is implemented since it takes into account the diverse sources of data that contribute to the inference process. Since each source may capture data of a different type at a different sampling rate, it is a challenge to combine the data to create a single enhanced dataset combining all records into one, with a column for each feature and a row for each timestamp. A row contains the values recorded for any source that corresponds with that timestamp. Examples of integration methods include sparse data, state machine, interpolation, and mapping to a coordinate model, each of which is presented briefly below:

Sparse data. If no value exists from a given source for a given timestamp, the cell remains empty. For instance, in the first row of Table 11.8, for the timestamp 141014135755652, there was no data from the eye tracker but there was data from the Emotiv EPOC headset and the skin conductance sensor. A similar situation occurred in the second row, where only eye tracker data exists for the timestamp 141014135755659. The presence of numerous empty cells in the dataset is referred to as sparse data. Applying a high sampling rate usually results in sparse data.

TABLE 11.8 Integration of Measurements (From an Eye Tracker, an Emotiv EPOC Headset, and a Skin Conductance Device) Using a Sparse Data Approach

Timestamp	Fixation	GPX	GPY	Short-term excitement	Long-term excitement	Engagement/ boredom	Meditation	Frustration	Conductance
141014135755652				0.436697	0.521059	0.550011	0.335825	0.498908	0.401690628
141014135755659	213	573	408						
141014135755668				0.436697	0.521059	0.550011	0.335825	0.498908	
141014135755676	213	566	412						
141014135755692	213	565	404						
141014135755709	213	567	404						
141014135755714									
141014135755726	213	568	411						
141014135755742	213	568	409						
141014135755759	213	563	411						
141014135755761									
141014135755776	213	574	413						
141014135755792	213	554	402						
141014135755809	214	603	409						
141014135755824									
141014135755826	214	701	407						
141014135755842	214	697	403						

141014135755859	214	693	401						
141014135755876	214	700	402						
141014135755892	214	701	411						
141014135755909	214	686	398						
141014135755918									
141014135755926	214	694	399						
141014135755942	214	694	407						
141014135755959	214	698	404						
141014135755964									
141014135755976	214	704	398						
141014135755992	214	693	415						
141014135756009	214	696	411						
141014135756025	215	728	406						
141014135756027				0.436697	0.521059	0.150011	0.335825	0.998908	

State machine. Instead of considering an affective state to exist only in the period of time equivalent to the sample, this method assumes that the state persists until a new state is obtained; thus, a source is in a state until it changes from that state to another. An example is shown in Table 11.9. There it is evident that, due to the sampling rate difference between the eye tracker (60 Hz), the Emotiv system (8 Hz), and the skin conductance device (2 Hz) in the interval between 57:55.652 and 57:56.027, there are 23 values reported from the eye tracker, 3 from the Emotiv system, and only 1 from the skin conductance device. Using the state machine approach, the values are repeated for each time stamp until a new value is obtained and there are no empty cells. The state machine approach is a good choice for real-time integration analysis.

Interpolation. This is a method for constructing new data points within the range of a discrete set of known data points. It is commonly used to deal with missing values by predicting them. Missing value computation is an active research topic and there are several successful techniques for handling the missing values in the literature (Little and Rubin, 2002).

Mapping to a coordinate model. This works with affective constructs and consists of mapping constructs to a coordinate vector and then adding all the coordinate vectors together. This approach has been suggested for systems that perform emotion recognition in real-time and use their inferences to trigger adaptive changes in a target system. Gilroy et al. (2009) suggested the use of the PAD three-dimensional coordinate model. The process of mapping the data from row 2 in Table 11.9 to the PAD three-dimensional coordinate model is described later. To simplify our example, only constructs from the Emotiv EPOC headset and the skin conductance device are used, that is, data from the eye tracker is ignored. Also, assuming a focus on real-time systems, short-term excitement is used here and long-term excitement is ignored.

Each construct value is mapped to a PAD coordinate vector as follows:

1. Excitement corresponds to $[+P, +A, -D]$, thus a value of 0.436697 can be mapped to the coordinate vector [0.436697, 0.436697, −0.436697].
2. Engagement and boredom are reported in the same column. Engagement corresponds to $[+P, +A, +D]$ and boredom corresponds to $[-P, -A, -D]$. A value of 0.5 means equilibrium, values below 0.5 are indicators of boredom, and values above 0.5 are indicators of engagement; thus, engagement/boredom can be mapped, in a scale of −1 to 1, to [(value − 0.5)*2, (value − 0.5)*2, (value − 0.5)*2]. Then a value of 0.550011 is mapped to [0.100022, 0.100022, 0.100022], which is positive and thus corresponds to a low level of engagement.
3. Meditation corresponds to $[+P, -A, +D]$; thus, a value of 0.335825 can be mapped to the vector [0.335825, −0.335825, 0.335825].

TABLE 11.9 Integration of Measurements (From an Eye Tracker, an Emotiv EPOC Headset, and a Skin Conductance Device) Using a State Machine Approach

Timestamp	Fixation	GPX	GPY	Short-term excitement	Long-term excitement	Engagement/ boredom	Meditation	Frustration	Conductance
141014135755652	213	574	414	0.436697	0.521059	0.550011	0.335825	0.498908	0.401690628
141014135755659	213	573	408	0.436697	0.521059	0.550011	0.335825	0.498908	0.401690628
141014135755668	213	573	408	0.436697	0.521059	0.550011	0.335825	0.498908	0.401690628
141014135755676	213	566	412	0.436697	0.521059	0.550011	0.335825	0.498908	0.401690628
141014135755692	213	565	404	0.436697	0.521059	0.550011	0.335825	0.498908	0.401690628
141014135755709	213	567	404	0.436697	0.521059	0.550011	0.335825	0.498908	0.401690628
141014135755714	213	567	404	0.436697	0.521059	0.550011	0.335825	0.498908	0.401690628
141014135755726	213	568	411	0.436697	0.521059	0.550011	0.335825	0.498908	0.401690628
141014135755742	213	568	409	0.436697	0.521059	0.550011	0.335825	0.498908	0.401690628
141014135755759	213	563	411	0.436697	0.521059	0.550011	0.335825	0.498908	0.401690628
141014135755761	213	563	411	0.436697	0.521059	0.550011	0.335825	0.498908	0.401690628
141014135755776	213	574	413	0.436697	0.521059	0.550011	0.335825	0.498908	0.401690628
141014135755792	213	554	402	0.436697	0.521059	0.550011	0.335825	0.498908	0.401690628
141014135755809	214	603	409	0.436697	0.521059	0.550011	0.335825	0.498908	0.401690628
141014135755824	214	603	409	0.436697	0.521059	0.550011	0.335825	0.498908	0.401690628
141014135755826	214	701	407	0.436697	0.521059	0.550011	0.335825	0.498908	0.401690628

(Continued)

TABLE 11.9 Integration of Measurements (From an Eye Tracker, an Emotiv EPOC Headset, and a Skin Conductance Device) Using a State Machine Approach *(cont.)*

Timestamp	Fixation	GPX	GPY	Short-term excitement	Long-term excitement	Engagement/ boredom	Meditation	Frustration	Conductance
141014135755842	214	697	403	0.436697	0.521059	0.550011	0.335825	0.498908	0.401690628
141014135755859	214	693	401	0.436697	0.521059	0.550011	0.335825	0.498908	0.401690628
141014135755876	214	700	402	0.436697	0.521059	0.550011	0.335825	0.498908	0.401690628
141014135755892	214	701	411	0.436697	0.521059	0.550011	0.335825	0.498908	0.401690628
141014135755909	214	686	398	0.436697	0.521059	0.550011	0.335825	0.498908	0.401690628
141014135755918	214	686	398	0.436697	0.521059	0.550011	0.335825	0.498908	0.401690628
141014135755926	214	694	399	0.436697	0.521059	0.550011	0.335825	0.498908	0.401690628
141014135755942	214	694	407	0.436697	0.521059	0.550011	0.335825	0.498908	0.401690628
141014135755959	214	698	404	0.436697	0.521059	0.550011	0.335825	0.498908	0.401690628
141014135755964	214	698	404	0.436697	0.521059	0.550011	0.335825	0.498908	0.401690628
141014135755976	214	704	398	0.436697	0.521059	0.550011	0.335825	0.498908	0.401690628
141014135755992	214	693	415	0.436697	0.521059	0.550011	0.335825	0.498908	0.401690628
141014135756009	214	696	411	0.436697	0.521059	0.550011	0.335825	0.498908	0.401690628
141014135756025	215	728	406	0.436697	0.521059	0.550011	0.335825	0.498908	0.401690628
141014135756027	215	728	406	0.436697	0.521059	0.150011	0.335825	0.998908	0.401690628

4. Frustration corresponds to [−P, +A, −D]; thus, a value of 0.498908 can be mapped to the vector [−0.498908, 0.498908, −0.498908].
5. Skin conductance is not mapped to an affective construct but to arousal; thus, a normalized value of 0.401690628 can be mapped as zero pleasure, positive arousal, and zero dominance, that is, to the vector [0, 0.401690628, 0].

Adding the vectors gives a resultant vector of [0.373636, 1.101492628, −0.499758], a vector in the zone of [+P, +A, −D], that is, a level of excitement. This vector represents the cumulative affective state at timestamp 141014135755659 (Oct 14, 2014 13:57:55.659).

DATA ANALYSIS

Once time sampling has been accounted for, noise has been removed, missing data values have been addressed, and the data has been integrated, data analysis can start. There is a vast body of literature available on the topic of data analysis. This section describes the overall process, and exemplifies techniques for data analysis in general and for the correlation of affect measurements with stimuli. Two examples of correlating affect and stimuli off-line using data collected in experimental studies are presented using two free tools. Advanced methods for data analysis are surveyed in Novak et al. (2012).

The overall strategy for data analysis is depicted in Fig. 11.5 and can be summarized as follows:

1. *Feature extraction*. This step starts with a set of clean data and builds derived values (called features) intended to be informative and

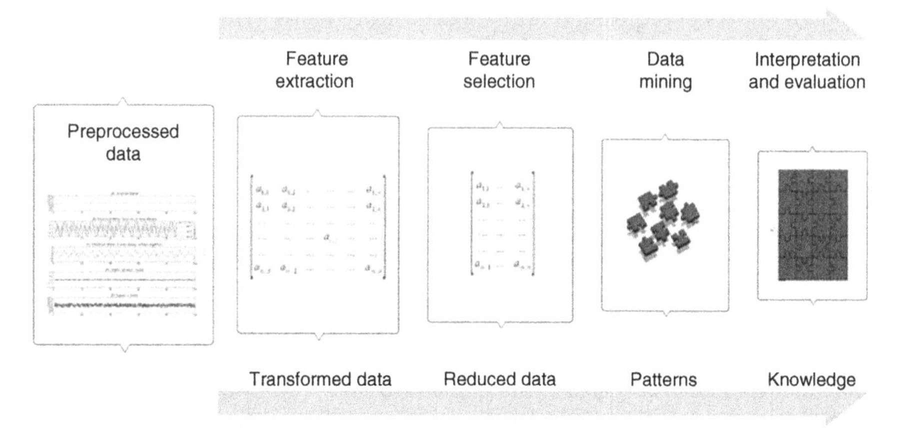

FIGURE 11.5 **Steps for data analysis: feature extraction, feature selection, data mining, and interpretation and evaluation.**

nonredundant. The extracted features are expected to contain the relevant information from the input data, so that subsequent steps can be performed using this reduced representation instead of the entire initial set of data. With dimensionality reduction or transformation methods, the effective number of variables under consideration can be reduced, or invariant representations for the data can be found.

2. *Feature selection.* This step entails the process of selecting a subset of relevant features for use in the construction of a model. The central premise of using a feature selection technique is that the data contains many features that are either redundant or irrelevant, and can thus be removed without incurring much loss of information. Feature selection techniques allow: (1) the simplification of models for ease of interpretation, (2) shorter training times, and (3) enhanced generalization by reducing overfitting.
3. *Data mining.* This step involves the selection of a mining approach to identify patterns of interest. Common approaches include classification rules or trees, regression, and clustering. The selection process takes into account: (1) the parameters that are going to be used and which model might be appropriate for them, for instance, models for categorical data are different to models for vectors; and (2) the overall criteria of the knowledge discovery process, for instance, understanding the model or achieving outstanding predictive capabilities.
4. *Interpretation and evaluation.* Interpretation involves the visualization of the extracted patterns and models, or the visualization of the data given the extracted models. Evaluation is the application of the knowledge by incorporating it into another system for further actions, or simply documenting it and reporting it to interested parties.

The following two subsections outline and exemplify two approaches for data analysis, regression and classification, using two freely available tools, *Eureqa* and *Weka*, respectively.

Regression Example

For regression or reverse engineering data searches, we use *Eureqa*[d]. *Eureqa* is used to compute mathematical expressions for structural relationships in the data records (Dubcakova, 2011). Typically, the records hold information about the actions or behaviors and affective states of a user who was engaged in an experimental setting. For instance, consider exploring the relationships between affective states and the tendency to make mistakes. We collected data from participants engaged in playing the video game Guitar Hero. The goal of the game is to press one or more colored buttons at the same time as moving target lights of the same color cross a line

[d]http://www.nutonian.com/products/eureqa/

on the screen. In each 1-h session, the first 15 min were allocated to practice, followed by a 45-min session in which participants played four songs of their choice, one of each level: easy, medium, hard, and expert. The data collected included the errors made and constructs for engagement, excitement, meditation, and frustration. For this example, a partial dataset was fed to the *Eureqa* tool and used to create a model expressed as a mathematical equation, showing possible relationships between attributes, where sensor readings are the independent variables and errors, the number of mistakes that are likely to happen in a timeframe, are the target variable.

Eureqa automatically splits the dataset in two for training and validation, respectively. The training data is used to build the models, and the validation data is used to test how well models generalize to new data. The models and their evaluation are displayed in the *Eureqa* graphical user interface, as shown in Fig. 11.6. Obviously there is no guaranty that a model with a good correlation exists or can be found for a given data set. However, the approach and this tool provide a good way to explore the collected data and identify possible relationships that can be investigated further later on, such as

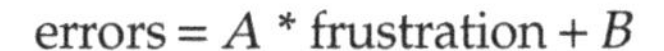

$$\text{errors} = A * \text{frustration} + B$$

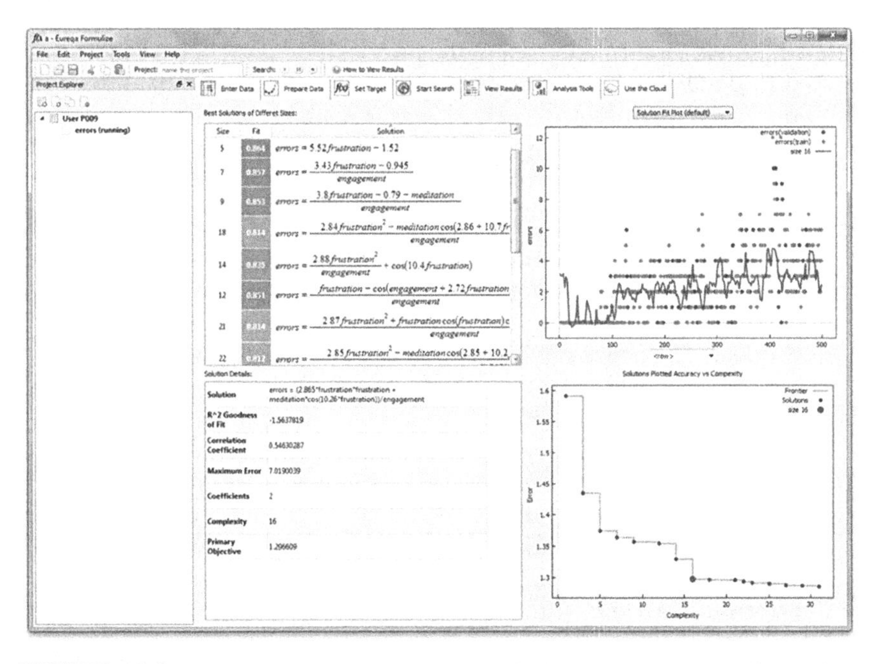

FIGURE 11.6 Eureqa tool during the analysis of a dataset investigating the relationship between affect measurements and the error-making tendency while playing a video game.

or

$$\text{errors} = (A * \text{frustration} - B * \text{meditation} - C)/\text{engagement}$$

The letters A, B, and C represent constant values weighting the variables frustration, meditation, and engagement. This nonconclusive exploration of mistakes made by a player of the Guitar Hero video game suggest a proportional relationship between the tendency to make a mistake and the levels of frustration and meditation, but an inversely proportional one with the level of engagement.

Classification Example

Weka[e] is a tool that implements a collection of machine learning algorithms for data mining tasks; these explore data composition and relationships and extract useful information by means of clustering and classification approaches (Hall et al., 2009). To exemplify the classification approach, let us consider a research study run to predict the level of difficulty for solving four puzzles. Each puzzle has a different difficulty level ranging from easy to hard. The Tobii Eye Tracker and the Emotiv headset were used to collect data from 44 users solving each puzzle; both raw EEG data and emotional constructs from the Emotiv headset were taken into account. With that data, the intention was to construct a prediction model to show if observed sensor data can provide information about the puzzle difficulty level. A model was created to understand and predict cognitive load (level of difficulty experienced in solving a problem), where sensor readings are the independent variables and puzzle difficulty level is the target variable (Joseph, 2013). Since both raw data and constructs were taken into account, this example allows the description of the complete set of steps presented before, as follows:

1. *Preprocessing.* For the eye-tracking data and Emotiv emotional constructs, no preprocessing was performed because it was assumed that they were already derived values and they are not as sensitive as EEG raw data. For the raw EEG data, median filtering was applied to reduce signal noise.
2. *Feature extraction.* For Emotiv constructs and pupil data, extracted features consisted of measures of variability, including variance, the minimum and maximum values, and skewness and kurtosis, as well as measures of central tendency, including mean and median. These features provide information about the characteristics of the signal distribution. For the raw EEG data, feature extraction consisted of transforming raw data into wave-frequency domains using a fast Fourier transform (FFT) algorithm and by band-pass filtering the signals within specific frequency ranges for alpha (8–12 Hz), beta

[e]http://www.cs.waikato.ac.nz/ml/weka/

TABLE 11.10 Features Calculated From Emotiv Affective Constructs and Their *P* Values for Predicting Puzzle Difficulty Level

Feature	Importance	*P* value
Long-Term Excitement_min	100%	<0.001
Long-Term Excitement_variance	98.13%	<0.001
Meditation_kurtosis	86.86%	<0.001
Engagement/Boredom_skew	60.82%	0.012
Engagement/Boredom_min	56.68%	<0.001
Frustration_variance	42.81%	0.004
Engagement/Boredom_mean	34.66%	<0.001
Engagement/Boredom_max	32.96%	<0.001
Short-Term Excitement_variance	25.66%	<0.001
Engagement/Boredom_variance	9.09%	<0.001

(13–30 Hz), and theta (4–7 Hz) waves. EEG alpha activity reflects attention demands, beta activity reflects emotional and cognitive processes, and theta activity has been related to consciousness.

3. *Feature selection.* From the 30 features extracted from the 5 Emotiv constructs and pupil data (5 measures of variability for each), 10 features were selected according to the process described by Tuv et al. (2009). Those features have *P* values less than 0.05; a small *P* value means corresponding variable values change significantly for different puzzle difficulty levels. Features, importance, and *P* values for each feature are listed in Table 11.10. Alpha, beta, and theta waves were retained as EEG features.
4. *Data mining algorithm.* The mining approach used for searching patterns that predict the puzzle being solved, that is, differentiating the puzzle type (difficult level), was classification with trees. A first approach was to train a decision tree using the *Weka* tool. The decision tree algorithm used was J48 (Quinlan, 1993). Fig. 11.7 shows a sample tree (not actual results) to exemplify a decision tree using only the features calculated from Emotiv constructs shown in Table 11.10. For instance, Fig. 11.7 shows that long-term excitement has taken a minimum value greater than 0.467 for 25 subjects while solving puzzle 4 (a higher level of difficulty), indicated by the branches on the right. Looking at the branches on the left, minimum long-term excitement values are less than 0.169 for 39 subjects while solving puzzle 1 (lower level of difficulty).

Trees are useful for understanding the patterns, but for some objectives a single tree suffers from problems with generalization—they tend to overfit. To avoid this, an approach with multiple trees can be introduced

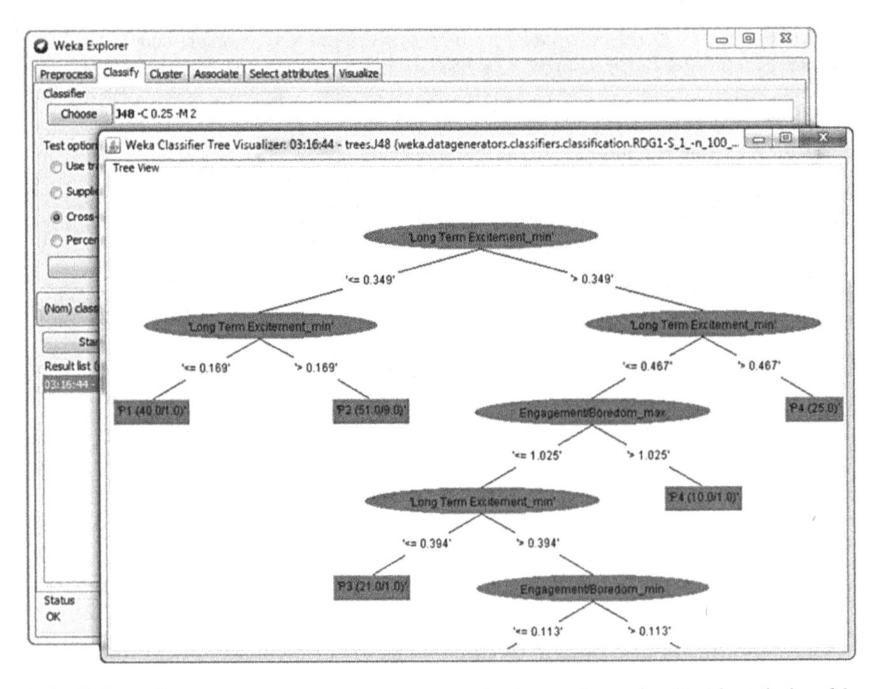

FIGURE 11.7 Weka tool during the analysis of a dataset investigating the relationship between affect measurements and the level of difficulty solving puzzles – a J48 decision tree generated by the tool.

in a randomized way; this approach is called Random Forest (Liaw and Wiener, 2002). Using trees for interpretation purposes and Random Forest for prediction is an option that we have explored with outstanding results.

SUMMARY

This roadmap aims to provide the reader with a better understanding of the technologies and approaches available for affect measurement, as well as an insight into how to process and analyze the data acquired. Some final thoughts are presented here along with a summary of each section.

The first section describes how a wide range of diverse approaches and technologies are being used and explored to recognize and quantify affective states. Emotions are complex constructs fused with physiological and electrical activity in our bodies. Sensing physiological and electrical activity, looking for changes, and finding patterns in them enables us to identify the affective state that is behind those changes.

The capabilities of the various approaches and technologies surveyed in the first section are summarized in Table 11.11. The data gathered by

TABLE 11.11 Summary of Approaches and Technologies for Affect Measurement

Device	Software	Input	Output (raw data and constructs)	Rate
Emotiv EPOC headset	Emotiv SDK	Brain waves	EEG activity: reported in 14 channels: AF3, F7, F3, FC5, T7, P7, O1, O2, P8, T8, FC6, F4, F8, and AF4 Face activity: blink, wink (left and right), look (left and right), raise brow, furrow brow, smile, clench, smirk (left and right), and laugh Emotions: excitement, engagement/boredom, meditation, and frustration	128 Hz (raw data) and 8 Hz (constructs)
Standard webcam	MIT Media Lab Mind-Reader	Facial expressions	Emotions: agreement, concentration, disagreement, interest, thought, and unsureness	10 Hz
Tobii Eye Tracker	Tobii SDK	Eye movements and pupil size	Gaze points (x, y), fixation, and pupil dilation	60 Hz
Skin conductance sensor	USB driver	Skin conductivity	Arousal	2 Hz
Heart rate sensor	USB driver	Heart rate	Arousal	Variable
Pressure sensor	USB driver	Pressure	One pressure value per sensor allocated into the input/control device, which can be related to frustration	6 Hz
Posture sensor	USB driver	Pressure	Pressure values in the back and the seat cushions (in the right, middle, and left zones) of a chair that can be related to interest	6 Hz
NA	Synesketch	Text	Happiness, sadness, anger, fear, disgust, and surprise	Variable

them is diverse and they can be used individually or in combination, depending on the goal of the study or the research questions being investigated. All technologies produce raw data but only some have a built-in capacity to infer constructs. In addition to each device's technical specifications, cost is also a key factor when deciding on a device or approach appropriate to a project. BCI devices, particularly the ABM B-Alert X-Series systems, are costlier in comparison to other systems discussed here. These are closely followed by eye-tracking systems embedded in glasses, the mobile and embedded eye-tracking systems, and the Emotiv EPOC headsets. For the face-based recognition systems the cost depends not on the hardware but the software, and there are a number options on the market. The least expensive systems are the skin conductance sensors, the pressure sensors, and the posture sensors, because these can feasibly be assembled in-house following blueprints available from many sources, such as those referenced above.

The second section describes the first steps in dealing with the data obtained, involving sampling, filtering, and integration. Diverse sources have diverse sampling rates; the first challenge is to synchronize these. A second challenge involves noise reduction or filtering, meaning the elimination of samples with no useful values (out of range as a result of faults in the hardware or the environment, low battery power, interference, glare, target out of the camera's line of sight, etc.). Last but not least, integration is the combination of data from diverse sources to improve the inference of affective states.

Finally, the last section explains an overall strategy for data analysis that consists of four steps: feature extraction, feature selection, data mining, and interpretation and evaluation. This section shows two examples with freely available tools, one using a regression and the other a classification approach, respectively dealing with the prediction of mistakes in videogames and puzzle solving. None of these examples are intended to present research results or conclusions. Instead, they provide an exemplary exploration of experimental date for demonstration purposes.

Acknowledgements

Our research in affect recognition and affect-driven adaptive systems is supported by the Office of Naval Research under Grants N00014-10-1-0143 and N00014-13-1-0438, awarded to Dr. Robert Atkinson. We thank Robert Christopherson, iMotion support specialist, for his advice in the initial stage of this manuscript. We would also like to thank our colleagues in the iLUX and ANGLE labs at Arizona State University for reading the manuscript and sharing their thoughts and responses, Irfan Kula in particular, for sharing his expertise and knowledge.

References

American Electroencephalographic Society, 1990. Guidelines for Standard Electrode Position Nomenclature. American Electroencephalographic Society.

Appelhans, B.M., Luecken, L.J., 2006. Heart rate variability as an index of regulated emotional responding. Rev. Gen. Psychol. 10 (3), 229–240, http://doi.org/10.1037/1089-2680.10.3.229.

Bradley, M.M., Miccoli, L., Escrig, M.A., Lang, P.J., 2008. The pupil as a measure of emotional arousal and autonomic activation. Psychophysiology 45 (4), 602–607, http://doi.org/10.1111/j.1469-8986.2008.00654.x..

Calvo, R.A., D'Mello, S.K., 2010. Affect detection: an interdisciplinary review of models, methods, and their applications. Trans. Affect. Comput. 1 (1), 18–37, http://doi.org/10.1109/T-AFFC.2010.1..

Cohn, J.F., De la Torre, F., 2014. Automated face analysis for affective computing. In: Calvo, R., D'Mello S., Gratch J., Kappas A. (Eds.), The Oxford Handbook of Affective Computing. Oxford University Press, USA.

Cooper, D.G., Arroyo, I., Woolf, B.P., Muldner, K., Burleson, W., Christopherson, R.M., 2009. Sensors model student self-concept in the classroom. Presented at the Proceedings of the 17th International Conference on User Modeling. Adaptation, and Personalization. Springer-Verlag, Berlin, Heidelberg, pp. 30–41. http://doi.org/10.1007/978-3-642-02247-0_6.

Dubcakova, R., 2011. Eureqa: software review. GPEM 12 (2), 173–178, http://doi.org/10.1007/s10710-010-9124-z..

Ekman, P., 1992. Are there basic emotions? Psychol. Rev. 99 (3), 550–553.

Ekman, P., Freisen, W.V., Ancoli, S., 1980. Facial signs of emotional experience. J. Pers. Soc. Psychol. 39 (6), 1125–1134, http://doi.org/10.1037/h0077722..

Gilroy, S.W., Cavazza, M., Benayoun, M., 2009. Using affective trajectories to describe states of flow in interactive art. Presented at the Proceedings of the International Conference on Advances in Computer Entertainment Technology. ACM, New York, NY, USA, pp. 165–172.

Hall, M.A., Frank, E., Holmes, G., Pfahringer, B., Reutemann, P., Witten, I.H., 2009. The WEKA data mining software: an update. SIGKDD Explor. 11 (1), 10–18, http://doi.org/10.1145/1656274.1656278..

Joseph, S., 2013. Measuring cognitive load: a comparison of self-report and physiological methods. ProQuest Dissertations and Theses. Arizona State University, Ann Arbor.

Kaliouby, El, R., Robinson, P., 2005. Generalization of a vision-based computational model of mind-reading. Presented at the Proceedings of the Third International Conference on Affective Computing and Intelligent Interaction. Springer, pp. 582–589.

Krcadinac, U., Pasquier, P., Jovanovic, J., Devedzic, V., 2013. Synesketch: an open source library for sentence-based emotion recognition. Trans. Affect. Comput. 4 (3), 312–325, http://doi.org/10.1109/T-AFFC.2013.18..

Liaw, A., Wiener, M., 2002. Classification and Regression by randomForest. R News 2 (3), 18–22.

Little, R.J., Rubin, D.B., 2002. Statistical Analysis with Missing Data, second ed. Wiley, New York, NY, USA.

Manolakis, D.G., Ingle, V.K., Kogon, S.M., 2000. Statistical and Adaptive Signal Processing: Spectral Estimation, Signal Modeling, Adaptive Filtering, and Array Processing. McGraw-Hill, Boston, MA, USA.

Mehrabian, A., 1996. Pleasure-arousal-dominance: a general framework for describing and measuring individual differences in temperament. Curr. Psychol. 14 (4), 261–292.

Mota, S., Picard, R.W., 2003. Automated posture analysis for detecting learner's interest level. Presented at the Proceedings of the 2003 Workshop on Computer Vision and Pattern Recognition, vol. 5. pp. 49–55.

Novak, D., Mihelj, M., Munih, M., 2012. A survey of methods for data fusion and system adaptation using autonomic nervous system responses in physiological computing. Interact. Comput. 24 (3), 154–172, http://doi.org/10.1016/j.intcom.2012.04.003..

Picard, R.W., 1997. Affective Computing, first ed. MIT Press, Cambridge, MA, USA.

Picard, R.W., 2010. Affective computing: from laughter to IEEE. Trans. Affect. Comput. 1 (1), 11–17, http://doi.org/10.1109/T-AFFC.2010.10..

Qi, Y., Picard, R.W., 2002. Context-sensitive bayesian classifiers and application to mouse pressure pattern classification. Presented at the Proceedings of the 16th International Conference on Pattern Recognition, vol. 3. pp. 448–451. http://doi.org/10.1109/ICPR.2002.1047973.

Quinlan, J.R., 1993. C4.5: Programs for Machine Learning. Morgan Kaufmann Publishers, San Francisco, CA, USA.

Russell, J., 1980. A circumplex model of affect. J. Pers. Soc. Psychol. 39 (6), 1161–1178.

Strauss, M., Reynolds, C., Hughes, S., Park, K., McDarby, G., Picard, R.W., 2005. The hand-wave bluetooth skin conductance sensor. In: International Conference on Affective Computing and Intelligent Interaction. Springer, Berlin, Heidelberg, pp. 699-706.

Tuv, E., Borisov, A., Runger, G., Torkkola, K., 2009. Feature selection with ensembles, artificial variables, and redundancy elimination. J. Mach. Learn. Res. 10, 1341–1366.

CHAPTER

12

The Role of Registration and Representation in Facial Affect Analysis

Evangelos Sariyanidi, Hatice Gunes**, Andrea Cavallaro**

*Queen Mary University of London, London, United Kingdom;
**University of Cambridge, Cambridge, United Kingdom

INTRODUCTION

Automatic facial expression analysis has attracted great interest due to potential applications in various domains, such as pain analysis in health care, drowsiness detection in automotive industry, facial action synthesis in animation industry, audience analysis in marketing, and novel human–computer interfaces in social robotics (Gunes and Schuller, 2013; Vinciarelli et al., 2009). Research is conducted not only in academia, but also by companies, such as Affectiva (Anon, in press a), Emotient (Anon, in press b), Kairos (Anon, in press c), Microsoft (Dooley, 2015), and Google (Hodson, 2014). For example, glasses that read emotions are one type of application (Dooley, 2015; Hodson, 2014).

There are two important properties that define how well a facial expression analysis system integrates with daily life applications. First, the ability of the system to recognize *spontaneous expressions*. Most research so far was conducted on datasets of *posed expressions*, which are datasets where participants were instructed to display expressions of particular emotions, possibly in an exaggerated manner; for instance, datasets where happiness is displayed with a big smile. In daily interactions, however, people often manifest their emotions with *subtle expressions* (Ambadar et al., 2005), which are expressions that are observed when a person is just starting to feel an

Emotions and Affect in Human Factors and Human–Computer Interaction. http://dx.doi.org/10.1016/B978-0-12-801851-4.00012-4
Copyright © 2017 Elsevier Inc. All rights reserved.

emotion, when an emotion is felt with low intensity or when an emotion is attempted to be covered up (Matsumoto and Frank, 2012). These expressions are very informative (Joormann and Gotlib, 2006; Matsumoto and Frank, 2012), however, they are also difficult to analyze due to the small deformations they cause in facial appearance (Sariyanidi et al., 2015).

The second important property for a facial expression analysis system is its ability to operate in *uncontrolled conditions*. Most research was conducted in environments where faces are illuminated uniformly using controlled lighting sources, and with people undergoing very limited head and body movement (Sariyanidi et al., 2015). However, in uncontrolled conditions faces seldom appear to be illuminated uniformly and daily life facial expressions are typically accompanied by head and body movements (Gunes and Schuller, 2013).

Facial expression analysis systems generally include two components that are critical for recognizing subtle expressions and performing analysis in uncontrolled conditions, namely *registration* and *representation*. The accuracy of the registration is critical for the recognition of subtle expressions, as the errors made during registration mislead the overall system by creating spurious expressions (Fig. 12.7) (Sariyanidi et al., 2015). The second component, representation, takes the face detected during registration and produces a numerical representation that is informative for recognizing the facial expression. This component has a direct influence on how the overall system performs in uncontrolled conditions. Ideally, the output of representation is expected to be sensitive to changes in facial expression while being insensitive to other factors that alter facial appearance, such as illumination variations or head rotations.

In this chapter we discuss the role of registration and representation in facial expression analysis. Fig. 12.1 shows an exemplar facial expression analysis process that illustrates the role of each component. The input to a facial expression analysis system is an image I or a video $\mathbf{I}$ that contains one face, which is typically cropped via automatic face detection. For clarity, in Fig. 12.1 we illustrate the overall analysis process for an image. The output of the system is an expression or emotion, which is represented with Y. The type of output depends on the affect model used. For example, Y can represent an emotion category (e.g., happiness) if

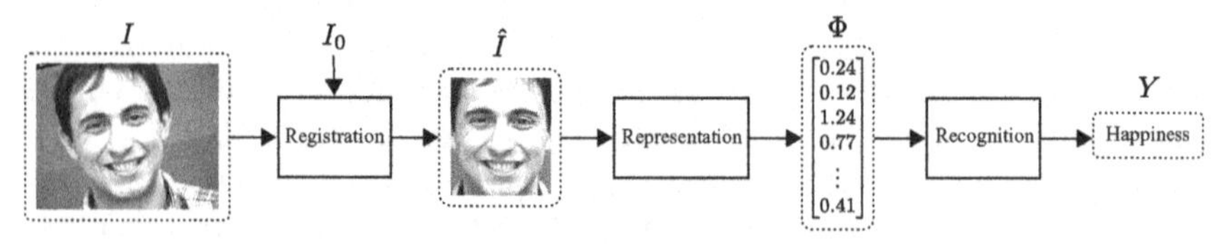

FIGURE 12.1 Illustration of a simple facial expression analysis pipeline with a still image as input and an emotion category as output. I_0 represents a frontal and upright facial pose that is taken as reference for registration.

a discrete affect model is used, or it can represent a set of numbers if a continuous affect model is used (e.g., valence-arousal space). The overall system comprises three layers of processing: registration, representation, and recognition. Registration is the process that scales and derotates the input face to bring it to an upright orientation, and then aligns the face after identifying more precise boundaries for the facial region (Fig. 12.1). Representation takes the cropped face and produces a numerical representation that is informative for the computer system for recognizing the facial expression. Finally, the recognition process takes the output of the representation and yields the predicted expression or emotion Y.

The registration and representation techniques used for image analysis are usually different from those used for video analysis. With images, we can use only *spatial* analysis techniques. With videos, we can use *spatiotemporal* techniques, which make use of the temporal variation of expressions during analysis. As we will discuss, the main advantage of spatiotemporal analysis is improved performance in the recognition of subtle expressions. This advantage, however, can be exploited only if more accurate registration techniques are employed.

In what follows, we will discuss registration and representation techniques first for spatial analysis and then for spatiotemporal analysis. We will see that research on human vision had significant influence on automatic facial affect analysis. One of the most important future directions is the development of spatiotemporal representation techniques. We will elaborate more on this in the end of our discussion on spatiotemporal analysis.

SPATIAL ANALYSIS

The research on automatic analysis of facial affect started with analysis on images. The literature on spatial analysis is more mature than that on spatiotemporal analysis. Moreover, spatial analysis can be conducted with simpler techniques than spatiotemporal analysis.

Registration

The standard method to register images is based on identifying landmark points in the face. There exist a plethora of methods that automatically localize fiducial points like the corners of the eyes or the corners of the mouth (Çeliktutan et al., 2013). Some methods are effective even in challenging imaging conditions where illumination is changing and the face may be partly occluded. The localization accuracy is relatively high (e.g., 2 pixels error per point; Çeliktutan et al., 2013). Most spatial representations can tolerate such small amounts of registration error, therefore, registration for spatial analysis can be considered as a solved problem.

Representation

Before discussing the progress in representations, let us first elaborate on what qualities make a representation useful for facial affect analysis. As mentioned in previous section, by representation we refer to the system component that takes as input a registered image and outputs a numerical array. The essential quality that makes a representation useful is to enable the discrimination of different facial expressions. This implies that the representation should be *sensitive* to the expression in an image, that is, the representation output for two images with different facial expressions should be different.

An even more useful representation is one that is sensitive *only* to facial expressions, and remains unaffected by irrelevant variations in facial appearance, such as variations caused by changes in illumination, by registration errors, or by head rotations. Such a representation is typically referred to as a *robust* representation, more specifically, robust with respect to the specific variation(s) (e.g., illumination, registration errors).

The last quality we list for a representation is computational efficiency, which has two aspects: speed and compactness. The former becomes particularly critical if the affect analysis system is desired to operate in real time. The latter relates to the length of the numerical array produced by the representation. All other things being equal, we prefer shorter representations, which not only require less computer memory, but may also improve recognition performance (Sariyanidi et al., 2015).

We shall start our inquiry with how to obtain a representation that is sensitive to facial expressions. Expression variations are most visible around the facial features; for example, lip corners are being pulled in the expression of happiness, and the inner parts of the eyebrows are raised upward in the expression of sadness. Therefore, we may say that a representation that is sensitive to facial expressions is one that is sensitive to the changes in the shape of facial features. A good way to describe the shape of facial features in an image is to describe the *edges*: The boundaries of facial features can be considered as edges, like the contours that enclose the eyes, eyebrows, and mouth.

How to obtain useful representations has been long investigated by computer scientists. The most satisfactory answers found so far are representations that resemble the low-level layers of the human vision system (Sariyanidi et al., 2015). These representations are sensitive to expressions and offer some robustness in uncontrolled conditions; specifically, robustness against illumination variations and registration errors.

The early layers of the human vision system contain cells that are sensitive to edges in visual scenes (Hubel and Wiesel, 1962) that provide us with the ability to perform complex tasks that seem simple to us, such as identifying the boundaries of objects or object parts, or reading letters and

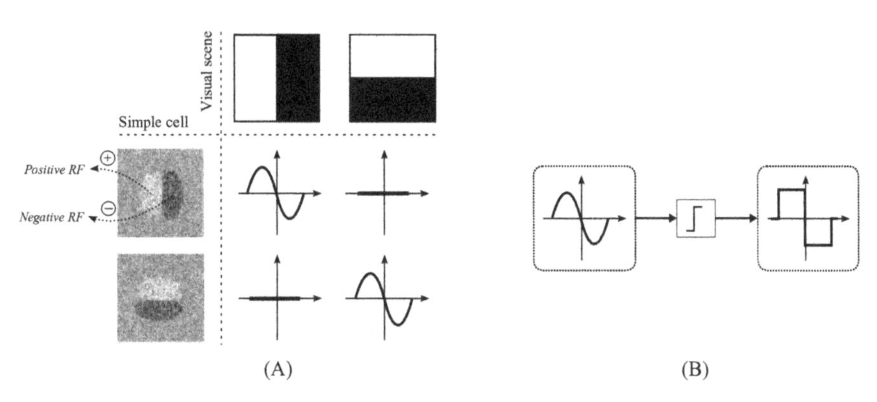

FIGURE 12.2 **Behavior of low-level human vision system.** (A) Illustration of simple cells in the human visual cortex, exemplar visual scenes and the (idealized) response of the cells to each scene. The cell on top is tuned to vertical edges and therefore provides response to visual scenes with vertical edges, while providing no response to horizontal edges. A similar discussion applies to the cell that is tuned to horizontal edges. (B) The nonlinear transformation ignores the precise granularity of the response and retains only one piece of information: whether the response is above or below a threshold. *Source: Part A, the illustrations of cells are produced based on the source images in Freeman, R.D., Ohzawa, I., 1990. On the neurophysiological organization of binocular vision. Vis. Res. 30 (11) 1661–1676.*

interpreting symbols. Each of those edge-sensitive cells is *oriented*, that is, it provides high response when the visual scene contains edges whose orientation is close to a particular angle—the angle that the cell is tuned for. Furthermore, each cell is *localized* in space, which makes it responsive only to edges that are situated at a particular location in the visual scene. These two properties are highly relevant to how we desire to describe edges for facial expression analysis. The orientation of edges is important to distinguish expressions; for example, the eyebrows of a sad-looking face are more slanted than those of a neutral face. The need for *localized* edge detection is more obvious: the location of an edge provides information about which facial feature the edge belongs to (e.g., an eye or a mouth), and further, which part of the facial feature (e.g., left or right corner of the mouth).

Fig. 12.2A illustrates the simple cells in the human visual cortex discussed in the previous section, along with their behavior to different visual stimuli. Each cell is tuned to detect edges in a specific orientation; for instance, the cell at top is tuned to vertical edges. Each cell contains two *receptive fields* (RFs) that enable the detection of edges: the positive RF displayed with light color, which is activated with bright visual elements, and the negative RF displayed with dark color, which is activated with dark visual elements—the gray parts are not responsive to any visual stimuli. The spatial order and orientation of these RFs define the orientation to which the cell is tuned for. In the top cell, the positive and negative RFs are ordered from left to right, which makes the cell sensitive to

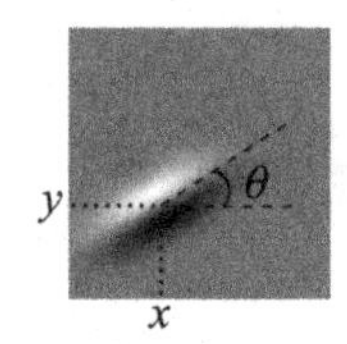

FIGURE 12.3 The illustration of a Gabor filter. The coordinates of the filter center are denoted with x, y, and the orientation angle of the filter is denoted with θ.

vertical edges, and in the bottom cell the RFs are ordered from top to bottom, which makes the cell sensitive to horizontal edges. At the early layers of the visual system, the precise granularity of the response is not very important (Wandell, 1995); what is important is whether the response is above or below a certain threshold. Therefore, the visual cortex performs a nonlinear transformation that saturates the output of the response to a fixed level as long as the response is above a threshold. An idealized version of such a nonlinear transformation is illustrated in Fig. 12.2B.

These findings about the human visual cortex had a major impact in automatic visual analysis. Computer scientists emulated these low-level layers of the visual cortex and created successful object, scene and face recognition systems. The essential task needed for this emulation is to find digital equivalents of the cells described in the previous paragraph. The *Gabor filters* are the widely popular mathematical tools that served for this purpose. A Gabor filter is a 2-dimesional (2D) array that can be visualized as in Fig. 12.3, where each pixel is colored based on the numerical value at the corresponding pixel. The gray color that is seen in the most parts represents zero, and the brighter and darker colors represent respectively positive and negative values, akin to the RFs of the visual cells (Fig. 12.2A). Recall that the visual cortex contains a population of localized and oriented visual cells; to emulate this, we can create a population by adjusting two parameters of the filter: the offset coordinates and the orientation (Fig. 12.3). Effectively, this population of filters operates like an edge detector which, when applied to faces, highlights the shape of facial features, as illustrated in Fig. 12.4.

The representation obtained through Gabor filters achieved a notable affect recognition performance; for example, 91.8% accuracy in the recognition of the six basic emotions (see Section "Spatial Analysis"). Many studies in the literature followed the basic idea of creating a representation that resembles an edge detector. The major drawback of the Gabor filters is their computational complexity. Gabor-based representation requires intensive computation, which hardly enables real-time speed on regular computers; also, the resulting representation is high dimensional and requires an additional *dimensionality reduction* component (Sariyanidi et al., 2015). The studies that followed the Gabor representation targeted

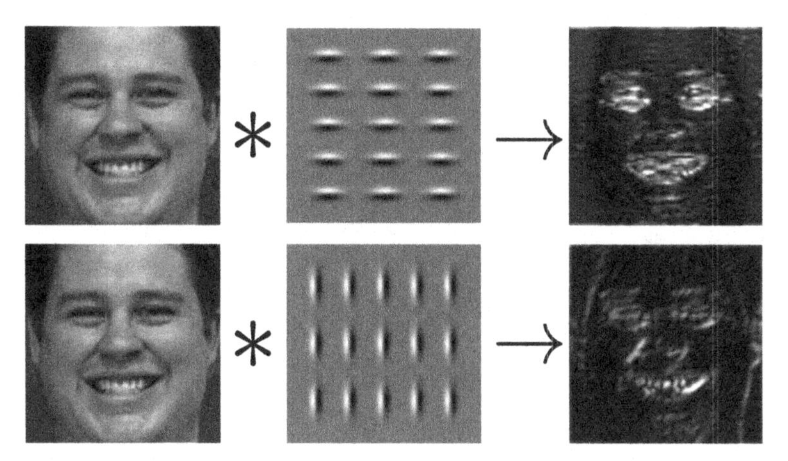

FIGURE 12.4 **A sample response of Gabor filters applied on a facial image.** Left column: input images. Central column: Gabor filters in a particular orientation (multiple filters at different locations highlight the localized nature of the Gabor filters). The filters are tuned to vertical edges (top) and horizontal edges (bottom). Right column: the response images highlight the vertical/horizontal edges of the facial features.

creating faster representations that are inherently compact. One of the first alternatives that proved successful in this search was the local binary patterns (LBP) representation (Shan et al., 2009). Here we discuss the quantized local Zernike moments (QLZM) representation (Sariyanidi et al., 2013), which showed better expression recognition performance than Gabor and LBP while also being computationally more efficient.

Quantized Local Zernike Moments

The aim with QLZM is to create a pipeline that mimics the main quality of the Gabor filters; that is, to represent the image edges in a localized and oriented manner. For this purpose, we must first create digital entities that resemble the Gabor filters. The *Zernike polynomials* (ZPs) (Teague, 1980) are the arrays that serve for this purpose. As illustrated in Fig. 12.5, ZPs are 2D arrays whose behavior resembles that of the visual cells with multiple positive and negative RFs. Each ZP encodes edges in a particular orientation, and ZPs have the advantage of being *orthogonal*; that is, the information provided by different polynomials does not overlap, promoting compactness (Sariyanidi et al., 2013).

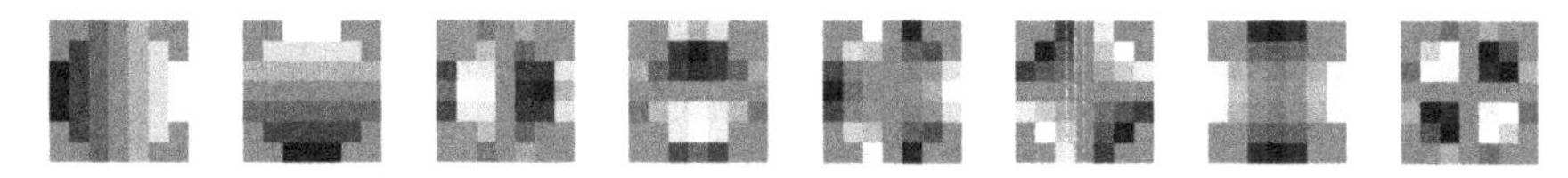

FIGURE 12.5 **Illustration of Zernike polynomials.**

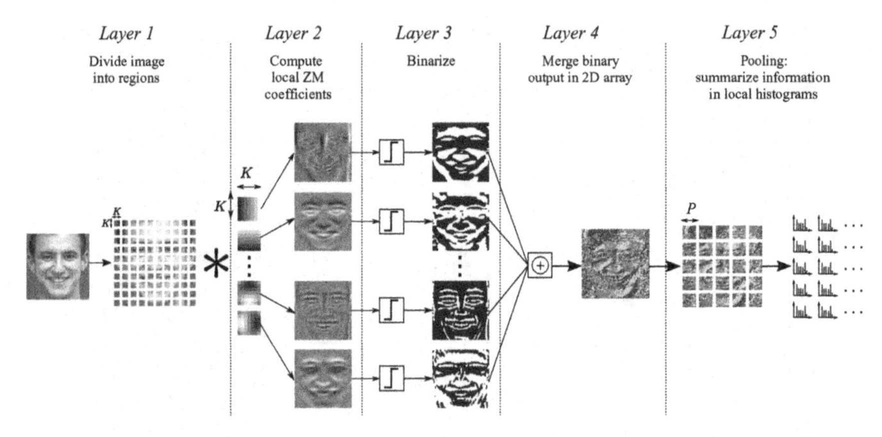

FIGURE 12.6 The illustration of the overall process of encoding an image with quantized local Zernike moments (QLZM).

The ZPs provide tolerance against illumination variations due to the following three properties:[a] first, ZPs are zero-mean, as the polynomial elements sum up to zero. This renders the representation robust to *uniform* illumination variations, which are the variations that cause the brightness in an image to change uniformly. Second, ZPs are localized in space and each local patch is treated independently from the others, providing some tolerance to illumination variations that affect an image *nonuniformly*; although such variations may have a nonuniform effect across the entire image, their effect can be assumed to be constant within a patch, and by treating each part independently and using zero-mean polynomials, QLZM provides partial tolerance against nonuniform illumination variations. The third property is performing illumination normalization to a spatial group of QLZM elements, as discussed next.

Fig. 12.6 illustrates the overall process of computing a QLZM representation. We first divide an input image to small, $M \times M$-sized patches and encode the local image variation (i.e., edges) across each patch by point-multiplying the patch with each of the K polynomials. The outputs of these multiplications are illustrated in Layer-2 of Fig. 12.6: each ZP encodes the variation across the image in a unique way. Then, similarly with the human vision system behavior in Fig. 12.2B, the responses in Layer-2 are converted into binary representations by yielding 1 if the response is positive or yielding 0 otherwise. These multiple binary responses are merged into one single 2D array as in Layer-3. This 2D array consists of

[a]Some of these properties are shared by other low-level representations as well, please see (Sariyanidi et al., 2015) for more details

K-bit values[b] that provide a summary of all the binary ZP representations; for instance, a $(00...00)_2$ value in a particular location of the array means that all binary responses for all ZPs were 0 at that location. A $(00...01)_2$ means that all binary responses were 0 except the last one. The final step, Layer-4, aims at reducing sensitivity to registration errors. We divide the 2D array to $P \times P$-sized subarrays, and produce a single *local histogram* for each subarray. A local histogram provides a statistical summary about the image variation across the $P \times P$-sized patch but neglects the precise location of the elements inside the patch—such techniques are typically referred to as pooling in the computer vision literature (Boureau et al., 2010) and have proved very useful in improving robustness against registration errors. Each histogram bin counts the number of occurrences of a particular K-bit integer; for instance, the first histogram bin counts the number of patterns with value $(00...00)_2$, and the second histogram bin counts the number of patterns with value $(00...01)_2$. Also, to reduce the sensitivity against nonuniform illumination variations, each of the local histograms is normalized to sum up to 1. This prevents potential imbalances that can occur due to nonuniform illumination variations (Sariyanidi et al., 2015). To conclude the section, QLZM achieves the desired numerical representation that summarizes expression-related information—that is, edges—across the image while being partly against illumination variations and registration errors.

Experimental Evaluation

Using the previously described QLZM representation, we evaluate facial expression recognition performance on two datasets: The CK+ dataset (Lucey et al., 2010), which consists of posed expressions, and the AVEC'12 dataset, which contains spontaneous expressions (Schuller et al., 2012). The task in both datasets is to recognize emotions. In the CK+ dataset, emotions are modeled with the 6-basic emotion model (Ekman et al., 2003). The evaluation metric in CK+ is classification accuracy; that is, the proportion of images where emotions are predicted correctly. In the AVEC'12 dataset emotions are modeled with a continuous affect model and we report results on the valence and arousal dimensions. The evaluation metric for the AVEC'12 dataset is the average of Pearson's cross correlation on the 32 sequences of the development partition (Schuller et al., 2012).

Table 12.1 shows the performance of QLZM on CK+ dataset and Table 12.2 shows the performance of QLZM on AVEC'12 dataset. The classifiers used when obtaining the results on CK+ are listed in the middle column of Table 12.1. The results on AVEC'12 dataset are obtained by

[b]A K-bit representation is a binary sequence of K values, such as $\underbrace{(010\ldots101)_2}_{K}$. Each K-bit value can be converted to a decimal integer in the range of $[0,2^K]$ using simple binary arithmetic; for instance $(0101) = 5$

TABLE 12.1 Facial Expression Recognition Performance on the CK+ Dataset Obtained With Several Representations

Representation	Classifier	Recognition rate (%)
Canonical appearance representation (CAPP) (Chew et al., 2012)	SVM Lin.	70.1
	Uni-Hyp.	89.4
Local directional numbers (LDN) (Ramirez Rivera et al., 2012)	SVM Lin.	89.3
	SVM RBF	89.3
	SVM Pol.	81.7
Gabor representation (Sikka et al., 2012)	SVM Lin.	91.8
Local binary patterns (LBP) (Sikka et al., 2012)	SVM Pol.	82.4
Bag-of-words (BoW) (Sikka et al., 2012)	SVM Lin.	95.9
Quantized local Zernike moments (QLZM)	SVM RBF	**96.2**

Bold number indicates the highest recognition rates.

TABLE 12.2 Facial Emotion Recognition Performance on the AVEC'12 Dataset

Representation	Arousal	Valence
QLZM	**0.195**	**0.229**
LBP (baseline)	0.151	0.207

Results are reported for arousal and valence dimensions in terms of cross correlation between the predicted emotion values and the ground truth. Both results are obtained with SVR regressor using the histogram intersection. Bold numbers indicate the highest cross correlation values for each affect dimension.

using SVR regressor with histogram intersection kernel both for QLZM and LBP. QLZM shows a considerable improvement on CK+ in comparison with other low-level representations, such as canonical normalized appearance features (CAPP) (Chew et al., 2012), LBP, Gabor and local directional number (LDN) (Ramirez Rivera et al., 2012) representation, and also on the AVEC'12. The only representation whose performance is close to that of QLZM is the bag-of-words (BoW) representation; however, this representation requires an offline training stage, and also a visual vocabulary search that is computationally costly, whereas QLZM is computationally simple—the computation time for a 200×200-sized image is very short, ranging between 1 and 37 ms depending on the parameter setting (Sariyanidi et al., 2013). QLZM offers also a considerable gain on AVEC'12 in comparison with the LBP: a cross correlation increase of 0.044 for the arousal dimension, which corresponds to an improvement of nearly 29%, and a cross correlation increase of 0.022 for the valence dimension, which corresponds to an improvement of nearly 10%. Overall, QLZM performs facial expression recognition with high accuracy both on CK+ and

AVEC'12 datasets. Moreover QLZM is computationally efficient not only in terms of computation time, but also in terms of memory requirement. The size of the local histograms for QLZM is 16, whereas, for comparison, the size of LBP's local histograms are 59 (Ahonen et al., 2006).

SPATIOTEMPORAL ANALYSIS

Facial actions are spatiotemporal events, and the evolution of a facial action is typically analyzed by breaking it into four *temporal phases* (Ekman et al., 2003): *neutral* is the expressionless phase where there is no muscular activity, *onset* is where muscular contraction begins and grows in intensity, *apex* is where muscular contraction reaches to a stable level and *offset* is the phase where muscles relax.[c]

Spatiotemporal analysis techniques are more suitable to analyze facial expressions in videos. The key difference between spatial and spatiotemporal methods is that the latter enable the analysis of the temporal evolution of the expression: Spatiotemporal analysis techniques may provide information about the duration of the temporal phases of the facial actions that constitute the expression, and information of whether there is a symmetry in the evolution of the different expressions; for instance, if the left and the right corners of the mouth are moving symmetrically during a smile. These cues are valuable to discriminate between expressions that cause a similar change in facial appearance (e.g., closing eyes vs. blinking) (Kaltwang et al., 2012; Koelstra et al., 2010), or in obtaining inference about complex tasks like distinguishing between genuine and posed affective behavior (Valstar et al., 2006, 2007).

Spatiotemporal analysis enables higher performance in the recognition of subtle expressions too. Research showed that the human vision is limited in its ability to recognize subtle expressions when the natural continuity of a facial video is interrupted, or when the videos are shown to subjects frame-by-frame (Ambadar et al., 2005). This can be verified through the simple illustration in Fig. 12.7. Two consecutive frames from a facial expression are shown in Fig. 12.7A. Looking at the frames separately, it is hard to notice any expression variation. A simple way of illustrating the temporal variation is to compute the *difference image*, that is, to subtract one image from the other. The difference image shown in Fig. 12.7B highlights that an expression *does* occur (around the eyelids), thus confirming the importance of temporal variation for recognizing subtle expressions.

[c]Strictly speaking, a facial action does not have to start from a neutral display, as there may be transition from one action to the other. Also, there may be multiple apices throughout a facial display (Valstar and Pantic, 2012). However, it is customary to use the neutral-onset-apex-offset order while introducing the definitions of these temporal phases.

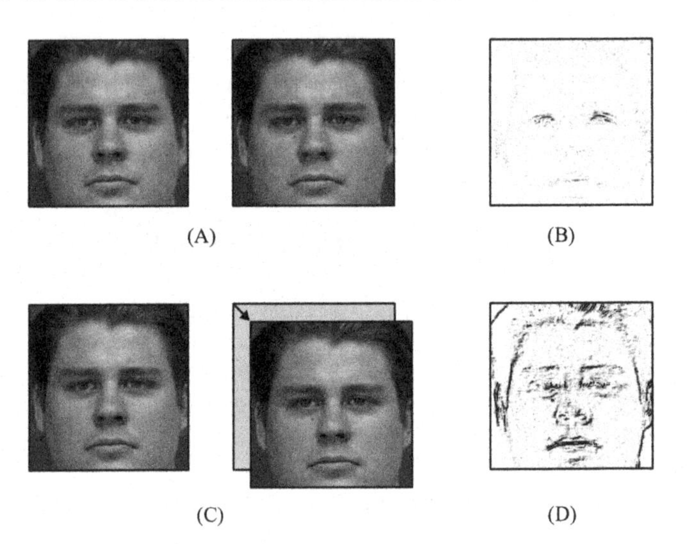

FIGURE 12.7 An illustration depicting the importance of accurate registration when analyzing subtle expressions. (A) Two consecutive images from a sequence that contains a subtle expression around the eyelids, which is hard to notice when looking at still images only. These two images are perfectly registered. (B) The temporal variation (i.e., difference image) between the perfectly registered images shows correctly the facial activity around the eyelids. (C) The same pair of consecutive images but the second image has been displaced by 0.5 pixels (this displacement is visualized with some magnification to facilitate the interpretation). (D) The temporal variation of the unregistered images is highlighting the registration errors rather than the facial activity even for registration errors as low as 0.5 pixels.

However, the analysis of temporal variation highlights accurate registration as an additional requirement to the design of the automatic analysis system. When we obtained the difference image in Fig. 12.7B, we used two images that are registered perfectly. Registration errors between consecutive frames create a jittering that may overshadow the expression. Even registration errors as small as 0.5 pixels may be detrimental to analysis, as shown in Fig. 12.7C–D.

Registration

The simplest and most popular (Sariyanidi et al., 2015) way to register a video is based on extending the typical spatial registration scheme discussed in Section "Spatial Analysis"; that is, registering each frame of the video based on facial landmarks. This scheme is effectively registering a frame with respect to a prototypical face model, as it aims at bringing some facial features (e.g., eyes) to a fixed position (Fig. 12.8A). However, this scheme is inherently limited in its ability to eliminate registration errors among consecutive frames, as it registers each frame independently from neighboring frames, and even state-of-the-art landmark localizers

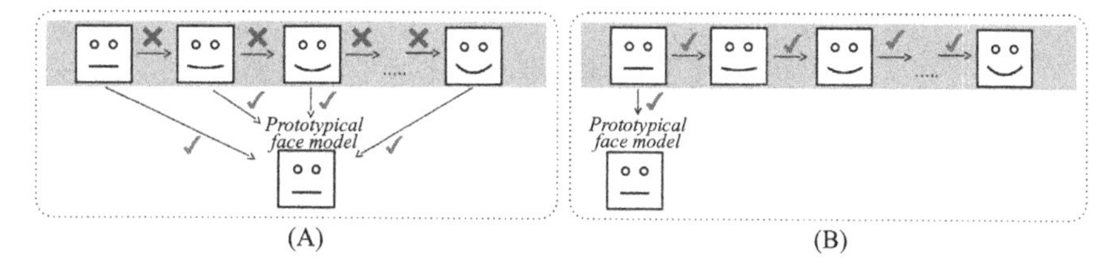

FIGURE 12.8 Visual illustration of frame registration and sequence registration. (A) In frame registration each frame is registered with respect to a prototypical face model, independently from neighboring frames. While this approach can minimize the registration errors with respect to the prototypical face, it does not necessarily minimize the errors among consecutive frames. (B) In sequence registration, only one of the frames (typically the first) is registered to a prototypical face, and the remaining frames are registered to the neighboring frames. This approach aims to model the temporal variation in facial expression more accurately by removing the registration errors among consecutive frames.

have an expected error of 2–3 pixels (Çeliktutan et al., 2013). As a result, the extension of the spatial registration method would inevitably cause jittering when used to register a video.

To eliminate jittering, the registration method must explicitly aim to reduce the registration errors among consecutive frames. This can be achieved by registering each frame of the video with respect to other frames in the video. A typical scheme would register the first frame with respect to a prototypical face (e.g., by localizing facial points), then the second frame with respect to the first, then the third frame with respect to the second and so on (Fig. 12.8B). In this process we only need a method that registers a given pair of frames with very high accuracy; then, we can use this method to register all the consecutive pairs of images sequentially.

Note that the jittering we try to eliminate with registration is a form of (spurious) motion and, what we aim to analyze, facial expressions, are also a form of motion. At this point it is useful to categorize motion in facial sequences into two types: *global motion* is observed throughout the image plane, typically caused by camera or head/body movement, and *local motion* is caused by facial activity and is observed only in parts of the image plane. The former is what we aim to eliminate via registration. To eliminate it, we must first measure it. There are three basic forms of global motion: rotation, scaling, and shifting. The global motion between two frames can be any mixture of these three basic forms, and what we mean by measuring global motion is discovering the parameters of this mixture (e.g., +15 degrees rotation, 110% scaling, −1 pixels of vertical translation).

How can we measure the global motion between two images? The progress in human vision research has been influential in answering this question too. The related field of study is *motion perception*: although it seems straightforward to us, the perception of motion speed and direction

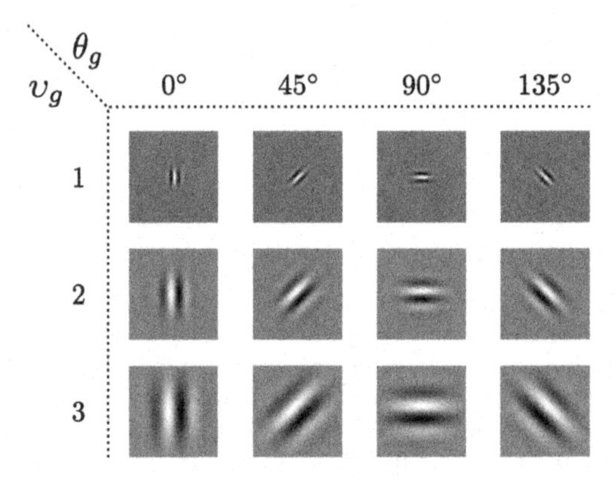

FIGURE 12.9 **Illustration of spatiotemporal Gabor filters that resemble motion-responsive filters in the human visual cortex.** Filters that are tuned to different speeds v_g have different sizes (i.e., spatial support). Filters that differ only in their orientation θ_g resemble rotated versions of one another.

of objects in a dynamic scene is a complicated neural process and an active research problem.

It is known that the early layers of the visual cortex contain cells that are responsive to motion (Adelson and Bergen, 1985). To enable inference about the speed and direction of motion, each of those cells is tuned to a particular speed and direction. Adelson and Bergen (1985) showed that the behavior of these cells can be well approximated with the Gabor filters that are similar to those discussed in Section "Spatial Analysis," but also have a third dimension that enables inference over time. One Gabor filter is denoted with $g^{v,\theta}$, where v and θ denote respectively the motion speed and direction that the filter is tuned for (Fig. 12.9). To emulate the early layers of the visual cortex, which contain a large number of motion-responsive cells, we must use multiple filters—each tuned to a particular speed and orientation.

We can understand why the 3D Gabor filters are useful by analyzing their output for a simple problem: discovering the speed and orientation of a moving line. On the left of Fig. 12.10 we show a moving line along with the response of four spatiotemporal Gabor filters to this line (the four 2D black-and-white arrays). The higher the response of a filter the brighter values in the array. The filter that produces the highest response is the one that is tuned to the orientation of the line. Therefore, by looking only at the filter responses, we are able to infer the orientation of the line. A similar discussion would apply for inferring the speed of the line—the filter that is best tuned to the speed of the line would produce the highest response.

To recall, our ultimate goal is to estimate the global motion between two consecutive images I_t, I_{t+1}, which we can denote with a symbol $\mathbf{y}_t = [\tau_x, \tau_y, \sigma, \theta]$, where τ_x, τ_y are the horizontal and vertical shifts (in pixel units),

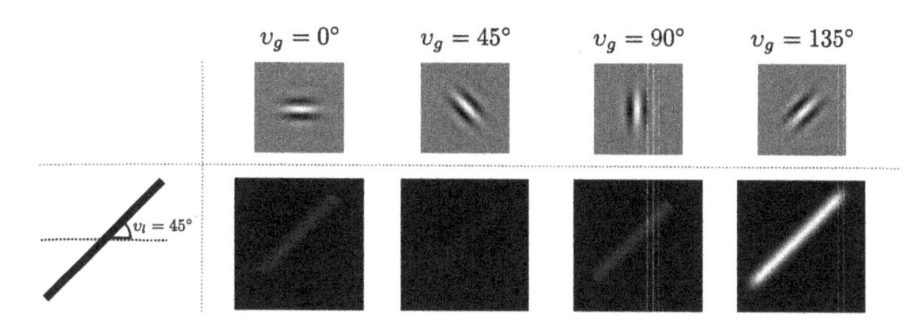

FIGURE 12.10 Illustration of how Gabor motion energies computed from multiple pairs of spatiotemporal Gabor filters enable the identification of the orientation of moving elements (a line with a 45 degree orientation). Second row: response of four spatiotemporal Gabor filters. The maximal energy is produced by the filter tuned to the motion of the line, that is, the one whose orientation is perpendicular to that of the line.

θ is rotation in degrees and σ the scale as a ratio. However, the numerical output of the filter responses (like the 2D black-and-white arrays in Fig. 12.10) provides only implicit information about motion. Therefore, we must somehow convert this implicit information into explicit parameters, $\mathbf{y}_t$.

Similar problems are encountered in other disciplines, such as meteorology, marketing, health care and engineering. An exemplar problem in meteorology would be to predict a future-day temperature using quantitative data about the current state of the atmosphere: We do know that the two are related, but we must still convert the quantitative data into a value in temperature unit (e.g., Celsius degrees). A well-established research field that deals with problems of this kind is *statistical learning* (Bishop, 2006). Let us denote the value to be predicted with **y** (in our problem, the global motion between two images) and the data to be used for prediction with **x** (in our problem, Gabor motion energy). Statistical learning assumes that there exist a number of pairs that contain the corresponding (**x**,**y**) values, that is, a *training dataset*, $\mathcal{D} = \{(\mathbf{x}^1, \mathbf{y}^1), (\mathbf{x}^2, \mathbf{y}^2), \ldots, (\mathbf{x}^n, \mathbf{y}^n), \ldots, (\mathbf{x}^N, \mathbf{y}^N)\}$; in the temperature prediction problem, we can think $\mathcal{D}$ as a collection of past observations. Then, statistical learning uses $\mathcal{D}$ to find a mathematical function f, typically termed a *statistical estimator* that maps any given **x** into a prediction **y**, that is:

$$\mathbf{y} = f(\mathbf{x}; \Theta) \tag{12.1}$$

where Θ denotes the (input-independent) function parameters that are learnt from $\mathcal{D}$.

As even very small registration errors can be detrimental to spatiotemporal analysis (Fig. 12.7), the estimator f must be very accurate. The accuracy of any statistical estimator f depends partly on the dataset $\mathcal{D}$ (Bishop, 2006). One factor that is the number of sample pairs, N, which must not be too low for an estimator to perform accurately (Bishop, 2006). Another factor is the noise on the output values $\mathbf{y}^n$ (for example, potential errors in the measuring of quantities like temperature), and excessive noise is generally detrimental to statistical learning. Fortunately, in our problem of registration we can create the dataset $\mathcal{D}$ in a way that we can have as many pairs as we like without any noise. To create one pair $(\mathbf{x}^n,\mathbf{y}^n)$, we can first take a facial image I and then create a nonregistered version of this image, I', simply by applying a random global motion $\mathbf{y}^n$. We can then compute the Gabor motion energy $\mathbf{x}^n = E(I,I')$. Since we do know the exact parameters of the applied global motion, there is no noise in the sample pair $(\mathbf{x}^n,\mathbf{y}^n)$. Sample pairs are generated automatically via a computer program without requiring a physical measurement, which allows us to create as many sample pairs as we desire from a handful of images by applying a different global motion y^n each time. The registration technique that we described in this section is referred to as probabilistic subpixel temporal registration (PSTR). The technical details of PSTR are discussed further in (Sariyanidi et al., 2014).

Experimental Evaluation

To demonstrate registration performance, we perform experimental evaluation both with posed and naturalistic facial expressions. Similarly to Section "Spatial Analysis," we use the CK+ dataset and the AVEC'12 dataset. We illustrate the performance of the PSTR method by comparing it with another method, namely the Robust FFT method (R-FFT) (Tzimiropoulos et al., 2010), which is currently one of the most robust methods in the literature that can be used for comparison. Further comparison of PSTR with other techniques (Bay et al., 2006; Matas et al., 2004) is shown in (Sariyanidi et al., 2014).

We begin with a qualitative analysis where we show sequence clips before and after registration. In Fig. 12.11A–B we show exemplar registration results from the CK+ and AVEC'12 datasets. Fig. 12.11A shows the difference between two pairs of images that differ by a mouth expression, after registering the pairs of images with PSTR and FFT. While the differences provided by FFT hardly help identifying the location of the expression, PSTR clearly shows where the expression occurs. Identifying the *absence* of facial activity is as important as detecting facial activity. We applied a similar test but for a pair with no facial activity (Fig. 12.11B). The differences provided by FFT generate spurious activity. Instead, the difference image of PSTR shows no signs of facial activity except from minor artifacts introduced by interpolation.

In Fig. 12.11C we illustrate the performance of PSTR on longer (5-frame) sequences from CK+ dataset (top) and AVEC'12 dataset (bottom)—the unregistered versions of both sequences are illustrated in Fig. 12.12. The difference images below each sequence are obtained by taking the difference

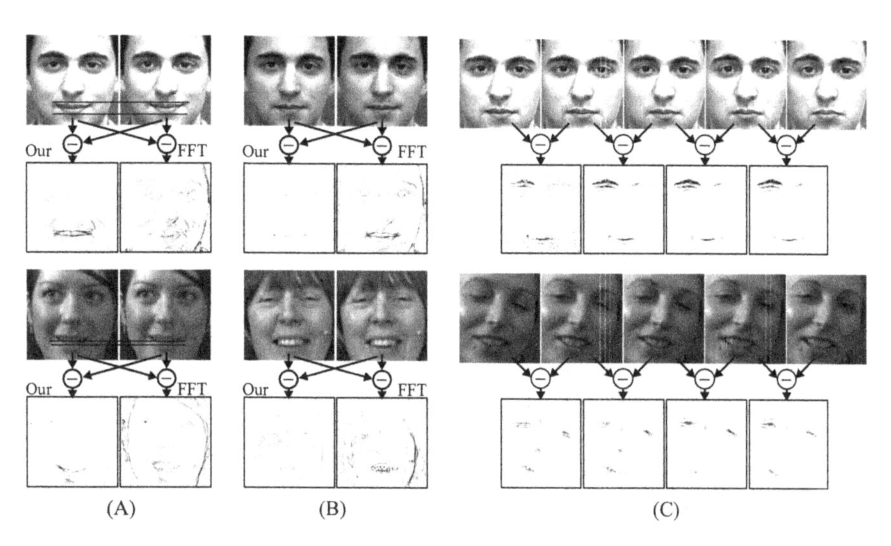

FIGURE 12.11 Illustration of sequence registration results for the subpixel temporal registration (PSTR) and FFT techniques. (A) Difference between pair of images with a subtle mouth expression, after registering the images with PSTR and FFT, (B) Difference between images without expression variation; (C) Difference of each pair in a sequence after registration. The images and sequence on top are from the (posed) CK+ dataset and the ones below are from the (naturalistic) AVEC'12 dataset.

of the consecutive image pairs. The CK+ sequence contains a subtle mouth expression and an eyebrow movement, and the AVEC'12 sequence contains eyelid and mouth movement. The resulting difference images clearly illustrate the usefulness of PSTR—no matter how slowly the expression evolves, the difference images capture face actions and *only* facial actions.

Representation

Spatiotemporal representations are less studied than spatial representations. We can describe the way that most spatiotemporal representations function by comparing them with spatial representations; while the latter encode the edges in a given frame, the former encode the variation of the edges over time in addition to encoding the edges at each frame. Most spatiotemporal representations are developed by extending existing spatial representations to the third dimension (i.e., time) rather straightforwardly (Sariyanidi et al., 2015) [e.g., LBP-TOP (Zhao and Pietikainen, 2007), LPQ-TOP (Jiang et al., 2011)]. The third dimension, however, creates a much less compact representation; such representations can be more than 10 times longer than their spatial counterparts (Sariyanidi et al., 2015). This not only makes them computationally more expensive, but also renders the task of automatic recognition more challenging due

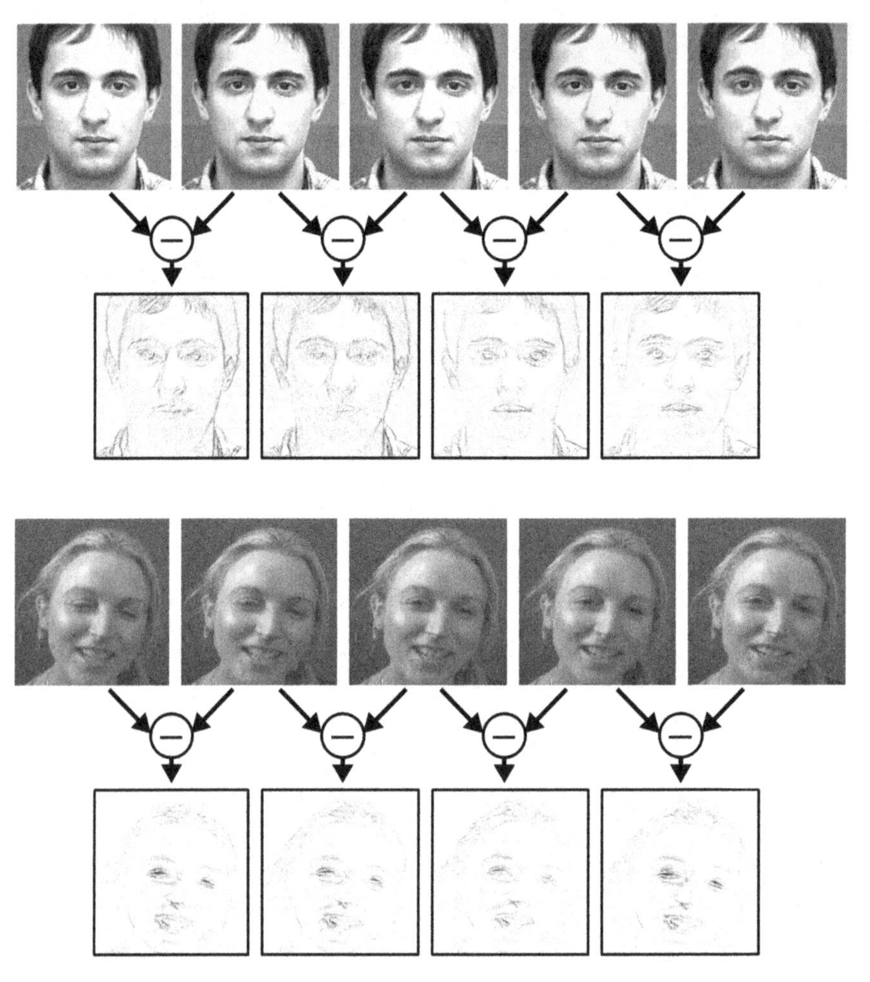

FIGURE 12.12 Exemplar sequences from the CK+ dataset (top) and the AVEC'12 dataset (bottom), and the difference images between consecutive frames. The sequences contain registration errors (which are hard to see when looking at static images), therefore the difference images are showing spurious motions. Fig. 12.11C illustrates these sequences and their difference images after registering them with PSTR.

to the curse of dimensionality (Friedman, 1997). Automatic facial affect analysis is essentially the task of finding the emotion label that best fits a given representation. As the representation gets longer (i.e., higher dimensional), finding this label becomes more challenging, akin to searching for a needle in the hay (Donoho, 2000).

So far, spatiotemporal representations are mostly used for proof of concept in relatively simple problems where illumination conditions are controlled and subjects display limited head movements. Spatial representations have been preferred in systems that are designed for real-world affect analysis where fewer assumptions can be made about illumination or head movements. Spatiotemporal representations have started being popular only more recently. This trend can also be noticed in the affect analysis competitions that were organized between the year 2011–14 and are known as the audiovisual emotion challenges (AVEC). None of the participants of the AVEC'11 and AVEC'12 relied on spatiotemporal representations, whereas nearly half of the participants of AVEC'13 and AVEC'14 used spatiotemporal representations. This indicates that further research needs to be conducted for spatiotemporal analysis of facial affect.

CONCLUSIONS

In this chapter we discussed recent advances in automatic facial expression analysis with emphasis on registration and representation. While research efforts have progressed so far mainly on spatial analysis, spatiotemporal techniques outperform spatial techniques when dealing with more challenging affect recognition problems that involve the analysis of subtle expressions. Therefore, although they require more sophisticated registration and representation, there has been recently a shift of interest toward spatiotemporal analysis, and the PSTR technique we discussed stands out as a candidate to provide the accurate registration required for spatiotemporal analysis.

Acknowledgement

The work of Hatice Gunes is funded by the EPSRC under its IDEAS Factory Sandpits call on Digital Personhood (grant ref: EP/L00416X/1). This work was partly done while Hatice Gunes was with the Queen Mary University of London.

References

Adelson, E.H., Bergen, J.R., 1985. Spatio-temporal energy models for the perception of motion,. J. Opt. Soc. Am. 2 (2), 284–299.
Ahonen, T., Hadid, A., Pietikäinen, M., 2006. Face description with local binary patterns: application to face recognition. IEEE Trans. Pattern Anal. Mach. Intell. 28 (12), 2037–2041.
Ambadar, Z., Schooler, J.W., Cohn, J., 2005. Deciphering the enigmatic face: the importance of facial dynamics in interpreting subtle facial expressions. Psychol. Sci. 16 (5), 403–410.
Affectiva. http://www.affectiva.com.
Emotient. http://www.emotient.com.
Kairos. http://www.kairos.com.
Bay, H., Tuytelaars, T., Van Gool, L., 2006. SURF: speeded up robust features. Proc. European Conf. Computer Vision. pp. 404–417.
Bishop, C., 2006. Pattern Recognition and Machine Learning. Springer, New York.
Boureau, Y.-L., Ponce, J., LeCun, Y., 2010. A theoretical analysis of feature pooling in visual recognition. Int'l Conf. Machine Learning. pp. 111–118.

Çeliktutan, O., Ulukaya, S., Sankur, B., 2013. A comparative study of face landmarking techniques. EURASIP J. Image Video Process. 2013 (1), 13.

Chew, S.W., Lucey, S., Lucey, P., Sridharan, S., Cohn, J.F., 2012. Improved facial expression recognition via uni-hyperplane classification. Proceedings of the IEEE Conference on Computer Vision and Pattern Recognition. pp. 2554–2561.

Donoho, D.L., 2000. High-dimensional data analysis: the curses and blessings of dimensionality. AMS Math Challenges Lecture. pp. 1–32.

Dooley, R., 2015. Microsoft glasses read your emotions. http://www.neurosciencemarketing.com/blog/articles/microsoft-glasses.htm; http://www.neurosciencemarketing.com/blog/articles/microsoft-glasses.htm.

Ekman, P., Campos, J., Davidson, R., Waals, F.D., 2003. Emotions Inside Out, ser., vol. 1000. Annals of the New York Academy of Sciences, New York Academy of Sciences.

Friedman, J.H., 1997. On bias, variance, 0/1loss, and the curse-of-dimensionality,. Data Min. Knowl. Discov. 1 (1), 55–77.

Gunes, H., Schuller, B., 2013. Categorical and dimensional affect analysis in continuous input: current trends and future directions. Image Vis. Comput. 31 (2), 120–136.

Hodson, H., 2014. Google glass, now in tune with your emotions. http://www.neurosciencemarketing.com/blog/articles/microsoft-glasses.htm; https://www.newscientist.com/article/dn26153-google-glass-now-in-tune-with-your-emotions/.

Hubel, D.H., Wiesel, T.N., 1962. Receptive fields, binocular interaction and functional architecture in the cat's visual cortex,. J. Physiol. 160 (1), 106.

Jiang, B., Valstar, M., Pantic, M., 2011. Action unit detection using sparse appearance descriptors in space-time video volumes. Int'l Conf. on Automatic Face Gesture Recognition and Workshops. pp. 314 –321.

Joormann, J., Gotlib, I.H., 2006. Is this happiness I see? biases in the identification of emotional facial expressions in depression and social phobia. J. Abnorm. Psychol. 115 (4), 705.

Kaltwang, S., Rudovic, O., Pantic, M., 2012. Continuous pain intensity estimation from facial expressions. Bebis, G., Boyle, R., Parvin, B., Koracin, D., Fowlkes, C., Wang, S., Choi, M.-H., Mantler, S., Schulze, J., Acevedo, D., Mueller, K., Papka, M. (Eds.), Advances in Visual Computing, ser. Lecture Notes in Computer Science, vol. 7432, Springer Berlin Heidelberg, pp. 368–377.

Koelstra, S., Pantic, M., Patras, I., 2010. A dynamic texture-based approach to recognition of facial actions and their temporal models,. IEEE Trans. Pattern Anal. Mach. Intell. 32 (11), 1940–1954.

Lucey, P., Cohn, J., Kanade, T., Saragih, J., Ambadar, Z., Matthews, I., 2010. The extended cohn-kanade dataset (CK+): a complete dataset for action unit and emotion-specified expression. IEEE Conf. on Computer Vision and Pattern Recognition Workshops. pp. 94–101.

Matas, J., Chum, O., Urban, M., Pajdla, T., 2004. Robust wide-baseline stereo from maximally stable extremal regions. Image Vis. Comput. 22 (10), 761–767.

Matsumoto, D., Frank, M.G., 2012. Nonverbal Communication: Science and Applications. Sage, California.

Ramirez Rivera, A., Rojas Castillo, J., Chae, O., 2012. Local directional number pattern for face analysis: face and expression recognition. IEEE Trans. Image Process. PP (99), 1.

Sariyanidi, E., Gunes, H., Gökmen, M., Cavallaro, A., 2013. Local Zernike moment representations for facial affect recognition. Proc. British Machine Vision Conf.

Sariyanidi, E., Gunes, H., Cavallaro, A., 2014. Probabilistic subpixel temporal registration for facial expression analysis. Proceedings of the Asian Conference on Computer Vision. pp. 320–335. http://dx.doi.org/10.1007/978-3-319-16817-3_21.

Sariyanidi, E., Gunes, H., Cavallaro, A., 2015. Automatic analysis of facial affect: a survey of registration, representation and recognition. IEEE Trans. Pattern Anal. Mach. Intell. 37 (6), 1113–1133.

Schuller, B., Valstar, M., Cowie, R., Pantic, M., 2012. AVEC 2012: the continuous audio/visual emotion challenge—an introduction. Proc. of ACM Int'l Conf. on Multimodal Interaction. pp. 361–362.

Shan, C., Gong, S., McOwan, P.W., 2009. Facial expression recognition based on local binary patterns: a comprehensive study,. Image Vis. Comput. 27 (6), 803–816.

Sikka, K., Wu, T., Susskind, J., Bartlett, M., 2012. Exploring bag of words architectures in the facial expression domain. Proceedings of the European Conference on Computer Vision Workshops and Demonstrations. pp. 250–259.

Teague, M.R., 1980. Image analysis via the general theory of moments. J. Opt. Soc. Am. 70 (8), 920–930.

Tzimiropoulos, G., Argyriou, V., Zafeiriou, S., Stathaki, T., 2010. Robust FFT-based scale-invariant image registration with image gradients. IEEE Trans. Pattern Anal. Mach. Intell. 32 (10), 1899–1906.

Valstar, M., Pantic, M., 2012. Fully automatic recognition of the temporal phases of facial actions. IEEE Trans. Syst. Man Cybern. B Cybern. 42 (1), 28–43.

Valstar, M., Pantic, M., Ambadar, Z., Cohn, J., 2006. Spontaneous vs. posed facial behavior: automatic analysis of brow actions. Proc. ACM Int'l Conf. Multimodal Interfaces. pp. 162–170.

Valstar, M., Gunes, H., Pantic, M., 2007. How to distinguish posed from spontaneous smiles using geometric features. Proceedings of the ACM Int'l Conf. Multimodal Interfaces. pp. 38–45.

Vinciarelli, A., Pantic, M., Bourlard, H., 2009. Social signal processing: survey of an emerging domain. Image Vis. Comput. 27 (12), 1743–1759.

Wandell, B.A., 1995. Foundations of Vision. Sinauer Associates, Massachusetts.

Zhao, G., Pietikainen, M., 2007. Dynamic texture recognition using local binary patterns with an application to facial expressions. IEEE Trans. Pattern Anal. Mach. Intell. 29 (6), 915–928.

CHAPTER

13

On Computational Models of Emotion Regulation and Their Applications Within HCI

Tibor Bosse

VU University Amsterdam, Amsterdam, The Netherlands

INTRODUCTION

Throughout history, the function of emotions in human beings has been a subject to much debate. In the early days of emotion psychology, many researchers treated emotions as (neural) states without a function (Hebb, 1949). However, there is increasing evidence that emotions are in fact functional (Damasio, 2000). For instance, they prepare a person for quick motor responses (Frijda, 1986), facilitate decision-making (Oatley and Johnson-Laird, 1987), and provide information about the ongoing match between organism and environment (Schwarz and Clore, 1983).

In addition to these functions, emotions also play an important role in interpersonal interaction. They provide us with the information about other people's intentions, and script our own social behavior (Gross, 1998). As a simple experiment, just try to count how many expressions of emotion you observe during a meeting with your colleagues at an ordinary working day. You will probably lose count within a few minutes. Mild smiles, frowny faces, hesitating speech, and enthusiastic arm gestures: all of these subtle expressions of emotion may occur during an average conversation. And what's more, they also convey important meaning to the person who observes them.

As a result of this central role that emotions play in human–human interaction, it is not surprising that they have recently received much attention in the area of human–computer interaction (HCI) (Brave and Nass, 2002).

Emotions and Affect in Human Factors and Human–Computer Interaction. http://dx.doi.org/10.1016/B978-0-12-801851-4.00013-6
Copyright © 2017 Elsevier Inc. All rights reserved.

After all, one of the main concerns of the HCI area is to facilitate the interaction between humans and computers. To achieve this goal, it makes sense to make this interaction more similar to the way humans interact with other humans, which can be achieved by including an element of affect.

There are many ways by which computational systems can be endowed with mechanisms to recognize, interpret, process, and simulate affective processes, and the interdisciplinary area that addresses this in general is called affective computing (Picard, 1997). In this chapter, I will focus on a specific subarea of affective computing, which can be referred to as *emotion modeling*. Although there is no complete consensus about the exact meaning of this term (Hudlicka, 2008), I will stick to the following definition proposed by Hudlicka (2014): "developing models of affective processes; that is, models of emotion generation in response to some triggering stimuli, and models of the effects of these emotions on cognition, expression and behavior."

The earlier definition does not explicitly include the word *computational*, but from the context it becomes clear that the author indeed assumes that emotion modeling is about developing computational models of affective processes. Emphasizing this additional criterion is important, as it distinguishes more "informal," conceptual models as (often developed in the human-oriented sciences) from formal models that can be processed by machines. As it will become clear from this chapter, representing models in a machine-readable format is a difficult challenge, but investing this effort usually pays off as the resulting models can actually be used by technical systems. More specifically, if the models are also *dynamic* (i.e., they address the development of a process over time), the possibility to represent them in a computational format typically enables computers to *reason about the dynamics* of such processes.

Applications that use computational models to reason about the dynamics of affective processes can roughly be divided into two categories, namely (1) systems that simulate the dynamics of affective processes within artificial agents and (2) systems that predict the dynamics of affective processes within humans. Most existing systems belong to the former category, for instance embodied conversational agents (Cassell et al., 2000) and social robots (Breazeal, 2001) that show emotions while interacting with humans. Examples of the latter category include adaptive empathetic interfaces (Klein et al., 2002) and stress regulation training systems (Bosse et al., 2014), where the system uses computational models to monitor and predict the development of the user's emotional state during interaction with the system. Hence, the difference between the two categories can be summarized as the difference between *artificial agent models* and *human agent models* of affect. However, for the most part of this chapter this distinction is ignored, thereby simply referring to the *agent* (i.e., artificial or human) as the entity of study.

As can be seen in the definition by Hudlicka (2014), there is another important distinction that can be made to the range of computational models of affect, namely the difference between models of emotion *generation* (or emotion elicitation), and models of the *effects* of these emotions on cognition, expression, and behavior. Obviously, for any serious HCI application, both of these elements are important. For example, an embodied conversational agent that plays the role of a virtual therapist, on the one hand, needs mechanisms to determine how its emotional state changes based on input provided by the human, but it also needs mechanisms that prescribe how this emotional state influences its behavior. Therefore, ideally, a computational model of affect should simulate the entire system of processes from (external and internal) stimuli via emotional states to the effects of these states.

The aim of this chapter is not to provide an overview of what such a "complete"—computational model of affect should look like, as this has already been done by others (Hudlicka, 2014; Marsella et al., 2010). Instead, the chapter will focus on a more specific, but very important aspect of emotions that is often overlooked in computational models, namely *emotion regulation* (or affect regulation). This is an area that has received considerable attention in the psychological literature over the past two decades (Gross, 1998, 2001; Ochsner and Gross, 2005; Thompson, 1994), and that can be defined as follows: "Emotion regulation includes all of the conscious and nonconscious strategies we use to increase, maintain, or decrease one or more components of an emotional response" (Gross, 2001). One of the most famous expressions that illustrate emotion regulation is the saying "better luck next time," which is typically used by people trying to reduce their disappointment after a failed attempt by assuring themselves that they might be successful in the future. However, as will become clear later on, this is just one out of countless examples that one can come up with.

To summarize, this chapter will focus on models that have the following characteristics:

- they are dynamic,
- they are computational,
- they are about affect, and
- they have an emphasis on emotion regulation.

The chapter will explain which elements are important when building dynamic computational models of emotion regulation, will show a concrete example of such a model, and will discuss the potential application areas within HCI. It is intended for any scholar with an interest in studying, using or designing emotion regulation models, both from a theoretical and from an applied perspective. Note that, although the focus is on emotion regulation (which can be seen as an effect of emotions), in several sections I will start by paying some attention to emotion generation. Despite the fact that researchers are not in complete agreement as to whether emotion generation and emotion regulation are two completely separate processes (Gross and Barrett, 2011), presenting them in this way makes it easier to study them from a computational perspective.

The remainder of this chapter is structured as follows. First, the psychological literature on emotion generation and emotion regulation is briefly reviewed from a computational modeling perspective. The next section explains how theories on emotion generation and regulation can be formalized into computational models. After that, a specific computational model of emotion regulation is presented in detail, and a number of resulting simulation runs are shown. Subsequently, a range of potential applications of emotion regulation models within HCI is discussed. The chapter concludes with a brief summary and some considerations for the future.

BACKGROUND

Before making a computational model of emotion regulation, the first question one needs to address is "what" to model. In other words, what are the elementary components of emotion regulation, and how do these components relate to each other? To get a better grip on these questions, this section will briefly review the psychological literature on emotion generation and regulation from a computational perspective. That means that I will not discuss the entire literature about these processes, but mainly list some of the existing theories that are most suitable (and indeed are most often used) for formalization.

As said before, it is too simple to state that emotion regulation is a completely separate process from emotion generation (both functionally and structurally). Indeed, there is still much debate about this question, and the different opinions can be represented as a continuum: on the one hand, there are scientists who treat them as two conceptually distinct processes, whereas on the other hand, there are people who treat them as completely intertwined processes (Gross and Barrett, 2011). In this chapter, I will not take a stand on any of these positions, but I will discuss emotion generation and emotion regulation in two separate steps, for the simple reason that this makes it easier to describe them from a computational perspective. Nevertheless, as will be shown in the section on simulations, the presented computational model simulates both processes simultaneously. This approach to make a conceptual distinction between the two processes, but still activate them in parallel in a computational environment, is used in several influential computational models of affect (Marsella and Gratch, 2009).

Following this dichotomy, the following subsections will discuss the literature on emotion generation and emotion regulation, respectively.

Emotion generation

According to Hudlicka (2014), three theoretical perspectives can be distinguished in the literature that may guide various aspects of emotion modeling: the discrete/categorical, dimensional, and componential perspective. These three perspectives are briefly discussed later.

Discrete or *categorical theories* are based on the assumption that there is a limited set of basic emotions categories, such as joy, sadness, fear, anger, and disgust (Ekman, 1992). Usually, these basic emotions are assumed to have a biological counterpart, that is, each emotion category can be loosely related to a specific corresponding brain region or circuit. Moreover, each emotion category is associated with its own stable pattern of triggers, behavioral expressions, and subjective experience.

In contrast, *dimensional theories* do not distinguish between discrete emotions, but view emotions as states that can be represented as points within a continuous space defined by two or three dimensions. Two-dimensional models of emotion typically make use of the dimensions, valence (or pleasure), and arousal (Russell, 2003; Russell and Mehrabian, 1977), where valence reflects the experienced level of pleasure (ranging from highly negative to highly positive), and arousal reflects a general degree of intensity of the emotion (ranging from very calm to very excited). Because this two-dimensional space cannot easily differentiate between emotions that share the same values of arousal and valence (e.g., anger vs. fear), sometimes a third dimension is added, called dominance. This results in a three-dimensional space that is often referred to as the pleasure-arousal-dominance (PAD) space (Mehrabian, 1995).

Finally, *componential theories* highlight the role of different components that play a role in the emotion generation process, such as the desirability and likelihood of the events that trigger the emotion. This perspective is compatible with appraisal theory, which roughly states that emotions are the result of an appraisal process in which human beings evaluate (internal and external) stimuli against their own goals (Lazarus, 1991; Scherer et al., 2001). Hence, the components are often referred to as appraisal dimensions or appraisal variables (Frijda, 1986). From a computational perspective, modeling the appraisal process involves assigning specific (numerical) values to the appraisal variables, and using these values to calculate the type and intensity of the agent's emotions.

A well-known model of emotion that roughly fits within this componential perspective is the OCC model, named after its authors Ortony, Clore, and Collins (Ortony et al., 1988). This model distinguishes different categories of emotions based on the type of stimuli by which they are triggered (events, actions by other agents, and objects) as well as some evaluative criteria associated with each category (e.g., desirability and familiarity). As these evaluative criteria are similar to the appraisal variables mentioned earlier, the emotion generation process described by the model indeed follows the pattern of cognitive appraisal. Because the OCC model was one of the first emotion generation models of which the dynamics were described in a very systematic way, it turned out to be extremely well suited as a basis for computational models. Indeed, nowadays the majority of the existing computational models of emotion is based on (or at least loosely inspired by) the OCC model.

In addition to the three types of theories mentioned earlier (discrete/categorical, dimensional, and componential), Marsella et al. (2010) mention three additional perspectives, namely *anatomic* (Panskepp, 1998), *rational* (Anderson and Lebiere, 2003), and *communicative* (Keltner and Haidt, 1999) theories. As these perspectives are less frequently used for the types of HCI applications that are the focus of this book, they will not be discussed further.

Emotion regulation

One of the most influential scientists on the topic of emotional (self-) regulation is the American psychologist, James J. Gross. Gross (2001) defines emotion regulation as "all of the conscious and nonconscious strategies we use to increase, maintain, or decrease one or more components of an emotional response." The process of decreasing components of an emotional response is often called downregulation of an emotion, whereas the opposite process is called upregulation. Although upregulation of one's emotional state may intuitively seem less useful (especially in case of emotions with a negative valence), there are numerous situations in which people do this intentionally (e.g., athletes often upregulate their state of anger prior to competition, with the aim to increase their performance).

According to Gross (2001), emotion regulation is concerned with three distinct components of the emotional response, namely (1) the *experiential component* (i.e., the subjective feeling of the emotion), (2) the *behavioral component* (i.e., behavioral responses), and (3) the *physiological component* (i.e., responses, such as heart rate and respiration). Human beings use a wide variety of strategies to influence their level of emotional response for a given type of emotion. On the highest level of abstraction, these strategies can be divided into *antecedent-focused strategies* and *response-focused strategies*. Antecedent-focused strategies are applied to the process preparing for response tendencies before they are fully activated. Response-focused strategies are applied to the activation of the actual emotional response, when an emotion is already underway.

In his process model of emotion regulation, Gross (1998) distinguishes four different types of antecedent-focused emotion regulation strategies, which can be applied at different points in the process of emotion generation: *situation selection, situation modification, attentional deployment,* and *cognitive change.* A fifth strategy, *response modulation,* is a response-focused strategy. See Fig. 13.4 in Gross (1998) for a graphical overview. In the following paragraphs, these five strategies will be explained in more detail (Bosse et al., 2010b).

The first antecedent-focused emotion regulation strategy in the model is situation selection. Someone who applies this strategy chooses to be in a situation that is expected to generate the emotional response level he or she wants to have. For example, a person can stay home instead of going to a party, because he wants to avoid being confronted with an annoying friend. This is an example of downregulating one's emotion (anger in this case). An example of situation selection to upregulate one's emotion (excitement in this case) is taking a roller-coaster ride.

The second antecedent-focused emotion regulation strategy in the model is situation modification. When this strategy is applied, a person modifies an existing situation so as to obtain a different level of emotion. For instance, when giving a speech, one could make some jokes to elicit laughter.

The third antecedent-focused emotion regulation strategy is attentional deployment. This strategy refers to shifting your attention toward or away from certain aspects of an emotional situation. Attentional deployment includes a number of substrategies, of which *distraction* is probably the most famous one (e.g., closing your eyes when you are watching an exciting penalty shoot-out). Other substrategies are *rumination* (i.e., repetitively focusing your attention on your symptoms of distress), *worry,* and *thought suppression.*

The fourth antecedent-focused emotion regulation strategy is cognitive change, which refers to the process of changing how one appraises a situation, with the effect to alter its emotional impact. A specific type of cognitive change, which is typically aimed at downregulating emotion, is *reappraisal.* This strategy involves a reinterpretation of the meaning of an event; an example is a case when a person loses a tennis match and blames the weather circumstances, instead of his own capacities. Other substrategies of cognitive change are *distancing* (i.e., taking a third-person perspective when evaluating an emotional situation) and using *humor.*

The fifth emotion regulation strategy, response modulation, is a response-focused strategy. As opposed to the other four categories, this strategy is applied after the emotion response tendencies have been generated: a person tries to influence the process of response tendencies becoming a behavioral response. A specific type of response modulation, again aimed at downregulation, is *suppression*. An example of suppression is a person that hides being nervous when giving a presentation. Other substrategies of response modulation are physical activity and drug use.

Emotion regulation strategies are used, both consciously and unconsciously, within a variety of domains, tasks, and situations, including sports (Hanin, 2007), demanding jobs (Grandey et al., 2004) and therapeutic settings (Campbell-Sills and Barlow, 2006). Nevertheless, some strategies are claimed to be more effective than others. For instance, Gross (2001) predicts that early emotion regulation strategies are more effective than the strategies that are applied at a later time point in the process. He provides evidence that reappraisal decreases both experiential and behavioral aspects of emotion, while suppression only decreases behavioral aspects, but fails to reduce the emotional experience. Moreover, in the reported experiment suppression resulted in an impaired memory, and an increased physiological response.

A concept that is similar to, yet slightly different from emotion regulation, is the notion of *coping*. Lazarus and Folkman (1984) define coping as the "constantly changing cognitive and behavioral efforts to manage specific external and/or internal demands that are appraised as taxing or exceeding the resources of the person." As can be seen from this definition, the term coping is usually reserved for the process of reducing negative emotions, in particular the emotion of stress. Instead, emotion regulation is assumed to be applicable to a much wider range of situations and emotions, hence covering both up- and downregulation of positive as well as negative emotions. Other differences between coping and emotion regulation are the fact that the latter includes both controlled and automatic processes, and that it includes both intrinsic processes (i.e., regulated by the self) and extrinsic processes (i.e., regulated by an outside factor). Instead, most people see coping as a controlled, volitional process that is performed by the person experiencing stress (Compas et al., 2014).

Nevertheless, also for coping a categorization in different strategies can be made, some of which are close to the emotion regulation strategies introduced by Gross. For example, in Billings and Moos (1984), a distinction is made between *problem-focused coping* (targeting stressors in the environment), *appraisal-focused coping* (targeting the agent's own cognition), and *emotion-focused coping* (targeting the agent's own emotional response). Among these categories, problem-focused coping roughly covers Gross' situation selection and situation modification, appraisal-focused coping roughly corresponds to cognitive change, and emotion-focused coping roughly covers attentional deployment and response modulation. For an extensive description of the distinguishing features between emotion regulation and coping, see Compas et al. (2014).

COMPUTATIONAL MODELING

The previous section provided some guidelines regarding the type of concepts that could be included in a computational model of emotion regulation. The next question is how to convert these concepts to a dynamic computational model.

Developing an accurate computational model of emotion regulation is not an easy task. It requires expertise from different disciplines, including emotion psychology on the one hand and computer science and mathematics on the other hand. Moreover, often a certain level of creativity is needed, in the sense that the steps to take are not fully specified in advance. The reason for this is that a model is by definition a simplification of reality, and designing and implementing a useful model involves making several choices about which elements to include, and about the level of granularity on which to represent these elements.

Despite the fact that computational modeling is a creative process, there are some guidelines that can be followed. Fig. 13.1 displays a general methodology for designing and analyzing models of human behavior: the modeling and simulation cycle (Gerritsen and Klein, 2014). In this cycle, two main phases can be distinguished, namely *design* (in which the model is built) and *analysis* (in which the model is analyzed). The methodology assumes that the modeler has decided which aspect of human behavior will be modeled (in our case: emotion regulation), visualized by the globe at the bottom of the scheme. To formalize this behavior, one should follow

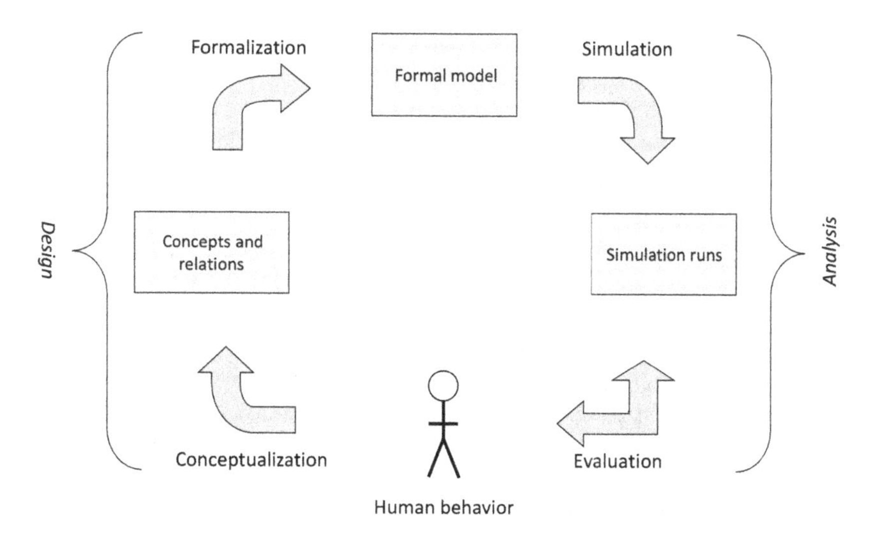

FIGURE 13.1 Modeling and simulation cycle. *Source: Adapted from Gerritsen, C., Klein, M.C.A., 2014. Dynamical simulation in criminology. In: Bruinsma, G., Weisburd, D. (Eds.), Encyclopedia of Criminology and Criminal Justice, Springer Verlag, pp. 1220–1231.*

a cycle that ideally consists of four steps (which will roughly be followed in this and the next sections):

1. *Conceptualization*: identifying the main concepts that play a role in the behavior to be modeled, as well as their mutual relationships.
2. *Formalization*: specifying the concepts and their relationships in a formal, machine-readable notation.
3. *Simulation*: executing the formal model on a computer, in order to show the resulting behavior over time.
4. *Evaluation*: verifying if the behavior produced by the model is as expected.

Following this methodology, any attempt to build a computational model of emotion regulation should start by considering which concepts should be included as its basic building blocks or *semantic primitives* (Wierzbicka, 1996). As we decided to build our model in two steps (starting with an emphasis on emotion generation and later extending this with a mechanism for regulation), let us start by identifying what could be useful semantic primitives for a model of emotion generation. The choice on which primitives to use is largely dependent on the choice, which of the three theoretical perspectives mentioned earlier will be used. If we look at the current state-of-the-art in computational models of emotion, we can see that each of the three perspectives typically provides its own semantic primitives. The situation can be summarized as follows (Hudlicka, 2014):

1. Emotion models that take the *discrete perspective* mostly use the distinct basic emotions themselves as basic building blocks.
2. Emotion models that take the *dimensional perspective* mostly use the two- or three-dimensions as basic building blocks.
3. Emotion models that take the *componential perspective* mostly use individual appraisal variables as basic building blocks.

As mentioned, the majority of computational models of emotion generation make use of the componential perspective, often taking the OCC model (Ortony et al., 1988) as a basis. Sometimes this OCC model is combined with the belief-desire-intention (BDI) framework (Rao and Georgeff, 1991) for representing rational agents, as these two paradigms are nicely compatible with each other (Reisenzein, 2009). For example, an often used approach is to represent an agent's goals in terms of desires, and to represent appraisal variables as beliefs about events. Then, the appraisal process can be simulated by introducing rules that match the beliefs and desires in order to calculate the resulting emotions and their intensities (Adam et al., 2006). Also, some authors use a combination of perspectives, for instance by using OCC for the appraisal process but representing the resulting emotions in terms of 3-tuples corresponding to the PAD dimensions (Dastani et al., 2014).

As explained earlier, the intention of this chapter is not to provide an extensive overview of computational models of emotion generation. For this purpose, the interested reader is referred to (Marsella et al., 2010), and in particular to Fig. 13.1 of that paper, which visualizes a history of computational models as well as the underlying theoretical perspectives.

In contrast, the current chapter has a specific emphasis on emotion regulation. If we look at the literature, we see that only a subset of the existing models addresses regulation in some detail. Probably, the most famous of these models is the emotion and adaptation (EMA) model (Marsella and Gratch, 2009). Like many others, this model also takes appraisal theory as point of departure, inspired in particular by the work of Smith and Lazarus (1990). Within EMA, intensities of emotional states are represented via real numbers in the [0..1] domain. Emotions arise when discrepancies between beliefs and desires (or other beliefs) are detected by automatic appraisal processes. Based on that perspective, a "content model" is used, in which appraisal operates on rich symbolic representations of the emotion-evoking situation. Different types of emotion regulation are simulated by manipulation of these representations. However, the regulation part of EMA mainly addresses coping, that is, downregulation of negative emotions.

Another computational model, which does address the entire scope of emotion regulation, is CoMERG (cognitive model for emotion regulation based on gross) (Bosse et al., 2010b). As suggested by the name, this model is a numerical formalization of the theory described informally by Gross (1998), based on a set of difference equations. CoMERG identifies a set of variables and their dependencies to represent quantitative aspects (e.g., levels of emotional response) as well as qualitative aspects (e.g., decisions to regulate one's emotion) of emotion regulation. These variables include the level of the current emotion, the optimal "desired" level of emotion, the personal tendency to adjust the emotional value, and the costs of adjusting the emotional value. A set of equations is used to simulate and evaluate the use of Gross' four antecedent-focus strategies of emotion regulation. Note that CoMERG does not explicitly address the underlying appraisal process that is required to generate specific emotions based on the observed world state. However, a similar model that does address this emotion generation process explicitly, and combines it with (two of) Gross' emotion regulation strategies, is presented in Martínez-Miranda et al. (2014).

To conclude, the number of computational models that address the complete process including emotion generation and regulation in detail is limited. Two candidate models that could be used as a basis for developing such a model are EMA (Marsella et al., 2010) and CoMERG (Bosse et al., 2010b). Moreover, in (Bosse et al., 2010a), the two models are compared, thereby making their main similarities and differences explicit, and a blueprint is provided for their integration into a "complete" computational model. In the next section, a simple version of such an integrated model for emotion generation and emotion regulation will be presented.

A COMPUTATIONAL MODEL OF EMOTION REGULATION

In this section the computational model for emotion regulation is described in detail. As mentioned earlier, it can be used to simulate emotion regulation of artificial agents as well as human agents. The model was first introduced in (Bosse et al., 2013). After that, a number of extensions have been proposed that make use of similar principles. These more recent models include a model that relates emotion regulation to specific social response patterns like autistic spectrum disorder (Treur, 2014), and a model that integrates emotion regulation with emotion contagion and decision making (Manzoor and Treur, 2015).

The computational model is formalized as a system of difference equations. It roughly consists of two parts:[a]

1. An *emotion generation model*. To simulate emotion generation, a componential perspective is taken, where emotions arise as the result of an appraisal process involving the agent's goals (or desires) and believed information about the world state as input elements. Hence, the model reuses some of the concepts from the BDI paradigm, such as beliefs and desires. However, the model does not represent the contents of the appraisal process in detail (e.g., using appraisal variables, such as desirability and likelihood, as in EMA). Instead, it abstracts from such aspects, and simply represents desires and beliefs as one-dimensional numerical variables. Similarly, the model assumes that one specific emotion type is simulated (e.g., sadness, fear, anger), but it abstracts from the specific characteristics of this emotion.
2. An *emotion regulation model*. To simulate emotion regulation, the theory by Gross (1998) is used as a basis, similar to CoMERG. Hence, the model distinguishes several strategies that are applied to different stages of the emotion generation process. However, in contrast to CoMERG, these strategies are now actually connected to the semantic primitives of the emotion generation model (such as beliefs and

[a]In Bosse et al. (2013), a third submodel is introduced, which enables learning of emotion regulation strategies. In contrast to most existing models, which assume that agents are equipped with static emotion regulation capabilities, that submodel includes mechanisms to represent an agent's ability to learn or strengthen emotion regulation strategies by training. These mechanisms are modeled based on a Hebbian learning principle, that is, the principle that "neurons that fire together, wire together" (Hebb, 1949).

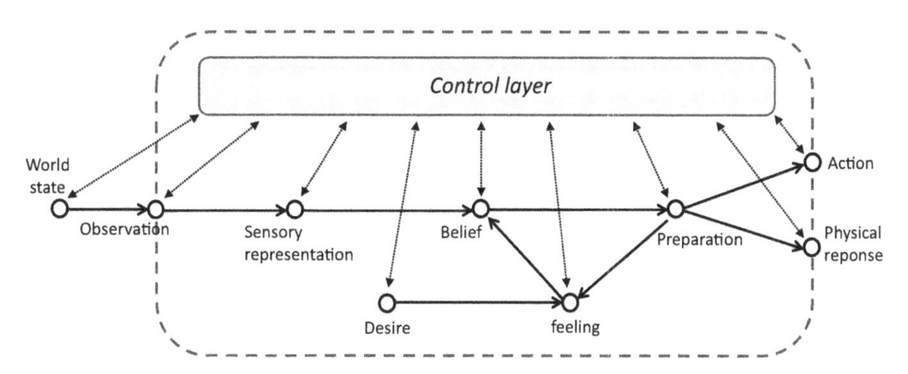

FIGURE 13.2 **Overview of the emotion regulation model.**

desires). They are formalized by means of separate *control states* that can express a suppressive effect on the different semantic primitives.

Before introducing the formal details, first, a general overview of the model is given at a conceptual level; see Fig. 13.2. Here, the circles denote different concepts (representing states of the world), and the influence of one state on another state is depicted by an arrow. The submodel that represents the emotion generation process is depicted by using solid arrows; the meaning of the relevant concepts within that process is explained in Table 13.1. This table shows an explanation for each concept at a generic level (i.e., independent from the specific type of emotion that is simulated), as well as two examples of how the concept can be filled in for a concrete situation. The first example explains the concept in the situation of an emotion with negative valence: here, the external stimulus is a dangerous animal and the resulting emotion type is fear. The second example addresses a situation of an emotion with positive valence: here, the external stimulus is a funny joke and the resulting emotion type is happiness. In both examples, the agent wants to avoid the emotion and the resulting action tendencies; that is, it has the desire to avoid (running away from) the animal or (laughing about) the joke. Hence, these examples address cases of emotional downregulation. Nevertheless, one can easily come up with similar examples involving upregulation, or even other emotion categories.

Each state is formalized in a numerical manner, in terms of a variable with a real value between 0 and 1, indicating the intensity or activation level of the corresponding state. For example, a world state with value 0.9 could represent a very scary animal, and a corresponding desire with value 0.5 could represent a moderately strong desire to avoid this animal.

Next, to represent emotion regulation, the *control layer* is introduced (see upper part of Fig. 13.2). The main idea is that each "basic" state of the emotion generation model can be regulated (individually) by the control layer.

TABLE 13.1 Explanation of the Concepts Used in the Emotion Generation Model

Concept	Explanation	Example (fear)	Example (happiness)
World state	External stimulus that triggers an emotion	Dangerous animal	Funny joke
Observation	The agent perceives the stimulus	Agent sees animal	Agent hears joke
Sensory representation	Brain areas representing the sensory input are activated	Mental image of animal is created	Mental image of joke is created
Belief	Agent forms a belief (or cognitive appraisal) about the stimulus	Agent believes the animal is scary	Agent believes the joke is funny
Desire	Agent has the desire to avoid the stimulus	Agent wants to avoid (running from) the animal	Agent wants to avoid (laughing on) the joke
Feeling	Agent experiences emotional response	Experience of fear	Experience of happiness
Preparation	Brain areas representing the tendency to act on the stimulus are activated	Agent prepares to run away	Agent prepares to laugh
Physical response	Physiological response to stimulus	Changed heart rate, sweating, etc.	Changed heart rate, sweating, etc.
Action	Agent acts upon the stimulus	Agent runs away	Agent laughs

This is indicated by the dashed arrows. As a result, the control layer has a suppressing (in case of downregulation) or reinforcing (in case of upregulation) effect on the basic states. Further, all of the basic states have a positive effect on the control layer, representing a kind of monitoring process: the more active the basic states are, the more likely the agent is to regulate them.

In the following two subsections, the formalization of the submodels for emotion generation and emotion regulation, respectively, is explained in detail.

Modeling emotion generation

The dynamics of the emotion generation submodel are as follows. As a starting point, the model assumes the presence of a world state with an emotional valence (either positive or negative) that is observed by the agent (e.g., a dangerous animal). This leads to a sensory representation and a belief (or cognitive appraisal) about this world state. Next, the agent prepares to act (e.g., to run away) and this preparation together with a desire (e.g., to avoid the stimulus) activate a feeling, which in turn may influence the belief state. This generation of feeling from preparation of emotional response follows the account based on an *as-if body loop* introduced by Damasio (1994). Following this theory, we use the term "feeling" to refer to the mental state that emerges when an organism senses that its body is preparing to act. In contrast, Damasio reserves the term "emotion" for the preparation state itself. Finally, the preparation state results in both a physical response (e.g., an increased heart rate) and an action (e.g., to run away).

To calculate the dynamics of the emotion generation process over time, a set of difference equations is used. For each of the states in the model, the same principle is used: the value of the state q at time point $t + \Delta t$ is calculated by taking the value of that state at time point t and adding a fraction of the *aggregated impact* on that state (from other states) minus the current value of the state. This is expressed in the following generic formula:

$$q_{state}(t+\Delta t) = q_{state}(t) + \eta_{state}\ [aggimpact_{state}(t) - q_{state}(t)]\ \Delta t$$

Here, η represents the speed of activation spread between states in the model. As an example, for the *belief* state, the formula would be as follows:

$$q_{belief}(t+\Delta t) = q_{belief}(t) + \eta_{belief}\ [aggimpact_{belief}(t) - q_{belief}(t)]\ \Delta t$$

Note that this formula can be rewritten in the following alternative form:

$$q_{belief}(t+\Delta t) = (1-\eta_{belief}\ \Delta t) * q_{belief}(t) + (\eta_{belief}\ \Delta t) * aggimpact_{belief}(t)$$

This shows how a new value of the *belief* state at time point $t + \Delta t$ is a weighted average of its old value (at time point t) and the *aggregated impact* on the *belief* state, with weights $\eta\ \Delta t$ and $1-\eta\ \Delta t$. So, suppose for instance that the *time step* Δt is taken 1 and the speed factor for activation spread η is taken 0.3, then the formula is instantiated as follows:

$$q_{belief}(t+1) = 0.7 * q_{belief}(t) + 0.3 * aggimpact_{belief}(t)$$

In other words, in this case the new value of the *belief* state is calculated by taking 70% of its old value and 30% of the aggregated impact on this state.

Next, the *aggregated impact* on a state is determined by the aggregated mean of the values of all states that have an influence on that state (which can be found by taking all incoming arrows in Fig. 13.2). For example, the aggregated impact on the *belief* state would be calculated by using the values of the states *sensory representation*, *feeling*, and (in case of emotion regulation, see next section) *belief control*:

$$aggimpact_{belief} = \omega_{sensoryrepresentation\_belief} * q_{sensoryrepresentation} + \omega_{feeling\_belief} * q_{feeling} + \omega_{bel.control\_belief} * q_{bel.control}$$

The values are weighted by the connection strengths between the two states, which are represented by ω. The generation of the other states is determined in a similar manner.

Modeling emotion regulation

To enable the agent to regulate its emotion levels in different ways, a specific *control state* has been added for each of the basis states in the model. These control states are assumed to represent the agent's efforts to perform various emotion regulation strategies. They are depicted together as one rounded rectangle (the *control layer*) in Fig. 13.2. For any basic state s of the model, its corresponding control state is represented by a variable $q_{s.control}$ with a value between 0 and 1. If all control states are 0, the agent does not perform any regulation (and consequently, if they would stay 0 during the entire simulation, the model only simulates a standard emotion generation process).

Additionally, the influence that each control state has on the related basic state in the emotion generation model is modeled by *downward connections* (the dotted arrows in Fig. 13.2 pointing downward). For any control state $q_{s.control}$, its corresponding downward connection is represented by a variable $\omega_{s.control\_s}$ with a value between −1 and 1. The value of the connection represents the type and strength of the regulation for each s: Positive values represent upregulation, negative values represent downregulation, and the value 0 represents no regulation.

Finally, the upward connections (the dotted arrows in Fig. 13.2 pointing upward) are used to monitor the activation levels of all basic states s. They are represented by variables $\omega_{s\_s.control}$ with values between 0 and 1. Their strength represents the extent to which the agent is able to monitor (and regulate) that particular state.

Based on these mechanisms, the model is able to simulate the emotion regulation strategies by Gross (1998) in the following way:

- For *situation selection* and *situation modification*, the value of the world state is altered (e.g., avoiding or changing a stimulus, such that its emotional valence is decreased or increased).
- For *attentional deployment*, two different variants can be simulated. First, the value of the observation state can be altered (e.g., by looking away from the stimulus). Second, the value of the sensory

representation can be changed, thereby regulating the internal focus of attention (e.g., thinking about something else).

- *Cognitive change* is possible when the value of the belief is modified (e.g., applying reappraisal of the situation: saying to yourself that the situation is not that bad). Additionally, it is also possible to change the value of the desire (e.g., by adjusting one's goals).
- Finally, the response-focused regulation strategy *suppression* is simulated by modifying the value of the feeling (e.g., suppressing feelings experienced), the physiological response (e.g., showing a pokerface), and the preparation/action states (e.g., staying at a location instead of running away).

SIMULATION RESULTS

To illustrate the working of the simulation model, a number of simulation experiments have been performed. The results are depicted in Fig. 13.3A–E. In these figures, time is shown on the horizontal axis and the activation values of (a selection of) the states of the model are shown on the vertical axis. All of these examples address a situation where the stimulus is assumed to have a negative valence (hence, the agent will need to perform downregulation).

Fig. 13.3A shows the simulation of the model without any regulation. Throughout this simulation, the activation value of the world state (q_{world}) and the desire (q_{desire}) have been set to 0.9 and 0.6, respectively. This corresponds to a case where the agent is exposed to a rather extreme stimulus (q_{world}), while having a reasonably strong desire to avoid this (q_{desire}). All other activation states start with a value of 0. In this simulation, the no regulation condition is realized by setting the $q_{s.control}$ and $\omega_{s\_control}$ parameters for all states s to 0. Moreover, all η_s are 0.1, representing a low speed of activation spread. The values of all positive connection strengths that lead toward the same state always sum up to 1, whereas the values of all negative connection strengths (i.e., the downward connections $\omega_{control\_s}$ from the control state) have been set to −0.5.

As shown in Fig. 13.3A, without regulation, the activation values of all basic states converge to a value somewhere in between the activation values of the *world state* and the *desire* state. Note that the exact equilibrium value of a state depends on the position of this state in the emotion generation process. For example, the *observation* state approximates the value of the *world state*, as it is directly influenced by this state. Instead, the *feeling* state approximates a lower value, as this state is also influenced by the *desire* state.

In Fig. 13.3B the results are shown for a situation in which all basic states are regulated in parallel. The parameters are equal to the parameters for the simulation in Fig. 13.3A, with the exception that the connection strengths from all basic states to the control layer (i.e., all $\omega_{s\_control}$) now have the value 0.5. This represents the case that the agent applies all emotion regulation strategies at the same time (with an average effort) during the entire simulation. As shown in Fig. 13.3B, these parameter settings result in a scenario where the activation of all states first increases, but is

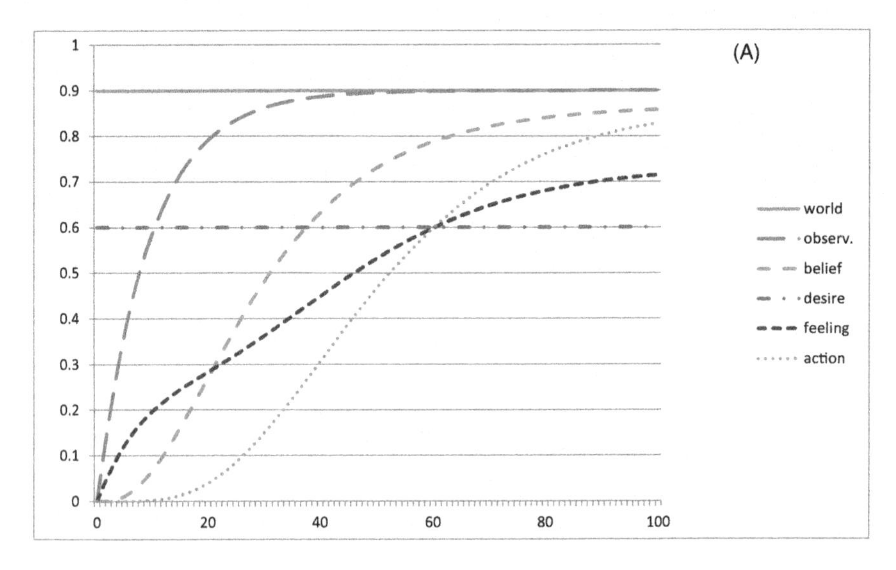

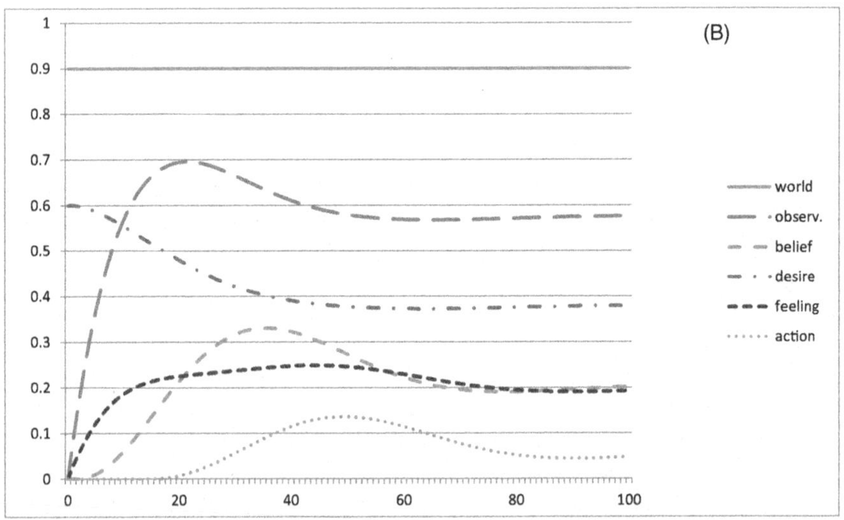

FIGURE 13.3 Simulation (A) without regulation; (B) simulation with all regulation strategies applied in parallel; (C) simulation with only *belief* regulated (i.e., an instance of *reappraisal*); (D) simulation with only *observation* regulated (i.e., an instance of *attentional deployment*); and (E) simulation with only *feeling* regulated (i.e., an instance of *suppression*).

then suppressed because of the regulation. The closer a state is to the *world state* (in Fig. 13.2), the higher the equilibrium value that it reaches.

Fig. 13.3C shows a simulation of the model in which only the *belief* state is regulated. This corresponds to an individual that applies the strategy *reappraisal*. This was realized by taking $q_{belief.control} = 0.5$, $\omega_{control\_belief} = -1$, and $\omega_{control\_s} = -0.1$ for all other *s*. The figure clearly shows that the *belief* state, as well as all states that "depend" on this state, end up at much lower values

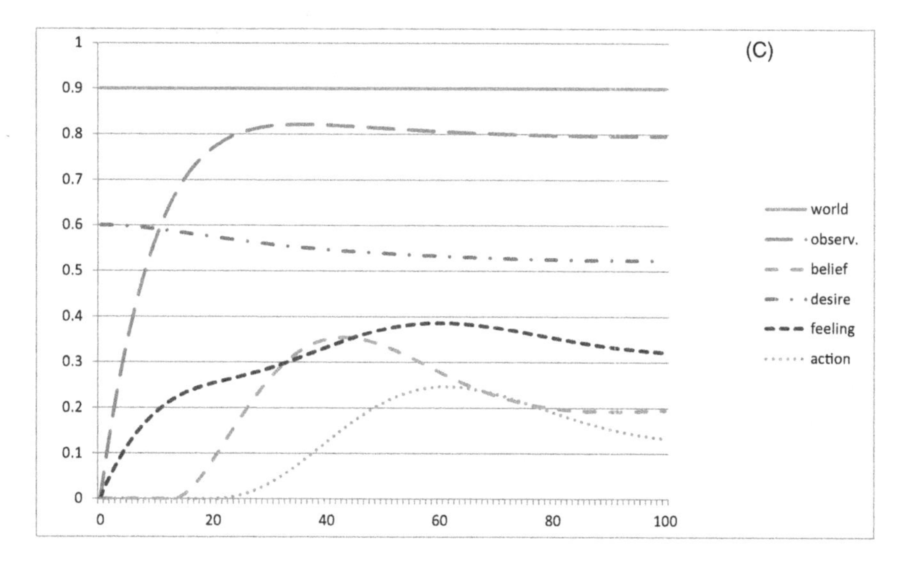

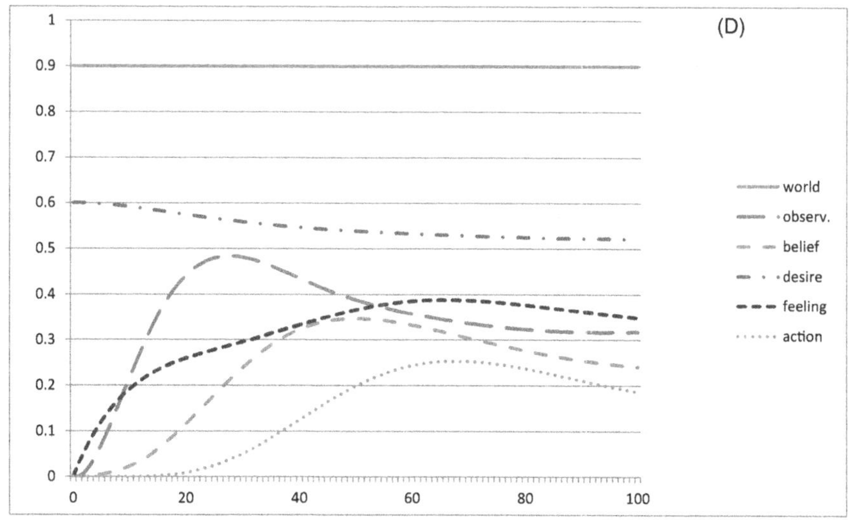

FIGURE 13.3 *(Continued)*

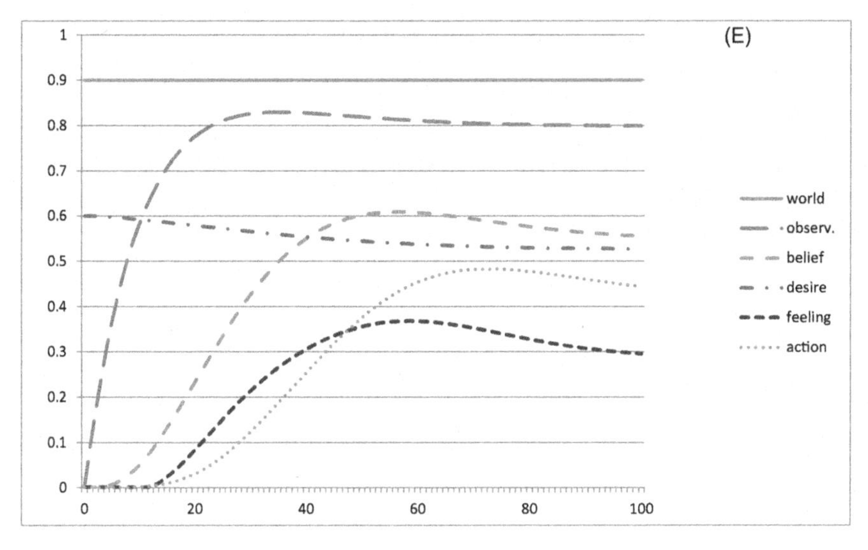

FIGURE 13.3 *(Continued)*

than in Fig. 13.3A. Instead, the *desire* state remains high, as this state is not affected by the reappraisal strategy.

Finally, Fig. 13.3D–E show simulations of the model in which only the respective states *observation* and *feeling* are regulated. This corresponds to individuals that apply the strategies *attentional deployment* and *suppression*, respectively. For Fig. 13.3D, this was realized by taking $q_{observation.control} = 0.5$, $\omega_{control\_observation} = -1$, and $\omega_{control\_s} = -0.1$ for all other s. For Fig. 13.3E, this was realized by taking $q_{feeling.control} = 0.5$, $\omega_{control\_feeling} = -1$, and $\omega_{control\_s} = -0.1$ for all other s. Like in Fig. 13.3C, Fig. 13.3D–E show that the state that is explicitly regulated (i.e., either the *observation* or the *feeling* state), as well as all states that "depend" on this state, end up at much lower values than in Fig. 13.3A, while the other states remain high.

APPLICATIONS WITHIN HCI

The simulations discussed earlier show some interesting patterns, and—assuming we can demonstrate that these patterns are, at least, somewhat accurate reflections of the dynamics of human emotion regulation (which is an entire topic in itself, see the discussion later)—the model might enable computational systems to reason about such patterns. However, the next obvious question is: which types of HCI applications would benefit from such an ability?

As outlined in the introduction, applications that make use of dynamic computational models of affect can be divided into (1) systems that simulate the dynamics of affective processes within artificial agents and (2) systems that predict the dynamics of affective processes within human agents. The former category includes, among others, embodied conversational agents (Cassell et al., 2000) and social robots (Breazeal, 2001), whereas the latter category includes adaptive empathetic interfaces (Klein et al., 2002) and stress regulation training systems (Bosse et al., 2014).

To make these examples more concrete, the current section will explain in brief how models of emotion regulation can have an added value for such applications. To start with the first category (i.e., using emotion regulation models within artificial agents), the research presented in (Bosse et al., 2007) is a good illustration. That paper proposes an approach to incorporate emotion regulation models into virtual characters, as part of an emergent narrative application. In particular, the emotion regulation model CoMERG (Bosse et al., 2010b), is used. The main idea is that, by endowing virtual characters with this model, they are able to autonomously perform actions within a virtual environment, by selecting those actions that are most effective in regulating their emotional state toward the desired level.

The approach is illustrated in the context of an emergent storyline involving three virtual agents that are trying to regulate their level of happiness. To achieve this, the particular topics they are allowed to talk about are football (more specifically, the Dutch football teams, Ajax and Feyenoord) and hockey. Regarding the specific emotion regulation strategies, the agents are able to perform *situation selection* by selecting different conversation partners, and withdrawing from conversations if desired. Moreover, they can perform *situation modification* by changing conversation topics, they can perform *attentional deployment* by changing the amount of attention they pay to a conversation, and they can perform *cognitive change* by changing the cognitive meaning, they assign to their thoughts (e.g., by stating to themselves that something is not very important).

A screenshot of a resulting scenario in the Vizard virtual environment is shown in Fig. 13.4. This figure shows a situation in which (on the foreground) two virtual characters are having a conversation. The right character is giving his opinion about his favorite football club, but since the left character does not agree with this, he tries to distract his attention from the conversation by moving his head away from the interlocutor (i.e., an instance of *attentional deployment*). Meanwhile, in the background a third character is standing alone. This character decided to not even join the conversation at all, as he does not like talking about football (i.e, an instance of *situation selection*).

An example of a system that falls within the second category of applications (i.e., using emotion regulation models to predict the emotional processes within humans) is the work reported in (Bosse et al., 2014). This paper sketches the design of a virtual reality based training environment

FIGURE 13.4 **Screenshot from an emergent storyline where the characters use emotion regulation.**

for professionals in public domains, such as law enforcement and public transport. The main idea is that trainees will be placed in a virtual scenario that may evoke negative emotions like fear and stress (e.g., they are confronted with a citizen that shows verbal aggression), which they have to solve as accurately as possible. Meanwhile, they are connected to various sensors measuring aspects of their physiological state that are indicators for negative affect (such as heart rate and skin conductivity). By combining these measurements with a dynamical model of emotion regulation (similar to the one shown in Fig. 13.2), the system is able to reason about the dynamics of the trainee's emotional state over time. As a result, it has the capacity to predict how this emotional state will develop in the near future, and to generate appropriate support actions if necessary. The types of support that can be provided include the following (Bosse et al., 2014):

1. *Situation modification*: modifying the stimuli displayed in the training environment (e.g., make a character show less aggressive behavior).
2. *Attentional deployment*: advising the trainee to focus his or her attention on something else (e.g., count to 10).
3. *Cognitive change*: advising the trainee to reinterpret the negative stimulus (e.g., trying to show more understanding for the aggressive person's situation).
4. *Suppression*: advising the trainee to suppress his or her negative feelings or tendencies.

TABLE 13.2 Four Possible Approaches to Validation of Computational Models of Emotion

	Subjective validation	Objective validation
Validation of model	Comparing predictions of the model with people's self reports about their emotional state (Bosse et al., 2013)	Comparing predictions of the model with physiological measurements that are indicative for people's emotional state (Bosse et al., 2011)
Validation of application	Asking people about their opinion on the effectiveness of the application (Bosse and Zwanenburg, 2013)	Testing the effectiveness of the application (quantitatively) (Bosse et al., 2015)

Hence, by applying these types of support, the system will "coach" the trainee in a way that is similar to the way human instructors would do this.

Taking the two applications explained earlier as examples, it seems evident that the use of emotion regulation models within HCI applications is promising. Nevertheless, we should not forget that such applications can only be effective if the underlying models are accurate. In other words, *validation* of emotion regulation models (and the systems that use them) is an important step. Although an extensive discussion about the different techniques to evaluate, validate, and even improve computational models (e.g., using parameter tuning) is beyond the scope of this chapter, let us state that there are, at least, four different categories of validation, which can be organized according to the structure shown in Table 13.2.

As shown, this table makes a distinction between validation of an emotion regulation *model* itself, and validation of an *application* that uses such a model. Moreover, a *subjective* type of validation (e.g., asking people about their subjective experience) can be distinguished from an objective type of validation (e.g., using objective criteria, such as physiological measurements or an explicit performance measure). The table includes references to some papers in which the different types of validation have been applied by the author. Another paper that provides an extensive discussion on validation of emotion models is Gratch and Marsella (2005).

CONCLUSIONS

For human beings, the ability to regulate their emotions is crucial to be able to function in daily life. Every day, there are numerous situations in which we attempt to suppress our smile, manage our anger, or canalize our anxiety. In this chapter, I made an attempt to provide more insight in the underlying mechanisms of emotion regulation, and to dissect them in such a way that the reader has a grasp on how to convert them into a dynamic computational model. As an illustration, I presented one specific instance of such a computational model, which consisted of two submodels based on difference equations: an emotion generation model based on (a simplified version of) appraisal theory, and an emotion regulation model based on the strategies proposed by Gross. As an illustration, a number of simulation results were shown, indicating that the model has the ability to reproduce the dynamics of the various regulation strategies in substantial detail. Validation of the model was not discussed in detail here, but is addressed in related papers, such as Bosse et al. (2011, 2013).

Finally, a number of application areas of computational emotion regulation models have been discussed. As mentioned, these applications use emotion regulation models either to regulate the emotions of an artificial agent or to regulate the emotions of the human user. As both tasks (i.e., regulating your own emotions and the emotions of others) are ubiquitous in human–human interaction, it is not far-fetched to predict that systems capable of these tasks will take a huge flight in HCI. If the research community manages to develop emotion regulation models that are increasingly accurate and better aligned with empirical data, this line of research has the potential to result in truly "empathetic" HCI systems that "understand" and regulate both their own emotions as well as the emotions of their human users.

Acknowledgments

Parts of this chapter are based on Bosse et al. (2010b, 2013). The author wishes to thank Charlotte Gerritsen, Jeroen de Man, Matthijs Pontier, and Jan Treur for various fruitful discussions.

References

Adam, C., Gaudou, B., Herzig, A., Longin, D., 2006. OCC's Emotions: A Formalization in BDI Logic. In: Euzenat, J., Domingue, J. (Eds.), Artificial Intelligence: Methodology, Systems, and Applications. Proceedings of AISMA'06. Springer LNAI, vol. 4183, pp. 24–32.

Anderson, J.R., Lebiere, C., 2003. The Newell Test for a theory of cognition. Behav. Brain Sci. 26, 587–640.

Billings, A.G., Moos, R.H., 1984. Coping, stress, and resources among adults with unipolar depression. J. Pers. Soc. Psychol. 46, 877–891.

Bosse, T., Zwanenburg, E., 2013. Do Prospect-Based Emotions Enhance Believability of Game Characters? A Case Study in the Context of a Dice Game. IEEE Transactions on Affective Computing, 2013.

Bosse, T., Pontier, M., Siddiqui, G.F., Treur, J., 2007. Incorporating Emotion Regulation into Virtual Stories. In: Pelachaud, C., Martin, J.C., Andre, E., Chollet, G., Karpouzis, K., Pele, D. (Eds.), Proceedings of the Seventh International Conference on Intelligent Virtual Agents, IVA'07. Lecture Notes in Artificial Intelligence, vol. 4722. Springer Verlag, 2007, pp. 339–347.

Bosse, T., Gratch, J., Hoorn, J.F., Pontier, M., Siddiqui, G.F., 2010a. Comparing Three Computational Models of Affect. In: Demazeau, Y., Dignum, F., Corchado, J.M., Bajo, J. (Eds.), Proceedings of the Eighth International Conference on Practical Applications of Agents and Multi-Agent Systems, PAAMS'10. Advances in Intelligent and Soft Computing, vol. 70. Springer Verlag, pp. 175–184.

Bosse, T., Pontier, M., Treur, J., 2010b. A computational model based on Gross' emotion regulation theory. Cogn. Syst. Res. 11 (3), 211–230.

Bosse, T., Brenninkmeyer, J., Kalisch, R., Paret, C., Pontier, M., 2011. Matching Skin Conductance Data to a Cognitive Model of Reappraisal. In: Hoelscher, C., Shipley, T.F., Carlson, L. (Eds.), Proceedings of the Thirty-Third Annual Conference of the Cognitive Science Society, CogSci'11.

Bosse, T., Gerritsen, C., Man, J. de, Treur, J., 2013. Learning Emotion Regulation Strategies: A Cognitive Agent Model. In: Proceedings of the Thirteenth International Conference on Intelligent Agent Technology, IAT'13. IEEE Computer Society Press, pp. 245–252.

Bosse, T., Gerritsen, C., Man, J. de., 2014. Agent-Based Simulation as a Tool for the Design of a Virtual Training Environment. In: Proceedings of the Fourteenth International Conference on Intelligent Agent Technology, IAT'14. IEEE Computer Society Press, pp. 40–47.

Bosse, T., Gerritsen, C., Man, J. de., 2015. Evaluation of a Virtual Training Environment for Aggression De-escalation. In: Proceedings of the Sixteenth International Conference on Intelligent Games and Simulation, GAMEON'16. Eurosis.

Brave, S., Nass, C., 2002. Emotion in human-computer interaction. In: Jacko, J., Sears, A. (Eds.), Handbook of Human-Computer Interaction. Lawrence Erlbaum Associates, New York, pp. 251–271.

Breazeal, C., 2001. Affective interaction between humans and robots. In:, J., Keleman, P., Sosik (Eds.), Proceedings of ECAL 2001, Springer Verlag, Lecture Notes in AI, vol. 2159, pp. 582–591.

Campbell-Sills, L., Barlow, D.H., 2006. Incorporating emotion regulation into conceptualizations and treatments of anxiety and mood disorders. In: Gross, J.J. (Ed.), Handbook of Emotion Regulation. Guilford press, New York.

Cassell, J., Sullivan, J., Prevost, S., Churchill, E., 2000. Embodied Conversational Agents. MIT Press, Cambridge, MA.

Compas, B.E., Jaser, S., Dunbar, J.P., Watson, K.H., Bettis, A.H., Gruhn, M., Williams, E., 2014. Coping and emotion regulation from childhood to early adulthood: points of convergence and divergence. Aust. J. Psychol. 66 (2), 71–81.

Damasio, A., 1994. Descartes' Error: Emotion, Reason and the Human Brain. Papermac, London.

Damasio, A., 2000. The Feeling of What Happens: Body, Emotion and the Making of Consciousness. MIT Press, Cambridge, MA.

Dastani, M., Floor, C., Meyer, J.-J. Ch., 2014. Programming Agents with Emotions. Emotion Modeling. Lecture Notes in Computer Science Springer Verlag, pp. 57–75.

Ekman, P., 1992. An argument for basic emotions. Cogn. Emot. 6 (3–4), 169–200.

Frijda, N.H., 1986. The emotions. Cambridge University Press, Cambridge, England.

Gerritsen, C., Klein, M.C.A., 2014. Dynamical simulation in criminology. In: Bruinsma, G., Weisburd, D. (Eds.), Encyclopedia of Criminology and Criminal Justice. Springer Verlag, New York, pp. 1220–1231.

Grandey, A.A., Dickter, D.N., Sin, H.-P., 2004. The customer is not always right: customer aggression and emotional regulation of service employees. J. Organ. Behav. 25, 397–418.

Gratch, J., Marsella, S., 2005. Evaluating a computational model of emotion. Auton. Agent Multi Agent Syst. 11 (1), 23–43.

Gross, J.J., 1998. The emerging field of emotion regulation: an integrative review. Rev. Gen. Psychol. 2 (3), 271–299.

Gross, J.J., 2001. Emotion regulation in adulthood: timing is everything. Curr. Dir. Psychol. Sci. 10 (6), 214–219.

Gross, J.J., Barrett, L.F., 2011. Emotion generation and emotion regulation: one or two depends on your point of view. Emot. Rev. 3, 8–16.

Hanin, Y.L., 2007. Emotions in sport: current issues and perspectives. In: Tenenbaum, G., Eklund, R.C. (Eds.), Handbook of Sport Psychology. Wiley, Hoboken, NJ, pp. 31–58.

Hebb, D.O., 1949. The Organization of Behavior: A Neuropsychological Theory. Wiley, New York.

Hudlicka, E., 2008. What Are We Modeling When We Model Emotion? AAAI Spring Symposium: Emotion, Personality, and Social Behavior (Technical Report SS-08-04, pp. 52–59). Stanford University, CA: Menlo Park, CA: AAAI Press.

Hudlicka, E., 2014. From habits to standards: towards systematic design of emotion models and affective architectures. Lecture Notes in Computer Science Springer Verlag, Switzerland, pp. 3–23.

Keltner, D., Haidt, J., 1999. Social functions of emotions at four levels of analysis. Cogn. Emot. 13, 505–521.

Klein, J., Moon, Y., Picard, R., 2002. This computer responds to user frustration: theory, design, results, and implications. Interact. Comput. 14, 119–140.

Lazarus, R.S., 1991. Progress on a cognitive-motivational-relational theory of emotion. Am. Psychol. 46 (8), 819–834.

Lazarus, R.S., Folkman, S., 1984. Stress, Appraisal, and Coping. Springer, New York.

Manzoor, A., Treur, J., 2015. An agent-based model for integrated emotion regulation and contagion in socially affected decision making. BICA J. 12, 105–120.

Marsella, S., Gratch, J., 2009. EMA: a process model of appraisal dynamics. Cogn. Syst. Res. 10 (1), 70–90.

Marsella, S., Gratch, J., Petta, P., 2010. Computational models of emotion. In: Scherer, K., Bänzinger, T., Roesch, E. (Eds.), A Blueprint for an Affectively Competent Agent: Cross-Fertilization Between Emotion Psychology, Affective Neuroscience, and Affective Computing. Oxford University Press, Oxford.

Martínez-Miranda, J., Bresó, A., García-Gómez, J.M., 2014. Modelling two emotion regulation strategies as key features of therapeutic empathy. Lecture Notes in Computer Science Springer Verlag, Switzerland, pp. 115–133.

Mehrabian, A., 1995. Framework for a comprehensive description and measurement of emotional states. Genet. Soc. Gen. Psychol. Monogr. 121, 339–361.

Oatley, K., Johnson-Laird, P.N., 1987. Towards a cognitive theory of emotions. Cogn. Emot. 1, 29–50.

Ochsner, K.N., Gross, J.J., 2005. The cognitive control of emotion. Trends Cogn. Sci. 9, 242–249.

Ortony, A., Clore, G.L., Collins, A., 1988. The Cognitive Structure of Emotions. Cambridge, NY.

Panskepp, J., 1998. Affective Neuroscience: The Foundations of Human and Animal Emotions. Oxford University Press, NY.

Picard, R., 1997. Affective Computing. MIT Press, Cambridge, MA.

Rao, A.S., Georgeff, M.P., 1991. Modeling Rational Agents within a BDI Architecture. In: Allen, J., Fikes, R., Sandewall, E. (Eds.), Proceedings of the Second International Conference on Principles of Knowledge Representation and Reasoning (KR'91), pp. 473–484. Morgan Kaufmann.

Reisenzein, R., 2009. Emotions as metarepresentational states of mind: naturalizing the belief-desire theory of emotion. Cogn. Syst. Res. 10, 6–20.

Russell, J., 2003. Core affect and the psychological construction of emotion. Psychol. Rev. 110 (1), 145–172.

Russell, J., Mehrabian, A., 1977. Evidence for a three-factor theory of emotions. J. Res. Pers. 11, 273–294.

Scherer, K.R., Shorr, A., Johnstone, T., 2001. Appraisal Processes in Emotion: Theory, Methods, Research. Oxford University Press, Canary, NC.

Schwarz, N., Clore, G.L., 1983. Mood, misattribution, and judgments of well-being: informative and directive functions of affective states. J. Pers. Soc. Psychol. 45, 513–523.

Smith, C.A., Lazarus, R., 1990. Emotion and adaptation. In: Pervin, L.A. (Ed.), Handbook of Personality: Theory & Research. Guilford Press, NY, pp. 609–637.

Thompson, R. A., 1994. Emotion regulation: a theme in search of definition. In: Fox, N.A. (Ed.), The Development Of Emotion Regulation: Biological And Behavioral Aspects. Monographs of the Society for Research in Child Development, vol. 59 (Serial No. 240), pp. 25–52.

Treur, J., 2014. Displaying and regulating different social response patterns: a computational agent model. Cogn. Comput. J. 6, 182–199.

Wierzbicka, A., 1996. Semantics: Primes and Universals. Oxford University Press, Oxford.

PART IV

APPLICATIONS: CASE STUDIES AND APPLIED EXAMPLES

CHAPTER

14

Evolution of Emotion Driven Design

Oya Demirbilek

University of New South Wales, Sydney, NSW, Australia

BACKGROUND

As per Solomon who suggested that emotions are the meaning of life (Solomon, 1976) and Van Gogh's quote (1889) "Let's not forget that the little emotions are the great captains of our lives and we obey them without realizing it," emotions influence our thoughts and behavior and are as such central to all human behavior and experience. Indeed, emotions form an essential part of our interactions with other people, environments, and products.

The origin of the word emotion has evolved from Latin and French, starting with the Latin verb *movere* implying "to move" with the prefix *ex*, meaning "out"—*exmovere*—*emovere*—*émouvoir* (12th century French meaning "stir up"), to *émotion* (16th century French word denoting social disturbance). All these permutations of the word denote some kind of action/reaction.

Meyers (2009) states that emotionality, the observable behavioral and physiological component of an emotion, is connected with an array of psychological phenomena, involving "physiological arousal, expressive behaviors, and conscious experience" (Meyers, 2009). Furthermore, according to Plutchik, during their evolutionary history, emotions have worked to support animals and humans deal with crucial survival issues. These emotions have evolved and derived into myriads of expression forms with a small number of these being "primary emotions" having identifiable basic common elements or patterns (Plutchik, 1980).

In the product design realm, one conference has been instrumental in shaping the direction of design and emotion research from 2002 onward:

Emotions and Affect in Human Factors and Human–Computer Interaction. http://dx.doi.org/10.1016/B978-0-12-801851-4.00014-8
Copyright © 2017 Elsevier Inc. All rights reserved.

the Design & Emotion Conference (www.de2016.org/). This initiative has originated from the combined disciplines of product design and design research and is now an interdisciplinary field of interest. This is noticeable in the themes of the 2016 Design & Emotion Conference that now encompass many more disciplines: *ambiguity* (designing rich experiences, irony, and uncomfortable interactions); *provocation* (design activism, future prototyping, and positive speculation); *well-being* (design for social behavior, ethics, happiness, and personal values); *beauty* (design aesthetics, materials, and the senses); *embodiment* (design ubiquity, internet, tangibility, and embedded computing); *poetry* (designing openness, drama, emotional durability, and storytelling); *empathy* (design for inclusion, participation, and codesign); and finally *spirituality* (design for mindfulness, memories, trust experiences, and awe) (Design, 2016).

HOW EMOTIONAL EXPERIENCE WORKS

Cherry (2010) classifies the main theories of emotion in three as, physiological, neurological, and cognitive. In brief, according to physiological theories, emotions are generated by reactions inside the body. Neurological theories on the other hand advise that emotions are engendered by activity inside the brain. Lastly, cognitive theories claim that emotions are mainly created by thoughts (Cherry, 2010). The way emotions work has been debated in different disciplines for centuries, and in the past two decades, there has been an increasing interest in researching emotions. A quick Google Ngram viewer search charts the frequencies of the words: *emotion driven*, *emotional design*, *emotional response*, *positive emotions*, and *emotional data*, found in printed sources between 1800 and 2012 in American English and British English (Fig. 14.1). The frequency in use for the terms *emotional response* and *positive emotions* has been increasing steadily, with positive emotions peaking after 2000.

Some of the latest research has synthesized this body of knowledge, and shown that our behavior is a direct consequence of our emotions, with the emotions directly affecting our perception, cognition, and personality system (Izard, 2001); that emotions and feelings affect one's perception and mind (Russell, 2003); that some emotions are learned and synthesized from accrued knowledge (Stets, 2010); that emotions are "forces" (Kövecses, 2003); and that body language is another outlet for communicating emotions (Carney et al., 2010; Cuddy et al., 2013; De Gelder, 2006). Other studies looked at the physical manifestation of emotions in the body, such a Finnish study that generated colored body maps showing the various regions of the body where (773) participants reported that activation was happening or not, when they were asked to feel certain emotions (Nummenmaa et al., 2014). In this study, Nummenmaa et al. (2014)

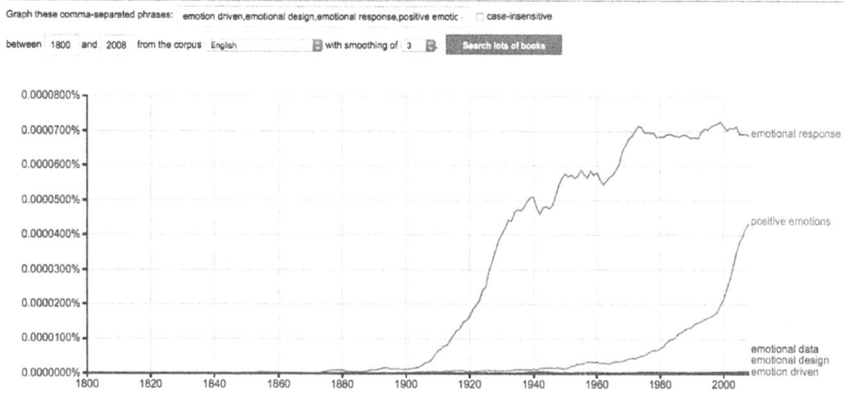

FIGURE 14.1 **Frequencies chart for: emotion driven, emotional design, emotional response, positive emotions, and emotional data (1800–2012).** Screen shot from Google Ngram viewer graph based on Google Books. https://books.google.com/ngrams

suggest that distinct emotional responses that are also somewhat overlapping maps of bodily sensations could be fundamental to the emotional experience (Fig. 14.2).

Decision-making relies heavily on emotional feedback. Damasio (2006) asserts that emotions dominate and guide the decision-making process by allowing us to discern if things are good, bad, or indifferent, and in this

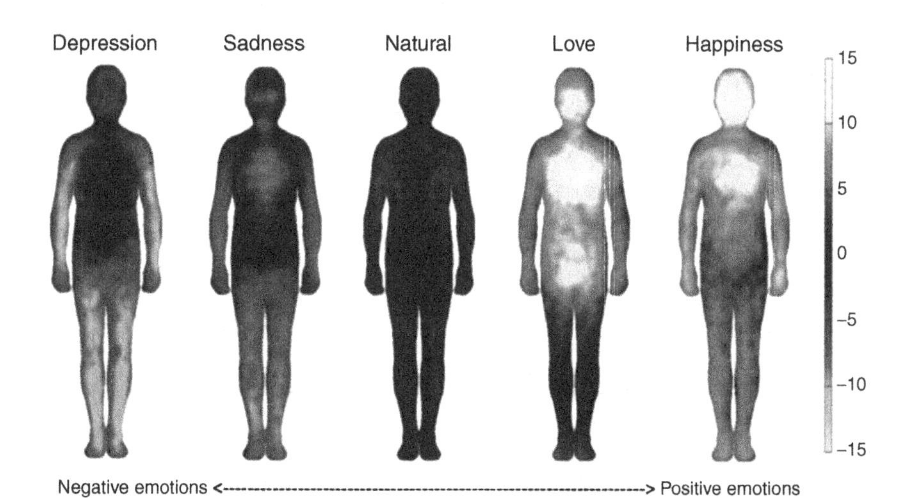

FIGURE 14.2 **Body maps of reported activity or inactivity felt during certain emotions.** *Source: Adapted from Nummenmaa, L., Glerean, E., Hari, R., Hietanen, J.K., 2014. Bodily maps of emotions. Proc. Natl. Acad. Sci., 111(2), pp. 646–651; image courtesy of the authors and PNAS.*

way, also strengthen and support our memory. He also states that humans "are not thinking machines." They "are feeling machines that think." This has also been supported by studies of MRI scans that allowed to demonstrate that the decisions we make are not solely made within the reason oriented cognitive parts of our brain, but that they are heavily influenced by both the amygdala in the limbic system (identified as the emotional governing center) and the prefrontal cortex for contemplation (Bechara et al., 2000; Graves, 2015).

Furthermore, the process of experiencing an emotion occurs outside our conscious awareness (Weinschenk, 2009) and is often complex and layered. This process is being simultaneously biochemical and neurological involving the heart, the brain, nervous and hormonal systems, and sensory organs (Barrett et al., 2007; Izard, 2013). Working together with the brain, the senses allow human beings to experience the events and environments they interact with, and mostly to determine how they will perceive any stimuli.

This is also described by Merleau-Ponty as *"être au monde,"* meaning simultaneously to be in the world, and to belong in it. This act of belonging involves having a connection and a rapport with the world, as well perceiving all of its dimensions (Merleau-Ponty, 1958). During our daily interactions with the world, we perceive its dimensions with the help of all our senses, our thinking, our memories, and skills. Phenomenologists and pragmatists describe personal knowledge creation as the result of "reflection upon action;" which means that reflection is preceding the action to be taken—both on the action and within it—a concept also described by Schön (1983) as fundamental to human-centered design practice.

Two studies (Gross and Thompson, 2007; Mauss et al., 2007) summarize the emotional response as follows:

1. Having contact with a stimulus (event, object, situation, sensation, or thought).
2. Focusing attention to this stimulus.
3. Interpreting the stimulus.
4. Having an emotional/physiological reaction to the stimulus (Fig. 14.3).

EMOTION DRIVEN DESIGN

Designers designing for emotional response are interested in designing the context in which emotions are perceived and the meanings they create or foster. To this end, they perform design research activities to understand the personal values and meaning for their target market of user. Desmet (2002) states that due to emotional reactions being personal and

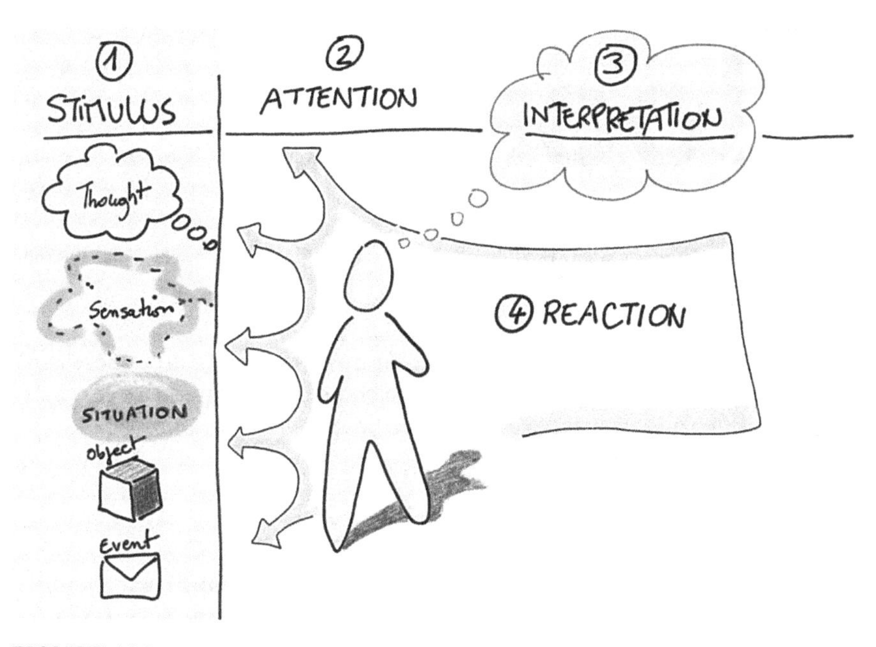

FIGURE 14.3 **Emotional response diagram.** *Source: Interpreted from Gross, J.J., Thompson, R.A., 2007. Emotion regulation: Conceptual foundations.; Mauss, I.B., Bunge, S.A., Gross, J.J., 2007. Automatic emotion regulation. Soc. Person. Psychol. Compass, 1(1), pp. 146–167.*

different for each individual, foretelling or even influencing the emotional impact of a design to induce targeted emotional responses is difficult.

The Three Layers of Human Nature and Abilities

As stated in an early work by Buxton (1994), human nature and abilities can be separated in three layers: physical (including human senses), cognitive, and social (relation to, and interaction with our social environment). These are also the layers where the interaction with things and environments happen. The physical layer has dominated this interaction, corresponding to the *look & feel* of products. To this, *sound* has been a more recent and still timid addition. Products' *look & feel* only engages two of our main senses and the addition of *sound* feedback activates a third human sense (Buxton, 1994). Buxton describes how most product interaction happens around these three senses, sight, touch, and hearing, using the font size to describe the magnitude of the interaction in Fig. 14.4.

Look, feel, and sound, proportionally represented in Fig. 14.4, are three important senses to consider in emotion driven design. They are emotion triggers allowing us to experience and connect with the products

look feel sound

FIGURE 14.4 **How we interact with products: through sight, touch, and hearing (Buxton, 1994: 2).**

and environments we interact with. For a more complete and successful user interface, the proportional unbalance between what our senses (sight, touch, and hearing) get exposed to has to be resolved (Buxton, 1994).

LOOK

The external appearance of products and environments dictates most of how we sense and control them. The look of something is defined by its form, geometry, color, proportion, material, and texture. These are all important visual design elements that dominate the look of things, which in turn affects the way we perceive all things and interact with them. The following quote "80% of what people see is behind their eyes" (Dan Buchner, Vice President Innovation and Design) is still valid for most products on the market and environments we live in.

Color

Color is one of the most important visual design elements. "Colors, like features, follow the changes of the emotions" Pablo Picasso (Spanish Artist and Painter, 1881–1973). This quote from Picasso points to the close connection between colors and emotions. Colors are also said to boost memory by allowing us to process and store visual images in a more effective way (Cherry, 2010; Kaya and Epps, 2004; von Goethe, 1810). Moreover, colors have physiological and emotional effects and can impact our emotional well-being. Colors can carry symbolic meanings and be perceived as having positive or negative traits, depending on who the target audience is, and also as specific emotions, such as happiness (yellow), invigorating (red), calming (light blue), exciting (orange), tired (purple), depressing (blue), irritating (red), and annoying (green-yellow) (Morris, 2006). Nevertheless, every person will perceive color differently, depending on their past experiences, culture, gender, nationality, race, and religion; which makes it hard to classify common responses to different colors.

FEEL

Touch

The very first sense to develop in the human baby embryo is touch; which then allows babies to learn about their environments through touch (Field, 2014; Montagu, 1971). In adulthood, tactility is important for us, as our physical interactions with most products and environments occur typically through our hands. Furthermore, Field (2014) says that touch is the *social sense*, and is different from all the other senses, as it is the only one that involves connecting with something or someone physically. All other senses can be experienced alone, whereas touch allows us to establish social connections.

How something feels in our hands is very important. This could be about the weight (a heavy car door would suggest good quality and luxury, for example), the material and temperature (a cold metal surface would be less inviting than a warm wooden one to seat on), the texture of a surface (which might indicate where and how we can interact with it), or the softness and smoothness of a surface finish. In relation to the last one, Tsakiris et al. (2010) reported a phenomenon called *the social softness illusion*, which is our innate ability to perceive other people's skin as softer than ours. Researchers explain this automatic and unconscious tactile mechanism as an important survival phenomenon (Silk et al., 2003) that promotes social bonds through touch, and demonstrate the hedonic benefits of social touch.

This phenomenon is also supported by psychophysical research done on stroking or gently caressing pets, which reports physical and mental health benefits to the touch providers (Cascio, 2010; Gallace and Spence, 2010). This also demonstrates that, compared to other tactile elements, softness and smoothness are considered as pleasant and rewarding tactile prompts (Gentsch et al., 2015). This is the main reason why we usually prefer soft and smooth textures over rough ones.

SOUND

The third vital sense for us is hearing, which is also an essential form of communication and human interaction. We hear through the information created by pressure variations in the air generated by vibrating objects (Plack, 2013). Özcan argues that product sounds are fundamentally ambiguous and that people would perceive what a sound signifies and make a cognitive evaluation of its meaning within its context (Özcan, 2008). Van Egmond (2008) classifies product sounds as *intentional* (distinct musical sounds designed into a product as part of its functionality or interface) or *consequential* (mechanical sounds as a result of a product's functioning). Intentional sounds communicate special messages and meanings. Jekosch (2005) says that appropriate sound design involves looking at it from an auditory communication and sign theory perspective—a field supported by semiotics. Jekosch (2005) also notes that sound design has gained importance in the last decade and has been promoted from being considered as something secondary and sometimes as unwanted noise; to something supporting the product's quality, and even becoming part of the product identity.

POSITIVE EMOTIONS

Norman's famous statement: "Attractive things work better" (Norman, 2005) has also been supported by many other researchers (Chawda et al., 2005; Crilly et al., 2004; Desmet, 2002; Petersen et al., 2004). One reason why attractive things would indeed work better could be that, as theorized by Aspinwall, positive emotions expand our attention and cognition on the one hand, and facilitate flexible and creative thinking on the other. This is an alpha state of mind that would also make it easier to deal with any stress and/or difficulty (Aspinwall, 1998: cited in Fredrickson and Joiner, 2002).

Both positive and negative emotions cause complex reactions (involving tension in muscles, release of hormones, cardiovascular fluctuations, facial expression, attention, and perception—among many others) that occur very fast. These emotional reactions help us create meaning, thus allowing us to connect to what makes sense in our lives; and to pay more attention to the things that impact our lives. Some emotions are deemed pleasant or positive and others as unpleasant or negative.

In one study that only looked at the appearance of certain products, Desmet (2002) developed a tool based on images, to assess seven unpleasant (indignation, contempt, disgust, unpleasant surprise, dissatisfaction, disappointment, and boredom) and seven pleasant (desire, pleasant surprise, inspiration, amusement, admiration, satisfaction, and fascination) emotions. Desmet demonstrated that physical products would elicit different layers of emotional responses to create meaning, some more predictable than others.

Positive emotions allow us to have better experiences overall and in this way, play an important role in our well-being. Fredrickson posits that experiencing positive emotions extends the inventory of our temporary reactions, which in turn strengthens our long-term physical, psychological intellectual, and social resources (Fredrickson, 2001); contributing in the long run to the improvement of our overall well-being (Fredrickson, 2004). Furthermore, Van Hout (2007) also supports the view that positive emotions elicited through design increase general well-being.

PLEASURE

Positive emotions are often confused with related affective states, such as sensory pleasure and positive mood (Fredrickson, 2004). Pleasure is a positive affective state, and sensory pleasures act as instigators for people to detect biologically useful stimulus for them (Cabanac, 1971; cited in Fredrickson, 2001). The Oxford English Dictionary defines pleasure as the "condition of consciousness or sensation induced by the enjoyment or anticipation of what is felt or viewed as good or desirable; enjoyment, delight, gratification. The opposite of pain" (Simpson and Weiner, 1989: 1031). In the context of artifacts and products, pleasure comes in three types of benefits: emotional, hedonic, and practical (Jordan, 2000). Moreover, pleasure comes through the relationship and the enjoyment taken by interacting with a product.

Lionel Tiger's four pleasure model (Tiger, 1992) describes a framework for the different types of pleasure enjoyed by people, such as physio-pleasure (touching, hearing, and smelling), psycho-pleasure (cognition, discovery, knowledge), socio-pleasure (feeling of belonging, signaling, conversation-starter, social self-identification, social awareness), and ideo-pleasure (signaling or reinforcing ideological standpoints or values).

As a real life example, some people love the smell of a new car, which is, in fact, nothing else but the scent of harmful volatile chemicals. The pleasure we might get from a smell is an olfactory pleasure classified under physio-pleasure, and the example earlier demonstrates that it is a *perceived* pleasure.

POSITIVE EMOTION TRIGGERS

Affordances, functionality, and ease of use are certainly very important in any design, yet the experience(s) and delight we get from the interaction would both support and strengthen the context for meaningful experiences to unfold and increase desirability. The same applies to the aesthetic appearance of a product, an environment, or an interface; all needing to respond exquisitely when we interact with them (Overbeeke et al., 2003). Design elements from the physical layer of human nature and activities, that can help trigger positive emotions and sensory pleasure are fun, cuteness, familiarity, and color (Demirbilek and Sener, 2003).

Embedding Fun in Design

Products, environments, and most interfaces are static, and on the other hand, experiences are fluid. A product, an environment, or an interface that we experience and perceive as funny would have particular characteristics. It would first be intuitive (helping us to retrieve cues from our memories or our past experiences), which means that it is easy to understand or discover without instructions. Second, it could have humor imbedded. Third, it could have elements of personification or expressiveness in human qualities (mirroring or abstracting the human body parts and posture, gestures, and facial expressions for familiarity), and as such, would convey enjoyment and/or feelings of happiness. Doyle says that funny, warm or friendly objects are engaging and would initiate a kind of "dialogue," creating an emotional connection with people (Doyle, 1998).

Fun is experienced and is situational, whereas happiness is built and can last longer, no matter what the situation is. Happiness and fun are not synonymous. The following quote from Aristotle reinforces the importance of focusing on feelings of happiness: "happiness is the meaning and the purpose of life, the whole aim and end of human existence." To this, Van Hout adds that "Design cannot *cause* happiness, but *good design* can be an occasion for and manifestation of happiness" (Van Hout, 2007).

Cuteness

Cuteness triggers feeling of warmth and protectiveness, feelings that are also associated with love. The best example for this is the way we react to baby features. Dissanayake writes about the power of a baby smile in giving happiness and strength to go on in life (Dissanayake, 2005). Baby features are considered as *naïve, honest,* and *helpless,* compared to more mature features (Lidwell et al., 2010) and this would trigger a desire to protect, nurture, and care for a product having baby features.

Familiarity

Familiarity involves anthropomorphic (human like) and zoomorphic (animal like) design, which is the creation of forms suggesting human or animal-like qualities. This is about humanizing products, websites, and environments by including human aspects that will allow people to relate and connect; in other words, to design "the relationship someone has with his or her environment" (Tromp and van den Berg, 2014: 15). This includes physical characteristics (posture, shape, and size), personality and behavioral qualities. Familiarity can also help in conveying human values.

The key to good anthropomorphic design is to keep the human characteristics abstract, simple, and subtle. "Abstraction reduces the chance of directly evoking negative emotions, while preserving the positive associations." Zoomorphic design on the other hand is easier to implement, as people would be more tolerant when a product does not perform as it should, and amazed when it exceeds their expectations (NextNature.net, 2011).

Facial expressions for six basic human emotions, including anger, fear, happiness, surprise, disgust, and sadness (Fig. 14.5) have a very high ratio of recognition (90%) (Goleman, 1995). This recognition happens very fast

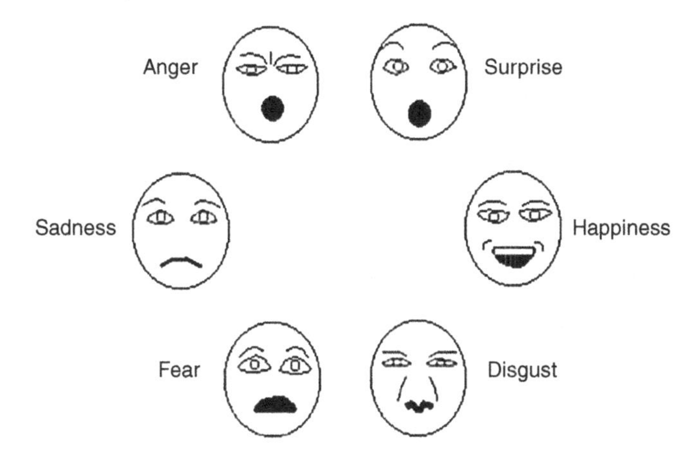

FIGURE 14.5 **Facial expressions for the six basic human emotions (author's sketches).**

through our visual system and is based on minimal cues, and would occur regardless of nationality, culture, or age group (Goleman, 1995; Hadjikhani et al., 2009). Facial expressions also involve more subtle cues as eye blinking patterns and changes in pupil size for the eyes. Our innate instinct for recognizing facial expressions as well as imitating them allows us to empathize with other people.

Smile

Humans are attracted to smiles, and smiling faces are perceived as friendly and unthreatening in most cultures. Furthermore, research has shown that smiling is in itself contagious, and that the use of genuine smiling (also called Duchenne smiling) in advertisement elicits positive moods and emotions in viewers (Scanlon and Polage, 2011; Soussignan, 2002; Surakka and Hietanan, 1998). This simple yet effective phenomenon has to do with the mirror neurons in our brain that help us empathize and experience emotions replicating the ones displayed in ads. The use of authentic smiling and even suggestions of smiles have been a particular focus in advertisement for a long time.

TECHNOLOGICAL ADVANCES AND THE COLLECTION OF EMOTIONAL DATA

Emotional Data

Whether we like it or not, we are rapidly heading toward a new era of insight into ourselves. For now, this data can be collected in two ways. One is through sentiment analysis software looking at linguistic patterns in written content and the other way is through data collected via wearable devices (*emotional arousal*—quicker pulse, self-reported mood). Emotions are contextual and complex, both psychological and biological; and as such, would require a combination of biometric data from pulse rates, hormone production, facial expressions, and self-reported moods and emotions—to truly reflect what is actually happening.

Emotional data covers everything that signifies state of mind. It has been used to track people's self-reported emotional reactions and mood to specific events; such as joy, delight, surprise, excitement, fear, and sadness. As new technology is developed, wearable technology and biofeedback will continue to flourish.

Recent technological advances, such as brain scans, endocrine systems, and wearable technologies have allowed the collection of vast amounts of biometric as well as emotional data. This has in return allowed uncovering a great deal about human emotions. However, some of these technologies have also transformed the way we do many things. A majority of people will agree that their smart devices can take over their lives. It all comes down to the experience one has with any such a device.

A badly designed user interface will require users to focus and concentrate too much of their attention, for too long, with information and feedback that compete for their attention—which has brought the concept of calm technology back on the table (Weiser and Brown, 1995; http://www.calmtech.com/). Bakker et al. (2010) talk about designing for the periphery of our focus, to encalm. Weiser was among the first to coin the term *calm technology*. He argues that technologies can first encalm when they allow easy movement for our focus, from center to periphery and back (Weiser and Brown, 1995); and second, by bringing more details into the periphery—without causing information overload. Calm technology does not demand more of our attention, as it works in the background. Calm technology brings back familiarity and ease of use to help us connect to the world and tune into what is occurring around us, without being overwhelmed by the amount of information.

Health related wearable devices now allow us to log how we feel at particular times of the day, after certain achievements (Fig. 14.6). This collected data is then analyzed by data scientists specializing in mood data, and the results show rough approximations of which sections of our daily routine influence how we feel (Fig. 14.7).

READING HUMAN EMOTIONS

Facial Expressions

Recent advances in the fields of machine learning (Microsoft Research, 2016) and artificial intelligence have allowed computer scientists to make

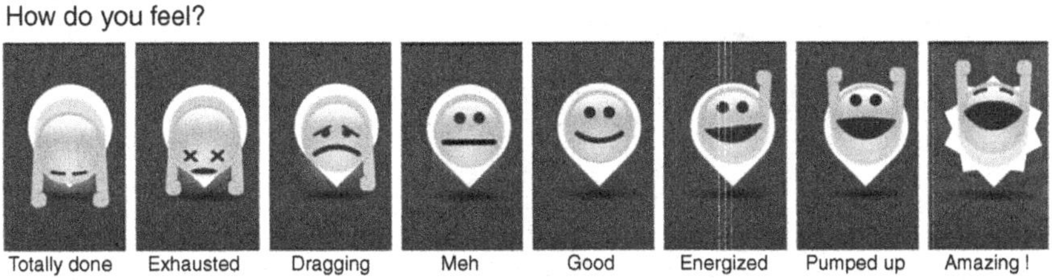

How do you feel?

Totally done Exhausted Dragging Meh Good Energized Pumped up Amazing !

FIGURE 14.6 Icons helping to collect self reported emotional data in a brand of wearable device (Available from: http://www.warriorwomen.co.uk/2013/05/28/jawbone-up-another-geeky-bracelet-review/, Warriorwoman, 2013).

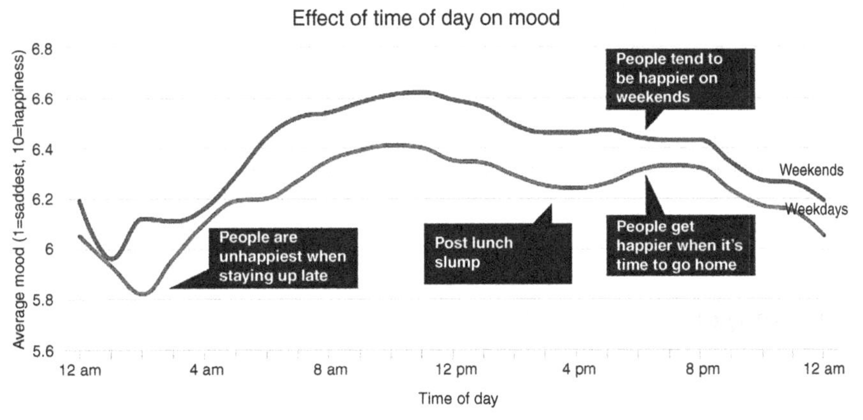

FIGURE 14.7 Emotional data plotted against the time of the day in a brand of wearable device. *Source: Mohan, S., 2015. What Makes People Happy? We Have The Data. Available from: https://jawbone.com/blog/what-makes-people-happy/*

smarter apps that can identify sounds, words, images, and even facial expressions.

An *Application-Programming Interface* (API) is a set of programming commands and standards for accessing a Web-based software application or Web tool. The Microsoft Project Oxford Emotion API identifies emotions in facial expressions in images, using machine learning techniques, and returns the confidence by allocating a set of emotions for each face in an image (Fig. 14.8). This can detect cross-cultural and universally communicated emotions, such as anger, contempt, disgust, fear, happiness, neutral, sadness, and surprise.

Voice Recognition

New software and application developments are also now targeting emotion recognition in voices, through speech. One such application is

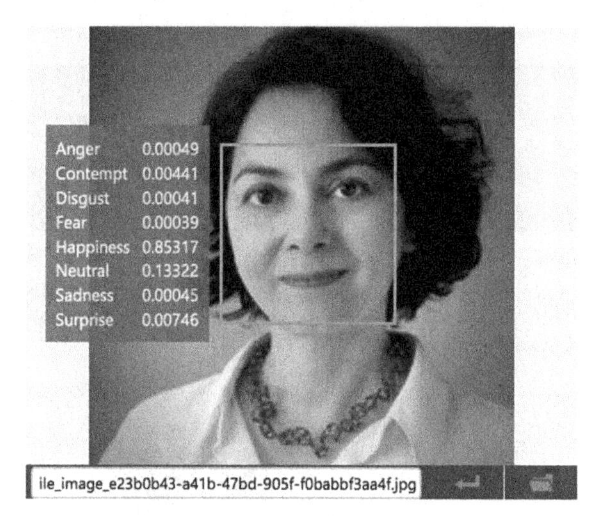


```
Detection Result:
JSON:
[
  {
    "faceRectangle": {
      "left": 62,
      "top": 53,
      "width": 83,
      "height": 83
    },
    "scores": {
      "anger": 0.000488546,
      "contempt": 0.004413109,
      "disgust": 0.000412132067,
      "fear": 0.000390713423,
      "happiness": 0.85316515,
      "neutral": 0.133217022,
      "sadness": 0.000450308376,
      "surprise": 0.007463029
    }
  }
]
```


FIGURE 14.8 **Emotions identified in the author's face on the Microsoft Project Oxford website (screen shot taken after uploading the image on https://www.microsoft.com/cognitive-services/en-us/emotion-api#detection).**

Moodies Emotions Analytics (ios app) that analyzes mood and emotions from voice. The application allows for recording real time speech and immediately extracts, decodes, and measures the "full spectrum of human emotions", one voice at a time.

CONCLUDING REMARKS

Design is omnipresent, as almost everything around us is designed. With the advances of technology, a range of the products we use will fade in the background and even disappear, and some will be reduced to an interface. All other devices will be connected via the Internet, which will allow them to learn from each other and even share functionality. This will also allow better ways to include people into any design process, as codesigners, in real time.

Design has an important role in the enhancement of well-being and the quality of life (Van Hout, 2007). Products and environments are usually specifically designed for a specific purpose, to solve a problem or meet an unmet need, and/or provide an answer or a solution to a real need. As Norman notes, any product is always greater than the sum of its parts or components, providing a "cohesive, integrated set of experiences" (Norman, 2009).

Important aspects of interaction design will still be usability, the way things work and fulfil needs, but also how it informs and guides, and above all, how encalming it is. To this end, when designing a new product, system, or interface, it is important to take into consideration how people experience, perceive, and connect with what they consider important and meaningful in their day-to-day environments. Emotion driven design revolves around our needs as humans to bond and create a connection between people, and man and machine (product, interface, space, etc.). At the end of the day, this is about establishing better relationships between humans and the world we live in. We need to promote the design of products and environments that enhance people's lives, inspire them, and elicit positive emotions.

All of this has the potential to be done in a way to provide quality of life and independent living for all ages and ranges of mobility, with age friendly accommodating and "caring" products and environments. Emotion driven design is also designing with empathy to create more responsive and appropriate products and environments in terms of people's cultural, emotional, spiritual, and practical requirements.

References

Aspinwall, L.G., 1998. Rethinking the role of positive affect in self-regulation. Motiv. Emot. 22(1), pp. 1–32 (cited in Fredrickson, B.L. and Joiner, T., 2002. Positive emotions trigger upward spirals toward emotional well-being. *Psychological science*, 13.(2), pp. 172–175).

Bakker, S., van den Hoven, E., Eggen, B., 2010. Design for the Periphery. EuroHaptics 2010, 71.

Barrett, L.F., Mesquita, B., Ochsner, K.N., Gross, J.J., 2007. The experience of emotion. Ann. Rev. Psychol. 58, 373.

Bechara, A., Damasio, H., Damasio, A.R., 2000. Emotion, decision making and the orbitofrontal cortex. Cereb. Cortex 10 (3), 295–307.

Buxton, W., 1994. The three mirrors of interaction: a holistic approach to user interfaces. In: MacDonald, L.W., Vince, J. (Eds.), Interacting with Virtual Environments. Wiley, New York.

Carney, D.R., Cuddy, A.J., Yap, A.J., 2010. Power posing brief nonverbal displays affect neuroendocrine levels and risk tolerance. Psychol. Sci. 21 (10), 1363–1368.

Cascio, C.J., 2010. Somatosensory processing in neurodevelopmental disorders. J. Neurodev. Disord. 2 (2), 62.

Chawda, B., Craft, B., Cairns, P., Heesch, D., Rüger, S., 2005. Do "attractive things work better"? An exploration of search tool visualisations.

Cherry, K., 2010. The Everything Psychology Book: Explore the Human Psyche and Understand Why We Do the Things We Do, Adams Media, Everything® Books.

Crilly, N., Moultrie, J., Clarkson, P.J., 2004. Seeing things: consumer response to the visual domain in product design. Design Stud. 25 (6), 547–577.

Cuddy, A.J., Kohut, M., Neffinger, J., 2013. Connect, then lead. Harvard Bus. Rev. 91 (7), 54–61.

Damasio, A.R., 2006. Descartes' error. Random House, London.

De Gelder, B., 2006. Towards the neurobiology of emotional body language. Nat. Rev. Neurosci. 7 (3), 242–249.

Demirbilek, O., Sener, B., 2003. Product design, semantics and emotional response. Ergonomics 46 (13–14), 1346–1360.

Design & Emotion Conference, 2016. Available from: www.de2016.org/

Desmet, P.M.A., 2002. Designing Emotions. Unpublished doctoral thesis ISBN 90-9015877-4.

Dissanayake, E., What is Art For? 2005, University of Washington Press, Seattle, WA.

Doyle, S., 1998. On humor in design. In: Heller, S., Pettit, E. (Eds.), Design Dialogues. Allworth Press, New York.

Field, T., 2014. Touch. MIT Press, Cambridge, Massachusetts.
Fredrickson, B.L., 2001. The role of positive emotions in positive psychology: the broaden-and-build theory of positive emotions. Am. Psychol. 56 (3), 218.
Fredrickson, B.L., 2004. The broaden-and-build theory of positive emotions. Philos. Trans. R. Soc. Lond. B Biol. Sci., 1367–1378.
Gallace, A., Spence, C., 2010. The science of interpersonal touch: an overview. Neurosci. Biobehav. Rev. 34 (2), 246–259.
Gentsch, A., Panagiotopoulou, E., Fotopoulou, A., 2015. Active interpersonal touch gives rise to the social softness illusion. Curr. Biol. 25 (18), 2392–2397.
Goleman, D.P., 1995. Emotional Intelligence: Why It Can Matter More Than IQ for Character, Health and Lifelong Achievement. Bantam Books, New York.
Graves, C., 2015. Part One: We Are Not Thinking Machines. We Are Feeling Machines That Think. Available from: http://www.instituteforpr.org/part-one-not-thinking-machines-feeling-machines-think/
Gross, J.J., Thompson, R.A., 2007. Emotion regulation: Conceptual foundations. In: Gross, James, J. (Eds.), Handbook of Emotion Regulation. Guilford Press, New York, NY, xvii, 3-24, 654.
Hadjikhani, N., Kveraga, K., Naik, P., Ahlfors, S.P., 2009. Early (N170) activation of face-specific cortex by face-like objects. Neuroreport 20 (4), 403.
Izard, C.E., 2001. Emotional intelligence or adaptive emotions? Available from: http://citeseerx.ist.psu.edu/viewdoc/download?doi=10.1.1.459.5602&rep=rep1&type=pdf
Izard, C.E., 2013. Human emotions. Springer Science & Business Media, New York.
Jekosch, U., 2005. Assigning meaning to sounds—semiotics in the context of product-sound design. Communication Acoustics. Springer Berlin Heidelberg, The Netherlands, pp. 193–221.
Jordan, P.W., 2000. Designing Pleasurable Products. Taylor & Francis, London.
Kaya, N., Epps, H. H., 2004. Relationship between color and emotion: a study of college students. Coll. Stud. J. 38(3), p. 396 (cited in Nijdam, N.A., 2009. Mapping emotion to color. (Eds.):'Book Mapping emotion to color'.(2009, edn.), pp. 2-9).
Kövecses, Z., 2003. Metaphor and emotion: Language, culture, and body in human feeling. Cambridge University Press, Cambridge.
Lidwell, W., Holden, K., Butler, J., 2010. Universal Principles of Design: 125 Ways to Enhance Usability, Influence Perception, Increase Appeal, Make Better Design Decisions, and Teach Through Design 2010, 2nd ed., Rockport Publications , Beverly, Massachusetts.
Mauss, I.B., Bunge, S.A., Gross, J.J., 2007. Automatic emotion regulation. Soc. Person. Psychol. Compass 1 (1), 146–167.
Merleau-Ponty, M., 1958. Phenomenology of Perception (1945), (C. Smith, Trans.). Routledge, London and New York.
Meyers, D.G., 2009. Exploring Psychology, eighth ed. Worth Publishers, New York.
Microsoft Research, 2016. Available from: http://research.microsoft.com/en-us/research-areas/machine-learning-ai.aspx
Montagu, A., 1971. Touching: The Human Significance of the Skin. Columbia University Press, New York.
Morris, J.A., 2006. The Purpose and Power of Color in Industrial Design: Encouraging the Meaningful Use of Color in Design Education, IDSA Design Conference.
NextNature.net, 2011. 11 Golden Rules of Anthropomorphism and Design: Introduction. Available from: https://www.nextnature.net/2011/12/11-golden-rules-of-anthropomorphism-and-design-introduction/
Norman, D.A., 2005. Emotional Design: Why We Love (or Hate) Everyday Things. Basic Books, New York.
Norman, D.A., 2009. The way I see it. Systems thinking: a product is more than the product. Interactions 16 (5), 52–54.
Nummenmaa, L., Glerean, E., Hari, R., Hietanen, J.K., 2014. Bodily maps of emotions. Proc. Natl. Acad. Sci. USA 111 (2), 646–651.

Overbeeke, K., Djajadiningrat, T., Hummels, C., Wensveen, S., Prens, J., 2003. Let's make things engaging. Funology. Springer, Netherlands, pp. 7–17.
Özcan, E., 2008. Product Sounds-Fundamentals and Applications. PhD Thesis, TU Delft.
Petersen, M. G., Iversen, O. S., Krogh, P. G., Ludvigsen, M., 2004. Aesthetic Interaction: a pragmatist's aesthetics of interactive systems. Fifth Conference on Designing Interactive Systems: Processes, Practices, Methods, and Techniques, pp. 269–276. ACM.
Plack, C.J., 2013. The Sense of Hearing. Psychology Press, New York.
Plutchik, R., 1980. A general psychoevolutionary theory of emotion. Theories Emot. 1, 3–31.
Russell, J.A., 2003. Core affect and the psychological construction of emotion. Psychol. Rev. 110 (1), 145.
Scanlon, A.E., Polage, D.C., 2011. The Strength of a Smile: Duchenne Smiles Improve Advertisement and Product Evaluations. Pacific Northwest Journal Undergraduate Research and Creative Activities 2, 3.
Silk, J.B., Alberts, S.C., Altmann, J., 2003. Social bonds of female baboons enhance infant survival. Science 302, 1231–1234.
Schön, D.A., 1983. The Reflective Practitioner: How Professionals Think in Action. Basic Books, 5126.
Simpson, J.A., Weiner, E.S.C. (Eds.), 1989. The Oxford English Dictionary. Clarendon, Oxford.
Solomon, R.C., 1976. The Passions: Emotions and the Meaning of Life. Hackett Publishing, Indiana, USA.
Soussignan, R., 2002. Duchenne smile, emotional experience, and autonomic reactivity: a test of the facial feedback hypothesis. Emotion 2 (1), 52–74.
Stets, J.E., 2010. Future directions in the sociology of emotions. Emot. Rev. 2 (3), 265–268.
Surakka, V., Hietanen, J., 1998. Facial and emotional reactions to Duchenne and non-Duchenne smiles. Inter. J. Psychophysiol. 29, 22–33.
Tiger, L., 1992. The Pursuit of Pleasure. Transaction Publishers, New Jersey, USA.
Tromp, N, van den Berg, D., 2014, Redesigning Mental Healthcare, Interviewed by Marte den Hollander. In: SF., Fokkinga, DPS., Ferrari, D., Peters, S., Ferrari, D., Pete, (Eds.) Design Vision—CrISP Magazine, #4: Well, Well, Well... TuDelft, Eindhoven, pp. 16–17.
Tsakiris, M., Carpenter, L., James, D., Fotopoulou, A., 2010. Hands only illusion: multisensory integration elicits sense of ownership for body parts but not for non-corporeal objects. Exp. Brain Res. 204 (3), 343–352.
Van Egmond, R., 2008. The experience of product sounds. In: Schifferstein, H.N.J., Hekkert, P. (Eds.), Product experience. Elsevier, Amsterdam, pp. 69–89.
Van Hout, M., 2007. Designing Happy. Design & Emotion Blog. Available from: http://www.design-emotion.com/2007/11/05/designing-happy/
von Goethe, J.W., 1810. Zur Farbenlehre (Vol. 2). JG Cotta. (cited in Nijdam, N.A., 2009. Mapping emotion to color. (Eds.):'Book Mapping emotion to color'.(2009, edn.), pp.2-9).
Warriorwoman, 2013. Jawbone Up - Another Geeky Bracelet Review, 28 May, 2013 Available from: http://www.warriorwomen.co.uk/2013/05/28/jawbone-up-another-geeky-bracelet-review/
Weinschenk, S.M., 2009. Neuro Web Design: What Makes Them Click? New Riders Publishing, Berkeley, CA.
Weiser, M., Brown, J.S., 1995. Designing Calm Technology. Available from: http://www.ubiq.com/weiser/calmtech/calmicch.htm Author Index c Caselli, Stefano 309: 165.

Further Reading

Microsoft Project Oxford and Cortana website. Available from: https://www.microsoft.com/cognitive-services/en-us/emotion-api#detection
Online Etymology Dictionary. Available from: http://www.etymonline.com/index.php?term=emotion

CHAPTER

15

Affective Human–Robot Interaction

Jenay M. Beer, Karina R. Liles, Xian Wu, Sujan Pakala

University of South Carolina, Columbia, SC, United States

INTRODUCTION

For decades, researchers, designers, and the general public have been fascinated with the possibility of developing able and intelligent machines that engage in social interaction. The crux for socially interactive robots is the ability to interact with humans in a meaningful manner. Thus, the field of human–robot interaction (HRI) over the past several decades has emerged as a crucial field for the design of social robots that are useful, intuitive, and user friendly. The field of HRI has been described as "dedicated to understanding, designing, and evaluating robotic systems for use by or with humans" (Goodrich and Schultz, 2007, p. 204). Successful HRI depends on the ability for the human and robot to interrelate and collaborate. Thus, consideration of both the robot and human's social capability is critical, in order to develop and design robots that fit into our social schemas.

In this chapter, we provide an overview of affective HRI as a field of study. We highlight important factors to consider in the study of socially interactive robots and discuss three common application areas: companionship, education, and aging-in-place. We draw from our own previously published empirical studies and those of our colleagues in providing a review of the considerations that are pertinent in the development and integration of social robots.

Emotions and Affect in Human Factors and Human–Computer Interaction. http://dx.doi.org/10.1016/B978-0-12-801851-4.00015-X
Copyright © 2017 Elsevier Inc. All rights reserved.

AFFECTIVE HRI AS A FIELD OF STUDY

Many researchers (Goodrich and Schultz, 2007; Murphy, 2000; Thrun, 2004; Steinfeld et al., 2006) have contributed significant insight into the emergence of HRI as a field of study—from fundamental concepts, to continually evolving frameworks and metrics, to exhaustive surveys—that aim to provide robotic solutions to real life problems.

Affective HRI is related to, but separate from, human factors and human–computer interaction (HCI). While HRI certainly benefits from considering the various HCI design principles and iterative design process (Rogers et al., 2011), affective HRI may be more complex than traditional HCI since the robot's social interaction and physical embodiment may take place while functioning in and interacting with a physical space (environment). For successful interaction to take place, various factors including the robot's emotion expression, user expectations, adaptability, the task performed, and the roles of the constituents in a system (the designer, human operator, robot, human participant) all become essential, interdependent elements of a dynamic affective interaction.

To understand HRI, we must first consider what is "interaction." Interaction is generally categorized to be either remote or proximate. Design for remote interaction typically emphasizes functionality, as to accomplish a specific task through appropriate interfaces. For example, controlling a robot from a remote location, such as in search and rescue. Proximate interaction, on the other hand, includes interaction where the human and robot are colocated. Proximate interaction, compared to remote interaction, more likely includes affect and sociability as part of the robot's functionality, where humans and robots may act as peers or companions. Furthermore, depending on the specific application, proximate interaction may consider social, emotive, and cognitive capabilities as essential elements of the interaction (Goodrich and Schultz, 2007).

The primary impetus for much HRI research, whether the interaction is remote or proximate, has been the application of robots in domains, such as search and rescue, assistive robotics, and government applications, which require interaction in dynamic environments. The nature of these applications has been fundamentally cross-domain, needing the collaboration of multidisciplinary scholars (Burke et al., 2004). The development of robots to socially interact with the general public is increasing the need for the robots to behave and interact with humans in an affective manner. To date, there is no single unifying framework or theory of HRI, but, simply put, factors that influence affective HRI may be discussed and categorized as factors related to the human, the robot, and the social interaction between these two social entities.

HUMANS AS SOCIAL ENTITIES

"Know thy user"—the cornerstone of human factors and HCI is also true for HRI. If the robot design does not match or consider the users' needs and perspectives, the robot is doomed to not be adopted by the user (Broadbent et al., 2012; Davis, 1989; Heerink et al., 2010a; Smarr et al., 2014; Venkatesh et al., 2003). Understanding the purpose and the mode of interaction for which to design the robot (i.e., before writing any code) will save time, effort, and money. Robots designed for human use should also consider the users' capabilities and limitations.

Human beings are considered to be social entities. In that, humans maintain social relationships, playing numerous roles: professional, familial, platonic, fraternal, and so on. However, human beings are also capable of social association with inanimate objects. In fact, it is generally accepted that people are willing to apply social characteristics to technology. Users have applied social characteristics to computers, even though they admit that these technologies do not possess actual human-like emotions, characteristics, or "selves" (Nass et al., 1994). Humans have been shown to display social behaviors toward computers mindlessly (Nass and Moon, 2000), as well as to treat computers as teammates with personalities, similar to human–human interaction (Nass et al., 1995, 1996).

The work of Clifford Nass highlighted the need to learn from the silent social rules of how people treat and work with technology and then apply those rules to better understand how people treat people (and technology) for healthier social relationships (Nass and Moon, 2000; Nass and Yen, 2010). The focus we draw here is the importance of unambiguous, rigorous, and straightforward rules in social behavior that should be considered in designing appropriate robot presentation (form, feature, and emotion) and personality that can directly increase the effectiveness of the robot. For example, to best match a social robot to meet human needs, the logical step is to enable the robot to provide services in an acceptable manner that follow the users' expectations for rules of social behavior.

Subject to feasibility and necessity, other characteristics of the user—such as physical capabilities, age, gender, cohorts, technology experience, profession, and other demographics—should be considered in the design of a dynamic personality for robots. For example, design considerations to address age-related differences in emotion recognition (Beer et al., 2015). Moreover, cultural diversity is an important factor (Šabanović, 2010, 2014). Some of the rules mentioned in the works of Clifford Nass, whose study participants were mostly from the individualistic United States, may not be applicable to eastern nations like Japan and China, which have collectivist cultures (Katagiri et al., 2001).

For social robots to be successful, they need to be accepted by humans. HRI researchers are aware of the need to develop the right kinds of robots with appropriate, natural ways for them to interact with humans. Effort must be made to assess the user's perspective and use it to reduce negative aspects of user experience while enhancing positive ones. Since no two users are alike, the success of a robot's acceptance then directly depends on its effectiveness and ability to perform as a social entity and its versatility in terms of demonstrating diverse social behavior within appropriate response times and tasks.

ROBOTS AS SOCIAL ENTITIES

Some personal robots, such as the Roomba vacuum, were not designed to elicit social or affective interactions with humans. Yet, users display social behaviors toward Roomba robots, such as naming and talking to it (Forlizzi, 2007; Forlizzi and DiSalvo, 2006; Sung et al., 2010). This suggests that humans may have an expectation that robots behave like social entities. In fact, a personal robot that was designed to demonstrate social capabilities, such as turn taking or gazing, elicited more social behaviors from the user and was rated more favorably compared to a robot without social abilities (Heerink et al., 2007, 2010b; Looije et al., 2010).

Some designers and researchers promote the design of robots with human-like sociability, thus social norms and abilities can be used as a natural interface to interact with robots (Breazeal, 2004; Dautenhahn, 2007a,b). In other words, then users could interact with a robot as if it was another person. That is not to suggest that users will necessarily think that the robot is actually human or a living entity, but rather that they may behave toward it that way.

Great strides are being made in developing robots that are similar in both appearance and behavior to humans. People tend to form quick first impressions about others based on appearance (Bar et al., 2006) and initial impressions influence behavior toward others (Eberhardt et al., 2006; Todorov et al., 2005). Similar processes might influence a user's first impressions of social robots. For successful HRI, it has been suggested that the robot's form should enable people to intuitively understand and interpret its behavior and affect (Kanda et al., 2008).

Furthermore, the appearance of the robot should match the users' expectations of what they might presume a robot to look like. Prakash and Rogers (2015) found that older adults do not prefer robots that have a combination of highly human-looking and mechanical features. A mix of human and mechanical features makes it difficult to develop expectations of the robot's social capability, because of inconsistent aesthetic characteristics. The construct's unity and prototypicality are central aesthetic aspects of product design (Veryzer and Hutchinson, 1998). *Unity* refers to how well the numerous elements of a product match each other. *Prototypicality* is the degree a product represents a category (Barsalou, 1985). There are examples of human-like robots that do not meet the criteria of unity and prototypicality; for example, realistic human arms attached to robots that lack heads or realistic faces on mechanical bodies (e.g., H-type and Albert Hubo). Although, if a robot becomes too human-like it may be perceived as strange in appearance (uncanny valley; Mori, 1970). Thus, it is important for designers to consider the role of aesthetics in the development of socially assistive robot faces and form, because users' first impressions influence their expectations and acceptance.

Given that humans are willing to apply social characteristics to inanimate objects, such as a Roomba, it is not surprising that much research has been conducted on the development of robots that communicate affectively. Much research to date has focused on the depiction of social cues. Facial expressions are a critical component in successful human–human social interaction and also important in the development of affective robots in particular. Emotional facial expressions may be defined as configurations of facial features that represent discrete emotional states, and some basic emotions have been shown to be recognizable across cultures and norms (Ekman and Friesen, 1975).

Many robot applications require some level of social interaction, and thus expression, with the user. The development of affective robots has a long history, with some early computational systems aimed at the development of generating facial emotion. Some of these robots (e.g., Breazeal, 2003; Fabri et al., 2004) have been modeled from psychological-based models of emotion facial expression, such as FACS (Ekman and Friesen, 1978) or the circumplex model of affect (Russell and Pratt, 1980). One approach posits the use of classic animation principles (e.g., exaggeration, slow in/out, arcs, timing; Lasseter, 1987) to depict emotional facial expression (e.g., Becker-Asano and Wachsmuth, 2010) as well as other nonverbal social cues, such as movement and body posture (Takayama et al., 2011). All of these approaches have been developed with the emphasis of designing believable and easily recognizable robot emotive expression.

An affective robot may make use of facial expressions and other nonverbal cues to facilitate social interaction, communication, and express emotional state. This requires the user to interpret its facial expressions accurately (Beer et al., 2015). To facilitate social interaction, a robot will need to demonstrate affective facial expression effectively for the user to accurately interpret the robot's intended message. Emotion is thought to allow a user to assume that the robot is capable of caring about its surroundings (Bates, 1994), as well as to promote an enjoyable interaction (Bartneck, 2003). The ability of a robot to express emotion may play a role in the development of intelligent technology. Picard (1997) suggested that affect is an active part of intelligence, and critical for technology to function with sensitivity toward humans.

The expression of social cues, as discussed previosuly, is important for depicting robots as social entities. However, it is commonly thought that long-term, consistent social mannerisms or trends will contribute to a user's interpretation of robot personality. Interestingly, incorporating adaptability of behavior into technology (like robots) to match the personality of the human user can favor their acceptance (Moon and Nass, 1996). In other words, even though a robot may start with a neutral personality, if it is able to slowly embody emotions similar to or compatible with the personality of the user, the user is bound to increase their social association with the robot. Considering such factors, designing for usability and user experience will likely further emerge as robots become more commonplace and commercialized. Similar to the millions of cars developed each year to suit the varied needs and lifestyles of people, it may be easy to imagine in the future a similar variety in personalized robot social personalities and applications.

APPLICATION OF AFFECTIVE HRI

In the next sections we provide an overview of how affective HRI has been applied. We highlight three categories, or applications, of social robotics: aging-in-place, companionship, and education. In the recent decades, much HRI research has been conducted in these three applications. We summarize the literature to date and provide an overview of future directions for each application.

Social Robots for Aging-in-Place

Older adults are an increasing proportion of our population, and this population requires unique considerations in the design of technology (Erber, 2011; Fisk et al., 2009). Based on World Health Organization, the proportion of the world's population over 60 years of age will double from about 11%–22% between 2000 and 2050. In the year of 2050, it is expected 2 billion people will be aged 60 and older. Global aging is, in part, a product of successes in medical, social, and economic advancements (Dobriansky et al., 2007). However, the aging of the population will also present challenges for families, health care providers, and medical care facilities (Dobriansky et al., 2007).

Aging societies are expected to see two trends in the near future—an increase in demand to care for our older adults and a decrease in number of caregivers (Tapus et al., 2007). Moreover, many older adults prefer to age-in-place (Farber et al., 2011), that is to age in their own home setting. Older adults' preferences to age-in-place are due to various reasons, such as attachment with their homes, familiarity with local friends and businesses, and a desire to remain independent. Consequently, a very poignant question was raised—how can older adults age-in-place gracefully with dignity while also maintaining healthy social connectedness and well-being? Social robotics has the potential to help older adults maintain their independence and age-in-place.

Looking to technology as a solution requires multidisciplinary research, especially for the HRI community. As a natural progression of research on assistive and interactive robots over the years, socially assistive robotics (SAR) has emerged as a potential tech-based solution to serve older adults and help maintain their independence. SAR is defined as an intersection of assistive robotics and socially interactive robotics (Feil-Seifer and Mataric, 2005). In other words, SAR maintains the goal of developing close and affective interactions with a human user for the purpose of providing assistance and achieving measurable progress in convalescence, rehabilitation, learning, and so on. Although the term SAR evolved later (Feil-Seifer and Mataric, 2005), several projects leading up to it endeavored to develop social robots that could provide assistance and interaction in various situations. For example, in the Nursebot project (Montemerlo et al., 2002), autonomous mobile robots were deployed to remind residents about events and guide them through their environments. The robot was able to guide older adults, without the need for assistance from a nurse.

While assistive robots can be broadly defined as providing aid or support to a human user (e.g., intelligent walkers, wheelchair robots, manipulator arms, and exoskeletons), adding an interactive and affective component when necessary becomes the domain of SAR, such as personal assistants in the home. Our work and others' (Beer et al., 2012; Broadbent et al., 2012; Bugmann and Copleston, 2011; Mast et al., 2012; Smarr et al., 2014) have been conducted to understand the potential of robot usage at home. In one study (Smarr et al., 2014), we reported the extent to which older American adults are willing to accept robot assistance in the home and the actions they want robots to perform. We determined preferences for robot assistance versus human assistance for 48 home-based tasks. Overall, participants reported an overall preference for robot assistance with repetitive chores like cleaning the kitchen and in other aspects like surveillance. In contrast, participants reported preferring human assistance for the domains of personal care, such as shaving followed by leisure activities, such as entertaining guests and health in regards to deciding what medication to take and when. However, the results of this study are specific to a mobile manipulator robot, which had been designed with functionality, rather than sociability, as the primary consideration.

Torta et al. (2014) report results of a short-term and a long-term evaluation of a small, socially assistive humanoid robot in a smart home environment. Multimodal communication channels were employed to provide a proactive interface for the smart home and help in generating positive feelings overall for a wide range of user needs being met (i.e., information sharing, health and home monitoring, exercise companion, etc.). Short-term usage participants experienced the effectiveness of the robot's capabilities, while long-term participants even demonstrated trust in the small humanoid robot in assisting with many activities of daily living. Both these factors contribute to a more effective usage. Also, a cross-cultural comparison showed that results were not due to the cultural background of the participants.

Lastly, older adults' attitudes toward robot appearance depended on the task (Goetz et al., 2003; Smarr et al., 2014). For socially oriented tasks, such as playing a game, learning new skills, or socially communicating, many older adults reported that they prefer a robot to have a human appearance (Prakash and Rogers, 2015). A robot with a human appearance was preferred because it evoked perceptions of human-like capabilities. However, some older adults indicated that the human appearance was also invasive for a highly personal task (e.g., bathing), suggesting the relationship between appearance and social tasks may be driven by personal preference.

Summary and future directions of robots for aging-in-place. Robots have the potential to help older adults remain independent and to age-in-place. However, there are many considerations for the development of robots for aging-in-place, ranging from understanding older adults' attitudes and acceptance to determining the proper robot appearance for assisting older adults with social home-based tasks.

There are several open areas for future research. First, there is a need to conduct research that makes comparisons across geographies to understand the role culture plays in technology acceptance (Šabanović, 2010, 2014). Second, most researchers have used some prototype or commercial robot right from their initial studies. This may have limited the user response based on the capabilities and limitations specific to the model itself. Thus, surveys related to such factors may help improve the next wave of research by better assessing, which findings are robot specific and which are generalizable to social robots in general. Behavioral adaptability and customizability of socially assistive robots has potential for immediate diverse research—improving proxemics, conversational skills, and so on.

More thoroughly though, various factors, like the level of autonomy of the robots (Beer et al., 2014), modes of communication, their learning and adaptability, should continue to be integrated together to holistically define and solve HRI challenges. For example, consider allowing a user to adjust a robot's social and affective settings. After interacting with a system, a user may develop a set of preferences for the height of the robot, volume, facial expressions, or even personality type. If sufficient capability is provided to the robot, these social characteristics could then autonomously adjust to the user.

In summary, while last few decades have seen the development of assistive or interactive robots, the need for more sophisticated sociability in their design is continually being seen as having immediate potential. The field of SAR for aging-in-place therefore stands at the forefront in attempting to serve some of those immediate critical needs.

Social Robots for Companionship

Traditionally, robots have been designed to build products or transfer objects, for example, in manufacturing. However, companion robots are designed to perform more complex tasks that offer behavior enrichment through the interaction with humans (Saint-Aimé et al., 2007). A robot companion can be defined as a service robot that is *specifically* designed for personal social use (Dautenhahn et al., 2005). While the line between socially assistive robots and companion robots is blurred, companion robots primary purpose is to satisfy social needs and serve as a social companion or friend. In most cases, robot companions are designed for nonexpert users. Thus, communication between users and robots needs to be intuitive and natural. Robot companions can be used for different age groups to achieve various tasks. In the following sections, we highlight different types of companion robots. We will also discuss select case examples on how companion robots have been designed for various age groups (e.g., from children to older adults) and for different applications (e.g., assistive living, therapy, or social interaction).

Much research has been conducted investigating the application of animal-like robot companions to interact with younger generations. For most children, a visit to the hospital can be a stressful and scary experience (Stiehl et al., 2009). Although hospitals make efforts to provide socioemotional support for young patients and their families, gaps between what the patients need and what the hospitals provide still exist (Jeong et al., 2015). A robot companion introduced to this scenario has potential to help mitigate stress, anxiety, or pain for the children in pediatric care. Huggable is a teddy bear robotic companion created by the MIT Media Lab. Huggable has over 1500 sensors on the "skin," back-drivable actuators, video cameras, microphones, an inertial measurement unit, a speaker, and an embedded PC with 802.11g wireless networking. All these sensors make Huggable "feel" the environment so user can interact with it just as they interact with another human.

The Huggable robot has been applied to a pediatric unit, where it not only facilitated interaction with children but also served as an important link between the patients, their families, and the medical staff. One study compared and contrasted three different intervention conditions: a plushy teddy bear, a virtual character on a screen, and Huggable. In this study, children aged between 3 and 10 years old, who were scheduled to receive

48+ h of inpatient care, were recruited. The researchers videotaped up to 8 h of the interaction between the kids and the three different interventions. The researchers provided a qualitative analysis on the time that kids spent interacting with each of the interventions by coding the children's facial affect under each condition. Children who interacted with Huggable appeared to be more physically and mentally motivated to engage with it compared to the other conditions. This finding suggested a preference for the Huggable robot (Jeong et al., 2015).

Another application for children is the use of affective robots for the development of language skills. For preschool children, oral language skill is an important part of development, which might affect their future study and career (Kory and Breazeal, 2014). Robots have the potential to help in this application by telling stories to children to improve their oral language skills. In addition, robot companions can also be potentially helpful in assisting therapy with children who have autism spectrum disorder (ASD). ASD encompasses a group of developmental disabilities that cause significant social, communication, and behavioral challenges (CDC, 2015b). About 1 in 68 children have been diagnosed with ASD (Baio, 2014). Children with ASD experience difficulties engaging in social interaction (Simut et al., 2011). By interacting with a robot storyteller, instead of with a human, children with ASD may feel less stress. One hypothesis is that robots, such as Huggable, represent a "niche" between inanimate toys and animate humans, thus offering a middle ground in which children with ASD may be more willing or able to elicit social behavior (Scasselati et al., 2012; Simut et al., 2011).

In another example, researchers videotaped and audio recorded the interaction between children and DragonBot, a cloud based social robot. Children's speech and interaction were coded through the video and transcript of the audio. The preliminary data of this study showed children's acceptance of DragonBot (Kory and Breazeal, 2014). Similarly, Probo is a huggable animal-like robot, which can be used as robot assisted therapy for children with ASD by storytelling (Simut et al., 2011). The robot has a fully actuated head, with 20 degrees of freedom, capable of showing facial expressions, a moving trunk, and a soft and huggable jacket (Probo, 2015). Probo has also been shown to provide an enriched social environment, which improved the effectiveness of social story therapy intervention (Simut et al., 2011).

Companion robots have the potential to assist older adults as well. One common problem for older adults is the onset of dementia and Alzheimer's disease (CDC, 2015a). In 2013, there were approximately 5 million Americans living with Alzheimer's disease, and by 2050, this number will increase by 3 times to 14 million (Hebert et al., 2013). One use of a robot companion is to assist with social therapy for older adults with dementia. Research suggests that communication with animals can increase older adults' social interaction and decrease feelings of loneliness (Churchill et al., 1999; Kanamori et al., 2001; Sellers, 2006). Robotic pets have been introduced as care companions for older adults with cognitive impairment or other disabilities (Libin and Cohen-Mansfield, 2004). For example, Paro, a seal-like socially companion therapeutic robot developed by AIST, is originally designed as a "healing pet," which might offer animal therapy without needing real animals. Paro has five kinds of sensors. With these sensors, Paro can perceive people and its environment. The light sensor allows the Paro to recognize dark and light; the posture sensor enables Paro to feel being held; tactile sensor allows Paro to recognize being stroked; the audio sensor comes give Paro the ability to detect voice and recognize words. With those different sensors, Paro can express its own feeling by moving its body (Paro).

One study aimed to compare the effect with and without Paro in an interactive reading group was conducted in one aged care facility in Australia. In this study, 18 residents with mid to late stages of dementia were recruited; 9 groups of older adult participated in 45-min long intervention (with Paro) and control activities 3 afternoons a week for 5 weeks. Comparing with reading group, the Paro group had higher QOL-AD (quality of life in Alzheimer's disease scale) and OERS-pleasure (observed emotion rating scale) scores (Moyle et al., 2013). In addition, the staffs worked in the facility stressed that participants in Paro group showed less anxiety comparing with the reading group and this point of view was later verified by analyzing the video data (Moyle et al., 2013). The findings from this study suggested Paro could enhance the lives of older people as therapeutic companions (Moyle et al., 2013).

Another 8-week observational study with 10 older adults with mild to severe cognitive impairment was conducted in a nursing home in the United States (Chang et al., 2013). The results of this study demonstrated that interaction with Paro resulted in an increased amount of physical activity in the context of group sensory therapy. In addition, participants experienced an increased willingness to interact with Paro, which the authors cited as a sign of success in terms of sensory therapy (Chang et al., 2013). A study conducted by Wada et al. (2004) also showed interaction with Paro not only generated social interactions among the nursing home residents, but also positively affected the older adults' emotional states (Wada et al., 2004).

Summary and future directions of companion robots. Robot companions can be used in many different applications for different age groups, ranging from children to older adults. However, depending on the different user groups, the nature of the robot's social and affective ability should be designed with each demographics' capabilities and limitations in mind. Additionally, the nature of the companion robot's emotional display should probably vary according to task or setting.

Furthermore, given that the populations of interest range in age, cohort effects on technology experience vary. Thus, it is important to evaluate the effect of technology experience on the acceptance of robot companions. Furthermore, a robot's affective behavior may have long-term effects on users' adoption of and emotional attachment to the robot, which has not been systematically evaluated.

The studies discussed in this section were conducted in developed countries. People's perception toward robots might vary depending on cultural background. In addition, consider that each individual is different from one another; people's need for a companion robot likely varies. In this case, customizable companion robots might be needed for future research to satisfy users' different needs and expectations.

Social Robots for Education

Education is a popular domain for robotics research. Researchers are able to model the behaviors of the humans (teacher, tutor, student, etc.), apply these behaviors to a robot, and evaluate HRI in educational settings. In many cases, the robot may serve the role of the teacher or facilitator in the interaction. We provide an overview of social robots for education, including the necessary affective characteristics of a teacher, characteristics of the robot mimicking teacher behaviors, children's perceptions of social robots, and case studies on robots for education.

First, let us consider what affective characteristics of the teacher are ideal for learning. This is important because to develop effective robot teachers a starting point is to mimic human teachers. For effective learning to occur, a social interaction between the teacher and student is important (Tiberius, 1993) because this display of emotion during a teaching/learning scenario contributes to a student's academic motivation and success (Graziano et al., 2007). In fact, a teacher's lack of emotional display can negatively influence many facets of the educational interaction (Tiberius, 1993). A lack of emotion can have a harsh impact on a student's relationship with the teacher. It can also be detrimental to a student's higher order cognition, which affects a student's mental processes, such as working memory, ability to pay attention to detail, and ability to plan (Blair, 2002). In contrast, a positive emotional climate can influence a student's cognitive, emotional, and social state, which enhances a student's well-being and performance (Castillo et al., 2013).

Tiberius and Billson (1991) describe attributes that contribute to social supportive behavior of a teacher. Of those, some key traits were acting as a role model, nonverbal communication, attention guiding, empathy, and communicativeness. Each of these traits contributes to the emotional climate for a teaching/learning scenario.

As a role model, the teacher sets the tone for the educational interaction that the student must follow by facilitating learning while guiding and supporting the student (Billson and Tiberius, 1991). By being a role model, the teacher also cultivates the student's desire to learn and apply new knowledge (Ficklin et al., 1988).

Nonverbal communication is the subtle yet effective act of responding or communicating without using words. This can be displayed in many ways including gestures, facial expressions, or eye gazes (Payrató, 2009).

Attention guiding keeps the student on task by actively moving them through the learning material. This may be achieved using actions like verbal cues, eye gaze, or deictic gestures, such as pointing (Levin et al., 2009).

Eisenberg and Miller define empathy as "an affective state that stems from the apprehension of another's emotional state or condition, and that is congruent with it. Thus, empathy can include emotional matching and the vicarious experiencing of a range of emotions consistent with those of others" (pp. 91, 1987). It is the act of showing sensitivity to the student's feelings and having a shared responsibility for learning and mutual commitment to educational goals (Tiberius and Billson, 1991).

Lastly, multiple opportunities for communication exist when teaching and learning occurs. Communicativeness between the teacher and student includes open, multichannel correspondence, timely feedback, and an open flow of ideas (Tiberius and Billson, 1991).

The same social interaction between a human teacher and student can be applied where the robot will assist in the teacher's role. That is, the social characteristics of a robot used for education can be modeled after human teachers.

It is important to note that the physical capabilities of the robot help determine how its affective aspects are implemented and displayed. The NAO, DARwIn-OP, SAYA, RoboThespian, interactive Cat (iCat), and DragonBot are all examples of social robots used for education. These robots have a wide range of features that demonstrate each robot's affective state. More specifically, social and affective behaviors that were used on the NAO and DARwIn-OP humanoid robots included natural, conversational body movements, hand gestures, and vocal cues (Brown and Howard, 2013; Liles and Beer, 2015a,b). With zero to limited movement of its arms and legs, the android robot SAYA and humanoid RoboThespian displayed emotion via facial expressions, eye movement, and speech (Hashimoto et al., 2013). Modeled after a human's face, the iCat has 13 servomotors that can move its head and different parts of its face. The iCat uses speech and facial expressions to exhibit humor, empathy, and enthusiasm (Saerbeck et al., 2010). Finally, the DragonBot has the appearance of a dragon. With limited movement, it uses verbal cues to show emotion (Short et al., 2014). All of these examples illustrate how human affective traits can be mimicked in educational robots and enhance the sociability of the robot.

Considering children's perceptions of robots in general is helpful and offers insight for studies that focus on social robots for education. Fortunately, it is probable that children are willing to engage with robots. Most children agreed that they would act friendly toward the robot and were able to recognize the friendly traits exhibited by the robot. As a result, they felt that they could play with the robot, talk to it, and confide in it. Additionally, they believed that the robot liked them and could cheer them up if necessary (Fior et al., 2010). Furthermore, children showed more expressions and behaved more positively when engaging with an affective robot than when interacting with a nonaffective robot without adaptive expressive behavior (Tielman et al., 2014).

A considerable amount of research has been conducted to encompass the social characteristics of the teacher in a robot and explore teacher (robot)–student interaction. Some studies have measured the students' opinions and attitudes about the social robot after interacting with it. Other studies have measured the students' academic performance as a result of their interaction with a social robot.

Studies have shown that students have positive reactions to robots due to the robot's social behavior. We (Liles and Beer, 2015a,b) found that students perceived the NAO robot Ms. An (meeting students academic needs) to be social and that students expressed a preference for working with the robot over other kinds of study support. In this study, students practiced multiplication with the robot. Before their interaction with the robot, students had a positive perspective of the robot, which increased after practicing math with the robot. Furthermore, we carefully depicted the role of the robot as a role model, which helped contribute to the robot's socially supportive identity. In this case, the robot may serve as a teaching assistant or tutor (Brown and Howard, 2013; Liles and Beer, 2015a,b; Saerbeck et al., 2010) in which it guides the student and facilitates learning. In other cases, the robot served as a teacher or instructor that maintains a healthy balance between being authoritative and having a pleasant demeanor (Hashimoto et al., 2013; Short et al., 2014).

In another study involving mathematics, a robot guided and encouraged students as they completed an algebra test. From this study, it was determined that having a robot present when a student is taking a test reduces boredom and increases the student's enjoyment. Although the students performed well on the test, there was no significant variance in performance between students who completed the test with a robot agent and those who did not (Brown and Howard, 2013).

Researchers used a DragonBot robot to teach students about nutrition. As a result of their interaction with the DragonBot, students had positive perceptions and reactions about the robot and found it to be useful, enjoyable, exciting, valuable, and attractive. Additionally, students believed the robot had a strong social presence (Short et al., 2014). Other studies have shown that using a social robot in an educational setting may increase a student's academic performance. Two very similar studies delivered a science lesson to students where one lesson was mediated by SAYA and the other by RoboThespian. In these studies, students were taught a lesson on levers and then guided through hands-on experiments using levers. After the lesson, students were tested and almost all students answered the assessment questions correctly. It was shown that there is potential for the social presence of a robot to increase student success (Hashimoto et al., 2013).

Nonverbal feedback, as discussed earlier in this chapter, is a common mode to depict robot affect. Nonverbal feedback is commonly implemented in studies where social supportive behavior is employed. In our previously discussed tutoring application where students practiced multiplication (Liles and Beer, 2015a,b), nonverbal feedback was given according to the correctness of the student's response. If the student answered a question correctly, the robot executed a gesture to praise the child. For example, the robot may have given the student a fist bump or clapped its hands. Similarly, if the student answered incorrectly, the robot executed an appropriate gesture. For example, the robot looked down or shook its head when a student answered incorrectly (Liles and Beer, 2015a,b). Other affective gestures include gazes to direct attention toward an object, head nods to signal the student to continue or indicate "okay" or "yes," and head scratches to show confusion (Brown and Howard, 2013).

However, not all robots are capable of depicting nonverbal feedback in exactly the same way as a human teacher. In these instances, it is important that the robots' non- or less-humanoid affective cues need to be easily recognizable by students. For example, due to its facial limitations and machine-like appearance, the RoboThespian robot used for the previously discussed science lesson changed its facial color, animated its eyelids, and dropped its jaw to display facial expressions. It also moved its arms and legs to create gestures. In contrast, SAYA had a human-like appearance and made facial expressions coupled with head and eye movements (Hashimoto et al., 2013). Many of the verbal and nonverbal affective cues discussed previously are thought to provide students with attention guiding. For example, the socially supportive iCat recommended for the students to look at the learning materials when necessary and used gazes to direct the student's attention (Saerbeck et al., 2010).

Lastly, empathy is an affective behavior that can be implemented in social robots. Ms. An displayed empathy for the student throughout the tutoring session. For example, if a student answered incorrectly, a typical response would be "Don't give up. Let's try that problem again." This emotional feedback informed the student of an incorrect answer while not discouraging the student (Liles and Beer, 2015a,b). The DARwIn-OP robot also used emotional responses to encourage students. If a student required more time to answer a question, the robot may have stated, "This is really making us think." Using pronouns, such as "we" and "us" showed the student that the robot shared the same zeal for the student's success as the student (Brown and Howard, 2013).

Similar to empathy, many of these educational robots also depicted socially supportive behavior to display its communicativeness. This is because communication is highly likely to occur in an educational setting. Many studies began each interaction with an introduction (Brown and Howard, 2013; Liles and Beer, 2015a,b; Short et al., 2014). Additionally, the robots' responses throughout the interactions with the students showed communicativeness. The social supportive communication was determined by some action of the student. For instance, if the student answered correctly or incorrectly, the robot responded accordingly (Liles and Beer, 2015a,b; Saerbeck et al., 2010).

Summary and future directions of educational robots. There are many intriguing questions and issues with regard to affective educational robots that need to be explored and addressed. Presently, many studies have focused on administering a lesson to a group of students (Hashimoto et al., 2013) or one student at a time (Brown and Howard, 2013; Liles and Beer, 2015a,b; Saerbeck et al., 2010; Short et al., 2014) in a controlled setting. In these studies, the students often had similar traits (i.e., grade level, age). These studies rarely were conducted in a real-time classroom environment during a typical school day (Brown and Howard, 2013; Liles and Beer, 2015a,b; Saerbeck et al., 2010; Short et al., 2014). Ideally, future research for social robots for education will utilize robots that will adapt to these dynamic classrooms and be integrated as a part of the students' normal class activities.

For studies that have engaged a full classroom of students where the robot was able to answer unscripted questions from the student, often a Wizard of Oz control was used (Hashimoto et al., 2013). In other words, the robots lacked autonomy and were remotely controlled by a researcher. Currently, most robots are programmed with preset responses and students are limited to what they can say to the robot as well as when they are able to say it (Brown and Howard, 2013; Liles and Beer, 2015a,b; Saerbeck et al., 2010; Short et al., 2014). It will be important to have educational robots that can accurately interpret a student's question or comment and provide an appropriate response. This will allow for the natural dialogue that normally occurs between the teacher and student (or robot and student), which will increase the opportunities for the robot to display appropriate affective behavior.

The range of social behavior exhibited by a robot can vary. These variations lead to differences in the social behavior that the robot exhibits.

Enhanced personalization will be beneficial as the robot will be able to modify its behavior based on the student's temperament. For example, if the student is feeling down, the robot will be able to detect this (e.g., facial emotion recognition) and respond accordingly. Future studies will include a robot that has a high level of sensing capability so that it can apply its social supportive behavior accordingly.

In some studies, the students interacted with the robot for only one session (Brown and Howard, 2013; Liles and Beer, 2015a,b) or a small number of sessions (Short et al., 2014). More long-term studies should be conducted to further determine the effect of socially supportive behavior on a student's academic performance and the student's perceptions of the robot given these behaviors.

Finally, many popular subjects for social robots for education include math (Brown and Howard, 2013; Liles and Beer, 2015a,b) or language learning applications (Saerbeck et al., 2010). Researchers should consider transferring the socially supportive behavior of robots to other learning domains and subjects. Previous and current research demonstrates the potential for a promising future for affective robots in educational settings. The considerations for future research presented here are just a few of many possible research directions.

CONCLUSIONS

In closing, affective HRI is emerging as a critical and necessary field of study as robots become more commonplace in operating in social environments and domains, such as caring for older adults, companionship, and education. In this chapter, we provide a review of affective HRI, highlighting the importance of considering both the human and robot as social entities. Through this review, we identify common affective themes important across application domains, such as the importance of robot facial expression and a match between robot social capability, user expectations, and user acceptance. However, research on individual differences, varying user needs, long-term evaluations, and culturally specific studies are needed. Through continued research and development of affective HRI, the application of social robotics will continue to evolve and promote assistance and enrichment in users' lives.

References

Baio, J., 2014. Prevalence of autism spectrum disorder among children aged 8 years autism and developmental disabilities monitoring network. MMWR Surveill. Summ. 28 (63), 1–21.

Bar, M., Neta, M., Linz, H., 2006. Very first impressions. Emotion 6 (2), 269–278.

Barsalou, L.W., 1985. Ideals, central tendency, and frequency of instantiation as determinants of graded structure. J. Exp. Psychol. Learn. Mem. Cogn. 11 (4), 629–654.
Bartneck, C., 2003. Interacting with an embodied emotional character. Proceedings of the Design for Pleasurable Product Conference. pp. 55–60.
Bates, J., 1994. The role of emotion in believable agents. Commun. ACM 37 (7), 122–125.
Becker-Asano, C., Wachsmuth, I., 2010. Affecting computing with primary and secondary emotions in a virtual human. Auton. Agents Multi-Agent Syst. 20 (1), 32–49.
Beer, J., Fisk, A.D., Rogers, W.A., 2014. Toward a framework for levels of robot autonomy in human-robot interaction. J. Hum. Robot Interact. 3 (2), 74–99.
Beer, J.M., Smarr, C.-A., Chen, T.L., Prakash, A., Mitzner, T.L., Kemp, C.C., Rogers, W.A., 2012). The domesticated robot: design guidelines for assisting older adults to age in place. Proceedings of the 7th ACM/IEEE International Conference on Human-Robot Interaction (HRI). ACM/IEEE, pp. 335–342.
Beer, J.M., Smarr, C.A., Fisk, A.D., Rogers, W.A., 2015. Younger and older users' recognition of virtual agent facial expressions. Int. J. Hum. Comput. Stud. 75, 1–20.
Billson, J.M., Tiberius, R.G., 1991. Effective social arrangements for teaching and learning. New Dir. Teach. Learn. 45, 87–109.
Blair, C., 2002. School readiness: integrating cognition and emotion in a neurobiological conceptualization of children's functioning at school entry. Am. Psychol. 57 (2), 111.
Breazeal, C., 2003. Emotion and sociable humanoid robots. Int. J. Hum. Comput. Stud. 59, 119–155.
Breazeal, C., 2004. Social interactions in HRI: the robot view. IEEE Trans. Syst. Man Cybern. C Appl. Rev. 34 (2), 181–186.
Broadbent, E., Tamagawa, R., Patience, A., Knock, B., Kerse, N., Day, K., MacDonald, B.A., 2012. Attitudes towards health-care robots in a retirement village. Australas. J. Ageing 31 (2), 115–120.
Brown, L.N., Howard, A.M., 2013. Engaging children in math education using a socially interactive humanoid robot. Proceedings of the 13th IEEE-RAS International Conference on Humanoid Robots (Humanoids). pp. 183–188.
Bugmann, G., Copleston, S., 2011. What can a personal robot do for you? Towards autonomous robotic systems. 12th Annual Conference. Springer, pp. 360–371.
Burke, J.L., Murphy, R.R., Rogers, E., Lumelsky, V.J., Scholtz, J., 2004. Final report for the DARPA/NSF interdisciplinary study on human-robot interaction. IEEE Trans. Syst. Man Cybern. C Appl. Rev. 34 (2), 103–112.
Castillo, R., Fernandez-Berrocal, P., Brackett, M., 2013. Enhancing teacher effectiveness in Spain: a pilot study of the RULER approach to social and emotional learning. J. Educ. Training Stud. 1 (2), 263–272.
Center for Disease and Control and Prevention (CDC), 2015a. Alzheimer's disease. http://www.cdc.gov/aging/aginginfo/alzheimers.htm
Center for Disease and Control (CDC), 2015b. Autism spectrum disorder. http://www.cdc.gov/ncbddd/autism/index.html
Chang, W.L., Šabanovic, S., Huber, L., 2013. Use of seal-like robot PARO in sensory group therapy for older adults with dementia. Proceedings of the 8th ACM/IEEE International Conference on Human-Robot Interaction. IEEE Press, pp. 101–102.
Churchill, M., Safaoui, J., McCabe, B.W., Baun, M.M., 1999. Using a therapy dog to alleviate the agitation and desocialization of people with Alzheimer's disease. J. Psychosoc. Nurs. Ment. Health Serv. 37 (4), 16–22.
Dautenhahn, K., 2007a. Methodology & themes of human-robot interaction: a growing research field. Int. J. Adv. Robot. Syst. 4 (1), 103–108.
Dautenhahn, K., 2007b. Socially intelligent robots: dimensions of human–robot interaction. Philos. Trans. Royal Soc. B Biol. Sci. 362 (1480), 679–704.
Dautenhahn, K., Woods, S., Kaouri, C., Walters, M.L., Koay, K.L., Werry, I., 2005. What is a robot companion-friend, assistant or butler? 2005 IEEE/RSJ International Conference on Intelligent Robots and Systems (IROS 2005). IEEE/RSJ, pp. 1192–1197.

Davis, F.D., 1989. Perceived usefulness, perceived ease of use, and user acceptance of information technology. MIS Quart. 13 (3), 319–340.

Dobriansky, P.J., Suzman, R.M., Hodes, R.J., 2007. Why population aging matters: a global perspective. National Institute on Aging, National Institutes of Health, US Department of Health and Human Services, US Department of State. https://www.nia.nih.gov/research/publication/why-population-aging-matters-global-perspective

Eberhardt, J.L., Davies, P.G., Purdie-Vaughns, V.J., Johnson, S.L., 2006. Looking deathworthy: perceived stereotypicality of black defendants predicts capital-sentencing outcomes. Psychol. Sci. 17 (5), 382–386.

Ekman, P., Friesen, W.V., 1975. Unmasking the face. Prentice-Hall, Englewood Cliffs, NJ.

Ekman, P., Friesen, W.V., 1978. Facial Action Coding System: A Technique for the Measurement of Facial Movement. Consulting Psychologists Press, Palo Alto, CA.

Erber, J.T., 2011. Aging and Older Adulthood. John Wiley & Sons.

Fabri, M., Moore, D.J., Hobbs, D.J., 2004. Mediating the expression of emotion in educational collaborative virtual environments: an experimental study. Int. J. Virtual Real. 7 (2), 66–81.

Farber, N., Shinkle, D., Lynott, J., Fox-Grage, W., Harrell, R., 2011. Aging in place: a state survey of livability policies and practices. A Research Report by the National Conference of State Legislatures and the AARP Public Policy Institute. http://www.aarp.org/home-garden/livable-communities/info-11-2011/Aging-In-Place.html

Feil-Seifer, D., Mataric, M.J., 2005. Defining socially assistive robotics. IEEE 9th International Conference on Rehabilitation Robotics. pp. 465–468.

Ficklin, F.L., Browne, V.L., Powell, R.C., Carter, J.E., 1988. Faculty and house staff members as role models. J. Med. Educ. 63 (5), 392–396.

Fior, M., Nugent, S., Beran, T., Ramirez-Serrano, A., Kuzyk, R., 2010. Children's relationships with robots: Robot is child's new friend. J. Phys. Agents 4 (3), 9–17.

Fisk, A.D., Rogers, W.A., Charness, N., Czaja, S.J., Sharit, J., 2009. Designing for Older Adults: Principles and Creative Human Factors Approaches, second ed. CRC Press, Boca Raton, FL.

Forlizzi, J., 2007. How robotic products become social products: an ethnographic study of cleaning in the home. Proceedings of the 2nd ACM/IEEE International Conference on Human-Robot Interaction.(HRI).

Forlizzi, J., DiSalvo, C., 2006. Service robots in the domestic environment: a study of the roomba vacuum in the home. Proceedings of the 1st ACM SIGCHI/SIGART Conference on Human-Robot Interaction (HRI).

Goetz, J., Kiesler, S., Powers, A., 2003. Matching robot appearance and behavior to tasks to improve human-robot cooperation. Proceedings of the 2003 IEEE International Workshop on Robot and Human Interaction Communication (RO-MAN). pp. 55–60.

Goodrich, M.A., Schultz, A.C., 2007. Human-robot interaction: a survey. Foundations Trends Hum. Comput. Interact. 1 (3), 203–275.

Graziano, P.A., Reavis, R.D., Keane, S.P., Calkins, S.D., 2007. The role of emotion regulation in children's early academic success. J. Sch. Psychol. 45 (1), 3–19.

Hashimoto, T., Kobayashi, H., Polishuk, A., Verner, V., 2013. Elementary science lesson delivered by robot. Proceedings of the 8th ACM/IEEE International Conference on Human-Robot Interaction. pp. 133–134.

Hebert, L.E., Weuve, J., Scherr, P.A., Evans, D.A., 2013. Alzheimer disease in the United States (2010-2050) estimated using the 2010 census. Neurology 80 (19), 1778–1783.

Heerink, M., Kröse, B., Evers, V., Wielinga, B., 2007. Observing conversational expressiveness of elderly users interacting with a robot and screen agent. Proceedings of the IEEE 10th International Conference on Rehabilitation Robotics. (ICORR 2007). June 13–15.

Heerink, M., Kröse, B., Evers, V., Wielinga, B., 2010a. Assessing acceptance of assistive social agent technology by older adults: the Almere model. Int. J. Soc. Robot. 2 (4), 361–375.

Heerink, M., Kröse, B., Evers, V., Wielinga, B., 2010b. Relating conversational expressiveness to social presence and acceptance of an assistive social robot. Virtual Real. 14, 77–84.

Jeong, S., Logan, D.E., Goodwin, M.S., Graca, S., O'Connell, B., Goodenough, H., Anderson, L., Stenquist, N., Fitzpatrick, K., Zisook, M., Plummer, L., Breazeal, C., Weinstock, P., 2015. A social robot to mitigate stress, anxiety, and pain in hospital pediatric care. Proceedings of the Tenth Annual ACM/IEEE International Conference on Human-Robot Interaction Extended Abstracts. pp. 103–104.

Kanamori, M., Suzuki, M., Yamamoto, K., Kanda, M., Matsui, Y., Kojima, E., Fukawa, H., Sugita, T., Oshiro, H., 2001. A day care program and evaluation of animal-assisted therapy (AAT) for the elderly with senile dementia. Am. J. Alzheimers Dis. Other Demen. 16 (4), 234–239.

Kanda, T., Miyashita, T., Osada, T., Haikawa, Y., Ishiguro, H., 2008. Analysis of humanoid appearances in human-robot interaction. IEEE Trans. Rob. 24 (3), 725–735.

Katagiri Y., Nass C., Takeuchi Y., 2001. Cross-cultural studies of the computers are social actors paradigm: the case of reciprocity. Ninth International Conference on HHCI. pp. 1558–1562.

Kory, J., Breazeal, C., 2014. Storytelling with robots: Learning companions for preschool children's language development. Proceedings of the 23rd IEEE International Symposium on Robot and Human Interactive Communication (RO-MAN). pp. 643–648.

Lasseter, J., 1987. Principles of traditional animation applied to 3D computer animation. Comput. Graph. 21 (4), 35–44.

Levin, D.M., Hammer, D., Coffey, J.E., 2009. Novice teachers' attention to student thinking. J. Teach. Educ. 60 (2), 142–154.

Libin, A., Cohen-Mansfield, J., 2004. Therapeutic robocat for nursing home residents with dementia: preliminary inquiry. Am. J. Alzheimers Dis. Other Demen. 19 (2), 111–116.

Liles, K.R., Beer, J.M., 2015a. Ms. An, feasibility study with a robot teaching assistant. Proceedings of the Tenth Annual ACM/IEEE International Conference on Human-Robot Interaction Extended Abstracts. Portland, Oregon, pp. 83–84.

Liles, K.R., Beer, J.M., 2015b. Rural minority students' perceptions of ms. an, the robot teaching assistant, as a social teaching tool. Proc. Hum. Factors Ergon. Soc. Annu. Meet. 59 (1), 372–376.

Looije, R., Neerincx, M.A., Cnossen, F., 2010. Persuasive robotic assistant for health self-management of older adults: design and evaluation of social behaviors. Int. J. Hum. Comput. Stud. 68 (6), 386–397.

Mast, M., Burmester, M., Krüger, K., Fatikow, S., Arbeiter, G., Graf, B., Kronreif, G., Pigini, L., Facal, D., Qiu, R., 2012. User-centered design of a dynamic-autonomy remote interaction concept for manipulation-capable robots to assist elderly people in the home. J. Hum. Robot. Interact. 1 (1), 96–118.

Montemerlo, M., Pineau, J., Roy, N., Thrun, S., Verma, V., 2002. Experiences with a mobile robotic guide for the elderly. Proceedings of the AAAI National Conference on Artificial Intelligence. pp. 587–592.

Moon, Y., Nass, C., 1996. How "real" are computer personalities? Psychological responses to personality types in human-computer interaction. Commun. Res. 23 (6), 651–674.

Mori, M., 1970. Bukimi no tani. Energy 7 (4), 33-35. (K.F. MacDorman, T. Minato, Trans. Proceedings of the Humanoids–2005 Workshop: Views of the Uncanny Valley, Tsukuba, Japan).

Moyle, W., Cooke, M., Beattie, E., Jones, C., Klein, B., Cook, G., Gray, C., 2013. Exploring the effect of companion robots on emotional expression in older adults with dementia: a pilot randomized controlled trial. J. Gerontol. Nurs. 39 (5), 46–53.

Murphy, R.R., 2000. Introduction to AI Robotics. MIT Press, Cambridge, MA.

Nass, C., Moon, Y., 2000. Machines and mindlessness: social responses to computers. J. Soc. Issues 56 (1), 81–103.

Nass, C., Yen, C., 2010. The Man Who Lied to His Laptop: What We Can Learn About Ourselves From our Machines. Penguin, New York, NY.

Nass, C., Steuer, J., Henriksen, L., Dryer, D.C., 1994. Machines, social attributions, and ethopoeia: performance assessments of computers subsequent to"self-" or "other-" evaluations. Int. J. Hum. Comput. Stud. 40 (3), 543–559.

Nass, C., Moon, Y., Fogg, B.J., Reeves, B., Dryer, D.C., 1995. Can computer personalities be human personalities? Int. J. Hum. Comput. Stud. 43 (2), 223–239.
Nass, C., Fogg, B.J., Moon, Y., 1996. Can computers be teammates? Int. J. Hum. Comput. Stud. 45 (6), 669–678.
Payrató, L., 2009. Non-verbal communication. In: Verschueren, J., Östman, J. (Eds.), Key Notions for Pragmatics. John Benjamins, Amsterdam, The Netherlands, pp. 163–194.
Picard, R.W., 1997. Affective Computing. MIT Press, Cambridge, MA.
Prakash, A., Rogers, W.A., 2015. Why some humanoid faces are perceived more positively than others: effects of human-likeness and task. Int. J. Soc. Robot. 7 (2), 309–331.
Probo., 2015. Paro, a huggable robotic friend. http://probo.vub.ac.be/
Rogers, Y., Sharp, H., Preece, J., 2011. Interaction Design: Beyond Human Computer Interaction, third ed. John Wiley & Sons Ltd, Chichester, United Kingdom.
Russell, J.A., Pratt, G., 1980. A description of the affective quality attributed to environments. J. Pers. Soc. Psychol. 38 (2), 311–322.
Šabanović, S., 2010. Robots in society, society in robots: mutual shaping of society and technology as a framework for social robot design. Int. J. Soc. Robot. 2, 439–450.
Šabanović, S., 2014. Inventing Japan's 'robotics culture': the repeated assembly of science, technology, and culture in social robotics. Soc. Stud. Sci. 44 (3), 342–367.
Saerbeck, M., Schut, T., Bartneck, C., Janse, M.D., 2010. Expressive robots in education: varying the degree of social supportive behavior of a robotic tutor. Proceedings of the SIGCHI Conference on Human Factors in Computing Systems. pp. 1613–1622.
Saint-Aimé, S., Le-Pévédic, B., Duhaut, D., Shibata, T., 2007. EmotiRob: companion robot project. Proceedings of the 16th IEEE International Symposium on Robot and Human Interactive Communication (RO-MAN). pp. 919–924
Scasselati, B., Admoni, H., Mataric, M., 2012. Robots for use in autism research. Annu. Rev. Biomed. Eng. 14, 275–294.
Sellers, D.M., 2006. The evaluation of an animal assisted therapy intervention for elders with dementia in long-term care. Act. Adapt. Aging 30 (1), 61–77.
Short, E., Swift-Spong, K., Greczek, J., Ramachandran, A., Litoiu, A., Grigore, E.C., Feil-Seifer, D., Shuster, S., Lee, J., Huang, S., Levonnisova, S., Li, J., Ragusa, G., Spruijt-Metz, D., Mataric, M., Scassellati, B., 2014. How to train your DragonBot: Socially assistive robots for teaching children about nutrition through play. Proceedings of the 23rd IEEE International Symposium on Robot and Human Interactive Communication (RO-MAN). pp. 924–929.
Simut, R., Pop, C., Vanderborght, B., Saldien, J., Rusu, A., Pintea, S., Vanderfaeillie, J., Lefeber, D., David, D., 2011. The huggable social robot probo for social story telling for robot assisted therapy with ASD children. Proceedings of the 3rd International Conference on Social Robotics (ICSR 2011). pp. 97–100.
Smarr, C.A., Mitzner, T.L., Beer, J.M., Prakash, A., Chen, T.L., Kemp, C.C., Rogers, W.A., 2014. Domestic robots for older adults: attitudes, preferences, and potential. Int. J. Soc. Robot. 6 (2), 229–247.
Stiehl, W.D., Lee, J.K., Breazeal, C., Nalin, M., Morandi, A., Sanna, A., 2009. The huggable: a platform for research in robotic companions for pediatric care. Proceedings of the 8th International Conference on Interaction Design and Children. pp. 317–320.
Steinfeld, A., Fong, T., Kaber, D., Lewis, M., Scholtz, J., Schultz, A., Goodrich, M., 2006. Common metrics for human-robot interaction. In: Proceedings of the 1st ACM/IEEE International Conference on Human-Robot Interaction (HRI). ACM/IEEE, pp. 33-40.
Sung, J.-Y., Grinter, R.E., Christensen, H.I., 2010. Domestic robot ecology: an initial framework to unpack long-term acceptance of robots at home. Int. J. Soc. Robot. 2 (4), 417–429.
Takayama, L., Dooley, D. Ju, W., 2011. Expressing thought: Improving robot readability actions with animation principles. Proceedings of the 6th ACM/IEEE International Conference on Human-Robot Interaction (HRI). pp. 69–76.
Tapus, A., Matarić, M.J., Scassellati, B., 2007. The grand challenges in socially assistive robotics. IEEE Rob. Autom. Mag. 14 (1), 35–42, (special issue on grand challenges in robotics).

Thrun, S., 2004. Toward a framework for human-robot interaction. Hum Comput. Interact. 19 (1), 9–24.
Tiberius, R., 1993. The why of teacher/student relationships. Essays on Teaching Excellence (POD) 5 (8), http://podnetwork.org/content/uploads/V5-N8-Tiberius.pdf.
Tiberius, R.G., Billson, J.M., 1991. The social context of teaching and learning. New Dir. Teach. Learn. 45, 67–86.
Tielman, M., Neerincx, M., Meyer, J.J., Looije, R., 2014. Adaptive emotional expression in robot-child interaction. Proceedings of the 2014 ACM/IEEE International Conference on Human-Robot interaction. pp. 407-414.
Todorov, A., Mandisodza, A.N., Goren, A., Hall, C.C., 2005. Inferences of competence from faces predict election outcomes. Science 308 (5728), 1623–1626.
Torta, E., Werner, F., Johnson, D.O., Juola, J.F., Cuijpers, R.H., Bazzani, M., Berzaucher, J., Lemberger, J., Lewy, H., Bregman, J., 2014. Evaluation of a small socially-assistive humanoid robot in intelligent homes for the care of the elderly. J. Intell. Rob. Syst. 76 (1), 57–71.
Venkatesh, V., Morris, M.G., Davis, G.B., Davis, F.D., 2003. User acceptance of information technology: toward a unified view. MIS Quart. 27 (3), 425–478.
Veryzer, Jr., R.W., Hutchinson, J.W., 1998. The influence of unity and prototypicality on aesthetic responses to new product designs. J. Cons. Res. 24 (4), 374–394.
Wada, K., Shibata, T., Saito, T., Tanie, K., 2004. Effects of robot-assisted activity for elderly people and nurses at a day service center. Proceedings of the IEEE. pp. 1780–1788.

Further Reading

Bemelmans, R., Gelderblom, G.J., Jonker, P., de Witte, L., 2012. Socially assistive robots in elderly care: a systematic review into effects and effectiveness. J. Am. Med. Dir. Assoc. 13 (2), 114–120.
Bennewitz, M., Burgard, W., Cielniak, G., Thrun, S., 2005. Learning motion patterns of people for compliant robot motion. Int. J. Rob. Res. 24, 31–48.
Bickmore, T., Caruso, L., Clough-Gorr, K., Heeren, T., 2005. It's just like you talk to a friend' relational agents for older adults. Interact. Comput. Interdiscipl. J. Hum. Comput. Interact. 17 (6), 711–735, (special issue: HCI and the older population).
Chen, T.L., Kemp, C.C., 2010. Lead me by the hand: evaluation of a direct physical interface for nursing assistant robots. Proceedings of the 5th ACM/IEEE International Conference on Human-Robot Interaction
Eisenberg, N., Miller, P.A., 1987. The relation of empathy to prosocial and related behaviors. Psychol. Bull. 101 (1), 91.
Fasola, J., Mataric, M., 2013. A socially assistive robot exercise coach for the elderly. J. Hum. Robot. Interact. 2 (2), 3–32.
Fong, T., Nourbakhsh, I., Dautenhahn, K., 2003. A survey of socially interactive robots. Rob. Auton. Syst. 42, 143–166.
Friedman, B., Kahn, Jr., P.H., 2002. Human values, ethics, and design. In: Jacko, J., Sears, A. (Eds.), The Human-Computer Interaction Handbook: Fundamentals, Evolving Technologies and Emerging Applications. Lawrence Erlbaum Associates, New York, NY, pp. 1177–1201.
Glover, J., Holstius, D., Manojlovich, M., Montgomery, K., Powers, A., Wu, J., Kiesler S., Matthews, J., Thrun, S., 2003. A Robotically-Augmented Walker for Older Adults (CMU-CS Report 03-170). Carnegie Mellon University, School of Computer Science Technical Report Collection: http://reports-archive.adm.cs.cmu.edu/anon/2003/abstracts/03-170.html
Hossain, M.A., 2014. Perspectives of human factors in designing elderly monitoring system. Comput. Hum. Behav. 33, 63–68.
Loper, M.M., Koenig, N.P., Chernova, S.H., Jones, C.V., Jenkins, O.C., 2009. Mobile human-robot teaming with environmental tolerance. Proceedings of the 4th ACM/IEEE International Conference on Human-Robot Interaction (HRI). ACM/IEEE, pp. 157–164.

Pang, W.C., Seet, G., Yao, X., 2014. A study on high-level autonomous navigational behaviors for telepresence applications. Presence (Camb) 23 (2), 155–171.
Russell, S., Norvig, P., 2003. Artificial Intelligence: A Modern Approach, second ed. Pearson, Upper Saddle River, NJ.
Sharkey, A., Sharkey, N., 2012. Granny and the robots: ethical issues in robot care for the elderly. Ethics Inf. Technol. 14 (1), 27–40.
Thrun, S., Bennewitz, M., Burgard, W., Cremers, A.B., Dellaert, F., Fox, D., Haehnel, D., Lakemeyer, G., Rosenberg, C., Roy, N., Schulte, J., Schulz, D., Steiner, W., 1999. Experiences with two deployed interactive tour-guide robots. Proceedings of the International Conference on Field and Service Robotic.
Torta, E., Cuijpers, R.H., Juola, J., van der Pol, D., 2011. Design of robust robotic proxemic behaviour. In: B., Mutlu, C., Bartneck, J., Ham, V., Evers, T., Kanda, (Eds.), International Conference on Social Robotics, vol. 7072. pp. 21–30.
Torta, E., Cuijpers, R., Juola, J., Van der Pol, D., 2012. Modeling and testing proxemic behavior for humanoid robots. Int. J. Humanoid Rob. 9 (4), 1–24.
Waldherr, S., Romero, R., Thrun, S., 2000. A gesture based interface for human-robot interaction. Auton. Robots 9 (2), 151–173.

CHAPTER

16

Computational Modeling of Cognition–Emotion Interactions: Theoretical and Practical Relevance for Behavioral Healthcare

Eva Hudlicka

Psychometrix Associates, College of Information and Computer Sciences, University of Massachusetts-Amherst, Amherst, MA, United States

INTRODUCTION

Recent years have witnessed an increasing interest in developing computational models of emotion and emotion–cognition interaction, within affective computing and the emerging area of computational affective science. Models of emotion–cognition interactions are typically developed in the context of modeling emotion generation via cognitive appraisal (e.g., Becker-Asano, 2008), typically based on the conceptual framework developed by Ortony, Clore and Collins (referred to as OCC) (Ortony et al., 1988). A few models have been developed that aim to model the effects of emotions on cognition (e.g., Hudlicka, 1998, 2003, 2008b) and some efforts exist that aim to model emotion regulation (Martinez-Miranda et al., 2015a). The majority of emotion models developed to date are "applied models," in that their primary objective is to increase affective realism and affective and social competence in virtual agents or social robots [e.g., the Wasabi architecture (Becker-Asano, 2008)]. A small number of models are "research models," whose objective is to elucidate the mechanisms mediating some aspect of emotion–cognition interactions; for example, modeling the mechanisms mediating affective biases (Hudlicka, 1998, 2008b).

Emotions and Affect in Human Factors and Human–Computer Interaction. http://dx.doi.org/10.1016/B978-0-12-801851-4.00016-1
Copyright © 2017 Elsevier Inc. All rights reserved.

At the same time, emotion theorists and clinical psychologists have begun to recognize the importance of moving beyond descriptive characterizations of psychopathology, and identifying the underlying mechanisms that mediate the etiology and maintenance of affective disorders, and developing mechanism-based approaches to diagnosis and treatment of affective disorders (e.g., Macleod and Mathews, 2012). Such mechanism-based understanding supports more accurate assessment and more differentiated and nuanced diagnosis of affective disorders, and provides a basis for more efficient and targeted approaches to treatment that directly target the underlying maladaptive or biased processes.

This trend is particularly evident in the recently emerging emphasis on the transdiagnostic models of psychopathology, and the associated Research Domain Criteria (RDoC) framework (Sanislow et al., 2010), that aim to shift from descriptive approaches to more mechanism- and first-principles oriented approaches to diagnosis, assessment and treatment of affective disorders (e.g., Kring, 2008; Nolen-Hoeksema and Watkins, 2011).

Computational models of cognition–emotion interactions within the context of *emotion generation, emotion effects on cognition,* and *emotion regulation,* have the potential to significantly contribute to this effort, by providing the ability to construct simulation-based models of hypothesized mechanisms that mediate specific affective disorders. Computational models have the additional advantage of requiring a degree of operationalization of psychological theories that helps identify gaps and inconsistencies, and providing a means of generating hypotheses for further empirical investigations.

This chapter discusses the state-of-the-art in modeling emotion–cognition interaction and the relevance of these models for understanding the mechanisms mediating psychopathology and therapeutic action. The chapter also discusses how these models can support the development of advanced behavioral healthcare technologies, which directly target the maladaptive processes, such as serious therapeutic games and virtual reality environments. Augmenting these technologies with models of cognition–emotion interaction would enhance their ability to accurately assess affective functioning and psychopathology, would facilitate the development and administration of individualized treatment protocols, and would provide a basis for more accurate assessments of outcomes.

The chapter is organized as follows. Section "Background Information" provides background information about relevant research in emotion, overview of affective disorders, brief summary of contemporary theories of therapeutic action, that is, the hypothesized mechanisms whereby symptoms of affective disorders are reduced via psychotherapy, and an overview of serious therapeutic games. Section "State-of-the-Art in Modeling Emotion-Cognition Interaction" discusses the state-of-the-art in computational models of emotion–cognition interaction, focusing on symbolic models. Section "Modeling Mechanisms of Psychopathology and Therapeutic Action " describes an example of a symbolic model capable of modeling the mechanisms mediating affective disorders and how this model could be augmented to model mechanisms of therapeutic action. Section "Potential of Model-Enhanced Therapeutic Games" discusses how these models could be integrated into serious therapeutic games and virtual reality treatment environments, to support enhanced assessment and diagnosis and customized treatment protocols. Section "Summary and Conclusions" concludes with a discussion of emerging ethical issues and challenges and open questions.

BACKGROUND INFORMATION

This section provides a brief overview of the research and technologies applicable for the development of computational models of emotion–cognition interactions and their relevance to an improved understanding of the mechanisms of affective disorders and therapeutic action, as well as their applications in the development of behavioral healthcare technologies.

Emotion Research

Given the significant activity and progress in emotion research over the past two decades, it would be impossible provide a comprehensive review of the state-of-the-art. Following a brief summary of some of the key characteristics of emotions and their roles, I therefore discuss the developments directly relevant to computational modeling of emotion–cognition interactions, focusing on the three dominant theoretical perspectives relevant to symbolic computational modeling of emotions, current theories regarding emotion generation, and examples of theories and data regarding the influences of emotions on cognitive processes.

Definition, Characteristics and Roles of Emotions

Definitions

When searching for a definition of emotions, it is interesting to note that most definitions involve descriptions of characteristics (e.g., fast, undifferentiated processing) or roles and functions (e.g., coordinating mechanisms for goal management in uncertain environments, communicative mechanisms for facilitating social interaction, hardwired responses to critical stimuli). The fact that we so often describe emotions in terms of their characteristics, rather than their essential nature, underscores our lack of understanding of these complex phenomena. Nevertheless, most emotion researchers would agree on a high level, working definition that views emotions as *states that reflect evaluative judgments of the environment, the self and other agents, in light of the agent's goals and beliefs, and which motivate and coordinate adaptive behavior* (e.g., Ekman, 1992; Frijda, 1986; Ortony et al., 1988). Note that the terms "goals" and "beliefs" are used in a generic sense: goals reflecting desirable states, and beliefs reflecting current knowledge. The term "goal" in this discussion therefore subsumes desired states and preferences, and includes conscious and unconscious, explicit or implicit, and innate or learned goals and preferences.

Multiple Modalities of Emotions

A characteristic feature of emotions is their multimodal nature. Emotions in biological agents are manifested across four distinct and interacting modalities. The most visible is the *behavioral/expressive* modality, where the expressive and action-oriented characteristics are manifested; for example, facial expressions, speech, gestures, posture, and behavioral choices. Closely related is the *somatic/physiological modality*—the neurophysiological substrate making behavior (and, of course, cognition) possible (e.g., physiological changes mediated by the autonomic nervous system and the neuroendocrine system). The *cognitive/interpretive* modality is most directly associated with the evaluation-based definition provided above, and emphasized in the current cognitive appraisal theories of emotion generation. The most problematic modality, from a modeling perspective, is the *experiential/subjective* modality: the conscious, and inherently idiosyncratic, experience of emotions within the individual.

While the current emphasis in emotion modeling is on the cognitive modality (in models of emotion generation) and the behavioral/expressive modality (in models of emotion effects), it is important to recognize that both the physiological and the experiential modalities also play critical roles (Izard, 1993). Recently, there has also been a renewed interest in the role of somatic modality of emotions, reflected in the emerging literature on embodied emotions (e.g., Niedenthal and Marcus, 2009).

Emotion Roles

The dominant contemporary view regarding the evolution and utility of emotions emphasizes their primary role as ensuring survival and improving the "adaptive fit" of an individual within his/her environment. This is accomplished by emotion-mediated rapid detection of survival-critical cues, including social cues, and by rapid preparation for the necessary behavioral responses, including expression of emotions for communication and social coordination (Ekman and Davidson, 1994; Frijda, 2008; Plutchik, 1984). Distinct emotions are linked to different patterns of stimuli and associated desired behavior, their "action tendencies" (Frijda 1986). Emotions can, however, also become highly maladaptive, even dangerous, both to the individual experiencing or manifesting the emotions, and to other agents in his/her social environment. These pathological

TABLE 16.1 Summary of Intrapsychic and Interpersonal Roles of Emotions

Intrapsychic roles

- Rapid detection and processing of salient stimuli (e.g., avoid danger, get food)
- Triggering, preparation for, and execution of fixed behavioral patterns necessary for survival (e.g., fight, freeze, flee)
- Rapid resource (re)allocation and mobilization
- Coordination of multiple systems (perceptual, cognitive, physiological)
- Implementation of systemic biases biasing processing in a particular direction (e.g., threat detection, self-focus)
- Interruption of ongoing activity and (re)prioritization of goals
- Motivation of behavior via reward and punishment mechanisms
- Motivation of learning via boredom and curiosity

Interpersonal roles

- Communication of internal state via nonverbal expression and behavioral tendencies (e.g., frown versus smile, inviting versus threatening gestures and posture)
- Communication of status information in a social group (dominance and submissiveness)
- Mediation of attachment
- Communicate of acknowledgment of wrongdoing (guilt, shame) in an effort to repair relationships and reduce possibility of aggression

manifestations of emotions will be discussed in more detail in Section "Affective Disorders: A Transdiagnostic Perspective."

Emotion roles can be grouped into two broad categories: *intrapsychic and interpersonal* (Table 16.1). *Intrapsychic emotion roles* are functions that emotions perform within the mind/brain, to ensure homeostasis and survival, by motivating adaptive behavior, in part via "reward" and "punishment" processes. Emotions play critical roles in the *perception and interpretation* of incoming stimuli, by implementing biasing functions that facilitate cognitive processing (e.g., threat bias associated with fear), in mediating *motivation* and *learning*, in goal management and prioritization, and in preparation for, and coordination of, *behavior.*

Interpersonal (also termed social) roles of emotions facilitate social interaction and successful social functioning by rapidly, typically nonverbally, communicating internal states, and associated behavioral intents; for example, communication of rising levels of frustration via facial expressions, gestures, and posture helps regulate the degree of frustration and prevents open expression of aggression in social interactions. In their social roles, emotions also facilitate the development of relationships (e.g., empathy) and social structures necessary for the individual's well-being and survival, including attachment. The social roles of emotion are evident in groups of varying sizes and types, ranging from intimate dyads to large organizations and even entire nations, and across cultures and species.

Theoretical Perspectives

Given the complexity of affective phenomena, it is not surprising that a number of distinct types of theories have evolved over time, to explain a particular affective phenomenon, or to account for some observed data. Below I describe three theoretical perspectives that are most relevant for computational affective modeling: *discrete or categorical theories, dimensional theories, and componential theories.*

Discrete or Categorical Emotion Theories

Discrete/categorical theories of emotions emphasize a small set of discrete emotions. The underlying assumption of this approach is that these fundamental emotions are mediated by associated dedicated neural circuitry, with a large innate (hardwired) component. Different emotions are characterized by stable patterns of triggers, behavioral expression, and associated distinct subjective experiences. The emotions considered in these theories are typically the "basic" emotions, which include joy, sadness, fear, anger, and disgust. Due to its emphasis on discrete categories of states, this approach is also termed the *categorical approach* (Panskepp, 1998). For modeling purposes, the semantic primitives representing emotions in affective models are the distinct "basic" emotions.

Dimensional Emotion Theories

An alternative method of characterizing affective states is in terms of a small set of underlying dimensions that define a space within which distinct emotions can be located. This dimensional perspective describes emotions in terms of two or three dimensions. The most frequent dimensional characterization of emotions uses two dimensions: valence and arousal (Russell, 2003; Russell and Barrett, 1999). Valence reflects a positive or negative evaluation, and the associated felt state of pleasure (vs. displeasure), as outlined in the context of undifferentiated affect above. Arousal reflects a general degree of intensity or activation of the organism, reflecting in large part the degree of activation of the autonomic nervous system. The degree of arousal reflects a general readiness to act: low arousal is associated with less energy, high arousal with more energy. Since this two-dimensional space cannot differentiate among emotions that share the same values of arousal and valence (e.g., anger and fear, both characterized by high arousal and negative valence), a third dimension is often added, termed dominance or stance. The resulting three-dimensional space is often referred to as the PAD space (Mehrabian, 1995; Russell and Mehrabian, 1977) [pleasure (synonymous with valence), arousal, dominance]. The representational semantic primitives within this theoretical perspective are thus these two or three dimensions.

Componential Emotion Theories

The third view emphasizes the distinct components of emotions, and is often termed the componential view (Leventhal and Scherer, 1987). The "components" referred to in this view are both the distinct modalities of emotions (e.g., cognitive, physiological, behavioral, subjective) and also the components of the cognitive appraisal process. The latter are referred to as appraisal dimensions or appraisal variables, and include novelty, valence, certainty, urgency, goal relevance, goal congruence, and coping abilities. The representational semantic primitives within this theoretical perspective are the individual appraisal variables. The roles of these variables in emotion generation are discussed in more detail below.

These theories should not be viewed as competing for a single ground truth, but rather as distinct perspectives, each arising from a specific research tradition (e.g., biological vs. social psychology), focusing on different sets of affective phenomena, considering distinct levels of resolution and fundamental components (e.g., emotions vs. appraisal dimensions as the distinct components), and using different experimental methods (e.g., factor analysis of self-report data vs. neuroanatomical evidence for distinct processing pathways). The different perspectives also provide different degrees of support for the distinct processes mediating emotion; for example, the componential theories provide extensive details about cognitive appraisal. Until such time as emotions are better understood it is best to view the three theoretical perspectives as alternative explanations, each with its own set of explanatory powers and scope, and supporting data; analogously, perhaps, to the wave versus particle theory of light, as suggested by Picard (1997).

Emotion Generation

Emotion generation is an evolving, dynamic process that occurs across the multiple modalities, discussed previously, with complex feedback and interactions among them. While all modalities are involved, contemporary understanding of these phenomena emphasizes the cognitive modality and emotion generation via cognitive appraisal. Cognitive appraisal involves the evaluation of the match between the agent's goals, needs and preferences and its current situation, which may further be categorized into events, actions by other agents and particular objects. The subsequent discussion is therefore limited to cognitive appraisal, recognizing that the current cognitive bias may well be an example of "looking for the key under the lamp because there is light there." For information about the role of noncognitive modalities in emotion generation the reader is referred to a summary of the emerging literature on embodied emotions that discusses the role of the somatic modality in affective processing (Niedenthal and Marcus, 2009), as well as an excellent summary of the state-of-the-art in understanding the fundamental processes mediating emotions by LeDoux (2014).

All cognitive appraisal theories emphasize the critical role that cognition plays in generating the subjective emotional experience, by mediating the interpretations required for the evaluative judgments involved in generating emotion. Appraisal theories have their roots in antiquity and have undergone a number of iterations since then. Many researchers over the past four decades have contributed to the current versions of cognitive appraisal theories (e.g., Arnold, 1960; Frijda, 1986; Lazarus, 1984; Mandler, 1984; Ortony et al., 1988; Roseman and Smith, 2001; Scherer et al., 2001; Smith and Kirby, 2001).

Cognitive appraisal theories provide a set of domain-independent evaluative criteria that capture the current interpretation of the agent's internal and external environment, as it pertains to the agent's current goals. The theories differ in terms of the specific criteria provided, their structure and organization [e.g., hierarchy (OCC) versus list (e.g., Scherer)], as well as in the number and type of processing stages involved in the appraisal process.

For example, the evaluative criteria defined within the componential theoretical perspective (e.g., Reisenzein, 2001; Roseman, 2001; Scherer, 2001; Smith and Kirby, 2001), and discussed previously, include novelty, valence, goal relevance and goal congruence, urgency, certainty, responsible agent, coping potential, and individual and social norms. A stimulus, whether real, recalled or imagined, is evaluated in terms of its meaning and consequences for the agent, to determine the appropriate affective reaction. This analysis involves assigning specific values to the individual appraisal variables. Once the appraisal variable values are determined by the agent's evaluative processes, the resulting vector is mapped onto a particular emotion, within the n-dimensional space defined by the set of appraisal variables.

A related and overlapping set of evaluative criteria has been proposed by Ortony and colleagues, in their seminal theory of the "cognitive structure of emotions" (Ortony et al., 1988). Their OCC model of appraisal provides a rich taxonomy of triggers and resulting emotions, and emphasizes fundamental distinctions among three types of triggers, and corresponding types of emotions: *events* (event-based emotions, such as desirability, hope, fear), *acts by other agents* (attribution emotions, such as anger), and *characteristics of objects* (attraction emotions, such as like, dislike). An OCC-based appraisal model proceeds through a sequence of evaluation steps as it classifies a trigger within this taxonomy, eventually generating a specific emotion.

Appraisal theorists recognize that appraisal processes vary in complexity and cognitive involvement, from: low-level, "hardwired," to complex, culture-specific, and idiosyncratic triggers. Three interconnected levels are typically proposed: sensorimotor, schematic, and conceptual. [Similar trilevel organization has also been proposed for cognitive-affective architectures in general (Ortony et al., 2005; Sloman et al., 2005)].

Emotion Effects on Cognition

Emotions exert profound influences on cognition in biological agents, influencing both the fundamental processes mediating information processing (attention, perception, memory), but also higher level processes, including situation assessment, decision-making, goal management, planning, and learning. Emotion effects, including affective decision biases and heuristics, can be adaptive or maladaptive, depending on their type, magnitude and context. For example, anxiety and fear are associated with preferential processing of threatening stimuli. This bias can be adaptive in situations where survival depends on fast detection of danger and protective behavior (e.g., avoid an approaching car that has swerved into your lane). The same bias can be maladaptive if neutral stimuli are judged to be threatening (e.g., a passing car is assumed to be on a collision course and causes the driver to swerve into a ditch), or if the threat level of a stimulus is exaggerated.

A number of emotion effects and biases has been identified by cognitive psychologists and emotion researchers, including in the context of psychopathology research (Table 16.2). For example, positive emotions induce a global focus and the use of heuristics, whereas negative emotions induce a more local focus and analytical thinking (Gasper and Clore, 2002); anxiety reduces attentional and working memory capacities, biases attention toward the detection of threatening stimuli, and biases interpretive processes toward higher threat assessments, anxiety also induces a self-bias; mood induces mood-congruent biases in recall, and negative affect reduces estimates of control, and induces more analytical thinking (Isen, 1993; Macleod and Mathews, 2012; Mineka et al., 2003).

Theories of emotion effects on cognition are not as well defined as those for emotion generation. While the dominant theoretical perspective on emotion generation is cognitive appraisal, and reasonable agreement exists among researchers regarding the evaluative criteria used by the appraisal processes, no single theoretical perspective exists regarding the effects of emotions on cognition.

Distinct theories have been proposed to explain specific observed effects. Examples include the theory of mood-congruent recall (material retrieved from memory matches the affective tone of current emotion) (Bower, 1981), Forgas' AIM theory addressing the influence of emotions on judgment and decision-making (Forgas, 1999), and influence of emotions on distinct cognitive and perceptual processes (Derryberry and Reed, 2002). The different theories emphasize different components of the information-processing apparatus (e.g., attention, memory, automatic vs. controlled processes) and researchers often group affective influences

TABLE 16.2 Effect of Emotions on Attention, Perception, and Decision-Making: Examples of Empirical Findings

Anxiety and attention	**Anger and attention, perception, decision-making and behavior**
(Williams et al., 1997; Mineka and Sutton, 1992) Narrowing of attentional focus Reduced responsiveness to peripheral cues Predisposing toward detection of threatening stimuli	Increases feelings of certainty, control, and ability to cope Induces shallow, heuristic thinking Induces hostile attributions to others' motives and behavior Induces an urge to act
Arousal and attention	**Affective state and memory**
(Edland, 1989) Faster detection of threatening cues Slower detection of nonthreatening cues	(Bower, 1981; Blaney, 1986) Mood-congruent memory phenomenon
Positive affect and problem solving	**Negative affect and perception, problem-solving, decision-making**
(Isen, 1993; Clore, 1994; Kahn and Isen, 1993; Mellers et al. 1998; Gasper and Clore, 2002) Promotes heuristic processing (Clore, 1994) increases likelihood of stereotypical thinking, unless held accountable for judgments (Mellers et al., 1998) increases estimates of degree of control Overestimation of likelihood of positive events/underestimation of likelihood of negative events increases problem solving Facilitation of information integration Promotes variety seeking Promotes less anchoring, more creating problem—solving (Mellers et al., 1998) Longer deliberation, use of more information, more reexamination of information Promotes focus on "big picture"	(Williams et al., 1997; Gasper and Clore, 2002) Depression lowers estimates of degree of control Anxiety predisposes toward interpretation of ambiguous stimuli as threatening Use of simpler decision strategies Use of heuristics and reliance on standard and well-practiced procedures Decreased search behavior for alternatives Faster but less discriminate use of information—increased choice accuracy on easy tasks but decreased on more difficult tasks Simpler decisions and more polarized judgments Increased self-monitoring Promote focus on details

into different categories, based on the cognitive structures and processes that are affected. Next, we discuss two types of theories that have been proposed as possible mechanisms mediating emotion effects on cognition: *spreading activation models* and *parameter-based models*.

Spreading activation has been proposed to explain several phenomena in emotion–cognition interaction, particularly *affective priming* (shorter response times required for identifying targets that are affect-congruent with the priming stimulus vs. those that have a different affective tone), and *mood-congruent recall* (the tendency to preferentially recall schemas from memory whose affective tone matches that of the current mood) (e.g., Bower, 1992; Derryberry, 1988). Bower's "Network Theory of Affect" assumes a semantic net representation of long-term memory, where nodes representing declarative information coexist with nodes representing specific emotions. Activation from a triggered emotion spreads to connected nodes, increasing their activation, thereby facilitating the recall of the associated information. Alternative versions of this theory place the emotion-induced activation externally to the semantic net. For example, Forgas (2003), focusing on emotion influences on attitudes and social judgments, suggests a distinction between memory-based influences and inference-based influences. An example of the former is network theories of affect, explaining mood-congruent recall via spreading activation mechanisms (Bower, 1981). Example of the latter is Schwartz and Clore's theory of affect-as-information (Schwarz and Clore, 1988). Derryberry and Reed (2002), focusing on personality and individual differences research, propose four categories of mechanisms whereby emotions influence cognition: automatic activation, response-related interoceptive information, arousal, and attention.

A number of researchers have independently proposed parameter-based mechanisms to represent a broad range of emotion effects on cognition. These theories suggest that emotions, and other affective states, can be encoded as sets of values across a range of parameters that induce variabilities across cognitive processes, and, ultimately, modify behavior (e.g., Hudlicka, 1998; Matthews and Harley, 1993; Ortony et al., 2005; Ritter and Avramides, 2000). Different theories specify different parameter sets, and distinct emotions are represented in terms of specific parameter configurations, which then influence the processing of both the fundamental cognitive processes (e.g., attention and working memory speed, capacity, and biasing), and high-level cognitive processes including situation assessment, decision-making, planning, problem-solving and learning), as well as the processes involved in cognitive appraisal. The latter in effect "closing the loop" and modeling the influence of existing emotions on the generation of new emotions via appraisal. It is tempting to consider the possibility that the parameter-based models are consistent with recent neuroscience theories which suggest that emotion effects on cognition are implemented in the brain in terms of systemic, global effects on multiple brain structures, via distinct patterns of neuromodulation, corresponding to different emotions (Fellous, 2004).

Affective Disorders: A Transdiagnostic Perspective

While emotions are fundamentally adaptive and facilitate information processing and behavior that are beneficial and essential for survival, the mechanisms mediating affective processes, both the generation of emotions and their effects on cognition, can become dysregulated, due to genetic predispositions as well as factors, such as trauma, early attachment failures, or chronic or acute stress. In these situations, the affective signals may become distorted [resulting in generation of inappropriate (maladaptive) emotions, which then trigger maladaptive behavior], too intense (e.g., causing an intense fear reaction in response to a low-threat stimulus), or not intense enough (e.g., failing to provide a danger signal and exposing the individual to harm). These biases and distortions in the affective processes can result in subjective distress, and the patterns of affective dysregulation and specific symptoms, including maladaptive and self-destructive behavior, that result in psychopathology and motivate individuals to seek psychotherapy.

Transdiagnostic Model of Psychopathology: From Descriptions to Mechanisms

Psychiatrists and psychotherapists have historically relied on the Diagnostic and Statistical Manual (DSM) to diagnose specific psychopathologies, including affective disorders. The DSM has undergone a number of revisions, but the underlying philosophy and approach to conceptualizing psychopathology has not changed and remains descriptive. Distinct disorders are characterized by lists of criteria describing specific behavior or subjective experience, their frequency and their history. Diagnosis consists of finding the closest match in the DSM that corresponds to the patient's condition, as evidenced by observable behavior and reported subjective experience.

While it continues to be widely used, and is the basis of diagnosis in medical and mental health settings, the descriptive approach has a number of shortcomings, the most critical one being the significant overlap among symptoms associated with distinct disorders. For example, social anxiety is a distinct disorder within the DSM, but social anxiety symptoms are also associated with a number of other distinct disorders, including depression, substance use, and schizophrenia (Charney, 2004). The nonspecificity of symptoms characterizing a particular disorder creates challenges both for diagnosis and for treatment. If a patient is diagnosed with major depressive disorder, but has significant social anxiety, should she/he be treated with protocols established for depression or those for social anxiety? Charney comments on this situation as follows: "The observations that the experience of social anxiety span DSM-IV-based conditions provide another example of the limitations of a diagnostic system not based upon etiology."

To address this issue researchers have begun to advocate for a classification system that is based on the underlying etiology of observed and subjective symptoms, rather than on cookbook-style lists of specific symptom configurations. This approach is termed the transdiagnostic model.

The transdiagnostic model of psychopathology, and the associated RDoC framework (Sanislow et al., 2010), aims to address psychopathology from a more mechanism- and first-principles oriented approach to diagnosis, assessment, and treatment (e.g., Kring, 2008; Nolen-Hoeksema and Watkins, 2011).

Computational emotion modeling in general, and models of emotion–cognition interactions in particular, are directly relevant to the transdiagnostic perspective on psychopathology. These methods provide the means of constructing mechanism-based models of psychopathology, and thus offer the hope that models based on first principles can be developed that emphasize the actual underlying processes contributing to the etiology and maintenance of specific visible symptomatology and subjective internal experience. Such models would then facilitate not only more accurate assessment and targeted treatment, but would also support research by enabling the generation of highly specific hypotheses regarding alternative mechanisms of both specific affective disorders, and the best therapeutic approaches to treat them, within a given individual. To be sure, this approach also has a number of challenges, since our ability to construct detailed computational models far outweighs our ability to validate them (Hudlicka, 2015a).

Nevertheless, computational models represent a promising new tool to help elucidate the mechanisms mediating affective disorders. The discussion of affective disorders in the subsequent section is therefore presented from the perspective of the unifying, mechanism-based transdiagnostic model of psychopathology. As such, it emphasizes the underlying affective, and cognitive, processes that contribute to psychopathology.

Mechanisms of Affective Disorders

Historically, the term affective disorders refers to mood disorders that include depression, bipolar disease and mania, and possibly schizoaffective disorders, with anxiety disorders being considered as a separate category, and including a range of disorders, such as generalized anxiety disorder (GAD), social anxiety, posttraumatic stress disorder (PTSD), obsessive–compulsive disorders (OCD), and a range of specific phobias (e.g., fear of flying, spiders, public speaking etc.).

However, given the significant involvement of cognition in affective and anxiety disorders, specifically, the role that affective biases play in the etiology and maintenance of these disorders, it would be more appropriate to (1) group these disorders into a single category that includes both mood and anxiety disorders, as is done in the transdiagnostic model, and (2) refer to them more accurately as cognitive-affective disorders, given the critical role that cognition plays in both categories of disorders. It is from this perspective that we consider the contributions of computational models of emotion–cognition interactions to both psychopathology research and the development of model-based behavioral technologies. The discussion of mechanisms of therapeutic action in Section "Mechanisms of Therapeutic Action" is also presented from this perspective.

A number of affective disorders have been identified, especially in relation to depressive and anxiety disorders, and frequently involving specific affective biases on cognitive processes. For example, depression is associated with preferential processing of negatively valenced cognitions, including negative self-evaluations, preferential recall of negative events in the past, and a bias toward anticipating negative events in the future. These are thought to be mediated in part by the phenomenon of mood-congruent recall, and in part on processes that maintain the activation of negative cognitions: rumination and worry.

Anxiety disorders represent another set of disorders where underlying affective biases on cognition are evident. These disorders are associated with attentional biases that selectively focus on negative and threatening stimuli, interpretive biases that bias interpretation of ambiguous stimuli and situations toward negative and threatening interpretations, biases that influence the appraisal processes to focus on the anxiety symptoms themselves, and overestimate their negative impact (Macleod and Mathews, 2012), as well as a self-bias and bias on affective states in general.

Mechanisms of Therapeutic Action

The phrase "mechanisms of therapeutic action" refers to a set of explanations or hypotheses regarding the means through which particular therapeutic interventions effect change in, or ultimately cure, a specific disorder. The term "mechanism" is used rather loosely in the psychotherapeutic community, depending on the theoretical orientation of the associated approach. In some cases, it refers to empirically validated and theoretically grounded hypotheses regarding the nature of the processes that effect change, specified at a level of detail that would support computational modeling; for example, systematic desensitization treatments of various anxiety disorders, including phobias and PTSD, which involve a variety of low-level conditioning and learning mechanisms. In other cases, the term refers to very high-level descriptions of change processes, which are not empirically grounded and whose theoretical foundations are not framed in terms of falsifiable hypotheses, as is the case with most psychoanalytic theories. In yet other cases, promising ideas are introduced which suggest interesting directions but do not provide sufficient level of detail to support modeling, for example, recent suggestions that a heightened level of arousal is required to transform cognitive structures from maladaptive to adaptive beliefs and interpretations, via processes analogous to annealing in physics.

Increasingly, however, there is emphasis on specifying mechanisms of therapeutic action in terms of the cognitive and affective processes, and associated structures, that lend themselves to symbolic modeling. The specific structures and processes proposed depend on the theoretical orientation of the associated approach. Given the emphasis on fundamental cognitive processes (e.g., attention, memory) and low-level learning mechanisms (classical and operant conditioning) in cognitive therapies (CT) and cognitive-behavioral therapies (CBT), it is not surprising that CT and CBT provide descriptions of mechanisms that are computation-friendly, and lend themselves to symbolic computational modeling. Of particular interest is the recent research on the role of cognitive biases in affective disorders. Specifically, threat biases in attentional and interpretive processing appear to mediate anxiety disorders. These biases operate on low-level processes and outside of conscious awareness (Mathews and MacLeod, 2005), inducing what MacLeod and Mathews refer to as "maladaptive patterns of selective information processing" (Macleod and Mathews, 2012, p. 197). Empirical studies have confirmed that individuals who suffer from anxiety disorders, as well as trait-anxious individuals, selectively focus on negative information (Bar-Haim et al., 2007) and experiments involving interventions targeted at directly reducing these biases demonstrated reductions in symptoms; for example, reducing attentional bias toward negative stimuli resulted in reduction of intrusive negative thoughts contributing to anxiety and depression (Hayes et al., 2010). Therapeutic approaches based on these mechanisms have been developed, the Cognitive Bias modification (CBM) approaches, and will be discussed in Sections "Modeling the Mechanisms of Therapeutic Action" and "Treatment."

Another example of recent focus on mechanisms is a process model of emotion regulation proposed by Gross (Gross, 1998), which emphasizes an information-processing framework and identifies five types of processes that contribute to emotion regulation. Gross' model includes processes controlling situation selection, situation modification, attention deployment, cognitive change, and response modulation. Although not addressing the mechanisms of actions of specific therapeutic approaches, the model provides a framework for integrating multiple processes involved in emotion regulation, thereby providing a basis for developing mechanism-based models of emotion regulation that span multiple therapeutic approaches, and involving multiple modalities of emotion.

Serious Therapeutic Games

Serious games are computer games developed for training and learning purposes, in contrast to games developed solely for entertainment. Games have a unique ability to engage the players and to provide highly immersive learning, training, and therapeutic environments, which can be customized to the user's specific learning needs or therapeutic goals. Their potential in education and training, coaching, rehabilitation, and psychotherapy has been increasingly recognized (e.g., Hudlicka, 2011, 2016) and in spite of their relatively recent emergence within the past two decades, serious games represent the fastest growing segment of the gaming market.

As is the case with games for entertainment, serious games typically provide a game "storyline," which evolves across distinct physical contexts within the simulated gameworld, and typically involves multiple nonplaying characters and distinct tasks which the player aims to achieve as s/he progresses through the game levels. The skills to be learned or practiced are embedded within the game tasks, and the game levels provide progressively more challenging tasks. Depending on the type of skill to be learned or the type of training or coaching, as well as on the age and abilities of the players, the gameplay may focus strictly on the "serious" task, or it may interleave these tasks with segments of gameplay designed for entertainment only. The latter is more typical for games aimed at children and younger users.

Games have the potential to create highly customized therapeutic assessment and treatment environments and facilitate the administration of a range of therapeutic protocols. The gameplay levels, reward structure, nonplaying characters, as well as the overall game storyline, can all be customized to provide an optimal learning and training experience for the user. The nonplaying characters can be implemented using technologies developed for affective intelligent virtual agents, and their appearance and behavior can be defined to match players' individual and cultural preferences and specific learning, training, and therapeutic needs.

Examples of how serious games could be used in psychotherapy include the following:

- a patient recovering from a major depressive episode could practice cognitive restructuring skills with the help of a virtual intelligent affective character in the context of a customized serious game;
- a child client on the autism spectrum could play with a virtual affective agent to learn social skills or interact with multiple such agents, in a customized learning environment embedded within a serious game;
- a patient with social phobia could attend "virtual parties" and practice approaching strangers and initiating and maintaining conversations.

While the majority of existing serious games in healthcare, whether for professional training or for patients, focus on physical health (e.g., surgery, emergency medicine, pain management), games focusing on behavioral care are beginning to emerge (Brezinka and Hovestadt, 2007). These include social skills training (Beaumont and Sofronoff, 2008); support for children experiencing divorce, based on family therapy (www.zipland-interactive.com); cognitive-behavioral treatment of obsessive–compulsive disorder (Brezinka, 2008); game to help veterans overcome PTSD (Rizzo et al., 2010) and a game designed to motivate adolescents for solution-focused therapy ["Personal Motivator" (Coyle et al., 2005)].

Ricky and the Spider represents a good example of the state-of-the-art in therapeutic serious games (Brezinka, 2008). The game was designed to treat OCD in children and is based on a cognitive-behavioral approach to OCD treatment. The players are provided psychoeducation about the condition, and are supported in creating a hierarchy of symptoms and provided with opportunities for simulated exposure and response prevention; these are established, evidence-based approaches to treating OCD. The players are also taught techniques for externalizing their symptoms, as a means of reducing anxiety. The game characters include "Dr. Owl," who provides advice and guidance (a stand-in for a therapist), "Spider," who represents the OCD condition and issues commands for the other game characters to engage in OCD behavior, and two characters with OCD symptoms: "Ricky the grasshopper," who must hop in a specific pattern, and "Lisa the Ladybug," who must count her polka dots before falling asleep. The Spider threatens Ricky and Lisa with terrible consequences if they do not follow his orders. The game is played by a child under the supervision of a therapist.

The game is being used by several thousand users across more than 40 countries. Data from an initial evaluation, conducted with 18 children and 13 therapists, were positive, with the children reporting satisfaction with the use of the game during treatment, and the therapist reporting that the children enjoyed playing the game and experienced increased motivation for treatment. The symptoms of OCD were significantly reduced in 15 children, and unchanged in one child.

It should be understood that it is not suggested that serious games should be replacing human therapists or face-to-face therapy. These technologies cannot function at the level of an experienced, empathic human therapist, and likely never will. Rather, they have a unique role in the delivery of behavioral healthcare, both supportive of, and distinct from, the roles of human therapists. These include the following capabilities:

- enhance dissemination of evidence-based treatment;
- make treatment more accessible (anytime/anywhere availability);
- support treatment between sessions (facilitate homework and skills practice);
- adapt to individual needs and cultural preferences;
- promote engagement and support motivation.

In some cases, technology may even be the preferred method of delivering therapeutic services, for example, for children or teens on the autism spectrum or for patients with social anxiety or agoraphobia. In addition, these technologies can also support enhanced diagnosis and assessment and play a role in research, where they can contribute to more mechanism-based diagnosis and treatment planning (e.g., Hudlicka, 2008b).

Section "Treatment" discusses how serious gaming technologies, augmented with models of the mechanisms that mediate psychopathology and therapeutic action, can facilitate assessment and diagnosis, outcome tracking, and implementations of customized treatments that target the mechanisms mediating specific affective disorders.

STATE-OF-THE-ART IN MODELING EMOTION–COGNITION INTERACTION

The past 15 years have witnessed a rapid growth in computational models of emotion and cognitive-affective architectures. Researchers in cognitive science, AI, HCI, robotics, and gaming are developing computational models of emotion primarily for applied purposes, to create more believable and effective virtual characters and robots, and to enhance human–computer interaction, but also for theoretical purposes, to elucidate the mechanisms mediating affective phenomena. This section provides an overview of the state-of-the-art in symbolic models of emotion–cognition interactions.

In spite of the many stand-alone emotion models, and the numerous affective agent and robot architectures developed to date, there is a lack of consistency, and lack of clarity, regarding what exactly it means to "model emotions." The term "emotion modeling" is used in the literature to refer to a wide range of processes, including: dynamic generation of emotion via black-box models, that map specific stimuli onto associated emotions; generation of facial expressions, gestures, or movements depicting specific emotions in synthetic agents or robots; modeling the effects of emotions on decision-making and behavior selection; including information about the user's emotions in a user model in tutoring and decision-aiding systems and in games. I have previously suggested that the term "emotion modeling" should be reserved for models that aim to model emotion generation and the subsequent effects of the generated emotions on cognition, expression and behavior (Hudlicka, 2008a). It is in this sense that the terms emotion modeling and affective modeling are used in this chapter.

There is also a lack of clarity regarding what affective factors are represented in a specific model. The term "emotion" itself is problematic. On the one hand, it depicts emotions in a generic, folk-psychology sense we all presume to understand, and which subsumes many types of affective factors. On the other hand, it has a specific meaning in the emotion research literature, referring to transient states, lasting for seconds or minutes, typically associated with well-defined triggering cues and characteristic patterns of expressions and behavior. (More so for the simpler, fundamental emotions than for complex emotions with strong cognitive components.) Emotions can thus be contrasted with other terms describing affective phenomena: *moods*, sharing many features with emotions but lasting longer (hours to months) and having more diffuse triggers and behavior patterns; *affective states*, undifferentiated positive or negative "felt states" and associated behavior tendencies (approach, avoid); and *feelings*, a problematic and ill-defined construct from a modeling perspective. [Averill points out that "feelings are neither necessary nor sufficient conditions for being in an emotional state" (Averill, 1994)]. Increasingly, emotion researchers are adopting the term affect as an umbrella term that covers all other affective states.

Some models also represent permanent affective personality *traits* (e.g., extraversion, neuroticism), or a variety of "mixed" mental states that involve both cognitive and affective components (e.g., attitudes). Emotion models also vary in terms of the specific roles of emotions that are included in a particular model; for example, which of the intrapsychic roles of goal management and goal selection, resource allocation and subsystem coordination, and the interpersonal roles of communication, coordination and attachment are included.

Although computational affective modeling is still in its infancy, efforts are beginning to emerge to introduce standards and design guidelines to facilitate model development, analysis, comparison and validation (Broekens et al. 2008; Hudlicka, 2011, 2015a; Reisenzein et al., 2013). This section therefore begins with a discussion of a computational analytical framework that aims to contribute to this effort, by defining a set of abstract computational tasks required to model emotion and emotion–cognition interactions (Section "Core Affective Processes and Generic Computational Tasks"). The objective of the framework is to facilitate a more systematic approach to affective model design, analysis, comparison, and validation. Sections "Emotion–Cognition Interaction in Models of Emotion Generation" and "Emotion–Cognition Interactions in Models of Emotion Effects" then provide an overview of existing approaches to modeling emotion–cognition interactions in symbolic affective models and architectures, focusing on interactions taking place during emotion generation and emotion effects, respectively.

It should also be noted that although historically cognition and emotion have been conceptualized as distinct domains, emerging neuroscience evidence suggests that what has traditionally been thought of as distinct cognitive and affective processing is in fact mediated by shared and overlapping neural circuitry. Thus, while it may be helpful to discuss cognition and emotion as distinct processes to help manage the complexity of the phenomena under investigation, particularly when considering the psychological versus neural level of analysis, it is important to keep in mind the close coupling of the underlying neural circuitry. In this regard it is helpful to consider the distinct levels of information processing, and the necessity to establish which level is being addressed in a particular modeling effort. Marr's levels are perhaps most often associated with identifying these distinct levels of analysis in the computational sciences (Marr, 1982) and distinguish among the *computational*, *algorithmic*, and *implementation* levels.

Core Affective Processes and Generic Computational Tasks

In spite of the progress in emotion research over the past 20 years, emotions remain elusive phenomena. While some underlying circuitry has been identified for some emotions (e.g., amygdala-mediated processing of threatening stimuli, the role of orbitofrontal cortex in emotion regulation), much remains unknown about the mechanisms of emotions. Given the multiple-modalities of emotion, the complexity of the cross-modal interactions, and the fact that affective processes exist at multiple levels of aggregation within the brain (neuronal, circuit-level, global neuromodulation), it may seem futile, at best, to speak of "core affective processes." Nevertheless, for the purpose of developing symbolic models of emotions, and for models of emotions in symbolic agent architectures, it is useful to cast the emotion modeling problem in terms of two broad categories of processes, and the generic computational tasks necessary to implement them.

I have previously suggested that for didactic and pragmatic purposes, computational affective modeling ought to be conceptualized in terms of two categories of core processes: those mediating *emotion generation,* and those that then mediate the *effects of the activated emotions* on cognition, expressive behavior (e.g., facial expressions, speech) and action selection (Hudlicka, 2008a, 2011). This temporally-based categorization (prior to, and following, the "felt" emotion) provides a useful perspective on modeling, by helping to manage the model complexity and by supporting a systematic deconstruction of the high-level processes into the underlying, generic computational tasks necessary to implement emotion generation and emotion effects.

These computational tasks can be thought of as the building blocks of emotion models and represent a candidate set of fundamental generic functions necessary to model the core affective processes. These building blocks can also provide a basis for the development of more systematic guidelines for emotion modeling, for a more exact specification of the associated theoretical and data requirements, and for the specification of representational and inferencing requirements and evaluation of possible alternatives. The building blocks can thus serve to support what Sloman calls "architecture based definition of emotion" (Sloman et al., 2005).

This *core processes/generic tasks* perspective provides a computational analytical framework that helps organize existing theories and data. The framework also facilitates more systematic approaches to model design, analysis and validation, by supporting a systematic comparison of different:

- theories for a particular model component (e.g., OCC appraisal vs. appraisal based on variables defined in the componential models, such as Scherer's);
- datasets from empirical studies;
- representational and inferencing formalisms used to implement a particular theory or a particular task (e.g., logic vs. production rules vs. Bayesian belief nets);
- algorithms for implementing a specific task (e.g., different functions used to model emotion onset and decay); and, ultimately
- cognitive-affective architectures aiming to model affective processing, whether for applied purposes, to enhance believability and social competence of synthetic agents, or for research purposes, to identify the underlying mechanisms and to help answer fundamental questions about emotions and their functions in biological agents.

By providing a hierarchical structure that links the high-level processes (emotion generation and emotion effects) to the generic computational tasks required to implement them, the framework also provides a hierarchical structure that connects the distinct affective processes to their underlying mechanisms, and, by extension, to the distinct roles they perform (i.e., the intrapsychic and interpersonal roles outlined above). The objective of this approach to analysis and design of emotion models is to move beyond the existing state of affairs, where individual models are used as the organizing dimension, and toward a more general approach, organized in terms of the individual computational tasks, and the associated representational and reasoning requirements. This approach provides a basis for both managing the complexity of affective modeling and for systematizing affective model development, by providing foundations for more concrete guidelines for both model design and for the analysis of existing models and theories, as well as for model validation. In addition, defining model structure in terms of these tasks promotes modularity, which in turn facilitates model reuse, model sharing, and the development of modeling tools.

This process-based approach to thinking about emotions also contributes to the emerging transdiagnostic approach increasingly emphasized in psychopathology research (e.g., Nolen-Hoeksema and Watkins, 2011), by supporting an understanding of psychopathology in terms of the underlying mechanisms, rather than in terms of descriptions of visible manifestations.

When we view affective processes from the core processes/generic tasks perspective, it is important to emphasize two points. *First*, it is not suggested that the two core processes, or the associated computational tasks, correspond to distinct neural circuits or processing mechanisms. Rather, they represent useful abstractions, and a means of managing the complexity associated with symbolic modeling of affective phenomena. *Second*, although the categorization emphasizes the chronological sequence of generation of emotion, followed by the various effects the generated emotion exerts, across multiple modalities, it must be understood that there are numerous complex feedback interactions among these processes, which are as yet poorly understood. In other words, the generated emotion influences the very processes involved in its generation, such as the cognitive appraisal processes (discussed next).

Generic Computational Tasks

The following distinct, generic computational tasks are required to implement *emotion generation via cognitive appraisal*:

- define and implement the {emotion elicitor(s)}–to–{emotion(s)} mapping. Depending on the theoretical perspective adopted, this may involve additional subtasks that map the emotion elicitor(s) onto an intermediate representation (e.g., PAD dimensions, appraisal variables vectors), and the subsequent mapping of these onto the final emotion(s);
- calculate the intensity of the resulting emotion(s);
- calculate the decay of these emotions over time;
- integrate multiple emotions, if multiple emotions were generated;
- integrate the newly generated emotion(s) with existing emotion(s) or moods.

The multimodal nature of emotion cannot as easily be ignored when considering models of emotion effects. This is particularly the case in models implemented in the context of embodied agents that need to manifest emotions not only via behavioral choices, but also via expressive manifestations within the channels available in their particular embodiment (e.g., facial expressions, gestures, posture etc.). The multimodal nature of emotion effects increases both the number and the type of computational tasks necessary to model emotion effects.

The following distinct computational tasks are necessary to implement *the effects of emotions* across multiple modalities:

- define and implement the emotion/mood–to–effects mappings, for the modalities included in the model (e.g., cognitive, expressive, behavioral, neurophysiological). Depending on the theoretical perspective adopted, this may involve additional subtasks that implement any intermediate steps, and are defined in terms of more

abstract semantic primitives provided by the theory (e.g., dimensions, appraisal variables);
- determine the magnitude of the resulting effect(s) as a function of the emotion or mood intensities;
- determine the changes in these effects as the emotion or mood intensity decays over time;
- integrate effects of multiple emotions, moods, or some emotion and mood combinations, if multiple emotions and moods were generated, at the appropriate stage of processing;
- integrate the effects of the newly generated emotion with any residual, ongoing effects, to ensure believable transitions among states over time;
- account for variability in the above by both the intensity of the affective state, and by the specific personality of the modeled agent;
- coordinate the visible manifestations of emotion effects across multiple channels and modalities within a single time frame, to ensure believable manifestations.

Abstract Domains Required for Modeling Emotions

The generic computational tasks outlined previously provide a basis for defining model design guidelines and for managing the complexity of affective modeling, but they provide only the first step in the top-down deconstruction of modeling requirements. Another example of a computational and design-oriented perspective on emotion modeling, one that focuses on emotion generation, is recent work by Broekens and coworkers, who developed a set-theoretic formalism, and an abstract framework, for representing and comparing appraisal theories.

Building on the work of Reisenzein (2001), Broekens et al. (2008) developed a set-theoretic formalism that depicts the abstract structure of the appraisal process, and represents both the processes involved, and the data manipulated. I have augmented their original framework with some of the abstract domains necessary to model emotion effects. The resulting structure is shown in Fig. 16.1. The framework illustrates the distinct processes involved in emotion generation and emotion effects modeling, and the data manipulated by these processes [e.g., perception (evaluative processes produce a series of mental objects), appraisal (processes that extract the appraisal variable values from the mental objects), and mediation (processes that map the appraisal values onto the resulting emotion(s)]. The distinct processes operate on distinct categories of data: their associated *domains*. Figure 16.1 illustrates the relationship among these domains and the abstract computational tasks.

This framework thus complements the computational task-based perspective with a set of domains required to implement both emotion generation and emotion effects, and helps define the constituent elements of

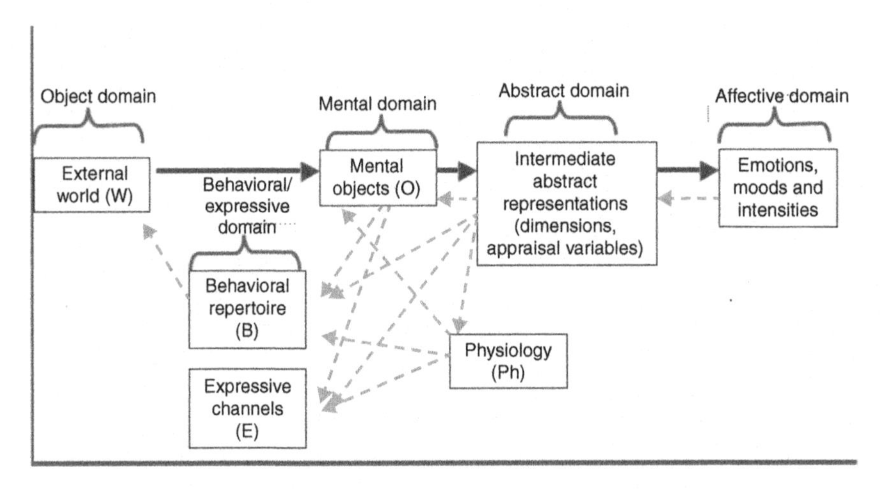

FIGURE 16.1 Distinct abstract domains necessary to implement the abstract computational tasks. *Note that this figure assumes the existence of some intermediate, abstract structures reflecting variables that mediate both emotion generation and emotion effects. Not all affective models necessarily require such an abstract domain. The solid (blue in the web version) arrows indicate paths mediating emotion generation; the dashed (green in the web version) arrows indicate paths mediating emotion effects.*

these domains. These definitions then form a basis for defining the mappings among these domains, which must be defined to model emotion generation and emotion effects.

Distinct Theoretical Perspectives and Core Affective Processes

Section "Theoretical Perspectives" above outlined three theoretical perspectives on emotions that represent the theoretical foundations for computational affective modeling. Many open questions remain regarding the most appropriate theory for a particular modeling task, and many opportunities exist for a systematic comparison of the alternative theories, and the associated representational and inferencing requirements, to identify the most appropriate theoretical framework for a particular modeling objective. Fig. 16.2 provides a high-level overview of the different theoretical perspectives, with respect to the core affective processes necessary to model emotions. As is clear from the figure, many opportunities exist for empirical studies, to collect the necessary data to fully populate the different cells, or to determine that for some cells theoretically plausible data don't in fact exist (e.g., the dimensional perspective and emotion generation).

Emotion–Cognition Interaction in Models of Emotion Generation

Two points should be noted. *First*, the majority of symbolic models focus on modeling only the cognitive modality, that is, modeling cognitive

	Generation	Effects ..on cognition	..on expression	..on behavior
Discrete	Fixed, simple triggers for basic emotions	Biases on attention, perception and congnitive processes (memory, learning, etc.)	Cross-cultural patterns of expression, esp. facial	Common patterns associated with basic emotions
Dimensional		Some known biases and broad effects on cognitive processes	Some data re: effects of arousal and valence on expressive behavior	Some dimension values associated with trends or broad categories of behavior
Componential (Including OCC) (cognitive appraisal)	Evaluative criteria to assess degree of congruence between situation and agent goals	Some data re: effects of appraisal variables on cognition	Some data mapping app. variable values to expression components	

FIGURE 16.2 **Summary of the degree of support provided by the distinct theoretical perspectives (*row labels*) for the different components of affective models (*column labels*).**

appraisal theories as the sole basis of emotion generation. There are two primary reasons for this choice: (1) cognitive appraisal theories are the most elaborated theories of emotion generation and thus a natural choice as a theoretical foundation of emotion generation models, and (2) since the majority of models are built for applied purposes (i.e., to increase believability and affective competence of an agent), cognitive appraisal is a reasonable choice since it enables the agent to dynamically generate emotions in response to events in the human-agent interaction. *Second,* most of the generic tasks involving affective dynamics (e.g., integrating multiple emotions and modeling onset and decay of emotions) are typically not adequately addressed in existing models. Thus the focus of the discussion below will be on the first two generic tasks outlined previously: mapping emotion elicitors onto emotions and calculating emotion intensity.

However, for research models, and particularly for models aiming to represent mechanisms mediating psychopathology and therapeutic action, it will be essential to enlarge the scope of existing models to include the additional modalities of emotions, somatic/physiological, expressive, and perhaps even the subjective/conscious modality, as these also influence emotion generation processes. The development of such multimodal models of emotions represents one of the core challenges in affective modeling, as neither the necessary data nor the theoretical foundations currently provide sufficient details to construct computational models.

Elicitor to Emotion Mapping

The most influential appraisal theories implemented in models of emotion generation are those that are cast in "computation-friendly" terms. The first of these was proposed by Ortony and coworkers, now referred to as the OCC model (Ortony et al., 1988). Numerous versions of this model have been implemented, and the majority of existing models of emotion generation use the OCC theory, and its associated categorization of emotions by the type of triggering stimuli (events, actions by other agents, and objects), and evaluative criteria associated with each category (e.g., desirability, praiseworthiness, familiarity) (e.g., Andre et al., 2000; Bates et al., 1992; Reilly, 2006). Examples of affective and cognitive-affective architectures using the OCC model include FAtiMA (Dias et al., 2015) and WASABI (Becker-Asano, 2008). More recently, appraisal models proposed by Scherer, and Smith and colleagues, have become the basis for computational appraisal models (Scherer et al., 2001; Smith and Kirby, 2000).

Implementations of the elicitor-to-emotion mappings typically involve two stages. In the first stage, the current stimuli (incoming cues, memories of previous situations, expectations) are analyzed with respect to the agent's goals and beliefs, to determine the values of the specific evaluative criteria/appraisal variables used in the model. This stage represents the more challenging phase, and may require significant amount of reasoning to evaluate the current situation with respect to the evaluative criteria, such as certainty and urgency, goal relevance, and congruence and coping strategies, depending on the complexity of the simulated environment and the modeled agent's beliefs and goals. The representational and inferencing formalisms used to implement this stage include production rules, Bayesian belief nets or predicate calculus.

In the second stage, the specific combinations of the values of the evaluative criteria (in effect a vector consisting of particular configuration of the evaluative criteria) are mapped onto a point in the affective space defined by these evaluative criteria, to identify the specific emotion corresponding to the current situation.

Fig. 16.3 illustrates this process with respect to the dominant theoretical perspectives outlined earlier. Note that the two-stage process is implemented in the OCC and the componential theories based models (lower two levels). The discrete/categorical model does not propose any intermediate, abstract evaluative criteria, and models of emotion generation based on this theory implement a direct mapping from a stimulus to the corresponding emotion (e.g., fearful object → fear). The dimensional model is included for completeness sake, but, as stated earlier, does not in

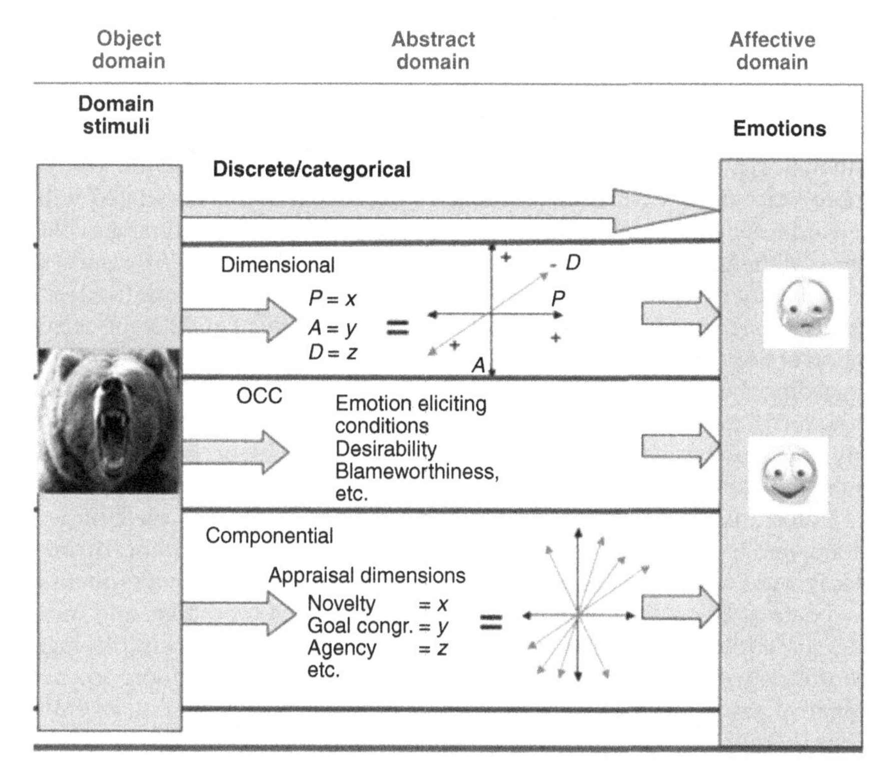

FIGURE 16.3 **Alternative theoretical perspectives on emotion generation.**

fact provide underlying theoretical foundations for cognitive appraisal, as it was developed to characterize felt mood states, rather than emotion generation via interpretive processes.

Emotion Intensity Calculation

Calculation of *emotion intensity* reflects a high degree of complexity. Not only must we define the fundamental formulae for the intensity calculation, based on the types, and characteristics, of the triggering stimuli, but we must then integrate the "current" intensities with those of existing emotions and moods, to ensure smooth and appropriate transitions among emotion states. This must take into account possible differences in the decay rates of different emotions, which are subject to a variety of influences that have not yet been identified or quantified to the degree required for computational modeling. Reilly (2006) discusses alternative approaches for modeling emotion dynamics.

Most existing models of appraisal use relatively simple formulae for calculating emotion intensity, typically focusing on desirability and likelihood;

for example, [desirability * likelihood] (Gratch and Marsella, 2004), [desirability * (change in) likelihood] (Reilly, 2006). A number of complexities are typically not addressed. For example, Reilly (2006) points out the need for representing asymmetry of success versus failure; in other words, for different types of individuals (and different goals) success may be less (or more) important than failure; for example, extraversion is associated with reward-seeking whereas neuroticism is associated with punishment avoidance. Modeling of these phenomena requires distinct variables for success (desirability of an event, situation or world state) versus failure (undesirability of the same). Directly related to the intensity calculation is the calculation of the emotion ramp-up and decay rates, which brings up a question regarding the extent to which emotions represent self-sustaining processes, that must "run their course." Reilly summarized current approaches to decay calculation as being linear, logarithmic, exponential, or "some arbitrary monotonically decreasing function over time" (2006).

Unfortunately for modelers, emotion dynamics are not well understood, and the data for precise calculations of intensities and ramp-up and decay rates are not available. Existing empirical studies provide qualitative data at best. Variability of these processes across emotions and individuals, while documented, has also not been quantified; for example, high neuroticism rate predisposes individuals toward faster and more intense negative emotions; anger appears to decay more slowly than other emotions (Lerner and Tiedens, 2006). Even more importantly, some researchers point out that the appraisal dimensions identified for emotion differentiation may not be the same as those that "allow prediction of duration and intensity," and that "the current set of appraisal dimensions may be incomplete" (Scherer, 2001, p. 375). Table 16.3 summarizes the computational

TABLE 16.3 Summary of Computational Tasks Required to Implement Affective Dynamics in Emotion Effects Modeling

Affective dynamics tasks	Examples of design alternatives
Intensity calculation	Step function (0 or 1); desirability * likelihood (of situation, event or relevant goal); modify desirability * likelihood to capture asymmetry in positive versus negative emotions
Intensity onset/decay	Step function, duration for time *t*; linear, exponential or logarithmic monotonically increasing/decreasing function
Integrating multiple affective states (similar)	Sum, maximum, average, logarithmic, sigmoidal
Integrating multiple affective states (dissimilar)	Max, precedence rules, mood congruent emotion selected

tasks required to implement affective dynamics, that is calculation of emotion intensity, its onset and decay, and integration of multiple emotions.

Emotion–Cognition Interactions in Models of Emotion Effects

As outlined in Section "Emotion Effects on Cognition" above, theories of emotion effects on cognition are not as elaborated, nor as uniform, as those aiming to explain emotion generation via appraisal. Nor are emotion effects on cognition modeled as frequently as emotion generation, in part precisely due to the lack of adequate theories. (In the majority of existing affective models the effects of emotions on cognition are not represented in any depth. Rather, these effects are manifested via the agent's expressive channels and behavioral choices.) However, if computational affective modeling is to contribute to improved understanding of the mechanisms mediating psychopathology, greater emphasis will need to be placed on developing models of emotion effects on cognition.

The primary generic computational task for emotion effects modeling is the mapping of a generated emotion onto the different modalities comprising emotion. Since the focus here is on emotion–cognition interactions, the subsequent discussion will be limited to the cognitive modality. Again, with the understanding that models of psychopathology and therapeutic action must eventually also include the other modalities of emotions.

The additional computational tasks regard affective dynamics: determining the magnitude of the effect as a function of the emotion intensity, with the possibility that distinct intensity intervals may exert distinct types of effects (e.g., Sonnemans and Frijda, 1994); determining the changes in the emotion-induced effects on cognition as the intensity decays over time; and determining any interactions when multiple emotions are generated or when existing emotion or mood interact with the newly generated emotion. As was the case with emotion generation, the affective dynamics of emotion effects are as yet poorly understood and the majority of emotion models do not address these in any significant detail.

Emotion to Effects on Cognition Mapping

In contrast to the core generic task in emotion generation, which maps emotion elicitors onto emotions, via evaluative criteria that are shared by multiple theories, there is greater variability in the "emotion to effects on cognition" generic task. This is due to the fact that there is currently no single (unified) theory of emotion effects on cognition. The available data regarding emotion effects on cognition are similarly fragmented, with distinct data representing different emphases on particular cognitive processes and structures (refer to Section "Emotion Effects on Cognition" and Table 16.1). Analogously, the existing models also vary in terms of the specific cognitive processes and structures represented, and thus implement different mappings between particular emotions and specific cognitive processes and structures that are influenced by these emotions.

Section "Emotion Effects on Cognition" above discussed several theories that have been proposed to explain emotion effects on cognition, including spreading activation over semantic net representations, explaining affective priming and mood-congruent recall (Bower, 1992) and Forgas' AIM theory (Forgas, 1999).

The parameter-based theories outlined in Section "Emotion Effects on Cognition" appear promising, in part due to their potential to encompass a broad range of effects across multiple structures and processes, and in part due to the possibility that this approach may be helpful for modeling neuromodulatory theories of emotion effects. The subsequent discussion therefore focuses on a parameter-based model, specifically, the MAMID modeling methodology and cognitive architecture (Hudlicka, 1998, 2003, 2007).

MAMID stands for methodology for analysis and modeling of individual differences. MAMID models emotion effects using a generic methodology capable of modeling multiple, interacting individual differences, both *stable traits and dynamic states.* Its focus is on modeling the effects of emotions (joy, fear, anger, and sadness) on the cognitive processes mediating decision-making (attention, situation assessment, expectation generation, goal management, and action selection), in terms of parameters that control processing within the individual modules of a cognitive-affective architecture; for example, the parameters control the speed and capacities of the different architecture modules, as well as the ranking of the individual constructs processed by these modules (e.g., cues, situations, goals), as they map the inputs (perceptual cues) onto the outputs (selected actions).

MAMID is a process-oriented model, focusing on emotion, and as such aims to explicitly represent the structures and processes mediating affective processing and emotion–cognition interactions. The MAMID model of emotion effects on cognition is implemented in the context of a symbolic, cognitive-affective domain-independent architecture, which was instantiated and evaluated in two domains [search-and-rescue operations (Hudlicka, 2005), and a peacekeeping scenario (Hudlicka, 2003, 2007)]. A high-level schematic of the MAMID cognitive-affective architecture is shown in Fig. 16.4. Fig. 16.5 illustrates the parameter-based modeling approach.

MAMID implements a sequential see-think-do processing sequence, consisting of the following modules: *Sensory Preprocessing* (translates incoming data into task-relevant cues), *Attention* (filters incoming cues and selects a subset for processing), *Situation Assessment* (integrates individual cues into an overall situation assessment), *Expectation Generation* (projects current situation onto possible future states), *Affect Appraiser* (derives a valence and four of the basic emotions from external and internal elicitors), *Goal Management* (identifies high-priority goals), and *Behavior Selection* (selects the best actions for goal achievement).

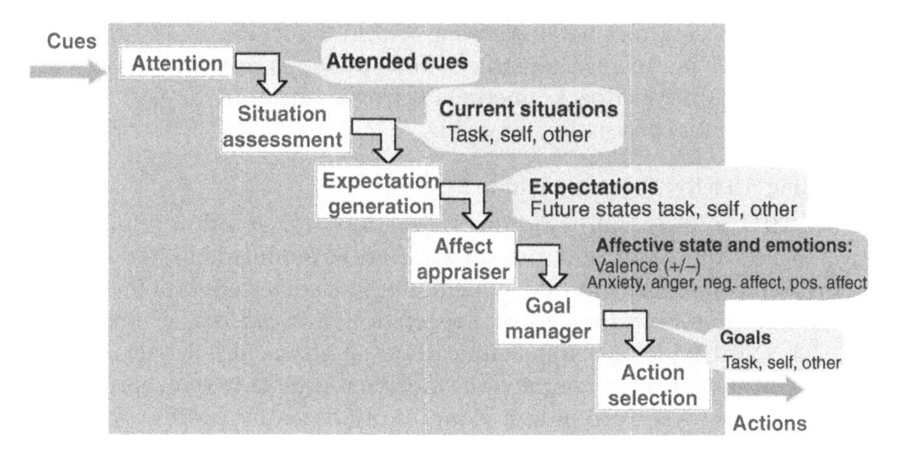

FIGURE 16.4 **MAMID cognitive-affective architecture.**

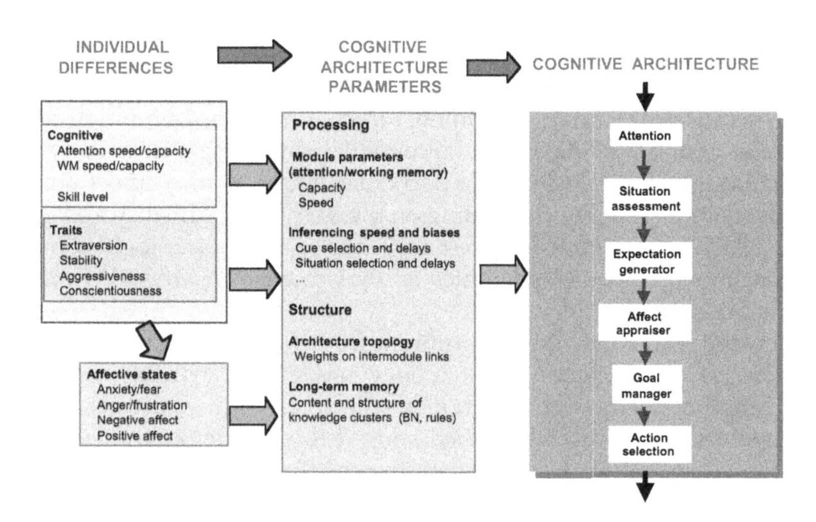

FIGURE 16.5 **Schematic illustration of MAMID methodology for state and trait modeling.**

These modules map the incoming stimuli (cues) onto the outgoing behavior (actions), via a set of intermediate internal structures (situations, expectations, and goals), collectively termed *mental constructs*. This mapping is enabled by long-term memories (LTM) associated with each module, represented by belief nets. *Mental constructs* are characterized by their attributes (e.g., familiarity, novelty, salience, threat level, valence, etc.), which influence their processing, that is, their rank and the consequent likelihood of being processed by the associated module within a given execution cycle (e.g., cue will be attended, situation derived, goal or action selected). All constructs derived in a given execution cycle are available to subsequent modules for processing, within that cycle. In addition, the availability of the mental constructs from previous execution cycles allows for dynamic feedback among constructs, and thus departs from a strictly sequential processing sequence.

Modeling Affective Processes Within MAMID

MAMID models both emotion generation and emotion effects, but emphasizes the latter. Emotion generation is modeled within a dedicated *Affect Appraiser* module, which integrates external data (cues), internal interpretations (situations, expectation) and desires and priorities (goals), with stable and transient individual characteristics (traits and emotional states), and generates an emotional state at two levels of resolution: a *valence* (corresponding to an undifferentiated positive or negative evaluation) and one of the four basic emotions (fear/anxiety, anger, sadness, joy).

Generation of basic emotions represents more differentiated processing, where the intensity of each emotion is influenced by task- and individual-specific criteria. This involves a consideration of a variety of idiosyncratic criteria that determine, for example, whether a high-threat situation or an impending goal failure will lead to anger or anxiety, in a particular agent; for example, a particular situation may affect anxiety positively or negatively, depending on the individual's history and experience. Such differentiated processing requires correspondingly complex inferencing and knowledge, which is implemented in MAMID in terms of belief nets.

Emotion intensities are determined from four contributing factors: *trait bias factor*—reflecting a tendency toward a particular emotion, as a function of the agent's trait profile (e.g., high neuroticism/low extraversion individuals are predisposed toward negative emotions). *Valence factor*—reflecting a contribution of the current valence, where negative valence contributes to higher intensities of negative emotions and vice versa. *Static context factor*—reflecting the agent's skill level and contributing to the anxiety level if skill level is low. *Individual factor*—weighted sum of the emotion intensities derived from the emotion-specific belief nets, reflecting the idiosyncratic contributions of specific elicitors.

The Affect Appraiser module incorporates elements from several appraisal theories: *domain-independent appraisal dimensions, multiple levels* of resolution, and *multiple stages* (Leventhal and Scherer, 1987; Smith and Kirby, 2000).

The effects of emotions (as well as traits and nonaffective states) are modeled by mapping a particular configuration of emotion intensities and trait values onto a set of parameter values, which then control processing within the architecture modules, as well as the data flow among the modules; for example, decrease/increase the modules' capacity and speed, introduce a bias for particular types of constructs, such as high-threat or self-related constructs (Figs. 16.5 and 16.7).

Functions implementing these mappings are constructed on the basis of the available empirical data. For example, the anxiety-linked bias to preferentially attend to threatening cues, and interpret situations as threatening, is modeled in MAMID by ranking high-threat cues and situations more highly, thereby making their processing by the Attention and Situation Assessment modules more likely. Currently, the parameter-calculating functions consist of weighted linear combinations of the factors that influence each parameter. For example, working memory capacity reflects a normalized weighted sum of emotion intensities, trait values, baseline capacity, and skill level.

MODELING MECHANISMS OF PSYCHOPATHOLOGY AND THERAPEUTIC ACTION

An ability to model mechanisms mediating affective disorders would provide a means of operationalizing high-level theories regarding their etiology and maintenance, and would provide a basis for more accurate, mechanism-based assessment and diagnosis, consistent with the aim of the transdiagnostic model of psychopathology. These models could also serve as a basis for modeling the mechanisms of therapeutic action and could contribute to a more personalized approach to treatment planning and assessment of outcomes.

As stated previously, the majority of existing models of emotion fall within the category of *applied models*, whose purpose is to enhance agent behavior, rather than *research models*, whose objective is to model the mechanisms mediating affective processing (Hudlicka, 2012). In addition, the majority of existing emotion models emphasize the cognitive modality in emotion generation, modeling cognitive appraisal, and the behavioral/expressive modality in models of emotion effects. Given the multimodal nature of emotions and the fact that all modalities likely contribute to psychopathology and play an important role in treatment, it is clear that the contemporary focus on only two of the modalities limits the ability of existing models to adequately represent the mechanisms mediating psychopathology and therapeutic action. (Fig. 16.6 provides a summary of the combinations possible when we consider the cross product of the distinct emotion modalities and the core affective processes, and lists some of the processes that need to be modeled, with the cells containing the most frequently modeled processes highlighted.)

Nevertheless, even models representing only the cognitive modality have significant potential to contribute to an improved understanding

	Generation	Effects
Cognitive	Cognitive appraisal	Attention, perception, cognitive processes, incl. memory
Physiological somatic	Environment and physio user factors	Autonomic nervous system manifestations
Expressive/ behavioral	Facial feedback/ Feeling theories	**Expr:** face, speech, gestures, movement **Behavior:** action selection
Subjective	?	?

FIGURE 16.6 **Summary of the processing required for modeling the core affective processes across the distinct emotion modalities (most common processing currently modeled is highlighted in yellow).**

of the mechanisms mediating affective disorders and their treatment. In addition, the fact that the most elaborated theories attempting to explain affective disorders and their treatment also emphasize the cognitive modality of emotion [the CBM theories (Macleod and Mathews, 2012; traditional CT protocols and their emphasis on distorted cognitions (Beck, 1967); metacognitive therapy (Wells, 2002); and to some extent emotion-focused therapy (Herrmann et al., 2016)] provides an additional argument for beginning the challenging enterprise of modeling psychopathology with an emphasis on cognition.

Below, I illustrate an approach to modeling the mechanisms of psychopathology and therapeutic action that is based on the MAMID modeling methodology and architecture outlined earlier in Section "State-of-the-Art in Modeling Emotion-Cognition Interaction." The approach emphasizes a parameter-controlled architecture that can produce a broad range of observable behaviors by manipulating the structures and processes within the individual architecture modules. Section "Modeling the Mechanisms Mediating Affective Disorders" below describes MAMIDs ability to generate a broad range of behaviors induced by manipulating the magnitude of attentional and interpretive biases, ranging from desirable adaptive vigilance through anxiety to a paralyzing panic attack. Section "Modeling the Mechanisms of Therapeutic Action" then discusses how the same modeling approach could be extended with additional modules and applied to model mechanisms of therapeutic action. It is important to note that these models represent early, exploratory stages of modeling psychopathology and therapeutic action, and no claims are being made regarding their validity.

Modeling the Mechanisms Mediating Affective Disorders

Contemporary theories of anxiety disorders emphasize the role of information-processing biases in contributing to, and maintaining, heightened anxiety levels; specifically, the role of a range of emotion-induced biases on attentional, interpretive, and memory processes (Macleod and Mathews, 2012). The MAMID modeling methodology and architecture described previously provides the representational and processing infrastructure that enables the construction of explicit models of these biases, within the context of a symbolic cognitive-affective architecture.

Two of the biases that have been extensively studied as mediators of anxiety disorders are the *attentional* and the *interpretive* bias. Attentional biases focus attention on stimuli with a particular affective content. In anxiety disorders, the biasing effects focus attention on negative and threatening stimuli. In other words, an individual in a state of anxiety selectively focuses on threatening stimuli and neglects nonthreatening stimuli, thereby maintaining or even increasing their state of anxiety. Interpretive biases selectively direct interpretation of stimuli to favor an interpretation with a specific affective tone (Hertel and Mathews, 2011). In anxiety disorders, this type of bias contributes to interpretations of ambiguous stimuli as dangerous, threatening, or negative; again, maintaining or increasing the individual's state of anxiety.

Both of these mechanisms can be explicitly modeled in MAMID, via parametric manipulations of the modeling processes and structures. MAMID is also well suited to modeling alternative mechanisms mediating anxiety disorders. Below, I describe how the MAMID methodology and architecture can represent both attentional and interpretive biases, and the resulting anxiety states, including the extreme state of a panic attack, through parametric manipulations of the underlying processing. The MAMID modeling approach demonstrates how the same set of underlying processes can generate a wide variety of effects, ranging from adaptive behavior through mild dysfunction to paralyzing pathology, depending on the values of the parameters controlling the processing: as the anxiety intensity increases, the processing becomes increasingly biased, demonstrating increasingly dysfunctional behavior.

Two key features of the MAMID model that make it suitable for modeling the mechanisms of psychopathology are the following: (1) high degree of parameterization, enabling manipulation of architecture topology and data flow, and processing within the individual modules, and (2) a testbed environment, within which the MAMID model is embedded, and which facilitates rapid model development and interactive model "tuning," by providing the modeler access to a range of model parameters, and control of the functions that derive their values. By manipulating these parameters, alternative hypotheses regarding the specific mechanisms of an observed phenomenon can be rapidly implemented and their behavior evaluated within the context of a specific simulated environment.

In the example below, the simulated environment represents a search-and-rescue game, where the MAMID architecture controls the behavior of simulated agents attempting to reach a "lost party" in an inhospitable terrain, and encountering a variety of anxiety-producing setbacks (e.g., various emergencies, inadequate resources, mechanical failures). An agent may need to obtain supplies from available "supply stations," to maintain adequate resources (fuel, first aid kits). In this example, the modeled agent approaches a difficult "emergency situation" but lacks the required resources. The agent's state of anxiety, dynamically calculated by the Affect Appraiser module within the MAMID architecture, is high; in part because of a trait-induced tendency toward higher anxiety, and in part because of the task difficulty level and a lack of adequate resources.

Panic attack is an interesting anxiety state to explore because its extreme nature provides a useful context in which to model the effects of anxiety on attentional and interpretive processes, and cognition–emotion interaction in general. Panic attack is a state where the confluence of multiple anxiety effects produces a type of a "perfect storm," frequently inducing behavioral paralysis. Three anxiety-linked effects are involved: *threat processing bias, self processing bias,* and *capacity reductions in both attention and working memory.* MAMID models all three of these effects, and provides parameters that control their relative contributions to the overall effect on information processing.

The MAMID processing parameter values are calculated from linear combination of the weighted factors influencing the parameter. A specific parameter-induced effect (e.g., reduced module capacity) can thus be obtained from multiple combinations of the individual factors that influence the final value of a given parameter, and their associated weights. These alternative configurations then provide the means of defining alternative mechanisms mediating specific effects. The MAMID testbed environment provides facilities that support the rapid construction of these alternative mechanisms, via interactive manipulation of the factors and weights, which allow the modeler to control the magnitude and contribution of each influencing factor.

Anxiety-induced threat bias is modeled by first calculating the threat level of each cue, situation and expectation (mental constructs), from factors that include an a priori "fixed" threat level (e.g., low level of resources is inherently more threatening than adequate resources), state and trait anxiety factors, and individual history (prior experience with a specific type of situation that has caused anxiety before). The threat level is then used

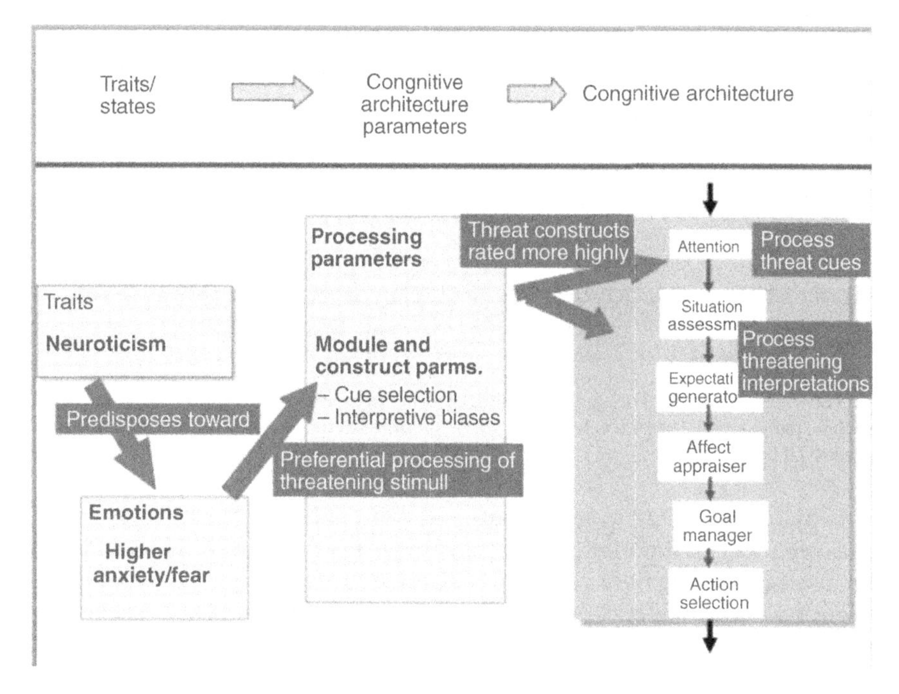

FIGURE 16.7 **Modeling threat bias within MAMID.**

as a weighted factor in the function calculating the overall construct rank, which determines the likelihood of its processing within a given execution cycle. In states of high anxiety, high-threat constructs have a higher ranking, and are thus processed preferentially. In other words, high-threat cues are given preference over low-threat cues and high-threat interpretations are therefore preferentially derived over low-threat interpretations in situation assessment (Figs. 16.7 and 16.8).

Self-bias is modeled by including a weighted factor reflecting the self versus nonself origin of each construct in its rank calculating function. High levels of state or trait anxiety then induce a higher ranking for self-related constructs, contributing to their preferred processing. In cases of high anxiety, this bias will produce a focus on the anxious state itself—a common feature of anxiety disorders that typically further increases the anxious state of the individual.

The *capacity reduction* effects on attention and working memory are modeled by dynamically calculating the capacity values of all modules during each simulation cycle, from weighted factors representing the emotion intensities, the four traits represented in MAMID, baseline capacity limits, and skill level.

MAMID models a panic attack state as follows. Stimuli, both external and internal, arrive at the Attention Module, whose capacity is

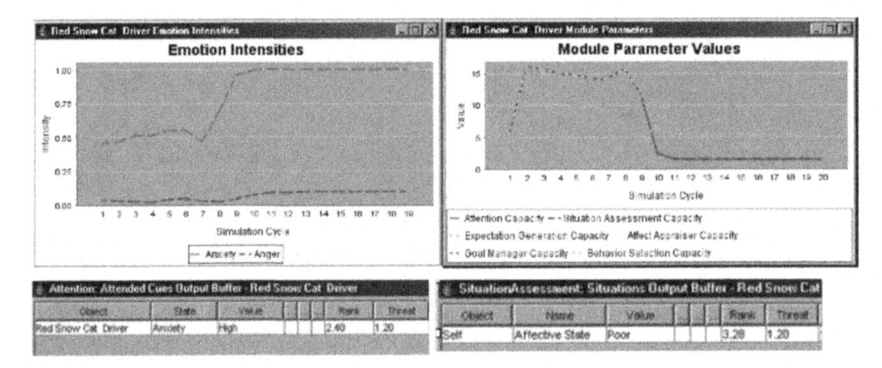

FIGURE 16.8 Affective dynamics and module capacity parameters (top) associated with a "panic attack" state (cycle 8), and subsequent reduction in the number of constructs processed (bottom).

alreadyreduced. Due to the threat- and self-bias, self-related high-threat cues are processed preferentially, in this case resulting in the agent's focus on a self-related anxiety cue (Fig. 16.8, lower left). This cue, reflecting the agent's anxious state, consumes the limited module capacity, leading to the neglect of external and nonthreatening cues (e.g., proximity of a supply station). This results in a continued self- and threat-focus in the downstream modules (Situation Assessment and Expectation Generation). No useful goals or behaviors can be derived from these constructs, and the agent enters a positive feedback-induced vicious cycle (an endless self-reflection), where the reduced-capacity and biased processing excludes cues that could lower the anxiety level and trigger adaptive behavior. Fig. 16.8 shows a diagram of the emotion intensities and module capacities, and representative contents of the cue and situation buffers, providing input to Attention and Situation Assessment modules, respectively.

A number of factors can be manipulated to induce the effects described previously, simultaneously or sequentially, reflecting multiple, alternative mechanisms mediating the anxiety biasing effects. In the case of the capacity parameters, alternative mechanisms can be defined from the agent's overall sensitivity to anxiety (reflected in the weights associated with trait and state anxiety intensity factors), the baseline, "innate" capacity limits (reflected in the factors representing the minimum and maximum attention and working memory capacities), and the anxiety intensity itself. This factor can be further manipulated via the set of parameters influencing the affect appraisal processes, including the nature of the affective dynamics (e.g., maximum intensity and the intensity ramp-up and decay functions).

The example outlined above illustrates how MAMID can represent alternative mechanisms mediating a range of anxiety behaviors, by explicitly representing the attentional and interpretive biases that contribute to the generation of anxiety, and the multiple, interacting causal pathways mediating these processes. The notion that the same underlying mechanism can result in distinct observable behavior and symptom severity, ranging from normal to severely disabling, depending on the values of the controlling parameters, is consistent with the emerging transdiagnostic model of psychopathology and MAMID thus lends itself to modeling the mechanisms mediating psychopathology from the transdiagnostic perspective. Again, it is important to emphasize that significant research would be required to validate the proposed model and that validation of emotion models represents a significant challenge in computational affective science.

Modeling the Mechanisms of Therapeutic Action

MAMID-based model of anxiety mechanisms can also be applied to modeling several proposed mechanisms of therapeutic action, focusing again on the information-processing theories of anxiety disorders and therapeutic approaches targeting specific biases; specifically, on mechanisms of action involving modifications of the attentional and interpretive biases, and those involving modifications in the cognitive appraisal process mediating emotion generation. Within the MAMID model, the former involve counteracting the effects of the biasing parameters, the latter involve reducing the parameters' values by reducing the contribution of the emotion intensity factor, and thereby reducing the magnitude of the bias.

Two of the possible pathways for reducing threat and self-bias are:

(1) directly (intentionally) reduce the biasing effects via a mechanism that either directly reduces the biasing parameter value, thereby counteracting its effect, or introduces an opposite bias (e.g., a bias to focus on nonthreatening and nonself-related cues), thereby indirectly counteracting the biasing effect; and

(2) reduce the level of anxiety that contributes to these biases, which will then reduce the value of the parameter inducing the bias, thereby reducing the biasing effect.

The MAMID models of these mechanisms are outlined below.

Reducing Attentional and Interpretive Anxiety Biasing Effects

As outlined previosuly, attentional and interpretive biases are modeled in MAMID via parameters that induce preferential processing of the mental constructs reflecting the particular bias, by increasing their rank in the processing queue. In the case of anxiety, this translates to preferential processing of threatening cues by the Attention Module, implementing *attentional biases,* and preference for threatening interpretations by the Situation Assessment and Expectation Generation modules, implementing *interpretive biases*. (Additional specific biases are implemented in an analogous manner, including self-bias, which is also associated with anxiety.)

To model intentional reduction in the anxiety (or any other) bias, the existing MAMID architecture would be augmented with an additional module, which would implement one or both of the antibiasing mechanism outlined previosuly: directly reduce the value of the biasing parameters, thereby reducing the likelihood of processing of threatening cues (in the case of attentional threat bias) or situations and expectations (in the case of interpretive threat biases), or indirectly counteract its influence by implementing an opposite biasing effect (e.g., by enhancing the ranking of the low-threat cues to counteract the attentional threat bias or by enhancing the ranking of low-threat situations and expectations to counteract the interpretive threat bias). Either approach, or both acting simultaneously, would reduce the likelihood that high-threat mental constructs will be selected for processing and thereby reduce their subsequent contribution to anxiety during emotion generation.

This module would in effect implement, in part, metacognitive processing that is a critical component of several therapeutic approaches. Initial steps toward implementations of metacognitive capabilities in MAMID are described in Hudlicka (2005).

Additional parameters would control which of the possible antibiasing effects would be implemented, as well as the threshold for the activation of the antibiasing module itself. The values of these parameters would reflect both individual differences in processing, as well as the effects of specific types of therapeutic interventions. Below, I outline how the effects of several types of psychotherapeutic interventions would be modeled in the augmented MAMID architecture.

Cognitive Bias Modification (CBM) therapy (Macleod and Mathews, 2012). A number of recent studies have demonstrated the effectiveness of retraining of the automatic processing that is thought to mediate the biasing effects, and subsequent reductions in anxiety. While speculative at this stage, it is intriguing to consider the possibility that these approaches can be modeled via direct reductions in the anxiety parameter values: the first approach outlined above.

Mindfulness-based cognitive therapy (MBCT) (Williams et al., 2008) and *metacognitive therapy* (Wells, 2002) aim to directly influence attentional deployment toward or away from particular cognitions, via techniques that train conscious attentional control. The effects of these therapeutic interventions would be modeled by the second method outlined previosuly: counteracting the anxiety biasing effects by an additional controlling parameter that would enhance the ranking of nonthreatening cues, thereby implementing conscious control of attentional deployment.

Cognitive therapy aims to identify and correct "distorted thinking styles" (e.g., catastrophizing, generalizing, etc.) and cognitive therapy interventions focus on retraining cognitive processing to correct these distortions and biases. In the augmented MAMID, these interventions would be modeled by first recognizing that biased interpretations are being generated (again, via a metacognitive mechanisms monitoring ongoing information processing), and subsequently manipulating the parameters controlling the interpretative processing in the Situation Assessment and Expectation Generation modules, by preferentially increasing the ranking of the nonthreatening situations and expectations, thereby limiting or eliminating their influence on the emotion generation processes and ultimately contributing to a reduction in anxiety.

Reducing Anxiety Intensity by Modifying the Cognitive Appraisal Process

An alternative means of reducing the anxiety biasing effects is to reduce the values of the biasing parameters themselves, by reducing the contributions of the individual factors that influence the final parameter values. The primary component of the parameter-calculating functions in MAMID is the intensity of the emotional state inducing a particular bias. In the case of anxiety, this is the intensity of the anxiety state generated via cognitive appraisal. If the anxiety intensity were reduced, the value of the biasing parameter would be lower and the consequent biasing effects would be reduced.

This mechanism would be modeled in MAMID by augmenting the existing architecture with a module that would control the cognitive appraisal process, and by modifying the currently implemented appraisal process to implement more closely the componential approach to cognitive appraisal, and its associated set of specific appraisal variables, including goal relevance, goal congruence, certainty, urgency, agency, coping, and social and individual norms. This metacognitive module would be activated when anxiety levels reached a specific threshold, and would selectively control the processes calculating the values of the appraisal variables, thereby modifying their values, which would in turn influence both the type of emotion derived and its intensity. For example, anxiety is often triggered by faulty assumptions about a lack of coping strategies, and cognitive therapy interventions often focus on helping clients recognize that they do in fact have the means to cope with anxiety-provoking situations (e.g., test anxiety, social anxiety). These types of interventions would be modeled in the augmented MAMID by triggering a reappraisal of a situation if the originally derived anxiety intensity were too high. Again, the threshold for triggering the reappraisal would be a function of individual differences, but could also be modified via psychotherapy, and could in fact be another target of therapeutic interventions. In other words, the threshold modification would aim to emulate the therapeutic approaches that focus on helping individuals recognize emotional states at lower intensities, when their regulation is easier (e.g., emotion-focused therapy). This type of therapeutic intervention would be modeled in MAMID by reducing the threshold for invoking reappraisal, thereby increasing the likelihood of successful reappraisal and subsequent reduction in anxiety.

The models outlined previosuly aim to represent the mechanisms of therapeutic action for a range of psychotherapies, including CBM, cognitive therapy, metacognitive therapy, and MBCT. More broadly, these mechanisms fall within the process model of emotion regulation proposed by Gross (1998), which identifies control of emotion deployment and modifications of cognitions mediating emotion generation via appraisal as two of the five processes involved in emotion regulation.

POTENTIAL OF MODEL-ENHANCED THERAPEUTIC GAMES

Section "Serious Therapeutic Games" outlined the potential of serious games to improve diagnosis and assessment, track treatment progress, and serve as adjuncts to traditional face-to-face psychotherapy, by supporting between-sessions homework exercises and by providing opportunities to practice specific skills. Section "Affective Disorders: A Transdiagnostic Perspective" discussed the potential of the emerging transdiagnostic model of psychopathology to provide a basis for diagnosis, assessment, treatment customization and outcome tracking, by identifying shared mechanisms mediating psychopathology and therapeutic action.

The synergy between the emerging gaming technologies, computational affective modeling, and the theoretical developments in the transdiagnostic model helps unify what have traditionally been distinct processes in the treatment of affective disorders: assessment and diagnosis, treatment planning and implementation, and outcome tracking. Serious therapeutic games, augmented with models of the shared mechanisms that mediate a variety of affective disorders, provide an immersive, engaging environment, within which the distinct processes of assessment and diagnosis, outcome tracking, and treatment can be integrated and implemented within the context of gameplay. Games increasingly incorporate elements of virtual reality, as well as affectively realistic virtual agents as the nonplaying characters. These features further enhance the games' immersive potential and thus the players' engagement and, in the case of serious games, their effectiveness.

Diagnosis and assessment can be conducted by presenting the player with a series of tasks that assess a specific aspect of processing contributing to a particular disorder; for example, degree of threat bias contributing to an anxiety disorder. Outcome tracking can be implemented by presenting the same set of tasks throughout treatment to determine whether progress is being made. Perhaps most importantly, serious therapeutic games can enhance treatment by providing an environment within which specific skills can be practiced, both high-level skill sets, such as social skills, assertiveness, and stress management, but also metacognitive skills aiming to control attentional and interpretive processes, as well as low-level tasks involving automatic processing that is outside of conscious awareness, such as some of the attentional and interpretive biases contributing to anxiety disorders. By integrating the therapeutic tasks within the gameplay, serious therapeutic games would provide environments where the actual training (treatment) would be conducted "under the radar," as it were, as the player engaged with the gameplay.

Given the key role that cognition–emotion interactions play in the mechanisms that mediate psychopathology, and the mechanisms of therapeutic action, explicit integration of these models within the therapeutic games would further enhance their effectiveness, by providing the means of mechanism-based modeling of the players' cognitive-affective processes. As mentioned previosuly, the development of valid models of cognition–emotion interactions in humans is not currently within the state-of-the-art. Nevertheless, part-task models aiming to represent specific components of these interactions, for example, specific attentional and interpretive biases, represent a promising beginning to the challenging enterprise of modeling affective mechanisms.

Below, I describe several examples of the use of model-enhanced therapeutic games to implement assessment, diagnosis and outcome tracking, and to serve as an adjunct tool to enhance traditional psychotherapy.

Assessment, Diagnosis, and Outcome Tracking

Continuing to focus on information-processing biases, the subsequent example illustrates how a set of diagnostic tasks would be incorporated within a therapeutic game to assess and quantify the type and magnitude of the biases associated with anxiety disorders, specifically, selective attention of stimuli with a particular affective tone, and subsequent biased interpretation of ambiguous stimuli to favor a particular affective tone. Attentional biases are identified via several techniques, including the *emotional Stroop task* (Williams et al., 1997) and *search tasks* involving the identification of stimuli with specific emotional tones (e.g., how rapidly are threatening stimuli located vs. neutral or positive stimuli) (Olatunji et al., 2010).

Two of the techniques used to identify interpretive biases include the following: (1) presentation of an ambiguous stimulus (e.g., ambiguous image), followed by a presentation of targets with alternative interpretations, with anxious individuals displaying faster interpretations of the meaning associated with a negative interpretation of the initial stimulus and (2) asking the subjects to first read an ambiguous scenario and then engage in a memory task testing for preferential processing of alternative interpretations of the initial scenario; again, the anxious individuals reveal a negative bias in the initial ambiguous scenario interpretation, as evidenced by faster processing of similarly toned stimuli in the subsequent memory task.

These diagnostic tasks could be readily incorporated within a gameplay to assess the type and magnitude of the associated biases, initially for assessment and diagnostic purposes and later during treatment for tracking progress and outcomes. For example, to assess attentional threat bias, the player would be presented with a set of stimuli, varying in danger or threat level, such as alternative paths to a treasure. The player's reactions to these stimuli would be tracked, including time it took to notice the stimulus as well as the time devoted to further analysis of the stimulus. As the treatment would progress, the same set of tasks would be presented to the player and any changes in the biased processing would be identified, to determine if the threat bias was changing.

The resulting data would be represented in an affective model of the player, using a modeling approach capable of representing these types of biases, such as the MAMID model outlined previosuly. An initial model would be constructed from the set of diagnostic tasks presented to the player at the beginning of treatment. As treatment would progress, the model would be updated to reflect any identified changes in processing. The model would in effect represent specific changes in the underlying mechanisms as different therapeutic interventions were explored. By enabling the modeling of alternative mechanisms of specific biases, the model could provide guidance regarding additional possible treatment interventions.

Treatment

Approaches to treating anxiety disorders by directly targeting the low-level biases are currently being developed and empirical data suggest that they represent an effective means of treatment (Macleod and Mathews, 2012). By directly targeting the underlying mechanisms that mediate psychopathology, these approaches have the potential to provide treatment that is also more efficient.

The CBM approaches mentioned earlier represent promising examples of these types of treatments and are being explored to implement targeted treatment protocols for a range of mood disorders. Therapeutic interventions based on the CBM approach aim to directly counter and correct affective biases, through targeted interventions on the biasing processing that operate outside of awareness and eventually result in changes in the higher-level cognitions, such as corrections in the distorted thought patterns that characterize anxiety disorders. By targeting interventions at the early processing stages, these approaches aim to prevent the generation of the distorted thinking styles that characterize many affective disorders, such as the catastrophizing and generalizing biases that are evident in anxiety disorders.

As outlined earlier, a variety of biases exist at multiple points in the information-processing sequence, including cognitive appraisal itself. CBM techniques have begun to be developed targeting these biases directly by augmenting existing assessment tasks with reward structures.

Anxiety disorders have been studied most extensively and positive impact of the antibias training has been demonstrated in terms of both symptom reduction in individuals with existing anxiety disorders (e.g., generalized anxiety disorder) as well as for prevention of anxiety symptoms with trait-anxious individuals who were later exposed to anxiety-inducing situations (Macleod and Mathews, 2012).

Several types of CBM training scenarios have been explored for the treatment, or prevention, of anxiety, with the most extensively evaluated training protocols targeting attentional biases (CBM-A), and interpretive biases (CBM-I). In each case, the training tasks are the same tasks that were used for assessment of these biases, but augmented with "training contingencies," that is with providing rewards with the desired choice—in the case of antibiasing training, with rewarding the nonbiased choices. The most frequently used CBM-A task used for treatment is the attentional probe task, augmented with reward structures to induce changes in processing in the desired direction (i.e., away from focus on threatening stimuli and negative interpretations), and data exist demonstrating the ability of this task to modify attentional biases to negative information (MacLeod et al., 2002; Hakamata et al., 2010). Existing studies suggest that multiple training sessions in the laboratory produce lasting positive effects and symptom reduction (Macleod and Mathews, 2012). The use of the visual search task has also demonstrated success in reducing attention to negative stimuli and symptoms of anxiety. Analogous training tasks have also been developed for CBM-I interventions, aiming to reduce the interpretive biases mediating anxiety disorders.

Games provide an ideal environment for implementing both the CBM-A and CBM-I interventions and provide opportunities both for initial systematic training, as part of targeted treatment, and for follow-up "refresher" training and ongoing practice. The high immersive potential of games, mediated in part by increasingly realistic virtual reality settings and increasing complexity and affective believability of the nonplaying characters, further enhance their therapeutic potential. The reward structures augmenting the original diagnostic tasks to implement the CBM-A and CBM-I interventions can be readily implemented within a reward structure that is embedded in the game plot. As the player acquires the targeted skill (e.g., reduced bias toward threatening stimuli), s/he accumulates rewards that allow him/her to progress through the game levels.

SUMMARY AND CONCLUSIONS

This chapter summarized the state-of-the-art in symbolic models of cognition–emotion interactions and discussed their relevance for research in psychopathology and the mechanisms of psychotherapeutic action, as well as their applicability to the development of technologies for behavioral health. The focus was on mechanisms of affective biases on attention and interpretive cognitive processes, which are thought to contribute to a range of affective disorders, including anxiety and depression. The MAMID methodology for modeling a broad range of affective biases on cognition was described, along with the associated cognitive-affective architecture. Examples of MAMID-based models of hypothesized mechanisms mediating psychopathology, as well as models of possible therapeutic actions involving bias modification, were then provided. The chapter concluded with a discussion of serious therapeutic games and the ability of these games to support improved assessment and diagnosis, outcome tracking, and support treatment as adjuncts to face-to-face psychotherapy. The focus was on discussing how games augmented with models of the mechanisms mediating psychopathology and therapeutic action, implemented in terms of symbolic models of cognition–emotion interactions, would enhance the effectiveness and efficacy of assessment, diagnosis, outcome tracking, and treatment itself. The emerging synergy between the increased focus on the mechanisms of psychopathology and therapeutic action, the transdiagnostic approach to psychopathology, cognitive-affective computational modeling, and gaming technologies was highlighted.

Below, I briefly discuss ethical concerns that emerge as advanced technologies are applied in behavioral health contexts, and conclude with a brief discussion of open questions and challenges in computational modeling of psychopathology.

Ethical Concerns

The use of technologies, such as serious games, virtual reality, and affective virtual agents in behavioral healthcare introduces a range of ethical concerns and dilemmas. The context of therapeutic technologies, in conjunction with the significant emphasis on the user's emotion, raises an entirely new set of ethical considerations. In addition to the fundamental issues of privacy, made more significant by the fact that affective data are concerned, which most people consider to be the most private aspect of their identity, there are potential ethical concerns regarding emotion induction by these technologies and the possibility of inadvertent interference with adaptive affective mechanisms during treatment. Although less likely at this point in time, there is also the possibility that individuals engaging with immersive games incorporating virtual reality and intelligent affective virtual characters may begin to develop relationships with these characters. The ethics of inducing "virtual" relationships with artificial agents is an unexplored area but one that will need to be addressed as agent technologies continue to proliferate.

Affective Privacy

Our emotions, perhaps even more so than our thoughts, are likely the most personal and private aspects of our lives. The development and use of applications that sense, infer, monitor, or aim to model our emotions therefore presents considerable and as yet unexplored ethical challenges. This is especially the case in any applications in behavioral health, where the users may be addressing a particularly painful experience or reveal emotions and thoughts that could have negative repercussions on their lives if they were made public, or revealed to other parties (e.g., employers, insurance companies). Affective modeling, a necessary component of therapeutic technologies, thus presents an ethical challenge, since these models may contain the most guarded personal information about the users: the emotions they feel, including "undesirable" emotions; the events that trigger those emotions, including triggers that may be considered inappropriate, etc. The existing concerns about data security and privacy protection are thus multiplied when we consider the personal nature of data now being collected. As remote sensing and emotion recognition technologies continue to advance, it will be increasingly possible to intrude into our private lives, including our internal affective experiences. It is therefore essential that both researchers and healthcare professionals actively address the ethical challenges associated with the privacy of affective data and become active and proactive, both within their specific professional organizations and more broadly at the political level, as government surveillance increases.

Emotion Induction

Serious games, particularly games with virtual affective agents as non-playing characters, not only have the potential to induce strong emotions but may often be designed with the express purpose of inducing specific emotions, including negative emotions. Virtual agents acting as coaches in serious therapeutic games may be designed to induce affection, so that they are viewed as empathic, that their message is trusted, and they can be more persuasive (e.g., induce behavior change, such as increasing exercise, reducing substance use). In some behavioral healthcare applications, games may be designed to induce a negative or unpleasant emotion; for example, when the technology aims to implement exposure-based treatments that require some degree of anxiety to be effective, such as PTSD or OCD treatments based on exposure.

Of course, computers already induce a range of negative emotions on a daily basis, without explicitly attempting to, and computer games designed for entertainment also induce a range of emotions, including frustration, disappointment, even rage. However, the use of these technologies in behavioral health, where emotion manipulation may be desirable and necessary to achieve the therapeutic goals, presents a unique set of ethical concerns.

How can we ensure that the induced emotions will not overwhelm the user or have a deleterious effect in the future? What if systematic desensitization occurs to the wrong stimulus (e.g., simulated violence or abusive behavior)? What emotions can ethically be induced in the user? In therapeutic contexts, it is generally accepted that the induction of some negative emotions is not only acceptable and unavoidable but even necessary for treatment (e.g., anxiety induction during exposure therapies; shame experience in a safe, supportive setting to help individuals reorganize some trauma-induced cognitive-affective schema).

Most existing active therapeutic systems are designed to be used under the supervision of a clinician (e.g., "Ricky and the Spider" OCD treatment game, The "Secret Agent Society" social skills and emotion regulation game, "Virtual Iraq" PTSD treatment environment). However, it is conceivable that users could gain access to these technologies without the supervision of a clinician, or that the supervising clinician might not be aware that a negative emotion of undesirable intensity was being induced.

There are no easy answers to these ethical dilemmas and carefully monitored application of these technologies, along with extensive education of both the end-users and the professionals administering the technologies regarding the possible risks, will be essential.

Virtual Relationships

Humans are "wired to relate" and one of the most powerful features of the affective agent technologies is precisely their ability to induce the "relational instinct" and attachment behavior in the human users (Reeves and Nass, 1996). Through the ensuing connection, and associated trust, agents can then provide support and coaching even social companionship. By inducing attachment and trust, virtual affective agents have the potential to mimic aspects of human relationships and humans can thus, at least theoretically, enter into relationships with agents. These virtual relationships present yet another ethical challenge.

How does a human user know when to trust a virtual agent? If the agent appears "confident" (e.g., a nonplaying character in a therapeutic game encouraging the player to engage in a particular behavior in his/her life), is this an indication that it should be trusted? What about the risk of virtual relationships replacing actual human relationships? What if a user of a therapeutic game finds the synthetic nonplaying character more compelling, and empathic, than his/her friends and family members and reduces contact with family, eventually investing emotions into a relationship with an entity not capable of experiencing real emotions or engaging in an emotionally responsible relationship? What if the virtual agent uses its persuasive capabilities to induce beliefs and behavior that are not in the player's best interests?

We may think that these are outlandish possibilities, given the state-of-the-art of agent technologies. However, we only have to consider the fact that DSM-5 has identified both "Internet Addiction" and "Internet Gaming Disorder" as possible new Axis I diagnostic categories to realize that as the virtual agent technologies advance we may soon be facing a variety of new potential disorders to consider, including, perhaps, a "Virtual relationship addiction disorder."

Some of these issues have begun to be raised by several agents and affective computing researchers (e.g., Castellano et al., 2010; Bickmore and Picard, 2005; Picard and Klein, 2002) but the research community has a long way to go in adequately understanding and addressing the ethics of virtual relationships.

Challenges and Open Questions

A number of challenges exist in all three of the key areas discussed in this chapter: identifying the mechanisms mediating psychopathology and therapeutic action, developing computational models of cognition–emotion interactions, and developing effective and engaging serious therapeutic games. However, given the significant progress in gaming technologies, along with advances in virtual affective agents and virtual reality to enhance the games' immersive potential, the gaming component is the least problematic.

Development and validation of computational models of affective processes represents the most significant challenge. Computational affective modeling is in its infancy, more so for research models that aim to emulate affective processing and cognition–emotion interactions in biological agents, than for applied models whose aim is to improve agent believability. While a number of models have been developed (refer to Section "State-of-the-Art in Modeling Emotion-Cognition Interaction"), the enterprise of computational affective modeling faces the expected challenges of most new discipline, foremost among these being a lack of established guidelines and standards for model design and development, and a lack of tools and shareable components to facilitate development (Hudlicka, 2015a). Promising efforts are being made in this area (e.g., Broekens et al. 2008; Hudlicka, 2012; Reisenzein et al., 2013) but significant cross-disciplinary collaborations among psychologists, neuroscientists, modelers, and clinicians will be necessary to advance the state-of-the-art.

Another challenge is the development of models that adequately represent the multiple modalities of emotions and their interactions. Our understanding of these interactions is limited, and most computation-friendly affective theories focus on the cognitive modality, as do most models of emotions. This limits the ability of contemporary emotion models to accurately model the mechanisms of psychopathology and therapeutic action, since emotions are inherently multimodal phenomena, and their dysregulation in psychopathology involves all four of the modalities, as does effective treatment. Thus more data, more refined theories, and likely novel computational approaches (e.g., dynamical systems) will be required to construct models that more accurately reflect the multimodal nature of emotions and provide more accurate representations of affective processing in humans.

The most significant challenge in affective modeling is model validation. The development of validation methods and criteria poses a significant challenge for research models aiming to emulate biological mechanisms mediating affective processing and cognition–emotion interactions, since our ability to construct detailed symbolic models far surpasses our ability to determine whether the corresponding model structures and processes actually exist in the mind, recent imaging studies in affective neuroscience notwithstanding.

However, the coupling of theoretical and empirical efforts to identify affective mechanisms, as represented by the recent emphasis on the transdiagnostic model, and computational modeling efforts aimed at modeling these processes, provides a unique environment within which these mechanisms can be explored, via coupled empirical and modeling approaches. These approaches would involve collaboration between emotion researchers in psychology and neuroscience, and computational modelers, which would enable research programs involving tight coupling between empirical studies with humans, providing initial data for modeling, and subsequent computational modeling of the hypothesized mechanisms. The models would then be used to generate more detailed hypotheses regarding the mechanisms, which could then be tested via further empirical studies, eventually, hopefully, yielding more accurate models of the mechanisms of psychopathology and therapeutic action via progressive iterative refinement of the computational models.

In conclusion, in spite of the ethical and technological challenges, as well as many theoretical open questions, the synergy between computational modeling, the increased focus on the mechanisms mediating psychopathology, embodied in the transdiagnostic model, and serious gaming technologies promise to significantly advance the state-of-the-art in understanding the mechanisms that mediate affective disorders, and in their treatment, via serious games that integrate diagnosis, intervention, and outcome tracking.

References

Andre, E. et al., 2000. Exploiting Models of Personality and Emotions to Control the Behavior of Animated Interactive Agents Proceedings of IWAI. Siena, Italy.

Arnold, M.B., 1960. Emotion and personality. Columbia University Press, New York.

Averill, J.R., 1994. I feel, therefore I am—I think. In: Ekman, P., Davidson, R.J. (Eds.), The Nature of Emotion: Fundamental questions. Oxford University Press, Oxford.
Bar-Haim, Y., et al., 2007. Threat-related attentional bias in anxious and nonanxious individuals: a meta-analytic study. Psychol. Bull. 133, 1–24.
Bates, J., et al., 1992. Integrating reactivity, goals, and emotion in a broad agent. Proceedings of the 14th Meeting of the Cognitive Science Society.
Beaumont, R., Sofronoff, K., 2008. A multi-component social skills intervention for children with Asperger syndome: the Junior Detective Training Program. J. Child Psychol. Psychiatry 49, 743–753.
Beck, A.T., 1967. Depression: clinical experimental aspects. Harper & Row, NY, New York.
Becker-Asano, C., 2008. WASABI: Affect Simulation for Agents With Believable Interactivity. IOS Press, Clifton, VA.
Bickmore, T., Picard, R.W., 2005. Establishing and maintaining long-term human-computer relationships. ACM Transactions on Computer-Human Interaction (TOCHI) 12 (2), 293–327.
Blaney, P.H., 1986. Affect and memory. Psychol. Bull. 99 (2), 229–246.
Bower, G.H., 1981. Mood and memory. Am. Psychol. 36, 129–148.
Bower, G.H., 1992. How might emotions affect memory? In: Christianson, S.A. (Ed.), Handbook of Emotion and Memory. Lawrence Erlbaum, Hillsdale, NJ.
Brezinka, V., 2008. Treasure hunt—a serious game to support psychotherapeutic treatment of children. In: Andersen, S.K. (Ed.), eHealth Beyond the Horizon—Get IT There. IOS Press, Clifton, VA.
Brezinka, V., Hovestadt, L., 2007. Serious Games Can Support Psychotherapy of Children and Adolescents. HCI and Usability for Medicine and Health Care. Springer, Berlin, vol. 4799.
Broekens, J., DeGroot, D., Kosters, W.A., 2008. Formal models of appraisal: theory, specification, and computational model. Cogn. Syst. Res. 9 (3), 173–197.
Castellano, G., et al., 2010. Affect recognition for interactive companions: challenges and design in real world scenarios. J. Multimodal. User Int. 3 (1–2), 89–98.
Charney, D.S., 2004. Discovering the neural basis of human social anxiety: a diagnostic and therapeutic imperative. Am. J. Psychiatry 161, 1.
Clore, G.L., 1994. Why emotions are felt? In: Ekman, P., Davidson, R.J. (Eds.), The Nature of Emotion: Fundamental Questions. Oxford University Press, Oxford.
Coyle, D., et al., 2005. Personal investigator: a therapeutic 3D game for adolescent psychotherapy. J. Int. Technol. Smart Education 2, 73–88.
Derryberry, D., 1988. Emotional influences on evaluative judgments: roles of arousal, attention, and spreading activation. Motiv. Emot. 12 (1), 23–55.
Derryberry, D., Reed, M.A., 2002. Anxiety-related attentional biases and their regulation by attentional control. J. Abnor. Psychol. 111, 225–236.
Dias, J., et al., 2015. FAtiMA modular: towards an agent architecture with a generic appraisal framework. In: Tibor Bosse, J.B., Joao Dias, Janneke van der Zwaan (Eds.), Towards Pragmatic Computational Models of Affective Processes. Springer, New York.
Edland, A., 1989. On cognitive processes under time stress: a selective review of the literature on time stress and related stress. Reports from the Department of Psychology. University of Stockholm, Stockholm.
Ekman, P., 1992. n argument for basic emotions. Cogn. Emot. 6 (3–4), 169–200.
Ekman, P., Davidson, R.J., 1994. The Nature of Emotion: Fundamental Questions. Oxford, New York, NY.
Fellous, J.M., 2004. From human emotions to robot emotions. AAAI Spring Symposium: Architectures for Modeling Emotion. Stanford University, AAAI Press, CA.
Forgas, J., 1999. Mood and judgment: the affect infusion model (AIM). Psychol. Bull. 117 (1), 39–66.
Forgas, J., 2003. Affective influences on attitudes and judgments. In: Davidson, K.R.S.R.J., Goldsmith, H.H. (Eds.), Handbook of Affective Sciences. Oxford University Press, New York, NY.

Frijda, N.H., 1986. The Emotions. Cambridge University Press, Cambridge.
Frijda, N., 2008. The psychologists' point of view. In: Lewis, M., Haviland-Jones, J.M., Barrett, L.F. (Eds.), Handbook of Emotions. The Guilford Press, New York, NY.
Gasper, K., Clore, G.L., 2002. Attending to the big picture: mood and global versus local processing of visual information. Psychol. Sci. 13 (1), 34–40.
Gratch, J., Marsella, S., 2004. A domain independent frame-work for modeling emotion. J. Cogn. Syst. Res. 5 (4), 269–306.
Gross, J.J., 1998. The emerging field of emotion regulation: an integrative review. Rev. Gen. Psychol. 2 (5), 271–299.
Hakamata, Y., et al., 2010. Attention bias modification treatment: a meta-analysis toward the establishment of novel treatment for anxiety. Biol. Psychiatry 68, 982–990.
Hayes, S., et al., 2010. Facilitating a benign attentional bias reduces negative thought intrusions. J. Abnorm. Psychol. 119, 235–240.
Herrmann, I.R., et al., 2016. Emotion categories and patterns of change in experiential therapy for depression. Psychother. Res. 26 (2), 178–195.
Hertel, P.T., Mathews, A., 2011. Cognitive bias modi cation: past perspectives, current findings, and future applications. Perspect. Psychol. Sci. 6 (6), 521–536.
Hudlicka, E., 1998. Modeling Emotion in Symbolic Cognitive Architectures. AAAI Fall Symposium: Emotional and Intelligent I. AAAI Press, Orlando, FL.
Hudlicka, E., 2003. Modeling Effects of Behavior Moderators on Performance: Evaluation of the MAMID Methodology and Architecture. BRIMS-12, Phoenix, AZ.
Hudlicka, E., 2007. Reasons for emotions. In: Gray, W. (Ed.), Advances in Cognitive Models, Cognitive, Architectures. Oxford, New York, NY.
Hudlicka, E., 2005. MAMID-ECS: Application of Human Behavior Models Capable of Modeling Individual Differences to Risk-Analysis and Risk-Reduction Strategy Development in Human-System Design. Psychometrix Associates, Inc, Blacksburg, VA.
Hudlicka, E., 2008a. What are we modeling when we model emotion? AAAI Spring Symposium: Emotion, Personality, and Social Behavior. Stanford University, CA, AAAI Press, Menlo Park, CA. Technical Report SS-08-04: 52–59.
Hudlicka, E., 2008b. Modeling the mechanisms of emotion effects on cognition. AAAI Fall Symposium: Biologically Inspired Cognitive Architectures. AAAI Press, Arlington, VA, Menlo Park, CA. TR FS-08-04 82–86.
Hudlicka, E., 2011. Affective gaming in education, training and therapy: motivation, requirements, techniques. In: Felicia, P. (Ed.), Handbook of Research on Improving Learning and Motivation through Educational Games: Multidisciplinary Approaches. IGI Global, Hershey, PA.
Hudlicka, E., 2015a. From habits to standards: towards systematic design of emotion models and affective architectures. In: Tibor Bosse, J.B., Dias, Joao, Zwaan, Janneke van der (Eds.), Towards Pragmatic Computational Models of Affective Processes. Springer, New York, pp. 1–21.
Hudlicka, E., 2012. Guidelines for Designing computational models of emotions. Int. J. Synthetic Emot. (IJSE) 2 (1), 26–79.
Hudlicka, E., 2016. Virtual Companions, coaches, and therapeutic games in psychotherapy. In: Luxton, D.D. (Ed.), Artificial Intelligence in Mental Healthcare. Academic Press/Elsevier, Waltham, MA.
Isen, A.M., 1993. Positive affect and decision making. In: Haviland, J.M., Lewis, M. (Eds.), Handbook of Emotions. Guilford, New York, NY.
Izard, C.E., 1993. Four systems for emotion activation: cognitive and noncognitive processes. Psychol. Rev. 100 (1), 68–90.
Kahn, G.E., Isen, A.M., 1993. The influence of positive affect on variety seeking among safe, enjoyable products. J. Cons. Res. 20 (2), 257–270.
Kring, A.M., 2008. Emotion disturbances as transdiagnostic processes in psychopathology. In: Lewis, J.H.-J.M., Barrett, L.F. (Eds.), Handbook of Emotion. The Guilford Press, New York, pp. 691–705.

Lazarus, R.S., 1984. On the primacy of cognition. Am. Psychol. 39 (2), 124–129.
LeDoux, J., 2014. Coming to terms with fear. PNAS 111 (8), 2871–2878.
Lerner, J.S., Tiedens, L.Z., 2006. Portrait of the angry decision maker: how appraisal tendencies shape anger's influence on cognition. J. Behav. Decis. Mak. 19, 115–137.
Leventhal, H., Scherer, K.R., 1987. The relationship of emotion to cognition. Cogn. Emot. 1, 3–28.
Macleod, C., Mathews, A., 2012. Cognitive bias modification approaches to anxiety. Annu. Rev. Clin. Psychol. 8, 189–217.
MacLeod, C., et al., 2002. Selective attention and emotional vulnerability: assessing the causal basis of their association through the experimental manipulation of attentional bias. J. Abnorm. Psychol. 111, 107–123.
Mandler, G., 1984. Mind and Body: The Psychology of Emotion and Stress. Norton, New York.
Marr, D., 1982. Vision. Freeman, San Francisco, CA.
Martinez-Miranda, et al., 2015a. Modelling two emotion regulation strategies as key features of therapeutic empathy. In: Tibor Bosse, J.B., Dias, Joao, Zwaan, Janneke van der (Eds.), Towards Pragmatic Computational Models of Affective Processes. Springer, New York.
Matthews, G.A., Harley, T.A., 1993. Effects of extraversion and self-report arousal on semantic priming: a connectionist approach. J. Pers. Soc. Psychol. 65 (4), 735–756.
Mathews, A., MacLeod, C., 2005. Cognitive vulnerability to emotional disorders. Annu. Rev. Clin. Psychol. 1 (1), 167–195.
Mellers, B.A., et al., 1998. Judgment and decision making. Annu. Rev. Psychol. 49, 447–477.
Mehrabian, A., 1995. Framework for a comprehensive description and measurement of emotional states. Genet. Soc. Gen. Psychol. Monogr. 121, 339–361.
Mineka, S., Sutton, S.K., 1992. Cognitive biases and the emotional disorders. Psychol. Sci. 3 (1), 65–69.
Mineka, S., et al., 2003. Cognitive biases in emotional disorders: information processing and social-cognitive perspectives. In: Davidson, R.J., Scherer, K.R., Goldsmith, H.H. (Eds.), Handbook of Affective Science. Oxford, NY.
Niedenthal, P.M., Marcus, M., 2009. Embodied emotion considered. Emot. Rev. 1 (2), 122–128.
Nolen-Hoeksema, S., Watkins, E.R., 2011. A heuristic for developing transdiagnostic models of psychopathology: explaining multifinality and divergent trajectories. Perspect. Psychol. Sci. 6 (6), 589–609.
Olatunji, B.O., et al., 2010. Efficacy of cognitive behavioral therapy for anxiety disorders: a review of meta-analytic findings. Psychiatr. Clin. North Am. 33, 557–577.
Ortony, A., Clore, G., Collins, A., 1988. The Cognitive Structure of Emotions. Cambridge, New York, NY.
Ortony, A., et al., 2005. Affect and proto-affect in effective functioning. In: Fellous, J.M., Arbib, M.A. (Eds.), Who Needs Emotions?. Oxford, New York, NY.
Panskepp, J., 1998. Affective Neuroscience: The Foundations of Human and Animal Emotions. Oxford University Press, New York, NY.
Picard, R., 1997. Affective Computing. The MIT Press, Cambridge, MA.
Picard, R., Klein, J., 2002. Computers that recognise and respond to user emotion: theoretical and practical implications. Int. Comput. 14, 119–140.
Plutchik, R., 1984. Emotions: a general psychoevolutionary theory. In: Scherer, K.R., Ekman, P. (Eds.), Approaches to Emotion. Erlbaum, Hillsdale, N.J.
Reeves, B., Nass, C., 1996. The media equation: how people treat computers, televisions and new media like real people and places. Cambridge University Press, Cambridge, UK.
Reilly, W.S.N., 2006. Modeling What Happens Between Emotional Antecedents and Emotional Consequents. ACE 2006, Vienna, Austria.
Reisenzein, R., 2001. Appraisal processes conceptualized from a schema-theoretic perspective: contributions to a process analysis of emotions. In: Scherer, K.R., Schorr, A., Johnstone, T. (Eds.), Appraisal Processes in Emotion: Theory, Methods, Research. Oxford University Press, New York, NY.

Reisenzein, R., et al., 2013. Computational modeling of emotion: toward improving the inter- and intradisciplinary exchange. IEEE Trans. Affect. Comput. 4 (3), 246–266.

Ritter, F.E., Avramides, M.N., 2000. Steps Towards Including Behavior Moderators in Human Performance Models in Synthetic Environments. State College, The Pennsylvania State University, PA.

Rizzo, A.S., Difede, J., Rothbaum, B.O., Reger, G., Spitalnick, J., Cukor, J., 2010. Development and early evaluation of the Virtual Iraq/Afghanistan exposure therapy system for combat related PTSD. Ann. N. Y. Acad. Sci. 1208 (1), 114–125.

Roseman, I.J., 2001. A Model of appraisal in the emotion system. In: Scherer, K.R., Schorr, A., Johnstone, T. (Eds.), Appraisal Processes in Emotion: Theory, Methods, Research. Oxford, New York, NY.

Roseman, I.J., Smith, C.A., 2001. Appraisal theory: overview, assumptions, varieties, controversies. In: Scherer, K.R., Schorr, A., Johnstone, T. (Eds.), Appraisal Processes in Emotion: Theory, Methods, Research. Oxford, New York, NY.

Russell, J., 2003. Core Affect and the psychological construction of emotion. Psychol. Rev. 110 (1), 145–172.

Russell, J., Barrett, L.F., 1999. Core affect, prototypical emotional episodes, and other things called emotion: dissecting the elephant. J. Pers. Soc. Psychol. 76 (5), 805–819.

Russell, J., Mehrabian, A., 1977. Evidence for a three-factor theory of emotions. J. Res. Pers. 11, 273–294.

Sanislow, C.A., Pine, D.S., Quinn, K.J., Kozak, M.J., Garvey, M.A., Heinssen, R.K., 2010. Developing constructs for psychopathology research: research domain criteria. J. Abnorm. Psychol. 119 (4), 631–639.

Scherer, K.R., 2001. Appraisal considered as a process of multilevel sequential checking. In: Scherer, K.R., Schorr, A., Johnstone, T. (Eds.), Appraisal Processes in Emotion: Theory, Methods, Research. Oxford, New York, NY.

Scherer, K., Schorr, A., et al., 2001. Appraisal Processes in Emotion: Theory, Methods, Research. Oxford, New York, NY.

Schwarz, N., Clore, G.L., 1988. How do I feel about it? The information function of affective states. In: Fiedler, K., Forgas, J.P. (Eds.), Affect, Cognition, and Social Behavior. Hogrefe, Toronto, pp. 44–62.

Sloman, A., Chrisley, R., Scheutz, M., 2005. The architectural basis of affective states and processes. In: Fellous, J.-M., Arbib, M.A. (Eds.), Who Needs Emotions?. Oxford University Press, New York, NY.

Smith, C.A., Kirby, L., 2000. Consequences require antecedents: toward a process model of emotion elicitation. Feeling and Thinking: the role of affect in social cognitionJ. P. Forgas, NY, Cambridge.

Smith, C.A., Kirby, L.D., 2001. Toward delivering on the promise of appraisal theory. In: Scherer, K.R., Schorr, A., Johnstone, T. (Eds.), Appraisal Processes in Emotion. Oxford, New York, NY.

Sonnemans, J., Frijda, N.H., 1994. The structure of subjective emotional intensity. Cogn. Emot. 8 (4), 329–350.

Wells, A., 2002. GAD, metacognition, and mindfulness: an information processing analysis. Clin. Psychol. Sci. Pract. 9 (1), 95–100.

Williams, J.M.G., et al., 1997. Cognitive Psychology and Emotional Disorders. John Wiley, NY.

Williams, J.M.G., et al., 2008. Mindfulness-based cognitive therapy: further issues in current evidence and future research. J. Consult. Clin. Psychol. 76 (3), 524–529.

CHAPTER

17

Emotions in Driving

Myounghoon Jeon
Michigan Technological University, Houghton, MI, United States

INTRODUCTION

Driving is a complex, ambiguous, and indeterminate task, where emotions and affect can have enormous consequences. For instance, a road-rage phenomenon provides a representative example of the impacts that emotions can have on driving safety. This extreme case, however, is not alone in describing affect-related driving situations. Consider a driver who decides to take the long route to work to avoid the dense traffic that agitated him or her the last time he or she tried the "short" route. This mundane example demonstrates that emotions provide an omnipresent backdrop to our everyday experience (Algoe and Fredrickson, 2011).

Despite the importance and prevalence of affective states in driving, a systematic approach to affect-related driving research has yet to be developed. Some driving models have contained basic affective elements (Van Elslande and Fouquet, 2007), but they are not that thorough. Most of the driving models have focused more on cognitive aspects of the drivers. Overall, affect has been considered peripherally and sporadically in driving behavior research. Given that affective elements allow for a systems approach (Czaja and Nair, 2006) with a more holistic view to understand the human–machine system, the inclusion of such necessary elements in behavior modeling will enrich driving research.

This chapter attempts to examine the roles and effects of emotions and affect on driving behavior in terms of both theoretical and practical aspects. First, driving behavior models are briefly reviewed with a focus on plausibility to include affective effects. Then, affect-related driving research is outlined including theoretical effects, empirical effects, vulnerable populations, and in-vehicle systems. Finally, implications for future works are presented.

Emotions and Affect in Human Factors and Human–Computer Interaction. http://dx.doi.org/10.1016/B978-0-12-801851-4.00017-3
Copyright © 2017 Elsevier Inc. All rights reserved.

DRIVING BEHAVIOR MODELS

Driving models have been developed to mainly account for primary tasks (i.e., direct driving maneuvers). Some models have focused on lower-level control behaviors, such as lane keeping and curve negotiation (Boer, 1996; Boer and Hildreth, 1999; Donges, 1978; Hess and Modjtahedzadeh, 1990; McRuer et al., 1977; Reid et al., 1981). Other integrated models have tried to combine lower-level control behaviors with higher-level information processes (Levison and Cramer, 1995; Wierwille and Tijerina, 1998). The higher-level elements of the integrated driver model include maintaining situation awareness (SA), determining strategies for navigation, deciding when to initiate and terminate maneuvers, and managing secondary (i.e., functions that increase driving safety) and tertiary (i.e., infotainment systems) tasks (Salvucci et al., 2001). These integrated driver models can be classified into affective models and cognitive models. Note that the driving models described here are examples of each category and not intended to be exhaustive.

Cognitive Driving Models

Among a number of cognitive driving models, a few more effective and widely used models are described here. I classified them into hierarchical operator models and dynamic systems models. The hierarchical operator models focus more on an individual driver's information processing, whereas the dynamic systems models extend the range to the relationship between the driver and tasks or the driver and environments. In both categories, a driving-specific model and a more generic model are delineated, each.

Hierarchical Operator Models

Michon (1985) introduced the hierarchical structure of driver behavior and cognitive architecture approaches. Both approaches assume interactive hierarchical elements with a focus on individual information process, relying on the traditional cognitive science framework. As such, almost all hierarchical driving models have adopted the three-level hierarchy structure. The Hierarchical Driver Model (Boer et al., 1998) is one of those examples, which consists of an operational, tactical, and strategic decision maker in a bottom-up fashion: the operational decision maker is responsible for low-level, skill-based controllers including actions, such as changing lanes and speeding up. The tactical decision maker is responsible for attention allocation, task scheduling, and communicating performance-related information to the performance mental model. The strategic decision maker is responsible for driver needs and individual differences and assesses whether the motivational utility value exceeds the constraining utility value. If this is the case, the current strategy can be applied, but if not, alternative strategies need to be explored. The model has the potential to embrace affective elements, such as a driver's mood or time pressure in a strategic decision-making layer (Boer et al., 1998). However, the model does not clearly identify those factors as its variables. Also, the other two levels have no room to include affective elements.

The Cognitive Architecture Model (Salvucci et al., 2001) is a computational driver model based on the traditional cognitive architecture, ACT-R (Anderson and Lebiere, 1998). The model also has three hierarchical components: *controlling*, *monitoring*, and *decision-making*, but its core foundation is the skill knowledge, specified as condition-action rules. For example, the model can examine the situation and decide whether or not to execute a lane change action based on the conditions made of controlling and monitoring behavior, in which visual attention is a core function. Clearly, this cognitive architecture model has advantages in integrating lower-level perceptual and motor control processes with higher-level cognitive and decision-making processes. However, in this model, there is no room for any affective elements to play a role with respect to driving behavior. With this type of computational model, initial parameters based on observation are critical. Thus, if the predefined parameters do not include any affective elements (Salvucci et al., 2001), the model's predictions would not be influenced by any affective states. Then, the model would explain a driver's behavior only with cognitive parameters, which can account for only partial behavior.

Dynamic Systems Models

Dynamic systems models go beyond the traditional individual information processing level, engaging more actively in the relationship between an operator, tasks, and contexts. This systems approach is expected to have more room to embrace affective elements in the model.

The Task–Capability Interface (TCI) Model describes a dynamic interaction between the driver capability and the determinants of task demand (Fuller, 2005; Fuller and Santos, 2002). In this model, task difficulty that arises out of the dynamic interaction is the key determinant to the driving behavior and control. According to the TCI Model (Fuller, 2005), driver capability has three elements: biological characteristics, knowledge and skills, and human factor variables. This human factor aspect can further include affective elements, but those are not specified in the model. Driving task demands are also identified by the interaction of various additional elements, such as environmental factors, other drivers, and operational features. The TCI Model seems to consider more affective elements than the hierarchical operator models. Nonetheless, it limits the scope of affect (e.g., arousal level) and its effects to only competence in capability and speed in task demand. The model does not account for the

TABLE 17.1 Cognitive Components in Driving Behavior Models

Model	Emphasized cognitive components
Hierarchical Driver Model	Attention[a], mental model (encoding, prediction, evaluation, and judgment[a]), decision-making[a]
Cognitive Architecture (ACT-R) Model	Visual attention[a], shift attention[a], perception, decision-making[a]
TCI Model	Knowledge and skills, decision-making[a]
SA Model	Attention[a], perception, comprehension, projection (judgment[a]), decision-making[a]

SA, Situation awareness; TCI, task–capacity interface.
[a]*Common cognitive factors in driving models, which are examined further.*

plausible impact of affective states on other critical components, such as decision-making.

The SA Model (Endsley, 1995) explains the components of operator SA, with a focus on dynamics of the operator and environments. The SA Model illustrates three states of SA formation: perception, comprehension, and projection. Specifically, projection state extrapolates all information pieces gathered from previous stages forward in time to determine how they will influence future states of the operational environment, which makes this model unique. In addition to three states, attention and working memory are treated as critical factors. They direct and limit operators in acquiring and interpreting information from the environment to form SA. Workload and stress are partially addressed in the model, but no other affective states are integrated into the model. Several researchers have revealed positive associations between SA and one or more dimensions of driving performance (Ma and Kaber, 2005, 2007) and developed their own driver SA Model (Gugerty, 2011; Ma and Kaber, 2005; Matthews et al., 2001). However, there is no standard SA Model in driving so far.

As shown in Table 17.1, integrated driver models commonly include attention and decision-making as their primary cognitive factors in dynamic driving situations. In addition, a couple of models include judgment, which is considered as a process prior to decision-making, but is often not differentiated from decision-making. Judgment seems to share functions with the mental model (e.g., predict and evaluate) in the Hierarchical Driver Model and serve as a part of comprehension and projection in the SA Model.

Affective Driving Models

There have been early driving models that attempt to reflect affective aspects of the driver. The majority of those models have assumed risk of collision as a core part of the decision-making process to achieve safe driving. This notion dates back to the 1960s (Taylor, 1964). In the same line, researchers (e.g., risk homeostasis theory: Adams, 1985; Fuller, 1984; Wilde, 1982) have proposed that drivers are motivated to reach an *optimal* level of risk. If the level of perceived risk is higher than the target risk, drivers try to reduce the perceived risk. From the slightly different perspective, but within the "risk" boundary, Näätänen and Summala (1976) suggested the zero risk theory, which assumes that drivers are *always* motivated to avoid risk. According to this theory, drivers are not likely to perceive risk in general, but only if the risk level reaches a certain threshold, they will perceive risk and respond to it.

Overall, it seems desirable that these models include affective elements in the driving model. However, those elements seem to be fairly limited to perceived "risk," which can be translated into "fear" or "anxiety" in affect terms (Fuller, 2005; Mesken, 2006). Fuller (2005), one of the proposers of the "risk" model, argued later that risk of collision is generally not relevant in the driver decision-making loop. Based on Fuller, what is relevant is the feedback regarding the difficulty of the driving task, just as described earlier in the cognitive model section. Moreover, these models addressed only *integral* affect (i.e., driving-relevant emotions: e.g., tailgating), which implies that effects of *incidental* affect (i.e., driving-irrelevant emotions: e.g., fighting with a friend before driving) on driving behavior should be addressed further.

LITERATURE ON AFFECT-RELATED DRIVING RESEARCH

Specific Affective Effects on Driving Behaviors

Road Rage and Aggressive Driving

One of the undeniable examples of the affective effects on driving performance and safety is road rage and aggressive driving (Galovski and Blanchard, 2004). Road rage has become one of the top three highway threats, along with drunken driving and failure to use a seat belt (Bowles and Overberg, 1998) and aggressive drivers are even more threatening to motorists than drunken drivers (Joint, 1995; Mizell et al., 1997).

In its broadest sense, road rage can refer to any display of aggression by a driver (Joint, 1995). However, given that road rage involves a large range of behaviors, there remain inconsistencies regarding an accurate definition of road rage and its frequency (Burns and Katovich, 2003). For example, in a study, 56% of respondents admitted to driving aggressively (Young, 1998). In another report of the same year, 90% of motorists said that they encountered road rage over the course of a year (Jouzaitis, 1998).

Road rage is extremely influential in driving behaviors and increases the risk of causing an accident (Wells-Parker et al., 2002). Even mild aggressiveness precludes the driver from concentrating on the traffic, increasing the risk of an accident (Deffenbacher et al., 1994). In fact, on average 1500 people are injured or killed each year in the United States as a result of "aggressive driving" (Mizell et al., 1997).

Generally, there has not been much research on the cause of road rage. One exception is the attempt of Burns and Katovich (2003). They showed the various causes of road rage and aggressive driving through a newspaper analysis in 1985 through 1999. They categorized the causes of road rage into two categories: those related to human behaviors or actions and those related to the structure of the environment. Human behaviors or actions (71.9%) seem to be the more dominant cause than environmental elements (28.1%), which demonstrates the high possibility of Human Factors approaches to successful intervention. Although the preferred approach to dealing with aggressive drivers largely includes punitive sanctions, researchers have given a skeptical response to the effectiveness of the punitive reactions to control or reform human behaviors (Currie, 1985; Kappeler et al., 2000). An alternative preventative approach can be to detect various drivers' affective states and to provide drivers with affect-regulating aids.

Anxiety and Nervousness

Anxiety is one of the most prevalent affective states during crisis situations and its effects have been extensively studied (Hudlicka and McNeese, 2002). The generic effects of anxiety on attention encompass a narrowing of attentional focus, difficulty focusing attention, and increased attention to threatening stimuli (Mineka and Sutton, 1992; Williams et al., 1997). This narrowing of attention may also lead to neglect of other critical tasks and a failure to detect other related cues. Anxiety also predisposes one toward the interpretation of ambiguous stimuli as threatening (Williams et al., 1997). Based on these principles, Hudlicka and McNeese (2002) identified plausible influences of the anxious state on pilot behaviors: anxiety-induced narrowing of attention (e.g., focusing on signals representing unknowns or threats) and anxiety-induced perceptual bias (e.g., misinterpreting ambiguous radar returns as threats). Likewise, in driving situations an anxious driver (e.g., a novice driver) is likely to focus only on the part of the entire task and neglect other crucial parts (i.e., cognitive processing capacity mechanism).

Nervousness seems to be a very close affective state to anxiety and can also result in poorer concentration. Li and Ji (2005) considered nervousness as one of the most dangerous states for drivers and suggested detecting it and providing adaptive assistance for nervous drivers. One study showed that nervousness induced by drugs, such as cannabis and cocaine, made driving performance worse (MacDonald et al., 2008). Anxiety and nervousness induced from driving itself may be moderated by training and instructions. However, for incidental affective states coming from driving-irrelevant events (e.g., fight before driving) or environmental factors (e.g., foggy weather), more dynamic coping strategies may be required.

Boredom and Fatigue

Even though majority of traffic psychologists have conducted research on strong affective states, such as anger or fear, weaker and less aroused affective states can also have a big role in everyday driving (Summala, 2007). The low level of activation can decrease driving performance by reducing attention and prolonging reaction time. Boredom and fatigue are representative of low-activated affective states while driving (Jeon and Walker, 2011b; Summala, 2007) that deteriorate alertness, quick and accurate perception, judgment, and action. Boredom often leads a driver to sleepiness, which is a critical state to threaten road safety.

A drowsy driver may increase speed to get out of the drowsy state, but the driver may not cope with time-critical operations, such as abrupt breaking due to slowing down of reaction or loss of attention. What makes the problem worse is that even when drivers recognize their sleepiness, they often force themselves to continue driving (James and Nahl, 2000). Surveys have demonstrated that many drivers experience excessive sleepiness while driving (Braver et al., 1992; Corfitsen, 1994; Marcus and Loughlin, 1996). In a study in New York, 55% of 1000 respondents indicated that they had driven while being sleepy during the year preceding the survey, 23% had fallen asleep while driving once or more than once in their life, and almost 5% already had an accident due to sleepiness (McCartt et al., 1996). According to a different study (Lyznicki et al., 1998), up to 3% of all motor vehicle crashes happen due to sleepiness. There are general causes of driver sleepiness: poor sleep, sleep disorder, stress, monotonous driving, shift work, and time of the day (Khare et al., 2009).

A number of fatigue or drowsiness detection techniques have been developed, which can be largely classified into three categories: sensor technology (e.g., vehicle lane position detection; Skipper et al., 1984); computer vision technology (e.g., opened eye detection using an infrared camera; Khare et al., 2009); and monitoring vehicle behavior (e.g., steering input detection; Ferrone and Sinkovits, 2005). However, it is hard to find an optimal regulation strategy beyond a simple alarm (Khare et al., 2009; Kyrtsos, 1998) whose effectiveness is in question.

Frustration and Sadness

When people's goals are blocked or interrupted with resultant inability to reach the satisfaction, they feel frustrated (Dollard et al., 1939;

Lazarus, 1991). For example, if slow traffic impedes people from getting somewhere they want to go, they may feel frustrated. According to an empirical study on a driving simulator (Lee, 2010), the induced frustration decreased drivers' awareness of potential distractions, mental state, and potential danger in the driving environment.

More importantly, frustration has high potential to develop into subsequent emotional reactions, which lead to more serious outcomes. For example, according to frustration-aggression hypotheses, the extent to which people feel frustrated predicts the likelihood that they will be aggressive (Dollard et al., 1939). Subsequently, a cognitive-neoassociationistic model proposed that frustration leads to aggression by eliciting negative emotions (Berkowitz, 1990). Indeed, when facing frustrating situations, anger led to unsafe driving (Deffenbacher et al., 2001, 2003; Johnson and McKnight, 2009). However, frustration is distinguished from anger by the degree of negativity and arousal (Grimm et al., 2007). Thus, it may be needed to consider the mechanism of preventing drivers from evolving frustration into more crucial reactions: anger or aggressiveness.

Sadness has double-sidedness. Affect researchers have widely shown that sadness generally increases systematic information processing and happiness decreases it (Schwarz and Clore, 2003). For example, sadness is associated with detail-oriented processing, whereas happiness is associated with global processing (Gasper and Clore, 2002). Sad participants engage more in message elaboration than happy participants, but happy participants do not (Schwarz and Bless, 1991). This tendency has similarly been reported in memory (Storbeck and Clore, 2005) or judgment (Alloy and Abramson, 1979) research and is typically deemed the "sadder but wiser" phenomenon. However, some researchers have questioned about this tendency (Bryson et al., 1984). Bryson et al. (1984) argue that sad people are not wiser, but more prone to self-attributions of incompetence in certain contingency learning tasks.

Some research has proposed that sadness, which is negative affect with low arousal, can have negative effects on driving (Dula and Geller, 2003). Just as some depression seems to be associated with a certain type of cognitive impairment (Dolan et al., 1992), sadness often occurs with a certain degree of passiveness or resignation and thus, may degrade drivers' attention and reaction time (Eyben et al., 2010). In fact, a recent study (Jallais et al., 2014) comparing anger and sadness in the localization task of road elements has shown that sadness increased the localization error rate, whereas anger slowed locating road elements. In sum, it would be interesting to investigate if sad drivers are actually wiser (i.e., showing refined processing, more accurate memory and judgment, which leads to better driving performance) or if they are slow and error prone and thus, show worse driving. This is addressed with empirical research in the next section.

Individual Differences Regarding Affective Effects on Driving Behaviors

There are some specific populations who are more vulnerable to issues regarding driving, affect, and affect regulation. Here are a few examples of populations, for whom affect-related driving research becomes more important than for others. Research has found that young drivers are overrepresented in various crashes involving excessive speeds, curves, alcohol, fatigue, distraction, and passengers (Ferguson, 2003; Gras et al., 2007; Lam, 2002; Poysti et al., 2005; Williams, 2003). What makes it worse is that young drivers are more biased to their assessment of their skill level (Holland, 1993; Matthews and Moran, 1986; Sivak et al., 1989). Consequently, young drivers overestimate their ability to drive and drive with *greater risks* of distraction-related crashes (Lam, 2002). Moreover, young drivers are more likely to exhibit aggressive driving behaviors compared to older drivers (Mathews et al., 1998). In an experiment, young drivers low in emotional adjustment and high in sensation seeking showed high levels of aggressive driving and speeding in competition with others and consequently, showed poor performance in a simulated driving (Deery and Fildes, 1999).

Novice drivers tend to feel nervous directly due to the driving task itself. Given that they have not yet honed their perceptual and motor skills nor decision-making abilities on the road (Borowsky et al., 2010; Lee, 2010), novice drivers may easily be frustrated and embarrassed at complex driving situations.

According to Senesac (2010), older adults may regulate their affective states with less cognitive resources than young adults. Therefore, affect regulation while driving might be an easier task for older drivers than for young drivers. However, they still have a significantly higher rate of accidents per mile driven than their younger counterparts (Cerrelli, 1989; Waller, 1991; Carr et al., 1992). This is expressed via specific tendencies that threaten the safety of older drivers: difficulties performing navigational tasks (e.g., searching for street signs or addresses on unfamiliar routes), maneuvering through complex intersections, and proactively avoiding roadway obstacles while simultaneously maintaining safe control of the vehicle (Burns, 1999; Rothe et al., 1990; Warnes et al., 1993). Moreover, older drivers are generally less familiar with in-vehicle technologies (IVTs) and are likely to be overwhelmed by the amount of information from the IVTs that they need to deal with simultaneously while driving (Ball and Rebok, 1994). Therefore, despite their defensive driving style and affect regulation skills, older drivers still have affective issues, such as confusion or embarrassment. In addition, not all older drivers are experienced drivers. Old but inexperienced drivers should be more seriously considered.

Even though affect-driving research may be able to help multiple classes of drivers; drivers with traumatic brain injury (TBI) are of primary interest. There are more than 5.3 million Americans with an identified TBI and 1.5 million new brain injuries are reported per year (Charle, 2011). They show persisting executive dysfunctions, such as attention/arousal, motor coordination, sleep and fatigue, and emotion regulation (ER) (Ewing-Cobbs et al., 1999; Gordon et al., 2006). In addition, a half of TBI survivors with drivers' licenses experienced behavior problems involving a "short fuse," uncontrolled aggression and irritability (Hawley, 2001). However, research reveals that many TBI patients hope to continue independent driving to facilitate community reintegration (Hawley, 2001; Lew et al., 2005). Based on the diverse needs, characteristics, and abilities and disabilities of each population, a specific approach might be required when designing an in-vehicle assistive system (Jeon et al., 2011a).

Affective In-Vehicle Systems and Research

Speech-Based In-Vehicle Systems

The main research interests of IVTs include the natural, intuitive interaction between a driver and a car (Eyben et al., 2010). Indeed, the concept of driving has been changed from an independent task of the driver into a collaborative work with an intelligent agent (Jeon, 2010) or with a passenger (Forlizzi et al., 2010). One of the most natural ways to communicate with an in-vehicle system is using speech just as with a human codriver. Several studies have identified affective design considerations for speech-based in-vehicle systems. Some tested more basic characteristics of the in-vehicle voice. For example, using a young adult voice made older drivers feel more confident while driving, need less time to complete the driving course, and have fewer accidents than using an old adult voice (Jonsson et al., 2005). Subsequent research (Jonsson, 2009) showed that using a familiar voice (famous TV and radio presenters) yielded a better performance, including accidents, traffic rules, and lane keeping on angry drivers than using an unfamiliar voice. Moreover, with the familiar voice, drivers perceived the in-vehicle system to have a more positive influence and rated themselves as more attentive while driving than the unfamiliar voice.

Others investigated more dynamic aspects of the speech-based in-vehicle systems. Nass et al. (2005) showed that when in-vehicle voice emotion matched to driver emotion (e.g., energetic to happy and subdued to upset), drivers had fewer accidents and attended more to the road (actual and perceived), and even spoke more with the car. In another study (Jonsson et al., 2004), drivers were given interspersed warnings about the drivers' performance while driving with three conditions: driver blame, driver and car blame, and environment blame. According to the results, with warnings associated with the environment, drivers felt most at-ease, liked the system, rated the quality of the car higher, and attended to the road better than with the other conditions.

Affect Measurement and Detection Systems for In-Vehicle Contexts

The ability to dynamically detect drivers' affective states is crucial in predicting drivers' behavior and performance and guiding them to safer driving. Besides, accurate affect detection would be helpful to design a more natural, intuitive driver–car communication. Mauss and Robinson (2009) categorized general affect measurement methods into self-report (for a dimensional framework), autonomic measures (e.g., electrodermal gland and cardiovascular for a dimensional framework), startle response magnitude (e.g., eye blink, EMG, etc. for valence), brain states (e.g., EEG, fMRI, PET for approach avoidance–related states), and behavior measures (e.g., vocal characteristics for arousal, facial behavior for valence, and whole-body behavior for discrete emotions). For additional human factors methods to measure affect, see Helander and Khalid (2006). Similarly, affect detection techniques in HCI can be classified into six categories (Hudlicka, 2003): self-report instruments, physiological sensing (e.g., heart rate, blood volume pulse, skin temperature and conductance, and breathing rate), facial expression recognition, speech analysis, diagnostic tasks, and expert observer evaluation (or knowledge-based assessment). Many affect detection applications have been developed using one or combinations of these methods for various purposes, such as learning, robots, games, and people with autism (see Hudlicka, 2003). Nevertheless, there has not been much research for in-vehicle situations, with a few exceptions outlined next.

Some researchers used speech analysis for affect detection. Grimm et al. (2007) evaluated the feasibility of detecting affective states using speech analysis in a variety of contexts (three road conditions: big cobbles, 30 km/h; city, 50 km/h; highway, 120 km/h) under challenging noise conditions. Human evaluators judged drivers' affective states from recordings along the valence, activation, and dominance dimensions. By using filtering techniques (e.g., high pass preprocessing), researchers could gain promising affect recognition accuracy, at least, for city and highway conditions. Jones and Jonsson (2008) also tried to design a driver affect detection and response system using speech analysis. In their experiment, older drivers were encouraged to converse with the car while they were driving on a simulator. The real-time detection results from their system were compared with the results from human judge for recorded speech. Investigators concluded that their in-vehicle affect detection system can recognize and track changes in the affective state of the driver by around 70% for the seven affective categories (see Table 17.2). However, further research is required for more reliable evidence because they analyzed only four participants' results, who actively talked with the in-vehicle system out of 18. A fairly recent study (Eyben et al., 2010) examined the acceptance

TABLE 17.2 Affect Detection Systems for In-Vehicle Contexts

Research	Target users	Detectable affect states	Detection techniques
Healey (2000)	General drivers	Predefined states (rest, city, and highway)	Physiological sensing (skin conductance, heart activity, respiration, and electrocardiogram)
Lisetti and Nasoz (2005)	General drivers	Panic/fear, frustration/anger, and boredom/sleepiness	Physiological sensing (galvanic skin response, heart rate, and temperature)
Hudlicka and McNeese (2002)	Fighter pilots	Anxiety	Knowledge-based, self-reports, diagnostic tasks, and physiological sensing
Jones and Jonsson (2008)	Older drivers	Boredom, sadness and grief, frustration and anger, happiness, surprise, not sure (multiple emotions), no decision (neutral/natural)	Speech analysis
Jeon and Walker (2011a)	Drivers with disabilities (e.g., TBI)	Currently positive, negative, and neutral (to be updated)	Facial recognition and speech analysis

TBI, Traumatic brain injury.

and feasibility of the speech-based in-vehicle system using a simulated virtual codriver in a Wizard-of-Oz experiment. Whereas their usability metrics showed promising results, the questionnaire results were controversial. For example, participants showed just neutral or less than neutral to the questions, such as "a car should react to my emotion" or "automatic evaluation of my stress level and emotional state is desired." These results reflect that users may still be reluctant to use such a speech-based system.

Others used physiological sensing. Healey (2000) tried to implement a wearable and automotive system for recognizing drivers' stress levels by measuring and analyzing their physiological signals (skin conductance, heart activity, respiration, and muscle activity). During the experiment, participants drove in a parking garage, in a city, and on a highway. Results showed that the drivers' stress could be recognized as resting in the parking garage, driving in Boston streets, and two-lane merge on the highway with 96% accuracy. Lisetti and Nasoz (2005) also showed the possibility of the affect-intelligent car interfaces using physiological signals (galvanic skin response, heartbeat, and temperature) to driving-related affective states (frustration/anger, panic/fear, and boredom/sleepiness). In the controlled driving simulation experiment, they focused more on algorithm performance. The result illustrated that the resilient backpropagation (RBP: 91.9%) could classify more accurately than the other two, *k*-nearest neighbor (KNN: 66.3%) and the Marquardt-backpropagation (MBP: 76.7%). Even though many researchers have attempted to use physiological signals for their assessment, the recognition of more natural affective states beyond the predefined two or three still remains as a challenge to be tackled. Thus, it is recommended that system designers use multiple methods in addition to physiological methods to detect various affective states more precisely just as Hudlicka and McNeese (2002) attempted. They proposed a system for the Air Force to compensate for performance biases caused by pilots' affective states and active beliefs. Researchers integrated a variety of affect assessment methods, including physiological sensing, self-reports, diagnostic tasks, and knowledge-based assessments. The system selected a compensatory strategy based on sensing and inferring the user's affective (especially, anxiety) and belief states, and tried to implement the mitigation strategy in terms of specific GUI adaptations. For a more accurate affect assessment, researchers included personality traits and personal history (e.g., training, recent events) as variables. This approach might work for a limited group, such as a few pilots, but it seems to be hard to generalize to drivers as a whole.

One of the most recent integrated approaches to affect detection includes the in-vehicle assistive technology (IVAT), which is an in-dash interface design project for people with TBI or posttraumatic stress disorder (PTSD) (Jeon and Walker, 2011a). In the IVAT project, researchers aim at enabling drivers with TBI to drive more safely by using a dynamic real-time affect detection system and providing adaptive user interfaces. To this end, they attempted to fuse facial expression and speech recognition to detect drivers' affective states, which could provide the highest predictive capability for affective state detection and thereby, result in better recognition accuracy (Cowie et al., 2001). Those two modalities are usually nonintrusive and more robust than other physiological measures and body motion detection. Currently, a prototype system can detect three discrete states: positive, neutral, and negative. More affective states are being updated by driving-specific affect dimension research (see Section "Empirical Research on State Affect and Interventions" for more details).

Table 17.2 shows a summary of affect detection systems for vehicles. Depending on a target user group, regulation strategies and adaptive user interface designs need to be specified.

Affect Regulation Research for In-Vehicle Contexts

The most well-accepted affect regulation model in psychology is a "process model" (Gross, 1998a,b, 2002; Gross and Thompson, 2006), in which affect may be regulated at five points in the affect generative process: situation selection, situation modification, attention deployment, cognitive change, and response modulation. Based on this model, drivers can select low-traffic routes instead of high-traffic routes (situation selection) or try to minimize its affective impact (situation modification). However, driving is a complicated, interactive situation so that often drivers are not able to control every variable. Attentional deployment includes not only distracting an individual from an affective source (Gross, 2002), but also concentrating intensely on a particular topic or task (Csikszentmihalyi, 1975). Accordingly, an effective strategy would allow the adaptive in-vehicle system to distract a driver from an affective source and enable the driver to concentrate on driving itself or driving environments (Jeon, 2012a). If the affective source is deeply associated with a driving task (i.e., integral affect), this strategy may not work effectively. Diverse empirical evidence has supported that cognitive change is cognitively and socially more effective than suppression, which is a response modulation at the final regulation point (Gross, 2002). Additionally, recent studies have found that engaging in cognitive reappraisal changes the activity of brain regions involved in the experience of emotion (Ochsner et al., 2002). Based on the cognitive reappraisal, Harris and Nass (2011) showed that drivers in a reappraisal-down speech condition (e.g., "heavy traffic results from limited routes, not the behavior of other drivers") had better driving behavior and reported less negative emotions than participants in a reappraisal-up speech condition (e.g., "the behavior of overly aggressive and inconsiderate drivers leads to traffic congestion") or a silent condition.

There are additional suggestions for a driver's affect regulation from road-rage literature (Mizell et al., 1997): consider altering your schedule; improve the comfort of your vehicle; while in traffic, concentrate on being relaxed; and don't drive when you are angry. These, however, look like more general advice on static preparation for driving and not like sophisticated strategies for dynamic driving environments. Further issues include how to tweak the affect regulation model fitting to the driving situation and the variations for specific driver populations, such as young drivers, older drivers, or TBI drivers. For motivated TBI drivers, direct mitigation of the speech-based system might be helpful, but for young drivers, it might result in a reverse effect (e.g., getting angrier because of the nagging system; Jeon et al., 2011a,b).

EMPIRICAL RESEARCH ON STATE AFFECT AND INTERVENTIONS

Trait Affect Research Versus State Affect Research

Affect-driving research has been mostly done on anger effects, but until recently research on angry driving has focused on trait anger (Deffenbacher et al., 2003; Nesbit and Conger, 2012). This trait affect research has tended to use respondents' subjective evaluation data, rather than using objective performance data. Research has shown that subjective perception does not necessarily predict or correspond to actual performance data (Durso et al., 1999; Walker et al., 2006). Moreover, the survey data inherently depend on respondents' memory. Therefore, we might not guarantee the accuracy or reliability of the self-report data. Further, drivers usually feel that they are safer drivers than similar drivers (McCormick et al., 1986; Svenson, 1981) and thus, the data can be easily biased. Fairly recently, researchers have started to conduct research on state affect using a driving simulator with affect-induction techniques (Abdu et al., 2012; Jeon, 2012a; Jeon et al., 2014b; Roidl et al., 2013). This state affect research is more reliable because it provides real-time performance data. With this type of state affect research, we can assess controlled, but more dynamic situations. Therefore, this state affect research is expected to be more useful in terms of policy making and prevention technology development (Abdu et al., 2012). Based on this background, this section presents an overview of recent state affect research studies using a driving simulation research protocol (Jeon, under review; Jeon and Croschere, 2015; Jeon and Zhang, 2013; Jeon et al., 2011b, 2014a,b, 2015; Sterkenburg, 2015).

Research Methods and Procedure

More than 300 young drivers (college students) have participated in more than 10 state affect experiments in multiple institutes in the United States. A couple of driving simulators [e.g., a mid-fidelity NADS MiniSim simulator (Fig. 17.1A) and a low-fidelity SimuRide simulator (Fig. 17.1B)] have been used. Both simulators included (existing or created) hazardous events across an urban and rural road, or a highway.

The between-subjects design was mainly used because having participants subsequently induce different affective states might negatively

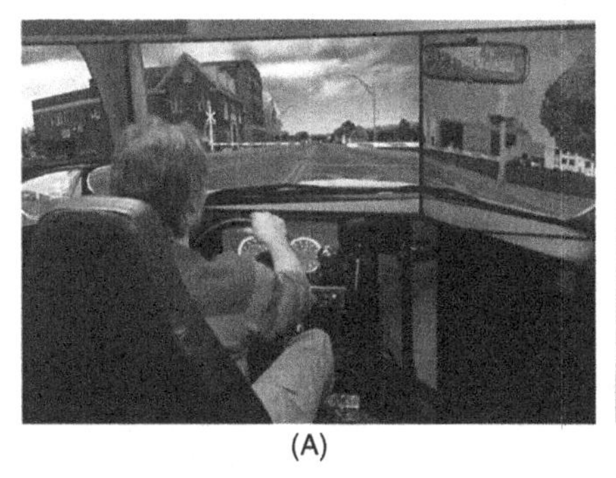

(A)

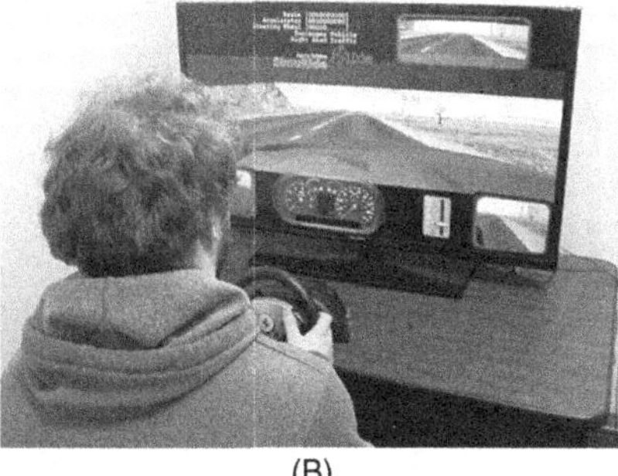

(B)

FIGURE 17.1 A participant drives in the NADS MiniSim driving simulator (A) and the SimuRide driving simulator (B).

influence the results. Before inducing affect, participants were asked to rate their current affective states (baseline) using a seven-point Likert scale. After the simulator sickness screening protocol, they had 12 min to write a past emotional experience (Ellsworth and Smith, 1988) associated with specific affect (e.g., anger, fear, happiness, sadness) depending on their affect condition. Participants in a neutral condition wrote their mundane events of the previous day. Participants were urged to refer to two sample paragraphs in the instruction sheet to help them write their own paragraphs (see Chapter 10). In some experiments, participants also watched validated affect-induction video clips. After affect-induction procedure, participants completed ratings on their present affective states. Then, they were instructed to drive as they would drive in the real world, following any traffic and safety rules. After the drive, participants completed the final affective state rating and the NASA-TLX (Hart, 2006) to provide measurements of perceived workload. Finally, participants filled out a short questionnaire for demographic information.

Affect Induction

In all experiments, specific affective states (e.g., anger, fear, happiness, sadness, and neutral, see Fig. 17.2) were induced using the combination of various affect-induction methods. Participants' affective states were checked 3 times: before affect induction, after affect induction, and after the experiment. In all experiments, regardless of types of affect, participants' designated affective state ratings significantly increased after affect induction and decreased after the experiment. In most cases, affective state ratings after the experiment were still higher than before induction, which accounts for the source of their performance difference between affect conditions. Successful results of this affect induction imply that we can more widely use this affect-driving experimental protocol.

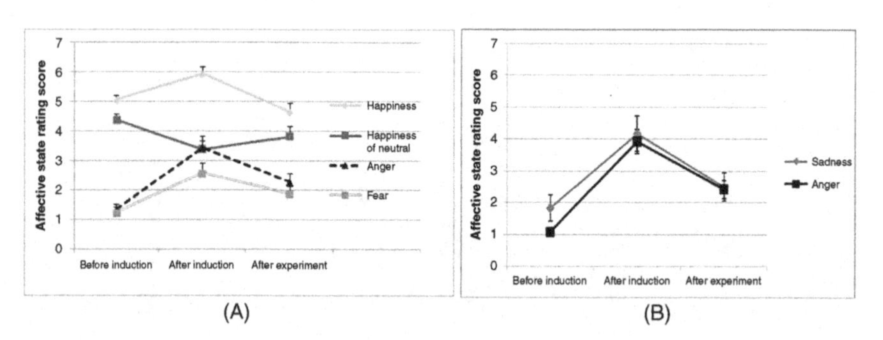

FIGURE 17.2 **Affective states rating scores across rating timings.** (A–B) The scores increased after affect induction. After the experiment, affect scores decreased. All seven experiments showed similar patterns. *Error bars* indicate standard errors of the means.

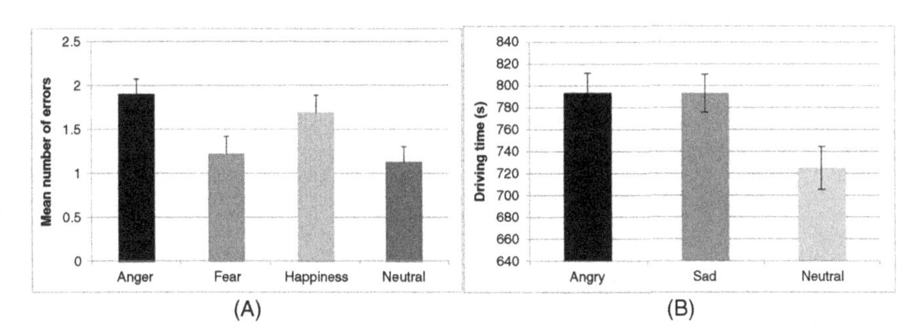

FIGURE 17.3 Mean number of errors across affective states (A) and driving time across affective states (B). *Error bars* indicate standard error of the mean.

Effects of Specific Affect on Driving Performance

Given that all experiments included an anger condition, we can address the effects of anger on driving more clearly than any other affective states. Anger degenerated driving performance (i.e., increased overall driving errors, overspeed, maximum speed, lane deviation, and aggressive driving behaviors, etc.) compared to the neutral condition (Deffenbacher et al., 2003; Jeon et al., 2011b; Roidl et al., 2013; Underwood et al., 1999).

Interestingly, happiness and sadness degenerated driving as much as anger, whereas fear did not show negative effects on driving (Fig. 17.3A). Driving errors more clearly appeared in risky driving behaviors (e.g., exceeding the speed limit or flooring the gas pedal), but the pattern was similar for other driving performance measures (e.g., lane deviation, violation of traffic rules, or collisions). In one experiment, participants with sadness even showed more consistent driving errors than those with anger (Jeon and Zhang, 2013). Two experiments consistently showed that the sadness and anger conditions showed significantly longer driving time compared to the neutral condition (Fig. 17.3B), with partial evidence of frequent lane departure (Jeon and Croschere, 2015; Jeon and Zhang, 2013).

Effects of Specific Affect on Neurophysiological Measures

Again, a couple of empirical studies using neurophysiological equipment have been conducted with induced anger. This type of objective measures is required because depending only on participants' subjective rating about their emotional states might not be perfectly reliable. In one study (Jansen et al., 2013), 10 participants drove with induced anger in a NADS MiniSim with a heart rate monitor [(electrocardiogram (ECG)]. Reading emotional passages and writing their emotional events were assumed to induce incidental anger, and hazards while driving were assumed to induce integral anger. Average heart rate increased after incidental affect

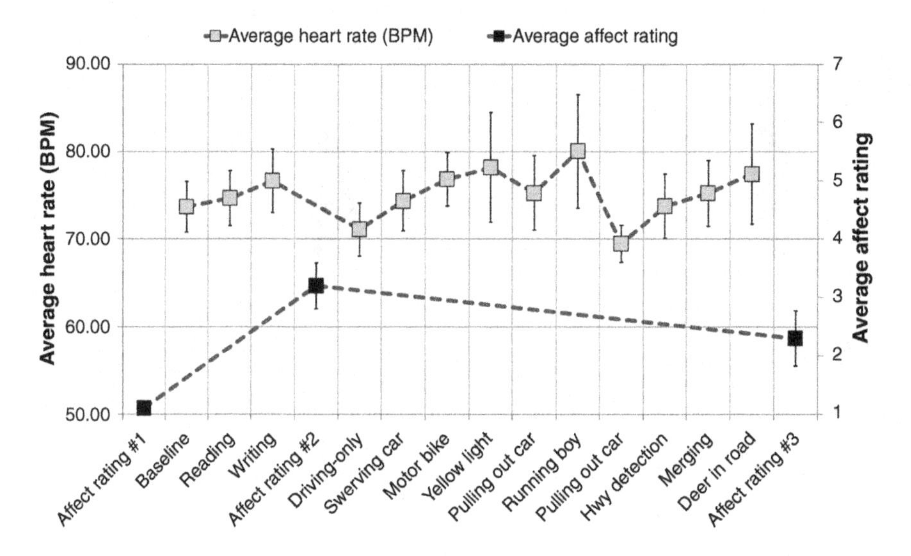

FIGURE 17.4 **Average heart rate (BPM) at each experimental event and average affect rating in three timings.** *Error bars* indicate standard error of the mean.

induction as compared to the baseline (Fig. 17.4). Moreover, all but one hazard event resulted in an increased heart rate as compared to both baseline and driving-only measurements. A decrease in heart rate was observed during the driving-only part of the experiment. This decrease is speculated to be a result of the participant's comfort with driving in the simulator, as writing about a past angry experience was a novel activity in comparison to previously driving in the simulator. This past simulator experience may have made the participant feel more comfortable than the novelty of writing and thus, their heart rate decreased while driving-only.

In a more recent study (FakhrHosseini et al., 2015), another 10 participants drove with induced anger in a NADS MiniSim with functional near-infrared spectroscopy (fNIRS). Two pairs of fNIRS probes were affixed to the forehead of the participants. The prefrontal cortex (PFC) is well known as being involved in cognitive and motor activities. Driving tasks require a driver's high-level cognitive functions and motor activities and so, the PFC is a good candidate for the investigation of driving tasks (Yoshino et al., 2013). It is also known that the frontal cortical areas are involved in perception of emotions (Rutkowski et al., 2011). Therefore, in this study, fNIRS was used as a noninvasive, lightweight imaging tool (Girouard et al., 2010) in a driving situation with some hazards. fNIRS measures hemodynamic and oxygenation changes in the PFC using near-infrared light. fNIRS has a better time resolution (e.g., 5 frames/s) than fMRI, but only measures the PFC rather than the entire brain. Literature shows that the right hemisphere is responsible for negative emotions and the left

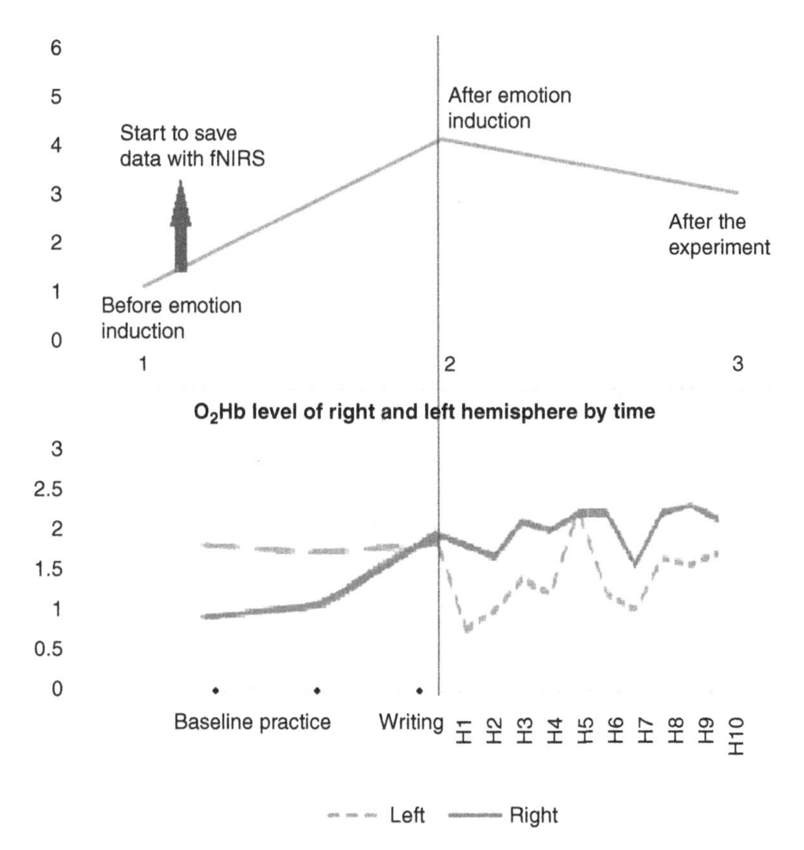

FIGURE 17.5 **A comparison between the self-report emotion survey results and changes in O_2Hb level during the experiment.**

hemisphere is responsible for positive emotions (Glotzbach et al., 2011; Rutkowski et al., 2011). Therefore, we hypothesize that the emotional driving period will show higher oxygen concentration levels than the baseline driving, specifically, in the right hemisphere.

Fig. 17.5 shows the trend of participants' self-report anger level and brain activities based on time. As is shown, the right hemisphere shows higher oxygen concentration levels than the left hemisphere after writing. Moreover, the level in the right hemisphere gradually increases as the hazardous events went by.

However, as expected, there is a large variance in brain activities among participants. Therefore, we focused on finding out the overall pattern rather than comparing in-between separate events. To classify events based on the changes in O_2Hb level and to find a pattern among them, a cluster analysis and factor analysis was conducted. Both methods clearly showed that fNIRS data support the differences between the baseline period and the anger period. More specifically, the data showed the difference between the right hemisphere and the left hemisphere with negative emotions.

Overall, data patterns indicate that neurophysiological measurement (i.e., average heart rate measured in BPM and oxygen concentration levels in PFC) could be used to identify the affective state, anger, compared to the baseline. Further, it demonstrates the benefits of physiological measurement in detection of both incidental and integral affect. However, it is not clear if incidental affect and integral affect can be differentiated by this type of measure.

Effects of Specific Affect on Other Measures

To see whether participants could expect before driving that their induced affect level might lead to degenerated driving, an experimenter asked them about their driving confidence level, safety level, and perceived risk perception right after affect induction procedure. Only in one experiment (Jeon et al., 2014b), angry participants showed significantly lower safety level, but other than that, there were no other results, which indicate that they might not be consciously aware of their plausible risk.

In two experiments (Jeon et al., 2014a, 2015), driver SA seemed to mediate between anger and driving performance either fully or partly. The association between affective states and speed-related performance variables (i.e., overspeed errors and average speed) was fully mediated by SA. These full mediation models imply that affective effects appear only via the effects of reduced SA. On the other hand, the association between affective states and the number of collisions was partially mediated by SA. This partial mediation model implies that affective states can have an impact on the number of collisions either with or without the effects of SA. In other words, affective states could influence collisions through other plausible channels (e.g., motor planning and controlling). However, we cannot simply attribute the entire affective effects to SA because there might be other steps where affect can still influence performance, such as decision-making or action selection. In fact, a recent study (Sterkenburg, 2015) showed that anger effects only appeared in a decision-making task while driving. In this experiment, among multiple hazards, participants with anger performed much worse than the other groups specifically on the traffic light hazard (abruptly changing into yellow).

For workload, we found mixed results. In most experiments, perceived workload was not different across affect conditions. Only in one experiment (Jeon and Croschere, 2015), anger showed significantly higher workload subscales (physical workload and frustration) than neutral. We could partly explain this by the cognitive appraisal mechanism. For example, according to the cognitive appraisal, anger is deeply associated with "self-control." Based on that, we cautiously infer that angry participants felt more workload during the new, frustrating (driving) task due to their failure of self-control than participants in other affective states. However, given that this workload difference was not shown in other experiments, it is hard to fully explain. One take-home message is that affective effects seem to be different from workload effects. Therefore, we need to have different measures and a different remedy for affect.

Plausible Mechanisms of Specific Affective Effects

Driving researchers have said that "happy drivers are a better driver." Also, affect researchers usually believe the "sadder, but wiser" phenomenon. Our empirical research shows that those notions are not working in that way in driving. Happy drivers were as bad as angry drivers. The difference might come from the degree of the affective state (e.g., normal happiness vs. excessive happiness). Sad drivers were also as bad as angry drivers. Regarding sadness, the different outcome might stem from the difference of the tasks. The traditional affect research on sadness has focused on the simple cognitive tasks, such as attention or decision-making, whereas driving is a compilation of different levels of tasks, including attention, comprehension, projection, decision-making, and action selection.

The previously mentioned results imply that we might not be able to explain affective effects on driving depending on traditional affect mechanisms. The valence dimension has functioned as a good axis to distinguish affective states and has worked well in general. However, beyond the valence dimension, affective states may differ in various aspects, such as their autonomic manifestations, the extent to which they produce rumination or distraction, their evolutionary connections to mental and behavioral proclivities, and their tendencies to endure across time (Bodenhausen, 1993). The arousal or activation has also served as a good axis. However, the optimal level of arousal may differ across task domains and situations. Given that affective states in the same valence or arousal dimension (e.g., both anger and fear belong to negative valence and high arousal) showed different performance results, we cannot account for affective effects by the valence or arousal dimensions.

The mood congruent effects or the cognitive appraisal mechanism partly explains the outcomes, but they cannot explain the entire results as well. Moreover, affective effects seem to be independent of perceived workload, which means that we need to adopt different approaches to affect other than traditional approaches. In sum, a more elaborated framework is required to disentangle complicated phenomena in driving-affect research.

Plausible Intervention Strategies to Mitigate Affective Effects

In our research, listening to validated emotional music pieces (either happy or sad) significantly mitigated anger effects on driving performance

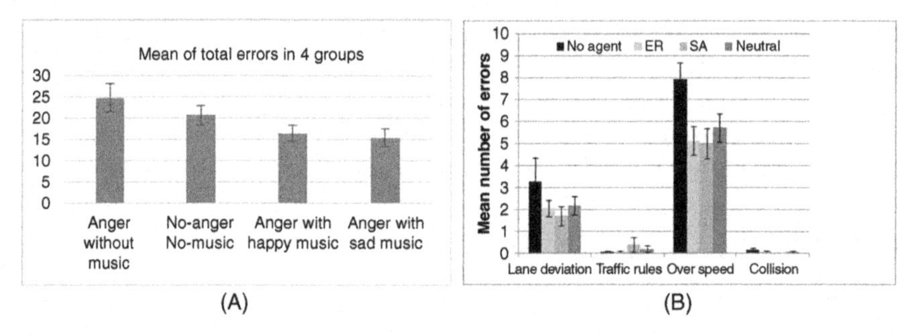

FIGURE 17.6 Number of driving errors across affect conditions. The anger conditions with either happy or sad music showed significantly fewer number of errors than anger condition without music (A). The anger conditions with either the emotion regulation *(ER)* speech condition or the situation awareness *(SA)* speech condition showed significantly fewer number of errors than the no-agent condition (B). *Error bars* indicate standard error of the mean.

(Fakhrhosseini et al., 2014) (Fig. 17.6A). However, there was no difference between happy and sad music conditions. The use of the in-vehicle speech-based systems not only reduced anger level and perceived workload, but also enhanced driving SA, which led to better driving performance (Jeon et al., 2015) (Fig. 17.6B).

Despite the similar promising results of the two types of speech-based system, the subjective assessment was different from those results. The emotion regulation (ER) type speech system was rated more "effective" than the situation awareness (SA) type speech system. Participants might think that way because directive messages felt more "helpful" in such a highly demanding situation, even with an angry state. Nonetheless, the ER system turned out to be more "annoying" and more "authoritative." Therefore, we might need to appropriately select either system depending on the target user.

RESEARCH AGENDA AND FUTURE WORKS

Theoretical Issues and Future Works

Theory-Driven Approach to Affect-Driving Research

Affect researchers have proposed considerable theories and hypotheses to identify the mechanisms of the relationship between affect and other psychological constructs. Those can be good theoretical ingredients that affect-related driving research can use and further develop. For example, if different affective states have different impacts on driving performance and safety even though they belong to the same valence, the cognitive appraisal mechanism might give an appropriate explanation. The somatic markers hypothesis could provide a proper account for affect induced from the driving tasks (i.e., integral affect). Regarding driver risk perception and automatized reactions, the experiential and analytic systems may function as a good taxonomy. In designing an adaptive affect regulation system, researchers may be able to consider the affect-as-information hypothesis as a design guide.

Driving is quite a complicated situation and consists of a number of subelements of human and environmental factors. On the human factors side, cognitive elements interact with each other and with performance as well. Affective elements also interact with cognitive elements and sometimes directly influence performance of action. Even within affective elements, positive and negative affective states sometimes appear independently (Watson et al., 1999). Such contradictory affective states may interact with each other and influence performance differently. Consequently, it is fairly difficult to identify each genuine effect separately or combine all of their effects with a single equation. Research has shown a tendency to label every effect as "distraction" or "reduced performance" (Dorn et al., 2013; Young et al., 2008) rather than revealing the specific effects of various affective states. However, if "affect A" results in high lane deviation, whereas "affect B" results in frequent traffic rule violations, the affect regulation strategy for each state is required to be distinctive. Relatively stable human factors, such as personality traits (Dorn and Matthews, 1995) or personal history (Hudlicka and McNeese, 2002) may also interact with affective states and also have different impacts on driving performance.

In fact, all of human factors may also interact with environmental factors, such as noise, temperature, weather, overcrowding, interpersonal communication, or even the status of the car. Integrating all of these factors may not be possible for now. Nonetheless, it has to be an ultimate long-term goal and researchers need to scrutinize the taxology—list and specify all the factors at play and their potential interactions—for a deeper understanding of the domain.

Driving-Specific Affect Taxonomy Research

During the past decade, technology development has spurred research on affect detection (Zeng et al., 2009). However, they have shown a typical technology-driven research on how to detect rather than what to detect and why. From the technology perspective, a recognition tool and its accuracy is certainly an important issue. However, above all, researchers need to ask what kind of emotions would be useful for their specific research domain and why. To date, most researchers have mainly depended on valence and arousal dimensions or Ekman's basic six emotions: happiness, sadness, anger, fear, surprise, and disgust (Ekman, 1992). As it is known to be robust across cultures, it may serve well for a general recognition task [e.g., emotion recognition of an agent's face (Beer et al., 2009) or a robot's voice (Jeon and Rayan, 2011)]. However, it may not work in a specific, complex context, such as driving or aviation. In such a case, researchers may need to tweak the existing taxonomy or create a new one. There has been some research on specific affective states for the driving contexts. For example, Gulian et al. (1989) developed three main affective factors related to driving: driving aggression, dislike of driving, and driving alertness. However, they mainly dealt with the issue in terms of driver's stress.

To understand the affect-related driving research domain in a sophisticated fashion and to develop a more effective and efficient affect detection and regulation system for drivers, identifying a driving-specific affect dimension is critical. To this end, Jeon and Walker (2011b) initially attempted to construct an affect dimension for driving contexts. In phase 1, they obtained 33 affect-inducing driving scenarios and 56 affective keywords from drivers' brainstorming. In phase 2 through drivers' ratings and factor analysis, they extracted nine affective factors to explain various driving situations: fear (anxiety), happiness, anger, depression (sadness), curiosity (confusion), embarrassment, urgency, boredom (sleepiness), and relief. This set included most of the critical affective states mentioned in the previous affect-driving literature review (Section "Literature on Affect-Related Driving Research") and showed some deviations from traditional affective taxonomies, such as Ekman (1992) or Plutchik (1994). Thus, researchers need to carefully scrutinize a driving-specific affect dimension rather than just follow an existing taxonomy. To achieve a more robust affect dimension for driving, this type of study should be replicated and expanded across more driving scenarios and more affective states, across age, driving experience, and various ethnic groups.

Inclusion of Positive Affect

Affect-related driving research has concentrated mostly on negative states. Driving literature has reported that "happy drivers are a better driver (Eyben et al., 2010; Grimm et al., 2007; Jones and Jonsson, 2005) and produce fewer accidents" (James and Nahl, 2000). However, affect research has demonstrated that positive emotions can also have diverse impacts on various cognitive aspects (Gable and Harmon-Jones, 2008; Isen, 1993; Storbeck and Clore, 2005). Happiness, for instance, has been shown to increase reliance on stereotypes (e.g., gender or accent of the speech-based interface) in judgment (Isen, 2000; Reeves and Nass, 1996; Schwarz and Bless, 1991). Thus, how positive affect influences driving performance and safety should be taken into consideration in driving research. In fact, as mentioned in the previous section, a study showed that happy drivers made as many errors as angry drivers (Jeon et al., 2014b). Therefore, "excessive happiness" or "excessive relief" can be investigated further. According to Yerkes and Dodson Law (Yerkes and Dodson, 1908), a moderate level of activation or arousal can bring out the best performance. Therefore, the affect detection system may also need to monitor happy and relieved states so that a driver would not stay in such states for too long. In terms of the regulation strategy, a good research question would be, "which emotional state should be the goal state for safer driving with regulation: happy (a strongly positive state), relieved (a counterpart of negative states), or neutral?"

Differentiation of Negative Affect

Traditional research addressing the impact of affect on cognitive processes has tended to focus on the effects of global mood rather than specific affect (Bodenhausen et al., 1994). Likewise, affect-related driving research has depended on the simple valence dimension (Megías et al., 2011a,b) and there has been few systematical comparisons of specific affective effects on driving performance measures. However, beyond valence, affective states may differ in various aspects, such as their autonomic manifestations, the extent to which they produce rumination or distraction, their evolutionary connections to mental and behavioral proclivities, or their tendencies to endure across time (Bodenhausen, 1993). Recent research has found differences of affective effects on attention, judgment, and decision-making even though they belong to the same valence. The cognitive appraisal has often been suggested as a mechanism to explain affective effects beyond the valence dimension. Ellsworth and Smith (1988) proposed eight appraisal dimensions that are important in differentiating emotional experiences: pleasantness, anticipated effort, attentional activity, certainty, responsibility, control, legitimacy, and perceived obstacle. In summary, appraisals of responsibility and control are central to the experience of anger, guilt, and sadness; appraisals of uncertainty are important in the experience of fear; high attentional activity is central to the experience of interest; effort, attention, and certainty are important to the experience of challenge/pride; and appraisals along the predictability dimension are important in the experience of resignation and surprise. These cognitive appraisals are involved in particular affect because appraisals have important implications for the individual's well-being and the affect associated with those appraisals prepares the person to adaptively cope with those implications (Ellsworth and Smith, 1988). However, cognitive appraisals do not always work for every case. The situation sometimes does not allow for cognition to be involved in affective processing. Affect sometimes works independently of cognitive process.

Based on the empirical evidence that goes beyond the valence dimension, research on the relationship between specified negative affective states and driving performance seems to be worth conducting to see if different negative states would yield different results just as in social judgment and decision-making tasks. For instance, a research question can be posited about "how does a fearful driver behave differently from an angry driver?" (Jeon et al., 2011b). If the valence is a valid dimension for them, the driving performance should be similar. However, if the empirical results are different from each other, another dimension may be needed to explain the results regardless of whether it is the cognitive appraisal or not. Moreover, the assistive system may need to consider detecting and mitigating those affective states separately.

Differentiation and Integration of Incidental and Integral Affect

Affect researchers have made a distinction between incidental and integral affect, depending on the relationship between the source of the affect and the task (Blanchette and Richards, 2010; Bodenhausen, 1993). Traditional research using mood-induction procedure falls into incidental affect, such as being late for an important appointment or driving after a fight or funeral. In contrast, affective states from driving relevant contexts, such as inexperienced driving, getting a ticket from the police, nearly running over a pedestrian, or tailgating belong to integral affect. Each example provides different strength and duration of the affective state. Therefore, more systematic approaches are needed to understand each effect (Jeon et al., 2011b) and incremental effects when both happen simultaneously (Jeon, 2012a).

Practical Issues and Future Works

Affect-Induction Methodologies

How to induce participants' affective states in driving simulation is a critical problem, which is directly related to a research goal (e.g., incidental or integral affect). Regarding incidental affect, researchers have had participants watch photos (Lang et al., 1990) or film clips (Fredrickson and Branigan, 2005), read scenarios or stories (Johnson and Tversky, 1983; Raghunathan and Pham, 1999), listen to music (Jefferies et al., 2008; Rowe et al., 2007), or write down their past experience (Gasper and Clore, 2002). Regarding integral affect, driving researchers have embedded frustrating or angry events on the driving scenarios (Jeon, 2012a; Lee, 2010). What is the best methodology of affect induction for driving environments? What is more adequate methodology for incidental affect? Since driving is a much more complicated and longer-lasting task than social judgment or decision-making tasks, the strength and duration of induced affective states are expected to be more important and to have a greater effect on driving performance. "How to guarantee that the induced affective states in a driving simulation are equivalent to affective states in real driving?" is another basic question to be addressed.

Simulated Driving Research

Driving simulation research has a number of advantages over real driving, including safety, cost and time efficiency, controlling over variables, and storing all the driving relevant data. However, using a simulator has clear disadvantages. Simulation sickness is one of the most prevalent issues to be tackled. A game-like simulator and an unrealistic scenario might have a disruptive effect on immersion during the experiment and might lead to some nonsignificant results. For example, in the cognitive appraisal mechanism, anger is characterized by other agency very consistently, but it might be hard for participants to attribute "rolling rocks" in a driving scenario to a specific agency (i.e., other person). If the simulator can have heavy traffic in front of the participant's vehicle, it could reinforce already triggered anger more directly and might influence driving behavior more severely. Physicality of the real road would also influence drivers' affective states and driving behavior. Given that physiological states are known to influence affective states, rumbling and body motion will likely have impact on drivers' affective states. A balanced approach is needed between minimizing simulation sickness and having a more realistic simulator and scenarios, which can keep more appropriate affective states and better immersiveness.

Affect Detection Methods and Modalities

To recognize more accurate affective states, researchers have tried to blend several methods and modalities (Hudlicka and McNeese, 2002; Jeon and Walker, 2011a). For example, Jeon and Walker (2011a) attempted to combine facial expression and speech analysis. Facial expressions could be detected by a small webcam and speech analysis could be conducted through a small microphone that is often embedded for an in-vehicle speech recognition system. If these data can optimally be integrated, the result may yield a relatively highly accurate recognition. However, a number of practical issues still exist. To list a few, some people have a poker face (Picard, 1997). As some affective states share similar facial expressions (e.g., anger and disgust, fear and surprise), they are commonly confusing (Leng et al., 2007). Moreover, detecting facial expressions at night may not work well with a webcam (without additional light or specifications that e.g., an infrared camera might offer). How about detecting different angles, gestures, or frequent movements of a driver's head? For speech analysis, what if a driver does not say anything when he or she is in an affective state? Discerning a driver's speech at a reliable level from a passenger's speech or the sounds of a radio is still impaired due to the in-car noise level even with a directional microphone.

For more accurate affect detection, diverse physiological measures could be added. However, physiological sensors are generally more obtrusive, which can lead to a driver's irritation or distraction. To date, the reliability and resolution of physiological sensing for a specific affect are still not at a satisfactory level. An alternative would be to directly match the results from physiological sensors to the percentage of risk or safety level without naming it as an affective state. However, this is a totally different approach from what has been described in this chapter so far (e.g., specification and extension of affect taxonomy).

Display Methods of Affect Detection Results

There are, at least, two possibilities of displaying the affect detection results in driving environments. On the one hand, making the covert affective process more visible could help drivers use their affective states as information to their judgment and decision-making. As people generally do not know well what they are feeling (even though they think they are well aware of it), objectively carried information on the current affective state can be much more powerful than self-reported subjective feelings (Picard, 2010). Thus, the affect detection system can have a type of gauge, indicating the driver's current state and acting just as a speedometer functions about vehicle's speed. This might enhance the driver's awareness about his or her internal state. Even just awareness of one's affective state or the source of the state can make the person less influenced by those affective states (Tiedens and Linton, 2001). In this type of display, a user can have a more active role in using the system. On the other hand, the system can be embedded into the dashboard or center fascia and run without the driver's awareness. Still, the system interface can adaptively change based on the driver's affective states in an unobtrusive way. This can be an automated, push-type interface. It might not facilitate driver's overt SA but might enhance "situation responsiveness" (Lee, 2006) and performance. As those two interface design strategies are on a continuum, it is possible to make some combinations of the two or a user may be able to customize the display methods.

Opportunities for Affect Regulation

There are several places for interventions to tackle a driver's affective issues. First, if the mechanisms are clearly known for the affective effects, direct mitigation of the affective states may be possible: change the induced affective state into a positive affective state using affect regulation techniques or adaptive user interfaces, or change the induced affective state into a neutral state. However, it is not clear which approach is better. So far, there is no empirical validation of those two possibilities. Based on the affect-as-information approach, the system can try to get rid of the affective effects or to utilize the effects reversely. In addition to Tiedens and Linton (2001), Isbell (1999) also showed that when the true cause of the participants' feelings was made salient before the task (undermining the information value of their feelings), the affective effects were reversed, suggesting that the affective effects were mediated by the apparent informativeness of affective states. This sheds light on the possibility that if the affect regulation system allows users to attribute their states properly and evaluate the informativeness of their affective states, it may be able to prevent the negative effects.

Second, attention deployment, which is one of the five steps in the process model for ER, can be used. For example, calling one's own name can sometimes break through to conscious awareness (Moray, 1959). This processing occurs preattentively, similarly to low-level stimuli that produce the pop-out effect in visual search tasks (Treisman and Gormican, 1988). It can directly have impact on driving performance and safety. It may work well for making a driver distracted from an affective source, but how the system can make a driver concentrate on driving per se is another question to answer. Simply telling a driver about driving or a driving task may make him or her feel that a system is a back-seat driver.

Finally, the system can be designed to increase performance, regardless of the regulation of the affective state. For high workload or confusing situations, the system can temporarily prevent incoming calls or emails. Making visual font larger (Hudlicka and McNeese, 2002) or adding auditory displays for a specific task may improve driving and other related task performance and reduce drivers' perceived workload (Jeon et al., 2009).

Modalities of Adaptive User Interfaces

Depending on the type and strength of induced affective states and their duration, an optimal interface strategy for each specific situation may vary. For some affective states, adaptive GUI modifications may be enough for a safe drive. Those enhancements can include changes in overall format or user interface elements, such as layout, font size, color, etc. (Hudlicka and McNeese, 2002). However, an adaptive GUI approach may not be the best solution because vision is the most heavily taxed sense in a driving situation.

Other than the visual modality, auditory or tactile displays can be appropriate alternatives. For auditory displays, they can include nonspeech sounds for attention-grabbing and speech prompts for cognitive reappraisal to help a driver rationalize the affect-inducing situations (Harris and Nass, 2011). Research also explores the possibility of using music to mitigate a driver's affective state (Brodsky and Kizner, 2012; Pêcher et al., 2009) although there are some practical issues, such as applicable affect, amount of time for listening, regularity, duration, and preference of music. For a systematic approach to the use of music for affect mitigation, see Jeon, 2012b. For some stronger and longer-lasting affective states, the in-vehicle interface may need to serve as an advisor or "coach" to help drivers regulate their affective states, which requires higher intelligence. The individual differences depending on cognitive and physical deficits or other characteristics should also be taken into account. Which strategy or modality can be the most effective should be empirically evaluated (Eccles et al., 2011).

CONCLUSIONS

This chapter researched the potential and empirical effects of various affective states on driving performance and road safety. Even though research on affect has not had a long history, affect researchers have proposed a number of constructs, taxonomies, and theories, which should provide a foundation for affect-related driving research. For a systematic approach I tried to offer multiple perspectives on the topic—diverse driving models and affect taxonomy, vulnerable populations, and in-vehicle applications. However, given the transdisciplinary characteristics of the domain, this chapter might not address all related issues. Thus, the components addressed here may not be exhaustive. Nonetheless, this review should help frame some useful, timely questions about the roles of affect in driving behavior—unfortunately, it cannot answer them just yet: specifically, scientific knowledge on optimal affect-induction procedure for research purpose and affect detection and mitigation for practical applications could be further updated based on empirical research studies.

The approach outlined in this chapter is expected to begin to close the existing gap between the traditional affect research and the emerging field of affect-related driving research. Once we integrate all necessary components—cognition, affect, and behavior, we could obtain a holistic map of what is going on in the complicated interaction scene. The evidence reviewed here suggests that this systematic approach may provide a legitimate framework of affect-related driving research. This approach can also be applied to any complex monitoring–controlling interaction situations, such as aviation or nuclear power plant.

References

Abdu, R., Shinar, D., Meiran, N., 2012. Situational (state) anger and driving. Transport. Res. F 15 (5), 575–580.

Adams, J.G.U., 1985. Risk and Freedom: The Record of Road Safety Regulation. Transport Publishing Projects, London.

Algoe, S.B., Fredrickson, B.L., 2011. Emotional fitness and the movement of affective science from lab to field. Am. Psychol. 66 (1), 35–42.

Alloy, L.B., Abramson, L.Y., 1979. Judgment of contingency in depressed and nondepressed students: sadder but wiser? J. Exp. Psychol. 108, 441–485.

Anderson, J.R., Lebiere, C., 1998. The Atomic Components Of Thought. Lawrence Erlbaum Associates, Hillsdale, NJ.

Ball, K., Rebok, G., 1994. Evaluating the driving ability of older adults. J. Appl. Gerontol. 13 (1), 20–38.

Beer, J.M., Fisk, A.D., Rogers, W.A., 2009. Emotion recognition of virtual agents' facial expression: the effects of age and emotion intensity. Proceedings of the Human Factors and Ergonomics Society 53rd Annual Meeting, October 19–23, San Antonio, TX.

Berkowitz, L., 1990. On the formation and regulation of anger and aggression: a cognitive-neoassociationistic analysis. Am. Psychol. 45, 494–503.

Blanchette, I., Richards, A., 2010. The influence of affect on higher level cognition: a review of research on interpretation, judgment, decision making and reasoning. Cogn. Emot. 24 (4), 561–595.

Bodenhausen, G.V., 1993. Emotion, arousal, and stereotypic judgment: a heuristic model of affect and stereotyping. In: Mackie, D., Hamilton, D. (Eds.), Affect, Cognition and Stereotyping: Interactive Processes in Intergroup Perception. Academic Press, San Diego, CA, pp. 13–37.

Bodenhausen, G.V., Sheppard, L.A., Kramer, G.P., 1994. Negative affect and social judgment: the differential impact of anger and sadness. Eur. J. Soc. Psychol. 24, 45–62.

Boer, E.R., 1996. Tangent point oriented curve negotiation. Proceedings of the IEEE Intelligent Vehicles 96 Symposium, September 19–20, Tokyo.

Boer, E.R., Hildreth, E.C., 1999. Modeling drivers' decision and control behavior on country roads. Proceedings of the Eight International Conference on Vision in Vehicles, December 17, Boston, MA.

Boer, E.R., Hildreth, E.C., Goodrich, M.A., 1998. A driver model of attention management and task scheduling. Proceedings of the 17th European Annual Conference on Human Decision Making and Manual Control, December, 14–16, Valenciennes, France.

Borowsky, A., Shinar, D., Oron-Gilad, T., 2010. Age, skill, and hazard perception in driving. Accid. Anal. Prev. 42, 1240–1249.

Bowles, S., Overberg, P., 1998. Aggressive driving: a road well-traveled. USA Today November 23, 17A.

Braver, E.R., Preusser, C.W., Preusser, D.F., Baum, H.M., Beilock, R., Ulmer, R., 1992. Long hours and fatigue: a survey of tractor-trailer drivers. J. Public Health Policy 13 (3), 341–366.

Brodsky, W., Kizner, M., 2012. Exploring an alternative in-car music background designed for driver safety. Transport. Res. F 15, 162–173.

Bryson, S.E., Doan, B.D., Pasquali, P., 1984. Sadder but wiser: a failure to demonstrate that mood influences judgements of control. Can. J. Behav. Sci. 16, 107–119.

Burns, P.C., 1999. Navigation and the mobility of older drivers. J. Gerontol. Ser. B 54, S49–S55.

Burns, R.G., Katovich, M.A., 2003. Examining road rage/aggressive driving media depiction and prevention suggestions. Environ. Behav. 35 (5), 621–636.

Carr, D., Jackson, T.W., Madden, D.J., Cohen, H.J., 1992. The effect of age on driving skills. J. Am. Geriatr. Soc. 40, 567–573.

Cerrelli, E., 1989. Older Drivers: The Age Factor in Traffic Safety. US Department of Transportation, National Highway Traffic Safety Administration, Washington, DC.

Charle, K., 2011. The right frame of mind. The Daily Checkup in the Daily News. http://www.mountsinai.org/vgn_lnk/Regular%20Content/File/Daily%20News/The%20Right%20Frame%20of%20Mind_Dr.%20Wayne%20Gordon.pdf.

Corfitsen, M.T., 1994. Tiredness and visual reaction time among young male nighttime drivers: a roadside survey. Accid. Anal. Prev. 26 (5), 617–624.

Cowie, R., Douglas-Cowie, E., Tsapatsoulis, N., Votsis, G., Kollias, S., Fellenz, W., Taylor, J.G., 2001. Emotion recognition in human–computer interaction. IEEE Signal Process. Mag. 18, 32–80.

Csikszentmihalyi, M., 1975. Beyond Boredom and Anxiety: The Experience of Play in Work and Games. Jossey-Bass, San Francisco, CA.

Currie, E., 1985. Confronting Crime: An American Challenge. Pantheon Books, New York, NY.

Czaja, S.J., Nair, S.N., 2006. Human factors engineering and systems design. In: Salvendy, G. (Ed.), Handbook of Human Factors and Ergonomics. John Wiley & Sons, Hoboken, NJ, pp. 32–49.

Deery, H.A., Fildes, B.N., 1999. Young novice driver subtypes: relationship to high-risk behavior, traffic accident record, and simulator driving performance. Hum. Factors 41, 628–643.

Deffenbacher, J.L., Deffenbacher, D.M., Lynch, R.S., Richards, T.L., 2003. Anger, aggression, and risky behavior: a comparison of high and low anger drivers. Behav. Res. Ther. 41 (6), 701–718.

Deffenbacher, J.L., Lynch, R.S., Deffenbacher, D.M., Oetting, E.R., 2001. Further evidence of reliability and validity for the driving anger expression inventory. Psychol. Rep. 89 (3), 535–540.

Deffenbacher, J.L., Oetting, E.R., Lynch, R.S., 1994. Development of a driving anger scale. Psychol. Rep. 74 (1), 83–91.

Dolan, R., Bench, C.J., Brown, R., Scott, L.C., Friston, K., Frackowiak, R., 1992. Regional cerebral blood flow abnormalities in depressed patients with cognitive impairment. J. Neurol. Neurosurg. Psychiatry 55 (9), 768–773.

Dollard, J., Miller, N.E., Doob, L.W., Mowrer, O.H., Sears, R.R., 1939. Frustration and Aggression. Yale University Press, New Haven, CT.

Donges, E., 1978. A two-level model of driver steering behavior. Hum. Factors 20 (6), 691–707.

Dorn, L., Matthews, G., 1995. Prediction of mood and risk appraisals from trait measures: two studies of simulated driving. Eur. J. Pers. 9, 25–42.

Dorn, L., Matthews, G., Victor, M.T.W., Lee, J.D., Regan, M.A., 2013. Driver Distraction and Inattention: Advances in Research and CountermeasuresAshgate Publishing, Ltd., Williston, VT.

Dula, C.S., Geller, E.S., 2003. Risky, aggressive, or emotional driving: addressing the need for consistent communication in research. J. Safety Res. 34 (5), 559–566.

Durso, F.T., Hackworth, C.A., Truitt, T.R., Crutchfield, J., Nikolic, D., Manning, C.A., 1999. Situation Awareness as a Predictor of Performance in En Route Air Traffic Controllers. Technical Report (DOT/FAA/AM-99/3). US Department of Transportation-Federal Aviation Administration.

Eccles, D.W., Ward, P., Woodman, T., Janelle, C.M., Scanff, C.L., Ehrlinger, J., et al., 2011. Where's the emotion? How sport psychology can inform research on emotion in human factors. Hum. Factors 53 (2), 180–202.

Ekman, P., 1992. An argument for basic emotions. Cogn. Emot. 6, 169–200.

Ellsworth, P.C., Smith, C.A., 1988. From appraisal to emotion: differences among unpleasant feelings. Motiv. Emot. 12 (3), 271–302.

Endsley, M.R., 1995. Toward a theory of situation awareness in dynamic systems. Hum. Factors 37 (1), 32–64.

Ewing-Cobbs, L., Prasad, M., Kramer, L., Landry, S., 1999. Inflicted traumatic brain injury: relationship of developmental outcome to severity of injury. Pediatr. Neurosurg. 31, 251–258.

Eyben, F., Wollmer, M., Poitschke, T., Schuller, B., Blaschke, C., Farber, B., Nguyen-Thien, N., 2010. Emotion on the road-necessity, acceptance, and feasibility of affective computing in the car. Adv. Hum. Comput. Interact. 2010, 1–17.

FakhrHosseini, S.M., Jeon, M., Bose, R., 2015. Estimation of drivers' emotional states based on neuroergonmic equipment: an exploratory study using fNIRS. Proceedings of the 7th International Conference on Automotive User Interfaces and Vehicular Applications (AutomotiveUI'15), September 1–3, Nottingham, UK.

Fakhrhosseini, S.M., Landry, S., Tan, Y.Y., Bhattarai, S., Jeon, M., 2014. If you're angry, turn the music on: music can mitigate anger effects on driving performance. Proceedings of the 6th International Conference on Automotive User Interfaces and Interactive Vehicular Applications (AutomotiveUI'14), September 17–19, Seattle, WA.

Ferguson, S.A., 2003. Other high-risk factors for young drivers-how graduated licensing does, doesn't, or could address them. J. Safety Res. 34 (1), 71–77.

Ferrone, C.W., Sinkovits, C., 2005. System and method for monitoring driver fatigue. US Patent 4602247.

Forlizzi, J., Barley, W.C., Seder, T., 2010. Where should I turn? Moving from individual to collaborative navigation strategies to inform the interaction design of future navigation systems. Proceedings of the SIGCHI Conference on Human Factors in Computing Systems (CHI2010), Atlanta, GA.

Fredrickson, B.L., Branigan, C., 2005. Positive emotions broaden the scope of attention and thought-action repertoires. Cogn. Emot. 19 (3), 313–332.

Fuller, R., 1984. A conceptualization of driving behaviour as threat avoidance. Ergonomics 27, 1139–1155.
Fuller, R., 2005. Towards a general theory of driver behaviour. Accid. Anal. Prev. 37, 461–472.
Fuller, R., Santos, J.A. (Eds.), 2002. Psychology and the highway engineer. Pergamon, Oxford.
Gable, P.A., Harmon-Jones, E., 2008. Approach-motivated positive affect reduces breadth of attention. Psychol. Sci. 19 (5), 476–482.
Galovski, T.E., Blanchard, E.B., 2004. Road rage: a domain for psychological intervention? Aggress. Violent Behav. 9 (2), 105–127.
Gasper, K., Clore, G.L., 2002. Attending to the big picture: mood and global versus local processing of visual information. Psychol. Sci. 13 (1), 34–40.
Girouard, A., Solovey, E.T., Hirshfield, L.M., Peck, E.M., Chauncey, K., Sassaroli, A., et al., 2010. From brain signals to adaptive interfaces: using fNIRS in HCI. Springer, London, pp. 221–237.
Glotzbach, E., Mühlberger, A., Gschwendtner, K., Fallgatter, A.J., Pauli, P., Herrmann, M.J., 2011. Prefrontal brain activation during emotional processing: a functional near infrared spectroscopy study (fNIRS). Open Neuroimag. J. 5 (1), 33–39.
Gordon, W.A., Cantor, J., Ashman, T., Brown, M., 2006. Treatment of post-TBI executive dysfunction—application of theory to clinical practice. J. Head Trauma Rehabil. 21 (2), 156–167.
Gras, M.E., Cunill, M., Sullman, M.J.M., Planes, M., Aymerich, M., Font-Mayolas, S., 2007. Mobile phone use while driving in a sample of Spanish phone university workers. Accid. Anal. Prev. 39, 347–355.
Grimm, M., Krroschel, K., Harris, H., Nass, C., Schuller, B., Rigoll, G., Moosmayr, T., 2007. On the necessity and feasibility of detecting a driver's emotional state while driving. In: Paiva, A., Prada, R., Picard, R.W. (Eds.), The 2nd International Conference on Affective Computing and Intelligent Interaction (ACII 2007), LNCS, vol. 4738, pp. 126–138.
Gross, J.J., 1998a. Antecedent-and response-focused emotion regulation: divergent consequences for experience, expression, and physiology. J. Pers. Soc. Psychol. 74 (1), 224–237.
Gross, J.J., 1998b. The emerging field of emotion regulation: an integrative review. Rev. Gen. Psychol. 2 (3), 271–299.
Gross, J.J., 2002. Emotion regulation: affective, cognitive, and social consequences. Psychophysiology 39, 281–291.
Gross, J.J., Thompson, R.A. (Eds.), 2006. Emotion Regulation: Conceptual Foundations. Guildford Press, New York, NY.
Gugerty, L., 2011. Situation awareness in driving. In: Lee, J., Rizzo, M., Fisher, D., Caird, J. (Eds.), Handbook for Driving Simulation in Engineering, Medicine and Psychology. CRC Press, Boca Raton, FL, pp. 1–25.
Gulian, E., Matthews, G., Glendon, A.I., Davies, D.R., Debney, L.M., 1989. Dimensions of driver stress. Ergonomics 32, 585–602.
Harris, H., Nass, K., 2011. Emotion regulation for frustrating driving contexts. Proceedings of the ACM SIGCHI Conference on Human Factors in Computing Systems (CHI 2011), Vancouver, BC, Canada.
Hart, S.G., 2006. NASA-Task Load Index (NASA-TLX); 20 years later. Proceedings of the Human Factors and Ergonomics Society 50th Annual Meeting, San Francisco, CA.
Hawley, C., 2001. Return to driving after head injury. J. Neurol. Neurosurg. Psychiatry 70, 761–766.
Healey, J., 2000. Wearable and automotive systems for affect recognition from physiology. PhD thesis. Massachusetts Institute of Technology, Boston, MA.
Helander, M.G., Khalid, H.M., 2006. Affective and pleasurable design. In: Salvendy, G. (Ed.), Handbook of Human Factors and Ergonomics. John Wiley & Sons, Inc., NJ, pp. 543–572.
Hess, R.A., Modjtahedzadeh, A., 1990. A control theoretic model of driver steering behavior. IEEE Control Syst. Mag. 10, 3–8.

Holland, C.A., 1993. Self-bias in older drivers' judgments of accident likelihood. Accid. Anal. Prev. 25, 431–441.
Hudlicka, E., 2003. To feel or not to feel: the role of affect in human–computer interaction. Int. J. Hum. Comput. Stud. 59, 1–32.
Hudlicka, E., McNeese, M.D., 2002. Assessment of user affective and belief states for interface adaptation: application to an air force pilot task. User Model. User-Adapt. Interact. 12, 1–47.
Isbell, L.M., 1999. Beyond heuristic information processing: Systematic processing in happy and sad moods. Doctoral dissertation. University of Illinois, Urbana-Champaign, IL.
Isen, A.M. (Ed.), 1993. Positive Affect and Decision Making. The Guilford Press, New York, NY.
Isen, A.M. (Ed.), 2000. Positive Affect and Decision Making. The Guilford Press, New York, NY.
Jallais, C., Gabaude, C., Paire-ficout, L., 2014. When emotions disturb the localization of road elements: effects of anger and sadness. Transport. Res. F 23, 125–132.
James, L., Nahl, D., 2000. Road Rage and Aggressive Driving. Prometheus Books, Amherst, NY.
Jansen, S., Westphal, A., Jeon, M., Riener, A., 2013. Detection of drivers' incidental and integral affect using physiological measures. Adjunct Proceedings of the 5th International Conference on Automotive User Interfaces and Vehicular Applications (AutomotiveUI'13), vol. 97–98, October 27–30, Eindhoven, The Netherlands.
Jefferies, L.N., Smilek, D., Eich, E., Enns, J.T., 2008. Emotional valence and arousal interact in attentional control. Psychol. Sci. 19 (3), 290–295.
Jeon, M., 2010. "i-PASSION": a concept car user interface case study from the perspective of user experience design. Proceedings of the 2nd International Conference on Automotive User Interfaces and Vehicular Applications (AutomotiveUI 2010), Pittsburgh, PA.
Jeon, M., 2012a. Effects of affective states on driver situation awareness and adaptive mitigation interfaces: focused on anger. Doctoral Dissertation. Georgia Institute of Technology, Atlanta, GA.
Jeon, M., 2012b. A systematic approach to using music for mitigating affective effects on driving performance and safety. Proceedings of the 14th ACM International Conference on Ubiquitous Computing (UbiComp'12), September 5–8, 2012. ACM Press, Pittsburgh, PA, USA.
Jeon, M., 2016. Don't cry while you're driving: sad driving is as bad as angry driving. Int. J. Hum. Comput. Interact. 32 (10), 777–790.
Jeon, M., Croschere, J., 2015. Sorry, I'm late; I'm not in the mood: negative emotions lengthen driving time. In: Harris, D. (Ed.), Engineering Psychology and Cognitive Ergonomics (EPCE) 2015, LNAI 9174. Springer International Publishing, Switzerland, pp. 1–8.
Jeon, M., Davison, B.K., Nees, M.A., Wilson, J., Walker, B.N., 2009. Enhanced auditory menu cues improve dual task performance and are preferred with in-vehicle technologies. Proceedings of the 1st International Conference on Automotive User Interfaces and Interactive Vehicular Applications (AutomotiveUI'09), Essen, Germany.
Jeon, M., Rayan, I.A., 2011. The effect of physical embodiment of an animal robot on affective prosody recognition. Proceedings of the 14th International Conference on Human–Computer Interaction, Orlando, FL.
Jeon, M., Roberts, J., Raman, P., Yim, J.-B., Walker, B.N., 2011a. Participatory design process for an in-vehicle affect detection and regulation system for various drivers. Proceedings of the 13th International ACM SIGACCESS Conference on Computers and Accessibility (ASSETS'11), Dundee, Scotland.
Jeon, M., Walker, B.N., 2011a. Emotion detection and regulation interface for drivers with traumatic brain injury. Proceedings of the SIGCHI Conference on Human Factors in Computing Systems (CHI11), Vancouver, BC, Canada.
Jeon, M., Walker, B.N., 2011b. What to detect? Analyzing factor structures of affect in driving contexts for an emotion detection and regulation system. Proceedings of the 55th Annual Meeting of the Human Factors and Ergonomics Society, Las Vegas, NV.

Jeon, M., Walker, B.N., Gable, T.M., 2014a. Anger effects on driver situation awareness and driving performance. Presence 23 (1), 71–89.
Jeon, M., Walker, B.N., Gable, T.M., 2015. The effects of social interactions with in-vehicle agents on a driver's anger level, driving performance, situation awareness, and perceived workload. Appl. Ergon. 50, 185–199.
Jeon, M., Walker, B.N., Yim, J.-B., 2014b. Effects of specific emotions on subjective judgment, driving performance, and perceived workload. Transport. Res. F 24, 197–209.
Jeon, M., Yim, J., Walker, B.N., 2011b. Fearful drivers are different from angry drivers: effects of different negative affective states on driving performance and workload. Proceedings of the 3rd International Conference on Automotive User Interfaces and Interactive Vehicular Applications (AutomotiveUI11), Salzburg, Austria.
Jeon, M., Zhang, W., 2013. Sadder but wiser? Effects of negative emotions on risk perception, driving performance, and perceived workload. Proceedings of the Human Factors and Ergonomics Society International Annual Meeting (HFES2013), September 30–October 4, San Diego, CA.
Johnson, M.B., McKnight, K.S., 2009. Warning drivers about potential congestion as a means to reduce frustration-driven aggressive driving. Traffic Inj. Prev. 10 (4), 354–360.
Johnson, E.J., Tversky, A., 1983. Affect, generalization, and the percedption of risk. J. Pers. Soc. Psychol. 45, 20–31.
Joint, M., 1995. Road Rage. AAA Foundation for Traffic Safety, American Automobile Association, Washington, DC.
Jones, C.M., Jonsson, I.-M., 2005. Automatic recognition of affective cues in the speech of car drivers to allow appropriate responses. Proceedings of the 17th Australia Conference on Computer–Human Interaction: Citizens Online (OZCHI'05), November 23–25, Canberra, Australia.
Jones, C., Jonsson, I.-M., 2008. Using paralinguistic cues in speech to recognise emotions in older car drivers. In: Peter, C., Beale, R. (Eds.), Affect and Emotion in Human–Computer Interaction: From Theory to Applications, LNCS vol. 4868. Springer-Verlag, Berlin, pp. 229–240.
Jonsson, I.-M., 2009. Social and emotional characteristics of speech-based in-vehicle information systems: impact on attitude and behaviour. PhD Dissertation. Linkoping University, Linkoping, Sweden.
Jonsson, I.-M., Nass, C., Endo, J., Reaves, B., Harris, H., Ta, J.L., et al., 2004. Don't blame me I am only the driver: impact of blame attribution on attitudes and attention to driving task. Proceedings of the Conference on Human Factors in Computing Systems (CHI'04), April 24–29, Vienna, Austria.
Jonsson, I.-M., Zajicek, M., Harris, H., Nass, C., 2005. Thank you, I did not see that: In-car speech based information systems for older adults. Proceedings of the SIGCHI Conference on Human Factors in Computing Systems (CHI05), April 2–7, Portland, OR.
Jouzaitis, C., 1998. AAA will ask drivers to cool it on 'road rage'. USA Today, 3A.
Kappeler, V., Blumberg, M., Potter, G., 2000. The Mythology of Crime and Criminal Justice, third ed. Waveland, Prospect Heights, IL.
Khare, A., Ghosh, H., Wattamwar, S.S., Sinha, A., Bhowmick, B., Kumar, K.S.C., Kopparapu, S.K., 2009. Multimodal interaction in modern automobiles. Proceedings of the 1st International Workshop on Multimodal Interfaces for Automotive applications (MIAA), FL.
Kyrtsos, C.T., 1998. Drowsy driver detection system, assigned to Meritor Heavy vehicle systems. US Patent 5900819.
Lam, L.T., 2002. Distractions and risk of car crash injury: the effects of drivers' age. J. Safety Res. 33, 411–419.
Lang, P.J., Bradley, M.M., Cuthbert, B.N., 1990. Emotion, attention, and the startle reflex. Psychol. Rev. 97 (3), 377–395.
Lazarus, R.S., 1991. Emotion and Adaptation. Oxford University Press, New York, NY.

Lee, J.D. (Ed.), 2006. Affect, Attention, and Automation. Oxford University Press, New York, NY.
Lee, Y., 2010. Measuring drivers' frustration in a driving simulator. Proceedings of the Human Factors and Ergonomics Society 54th Annual Meeting, CA, USA.
Leng, H., Lin, Y., Zanzi, L.A., 2007. An experimental study on physiological parameters toward driver emotion recognition. In: Dainoff, M.J. (Ed.), Ergonomics and Health Aspects, HCII 2007, LNCS, vol. 4566. Springer-Verlag, Berlin, pp. 237–246.
Levison, W.H., Cramer, N.L., 1995. Description of the Integrated Driver Model (Tech. Rep. No. FHWA-RD-94-092). Federal Highway Administration, McLEan, VA.
Lew, H.L., Poole, J.H., Lee, E.H., Jaffe, D.L., Huang, H.C., Brodd, E., 2005. Predictive validity of driving-simulator assessments following traumatic brain injury: a preliminary study. Brain Inj. 19, 177–188.
Li, X., Ji, Q., 2005. Active affective state detection and user assistance with dynamic bayesian networks. IEEE Trans. Syst. Man Cybernet. A 35 (1), 93–105.
Lisetti, C.L., Nasoz, F., 2005. Affective intelligent car interfaces with emotion recognition. Proceedings of the 11th International Conference on Human Computer Interaction, Las Vegas, NV.
Lyznicki, J.M., Doege, T.C., Davis, R.M., Williams, M.A., 1998. Sleepiness, driving, and motor vehicle crashes. J. Am. Med. Assoc. 279 (23), 1908–1913.
Ma, R., Kaber, D.B., 2005. Situation awareness and workload in driving while using adaptive cruise control and a cell phone. Int. J. Ind. Ergon. 35, 939–953.
Ma, R., Kaber, D.B., 2007. Situation awareness and driving performance in a simulated navigation task. Ergonomics 50 (8), 1351–1364.
MacDonald, S., Mann, R., Chipman, M., Pakula, B., Erickson, P., Hathaway, A., MacIntyre, P., 2008. Driving behavior under the influence of cannabis and cocaine. Traffic Inj. Prev. 9 (3), 190–194.
Marcus, C.L., Loughlin, G.M., 1996. Effect of sleep deprivation on driving safety in housestaff. Sleep 19 (10), 763–766.
Matthews, M.L., Bryant, D.J., Webb, R.D.G., Harbluk, J.L., 2001. Model for situation awareness and driving. Transport. Res. Rec. 1779, 26–32.
Mathews, G., Dorn, L., Hoyes, T.W., Davies, D.R., Glendon, A.I., Taylor, R.G., 1998. Driver stress and performance on a driving simulator. Hum. Factors 40, 136–149.
Matthews, M.I., Moran, A.R., 1986. Age differences in male drivers' perception of accident risk: the role of perceived driving ability. Accid. Anal. Prev. 18, 299–313.
Mauss, I.B., Robinson, M.D., 2009. Measures of emotion: a review. Cognit. Ther. Res. 23 (2), 209–237.
McCartt, A.T., Ribner, S.A., Pack, A.I., Hammer, M.C., 1996. The scope and nature of the drowsy driving problem in New York state. Accid. Anal. Prev. 28 (4), 511–517.
McCormick, I.A., Walkey, F.H., Green, D.E., 1986. Comparative perceptions of driver ability—a confirmation and expansion. Accid. Anal. Prev. 18 (3), 205–208.
McRuer, D.T., Allen, R.W., Weir, D.H., Klein, R.H., 1977. New results in driver steering control models. Hum. Factors 19 (4), 381–397.
Megías, A., Maldonado, A., Cándido, A., Catena, A., 2011a. Emotional modulation of urgent and evaluative behaviors in risky driving scenarios. Accid. Anal. Prev. 43, 813–817.
Megías, A., Maldonado, A., Catena, A., Stasi, L.L.D., Serrano, J., Cándido, A., 2011b. Modulation of attention and urgent decisions by affect-laden roadside advertisement in risky driving scenarios. Safety Sci. 49, 1388–1393.
Mesken, J., 2006. Determinants and consequences of drivers' emotions. Unpublished Dissertation, Rijksuniversiteit Groningen, Leidschendam, Nederland.
Michon, J.A., 1985. A critical review of driver behaviour models: what do we know, what should we do? In: Evans, L., Schwing, R.C. (Eds.), Human Behaviour and Traffic Safety. Plenum, New York, NY, pp. 485–520.

Mineka, S., Sutton, S.K., 1992. Cognitive biases and the emotional disorders. Psychol. Sci. 3 (1), 65–69.
Mizell, L.B., Joint, M., Connell, D., 1997. Aggressive Driving: Three Studies. AAA Foundation for Traffic Safety, New York, NY.
Moray, N., 1959. Attention in dichotic listening: affective cues and the influence of instructions. Q. J. Exp. Psychol. 11, 56–60.
Näätänen, R., Summala, H., 1976. Road-User Behavior and Traffic Accidents. North-Holland Publishing, Amsterdam.
Nass, C., Jonsson, I.-M., Harris, H., Reaves, B., Endo, J., Brave, S., Takayama, L., 2005. Improving automotive safety by pairing driver emotion and car voice emotion. Proceedings of the SIGCHI Conference on Human Factors in Computing Systems (CHI05), April 2–7, Portland, OR.
Nesbit, S.M., Conger, J.C., 2012. Predicting aggressive driving behavior from anger and negative cognitions. Transport. Res. F 15 (6), 710–718.
Ochsner, K.N., Bunge, S.A., Gross, J.J., Gabrieli, J.D.E., 2002. Rethinking feelings: an fMRI study of the cognitive regulation of emotion. J. Cognit. Neurosci. 14, 1215–1299.
Pêcher, C., Lemercier, C., Cellier, J.-M., 2009. Emotions driver attention: effects on driver's behaviour. Safety Sci. 47, 1254–1259.
Picard, R., 1997. Affective Computing. MIT Press, Cambridge.
Picard, R., 2010. Affective computing: from laughter to IEEE. IEEE Trans. Affect. Comput. 1 (1), 11–17.
Plutchik, H., 1994. The Psychology and Biology of Emotion. Harper Collins, New York, NY.
Poysti, L., Rajalin, S., Summala, H., 2005. Factors influencing the use of cellular (mobile) phone during driving and hazards while using it. Accid. Anal. Prev. 37, 47–51.
Raghunathan, R., Pham, M.T., 1999. All negative moods are not equal: motivational influences of anxiety and sadness on decision making. Organ. Behav. Hum. Decis. Process. 79, 56–77.
Reeves, B., Nass, C.I., 1996. The Media Equation: How People Treat Computers, Televisions, and New Media Like Real People and Places. Cambridge University Press, New York, NY.
Reid, L.D., Solowka, E.N., Billing, A.M., 1981. A systematic study of driver steering behaviour. Ergonomics 24, 447–462.
Roidl, E., Siebert, F.W., Oehl, M., Höger, R., 2013. Introducing a multivariate model for predicting driving performance: the role of driving anger and personal characteristics. J. Safety Res. 47, 47–56.
Rothe, P.J., Cooper, P.J., DeVries, B., 1990. The Safety of Elerly Drivers: Yesterday's Young in Todays Traffic. Transaction Publishers, London.
Rowe, G., Hirsh, J.B., Anderson, A.K., 2007. Positive affect increases the breadth of attentional selection. Proc. Natl. Acad. Sci. USA 104 (1), 383–388.
Rutkowski, T.M., Zhao, Q., Cichocki, A., Tanaka, T., Mandic, D.P., 2011. Towards affective BCI/BMI paradigms–analysis of fEEG and fNIRS brain responses to emotional speech and facial videos. Springer, Dordrecht, The Netherlands, pp. 671–675.
Salvucci, D.D., Boer, E.R., Liu, A., 2001. Toward an integrated model of driver behavior in a cognitive architecture. Transport. Res. Rec. 1779, 9–16.
Schwarz, N., Bless, H. (Eds.), 1991. Happy and Mindless, But Sad and Smart?: The Impact of Affective States on Analytical Reasoning. Pergamon Press, Oxford.
Schwarz, N., Clore, G.L., 2003. Mood as information: 20 years later. Psychol. Inq. 14 (3 and 4), 296–303.
Senesac, E., 2010. Cognitive depletion in emotion regulation: age differences depend on regulation strategy. Masters Thesis. Georgia Institute of Technology, Atlanta, GA.
Sivak, M., Soler, J., Trankle, U., 1989. Cross-cultural differences in driver self-assessment. Accid. Anal. Prev. 21, 371–375.

Skipper, J.H., Wierwille, W., Hardee, L., 1984. An investigation of low level stimulus induced measures of driver drowsiness. Virginia Polytechnic Institute and State University IEOR Department Report 8402, Blacksburg, VA.

Sterkenburg, J., 2015. Impacts of distraction on driving: an analysis of physical, cognitive, and emotional distraction. Masters Thesis. Michigan Technological University, Houghton, MI.

Storbeck, J., Clore, G.L., 2005. With sadness comes accuracy; with happiness, false memory: mood and the false memory effect. Psychol. Sci. 16 (10), 785–791.

Summala, H., 2007. Towards understanding motivational and emotional factors in driver behaviour: comfort through satisficing. In: Cacciabue, C. (Ed.), Modelling Driver Behaviour in Automotive Environments. Springer Verlag, London, pp. 189–207.

Svenson, O., 1981. Are we all less risky and more skillful than our fellow drivers? Acta Psychol. 47, 143–148.

Taylor, D.H., 1964. Drivers' galvanic skin response and the risk of accident. Ergonomics 7, 439–451.

Tiedens, L.Z., Linton, S., 2001. Judgment under emotional certainty and uncertainty: the effects of specific emotions on information processing. J. Pers. Soc. Psychol. 81, 973–988.

Treisman, A., Gormican, S., 1988. Feature analysis in early vision: evidence from search asymmetries. Psychol. Rev. 95 (1), 15–48.

Underwood, G., Chapman, P., Wright, S., Crundall, D., 1999. Anger while driving. Transport. Res. F 2 (1), 55–68.

Van Elslande, P., Fouquet, K., 2007. Analyzing 'Human Functional Failures' in Road Accidents. Final report. Deliverable D5.1, WP5 "Human factors". TRACE European Project.

Walker, G.H., Stanton, N.A., Young, M.S., 2006. The ironies of vehicle feedback in car design. Ergonomics 49 (2), 161–179.

Waller, P.F., 1991. The older driver. Hum. Factors 33, 499–505.

Warnes, A.M., Fraser, D.A., Hawkin, R.E., Sievey, V., 1993. Elderly drivers and new road transport technology. In: Parkes, A.M., Franzen, S. (Eds.), Driving Future Vehicles. Taylor & Francis, London, pp. 99–117.

Watson, D., Wiese, D., Vaidya, J., Tellegen, A., 1999. The two general activation systems of affect: structuring findings, evolutionary considerations, and psychobiological evidence. J. Pers. Soc. Psychol. 76, 820–838.

Wells-Parker, E., Ceminsky, J., Halberg, V., Snow, R.W., Dunaway, G., Guiling, S., et al., 2002. An exploratory study of the relationship between road rage and crash experience in a representative sample of US drivers. Accid. Anal. Prev. 34 (3), 271–278.

Wierwille, W.W., Tijerina, L., 1998. Modelling the relationship between driver in-vehicle visual demands and accident occurrence. In: Gale, A.G., Brown, I.D., Haslegrave, C.M., Taylor, S.P. (Eds.), Vision in Vehicles VI. Elsevier, Amsterdam, pp. 233–243.

Wilde, G.J.S., 1982. The theory of risk homeostasis: implications for safety and health. Risk Anal. 2, 209–225.

Williams, A.F., 2003. Teenage drivers: patterns of risk. J. Safety Res. 34 (1), 5–15.

Williams, J.M.G., Watts, F.N., MacLeod, C., Mathews, A., 1997. Cognitive psychology and emotional disorders. John Wiley, NY.

Yerkes, R.M., Dodson, J.D., 1908. The relation of strength of stimulus to rapidity of habit-formation. J. Comp. Neurol. Psychol. 18, 459–482.

Yoshino, K., Oka, N., Yamamoto, K., Takahashi, H., Kato, T., 2013. Functional brain imaging using near-infrared spectroscopy during actual driving on an expressway. Front. Hum. Neurosci. 7, 882.

Young, M., 1998. Driven mad: aggression behind the wheel gains momentum on busy city freeways. The Dallas Morning News, 1A.

Young, K., Lee, J.D., Regan, M.A. (Eds.), 2008. Driver Distraction: Theory, Effects, and Mitigation. CRC Press, Boca Raton, FL.

Zeng, Z., Pantic, M., Roisman, G.I., Huang, T.S., 2009. A survey of affective recognition methods: audio, visual, and spontaneous expressions. IEEE Trans. Pattern Anal. Mach. Intell. 31 (1), 39–58.

PART V

EMERGING AREAS

CHAPTER

18

Positive Technology, Computing, and Design: Shaping a Future in Which Technology Promotes Psychological Well-Being

*Andrea Gaggioli**,**, *Giuseppe Riva**,**, *Dorian Peters*[‡], *Rafael A. Calvo*[‡]

***Catholic University of Sacred Heart, Milan, Italy; **Istituto Auxologico Italiano, Milan, Italy; [‡]School of Electrical and Information Engineering, University of Sydney, Sydney, NSW, Australia**

INTRODUCTION

More and more of our daily activities depend on some kind of interactive device or digital service. Furthermore, the use of information and communication technologies (ICTs) is not limited to the long hours that we spend at the office. Our free time, too, has been increasingly colonized by technology-mediated experiences delivered through smartphones, tablets, and other personal wearable devices. Even sport activities have been infiltrated by technologies from fitness trackers to wearable action cams.

Computers and other digital devices have become regular companions in our daily lives, but have they made us any happier? Interestingly, the majority of psychological studies on the impact of technologies on well-being have focused on their potential negative effects, including investigations into cyber addiction, techno-stress, violent videogames, privacy risks, etc. On the other side, less attention has been paid to the question of how interactive digital systems could be used to improve well-being of individuals and groups.

Emotions and Affect in Human Factors and Human–Computer Interaction. http://dx.doi.org/10.1016/B978-0-12-801851-4.00018-5
Copyright © 2017 Elsevier Inc. All rights reserved.

Designing for digital experience that promotes psychological flourishing is not a straightforward task as it requires the integration of scientific knowledge and theory relating to those factors known to make life more fulfilling, with the technological expertise required to turn these factors into practical services and applications.

In recent years, the human–computer interaction (HCI) community has given growing importance to interaction design for promoting mental health and well-being (Calvo et al., 2016). This increasing interest has given rise to new research and development areas within HCI, including "Positive Technology" (Botella et al., 2012; Riva et al., 2012) and "Positive Computing" (Calvo and Peters, 2014), which aim to integrate the scientific principles of well-being into the design of interactive systems.

The development of the Positive Technology/Computing approaches result from the convergence of two main trends. First, interaction designers have increased the focus on the concept of User Experience (UX), which has been accompanied by an increasing recognition of the importance of considering human values and ethical issues in the design, development and use of interactive systems (i.e., value-sensitive design, reflective design). Second, the emergence of positive psychology as "the study of the conditions and processes that contribute to the flourishing or optimal functioning of people, groups, and institutions" (Gable and Haidt, 2005) (p. 104) has opened the way for the scientific investigation of the conditions that promote happiness and well-being.

The aim of this chapter is to introduce the concepts of Positive Technology and Positive Computing and examine some illustrative applications of these approaches. We will first introduce the field of positive psychology and examine key theoretical models. Next, we will discuss the objectives and main application areas of these design approaches. Last, we will consider some key challenges and future research directions for this emergent interdisciplinary area (Table 18.1).

TABLE 18.1 Positive Computing Strategies

Positive computing strategies	
Not positive design	Well-being and human potential were not considered in the design of the technology
Preventative integration	Obstacles or compromises to well-being are treated as errors
Active integration	A technology that is designed to actively support components of well-being or human potential in an application that has a different overall goal
Dedicated integration	A technology that is purposefully built to and dedicated to fostering well-being and human potential in some way

Adapted from Calvo, R.A., Peters, D., 2014. Positive computing: technology for wellbeing and human potential. Cambridge, Massachusetts: MIT Press.

THE EMERGENCE OF POSITIVE PSYCHOLOGY AS THE NEW "SCIENCE OF HAPPINESS"

Traditionally, the main concern of psychology has been to understand and treat dysfunctional behaviors, adopting a perspective focused on mental illness, or other psychological problems. However, at the end of 1980s, several scholars dissatisfied with this pathology-centered model of psychology started to investigate alternative ways to look at mental health, which were more focused on positive aspects of mental functioning—such as hope, resilience, strength, creativity, and growth—than on disorders, damages, and suffering.

This growing interest toward the scientific study of well-being led to the birth of positive psychology, which was officially announced by Martin Seligman and Mihaly Csikszentmihalyi in the first 21st century issue of the *American Psychologist* (Seligman and Csikszentmihalyi, 2000). Since then, the field of positive psychology has developed rapidly, with the establishment of dedicated academic courses, international conferences, journals, and books.

However—as also acknowledged by the founding fathers of the movement—positive psychology is not a new idea. Over the course of the twentieth century, several authors focused on the factors that help individuals and groups to thrive, although most of their contributions were unrecognized by "mainstream" psychology. One of the most significant contributions to the development of the theoretical background of positive psychology was provided by humanistic psychology with the work of Carl Rogers (1961) and Abraham Maslow (Maslow, 1954), who introduced, respectively, the concepts of fully-functioning person and self-actualization. Maslow was also the first psychologist to use the term "positive psychology" in his book "Motivation and Personality" (Maslow, 1954) (p. 201).

According to Seligman and Csikszentmihalyi (2000), the main distinguishing feature of positive psychology with respect to previous approaches is its explicit focus on empirical research. Thus, it might be said that the main distinction between positive psychology and its predecessors, with particular reference to humanistic psychology, is more related to its methodological approach than to the topics and subjects investigated (Peterson and Seligman, 2004).

But what are the goals of positive psychology as a scientific discipline? Answering this question is not straightforward, as positive psychology is a multifaceted research area that encompasses a number of different topics related to "what makes life worth living."

However, it is possible to identify three broad levels of analysis within this approach: the subjective level, the individual level, and the group level (Seligman and Csikszentmihalyi, 2000):

- The *subjective level* focuses on positive subjective states, such as happiness, well-being, satisfaction with life, love, hope, and optimism, as well as on the conditions under which these positive experiences occur.
- The *individual level* is concerned with the identification and cultivation of positive individual traits (such as honesty, courage, future-mindedness, self-determination, forgiveness, originality, wisdom, interpersonal skills, and high talent) and engagement in absorbing activities (e.g., flow).
- Last, at the *group or societal level*, the emphasis is on the creation and cultivation of meaningful positive relationships and positive institutions (e.g., schools, legislative bodies, the press, public services) as well as on fostering those civic virtues that are capable of promoting better citizenship, by increasing responsibility, altruism, tolerance of diversity, equality, opportunity, civility, reciprocity, and moderation.

In summary, positive psychology is the scientific study of optimal human functioning. It is concerned with the discovery, understanding and promotion of positive emotions, positive character, and the institutions that enable them to flourish (Seligman et al., 2005). As observed by Delle Fave et al. (2011), rather than representing a new formal sector or a new paradigm, positive psychology is a novel approach to study human behavior, which encompasses all areas of psychological investigation, such as development, occupation, mental and physical health. As such, it can be regarded as a transversal approach to psychology, with applications in a number domains, including work, development, education, clinical, health, and community. However, in their recent review of positive psychology McNulty and Fincham (2012) indicate the need to think beyond positive psychology (McNulty and Fincham, 2012). In particular, they argue that positive psychology needs "to move beyond labeling psychological traits and processes as positive" (p. 107). To overcome this issue Riva (2012) suggested to shift its focus to personal experience: the goal should be the understanding of how it is possible to manipulate the quality of personal experience for increasing wellness, and generating strengths and resilience in individuals, organizations, and society.

PERSPECTIVES ON HAPPINESS

In positive psychology, it is common to distinguish between two broad traditions in the study of happiness, which have their roots in ancient Greek philosophy: the *hedonic* view and the *eudaimonic* view (Ryan and Deci, 2001). The first view conceives happiness as a person's subjective evaluation of his or her positive and negative emotional feelings. In contrast with the focus on emotional life, eudaimonic well-being links happiness with lifelong conduct aimed at self-development.

The Hedonic Perspective

The hedonic view was originated by the Greek philosopher, Aristippus of Cyrene, who considered the pursuit of pleasure as the ultimate goal of life. Accordingly, psychologists who have advocated the hedonic perspective tend to consider happiness as the result of subjective positive experiences, such as pleasure, comfort, and enjoyment (Kahneman et al., 1999). Proponents of this approach are especially interested in understanding how people experience the quality of their lives, both in terms of emotional reactions and cognitive assessment of life satisfaction.

In particular, most research and intervention within the hedonic psychology has focused on the assessment/improvement of subjective well-being (SWB), which is typically defined as constituted by three key components: the presence of positive affect, the absence of negative affect, and a cognitive evaluation of life satisfaction (Diener, 1984, 2000). These subdimensions are evaluated through self-report measures, such as positive and negative affect scale (PANAS) (Watson et al., 1988) that assesses a person's positive and negative affect states and satisfaction with Life Scale (SWLS) (Diener, 1994) that evaluates the degree of satisfaction with life. An important area of research on SWB has examined the predictive power of this construct on different desirable life outcomes, including health and longevity, financial success, social relationships, and effective coping. In these studies, large sample of participants are followed for several years (typically, a decade or more), to identify whether the happier ones live longer and healthier. Results of several meta-analyses have consistently shown that SWB has a positive causal effect on longevity and physiological health although it is still controversial how various types of SWB influence specific diseases, and about the role of the possible mediating processes (Diener and Chan, 2011).

In particular, the famous "nun study" (Danner et al., 2001) is considered a landmark contribution to the field. The study started in 1986 with the goal of examining the onset of Alzheimer's disease. It has involved 678 of the older School Sisters of Notre Dame, which provides an ideal population to study effects on health, because of the highly homogeneous living conditions of religious sisters. The methodology concerned the careful analysis of emotional content of autobiographical essays that nuns wrote upon joining the congregation, which was then related to survival during ages 75–95. Findings showed a remarkable direct relationship between positive emotional content in autobiographies written in early adulthood and longevity 6 decades later: nuns who perceived themselves to be happier died at a median age of 93.5 years. In contrast, those who considered themselves to be less happy died at a median age of 86.6 years (Danner et al., 2001).

A further often-cited research concerning the long-term effects of happiness is Harker and Keltner (2001) study of women's yearbook pictures at an elite college, which these authors related to a variety of life outcomes (e.g. health, personality, and marriage) up to 30 years later. Results showed that the more intense the subject's positive expression shown in her senior yearbook picture, the more likely that she would be married by age 27 and would have a more satisfying marriage in adulthood (Harker and Keltner, 2001).

Positive emotions are a key component of SWB. However, *how* do positive emotions contribute to happiness? One major contribution to the understanding of this question is Barbara Fredrickson's *broaden-and-build theory of positive emotions* (Fredrickson, 2001). As the name suggests, the model encompasses two main hypotheses: the broaden hypothesis and the build hypothesis.

The broaden hypothesis argues that positive emotions—such as joy, interest, contentment, and love—*broaden* people's momentary thought-action repertoires. In contrast with negative emotions, which narrow attention, cognition, and physiology by calling to mind an urge to act in a particular way when facing an immediate threat or problem (e.g., flight in fear, attack in anger), positive emotions prompt individuals to pursue a wider range of thoughts and actions (e.g., play, explore, savor, and integrate) than is typical (Fredrickson and Branigan, 2005).

Fredrickson and Branigan (2005) performed several experiments to test the broaden hypothesis. In one study, the authors used videos to induce positive emotions (amusement and contentment), negative emotions (anger and anxiety), and neutral states in student participants. After viewing the clips, participants were administered a series of global-local visual processing tasks (Fig. 18.1), which required them to judge which of the two comparison figures (bottom) is more similar to a standard figure (top).

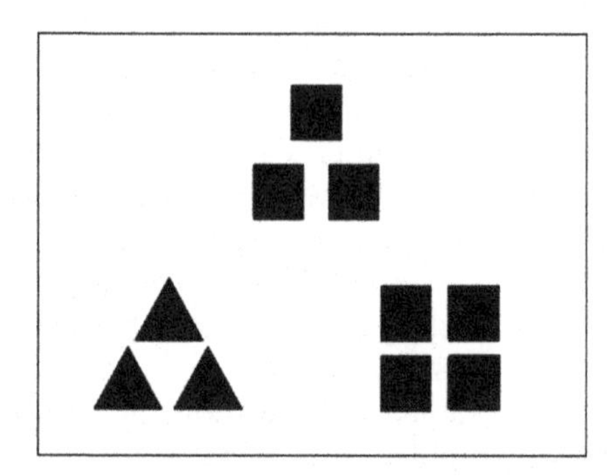

FIGURE 18.1 A sample item from a global-local choice task. *Adapted from Fredrickson, B. L., Branigan, C., 2005. Positive emotions broaden the scope of attention and thought--action repertoires. Cogn. Emot. 19, 313–332.*

Although neither choice is correct or incorrect, according to Fredrickson and Branigan the global response option (lower left) reflects more broadened thinking. The results of this experiment support the broaden hypothesis, as participants exposed to clips inducing positive emotions selected the global option significantly more often than those in the neutral or negative emotion conditions, thus suggesting that positive emotions broaden the scope of attention (Fredrickson and Branigan, 2005). Notably, this effect has also been shown through eye-tracking (Wadlinger and Isaacowitz, 2006) and brain-imaging (Schmitz et al., 2009; Soto et al., 2009).

In another study, Fredrickson and Branigan (2005) focused on the effects of positive emotions on momentary thought-action repertories. As in the global-local choice experiment, they exposed participants to film clips evoking the specific emotions of joy, contentment, fear, and anger, as well as to nonemotional videos as a neutral condition. Immediately after the induction procedure, the authors asked participants to write a list of all of the things they would like to do right now. Findings showed that participants assigned to the two positive emotion conditions (joy and contentment) listed more things that they would like to do as compared to those in the negative emotion conditions (fear and anger) and in the neutral control condition. Moreover, participants in the negative emotion conditions named fewer actions than those in the nonemotional condition. According to Fredrickson and Branigan, these results support the broaden hypothesis since they indicate that positive emotions expand the momentary thought and action options available to the individual, whereas negative emotions produce a narrower array of thought and action's repertoire. Moreover, these observations are complemented by findings from other studies that have documented the effect of positive emotions on creativity (Rowe et al., 2007), openness to new experiences (Kahn and Isen, 1993), and critical feedback (Raghunathan and Trope, 2002).

The second key hypothesis of Fredrickson's theory is that by encouraging a broadened range of actions, positive emotions *build* over time enduring psychological, intellectual, physical, and social resources. For example, the positive emotion of joy creates the urge to play and get involved; however, playing also allows to create long-lasting resources, which are the skills acquired through the experiential learning it prompts (Danner et al., 2001).

Cohn et al. (2009) carried out an experiment to test the hypothesis that positive emotions help the individual to build enduring personal resources. The authors measured emotions daily for 1 month in a sample of students and assessed life satisfaction and trait resilience at the beginning and end of the month. Findings showed that positive emotions were a significant predictor of gains in resilience and life satisfaction, whereas negative emotions had no or small effects. In another study, Fredrickson et al. (2008) carried out a randomized controlled trial to test the effects of loving-kindness meditation as a technique to induce positive emotions, comparing it with a wait-list control group. Findings showed that the treatment group reported greater positive emotions and lower depressive symptoms than participants in the wait-list control group. Further, gains in positive emotions observed in the meditation condition generated increases in several personal resources, including mindfulness, environmental mastery, positive relations with others, reduced illness symptoms, and increases in life satisfaction (Fredrickson et al., 2008).

Positive emotions have also shown beneficial effects at physical level. In particular, Fredrickson and colleagues (Fredrickson et al., 2000) argued that positive emotions have a homeostatic function by "undoing" the lingering aftereffects of negative emotional reactivity, returning an individual to cardiovascular equilibrium. In line with this hypothesis, Tugade and Fredrickson (2004) found that positive emotions were related to faster recovery from cardiovascular reactivity generated by negative emotions for resilient individuals. These findings support a potential health-promoting functions of positive emotions, with particular reference to the prevention of cardiovascular disease (Fredrickson and Levenson, 1998; Kok and Fredrickson, 2010; Tugade et al., 2004).

The Eudaimonic Perspective

In contrast to the hedonic view, the eudaimonic view considers happiness something different from the mere attaining of pleasure and enjoyment. Drawing on the Aristotelian definition of *eudaimonia*, this perspective identifies happiness with the full realization of true human nature, through the exercise of personal virtues and potentials in pursuit of complex goals that are meaningful to the individual and society (Ryff and Singer, 2008).

However, the eudaimonic perspective does not only encompass personal satisfaction, but it also pursues a development path toward the integration of the individual with the surrounding environment: it refers to the interaction between personal and collective space, which assumes that individual happiness is realized within interpersonal relationships (Nussbaum and Sen, 1993).

In the 20th century, the eudaimonic perspective has seen a significant contribution by pioneers of humanistic psychology, Abraham Maslow and Carl Rogers. According to humanistic psychology, individuals have the potential to express their authentic nature if they are able to support the natural inclination toward self-actualization, which exists in all people and is only "waiting" for the conditions to be realized.

A prominent model that has focused on the process of actualization of the self, and the ways it can be accomplished, is Deci and Ryan's self-determination theory (Deci and Ryan, 2000).

The main focus of self-determination theory is on the investigation of innate psychological needs and inherent growth tendencies that form the basis of self-motivation and integration of personality. Specifically, according to Deci and Ryan, in order to foster well-being and health, three basic needs must be satisfied:

- *autonomy*: this need refers to the urge to be causal and self-governing agents, who act in harmony with their integrated self;
- *competence*: this refers to the experience of behavior as effectively enacted; and
- *relatedness*: it concerns the universal need to interact with other human beings, be connected, and experience caring for others.

The fulfillment of these needs is essential for as a crucial condition for psychological growth (Deci and Ryan, 2000): several studies, across different life domains, have shown positive relations between need satisfaction and optimal functioning, both at the interpersonal and intraindividual level (Deci and Ryan, 2008). Also, self-determination theory holds that basic needs drive the goal-setting process: depending on the extent to which these needs are fulfilled, individuals set *intrinsic aspirations*, which encompass personal growth, affiliation and intimacy, contribution to one's community, and physical health. However, if basic needs are not satisfied, individuals develop *extrinsic aspirations*, such as financial success, social recognition and fame, and image or attractiveness. Since the satisfaction derived from the achievement of extrinsic goals has an ephemeral nature, the individual aims at creating novel and greater extrinsic aspirations. Furthermore, recent research findings have highlighted that people's well-being improves as they place relatively less importance on materialistic goals and values, whereas orienting toward materialistic goals relatively more is associated with decreases in well-being over time (Kasser et al., 2014). In contrast, the pursuit of intrinsic goals can generate an enduring well-being. Thus, it can be said that psychological need satisfaction plays a mediating role between the achievement of intrinsic goals and changes in well-being (Deci et al., 2008; Niemiec et al., 2009). In sum, Deci and Ryan's self-determination theory provides a comprehensive account of intrinsic motivation and its role in the development of psychological well-being. According to this model, the prototype of self-determined behavior is intrinsically-motivated action that one engages in because one enjoys it and is interested in it, not because of an "external" reward.

Mihály Csíkszentmihályi has defined the experience associated to such intrinsically-rewarding activity as "flow" (Csikszentmihalyi, 1975, 1990). This is a positive and complex state of consciousness, characterized by a perceived balance between high challenges in the task at hand and adequate personal skills in facing them. Additional characteristics of this optimal experience are positive affect, deep concentration, clear rules in and unambiguous feedback from the task at hand, loss of self-consciousness, and control of one's actions and environment (Csikszentmihalyi, 1975, 1990).

Previous research has shown that flow shows stable features at the cross-cultural level, and it can be associated with various contexts of activities, such as family, work, leisure, provided that individuals perceive these activities as complex opportunities for action in which to invest personal skills. From this perspective, an important aspect highlighted by previous research is that the association of flow with specific activities motivates people to replicate them, as well as to the preferential cultivation of individual skills (Csikszentmihalyi and Beattie, 1979). As the latter improves, the person will subsequently search for increasingly complex opportunities for action.

By virtue of this dynamic process of skills cultivation and challenge increase, optimal experience shapes the development of a *life theme,* namely the set of goals and interests a person preferentially pursues and cultivates in his/her life (Csikszentmihalyi and Beattie, 1979). Csikszentmihalyi and colleagues also introduced the concept of an *autotelic personality,* to describe people with several very specific personality traits which enable more frequent experiences of flow states than the average person (Csikszentmihalyi et al., 1993; Nakamura and Csikszentmihalyi, 2002). In Csikszentmihalyi's own words: "Autotelic is a word composed of two Greek roots: *auto* (self), and *telos* (goal). An autotelic activity is one we do for its own sake because to experience it is the main goal. Applied to personality, autotelic denotes an individual who generally does things for their own sake, rather than in order to achieve some later external goal" (Csikszentmihalyi, 1997, p. 117).

A further central concern in the eudaimonic approach to happiness is to explore the role of human character strengths and virtues. In the history of psychology, there have been several attempts to understand and classify mental illness using statistic criteria. These efforts eventually lead to the elaboration of the *Diagnostic and Statistical Manual of Mental Disorders* (DSM) which is today widely adopted by the clinical community. Following the DSM example, Park, Peterson, and Seligman proposed a taxonomy of *character strengths,* defined as "positive traits reflected in thoughts, feelings and behaviors" (Park et al., 2004, p. 603). Based on systematic review of psychological, philosophical, and religious literature, Peterson and Seligman (Peterson and Seligman, 2004) identified 24 measurable character strengths, which were further classified under six classes of core *virtues*: wisdom/knowledge (creativity, curiosity, open-mindedness, love of learning, perspective), courage (bravery, persistence, integrity, vitality), humanity (love, kindness, social intelligence), justice (citizenship, fairness, leadership), temperance (forgiveness and mercy, humility, prudence, self-regulation), transcendence (appreciation of beauty and excellence, gratitude, hope, humor, spirituality). These authors also developed a self-reported instrument, the *Values in Action Inventory of Strengths* (VIA-IS; 2004), in order to measure these character strengths.

Integrative Models

In his book "Authentic Happiness" Seligman (2002) attempted to provide a more integrated view of the concept of happiness, by bringing together the main theoretical traditions about this concept. In particular, Seligman identified three main components of the "good life": (1) the pleasant life: achieved through the presence of positive emotions, pleasure, and gratifications; (2) the engaged life: achieved through engagement in satisfying activities and utilization of one's strengths and talents; it involves engaging in activities that support flow; and (iii) the meaningful life: achieved through serving a purpose larger than oneself.

However, more recently Seligman realized these three components were not exhaustive elements of happiness. In particular, he acknowledged three main limitations of the Authentic Happiness theory: first, it did not take into consideration the role of other key variables, such as success and the sense of control. Furthermore, according to Seligman, the term happiness itself was misleading, because it posits a superficial identification with "feeling in a good mood," while neglecting the dimensions of engagement and meaning. A further weakness was that happiness was mainly operationalized as life satisfaction and measured subjectively by self-report thus overlooking the assessment of the other two components—engagement and meaning (Seligman, 2011).

To overcome the limitations of Authentic Happiness theory, in 2011 Seligman introduced the PERMA model (Seligman, 2011), an acronym for the five pillars of well-being: Positive emotions, Engagement, Relationships, Meaning, and Accomplishment. Of these components, three were inherited from the former Authentic Happiness model. The additional ones—relationships and accomplishment—refer, respectively, to the role played by significant others in helping us to face the challenges in life, and to the motivation to achieve, to have mastery and competence.

USING POSITIVE PSYCHOLOGY PRINCIPLES IN TECHNOLOGY DESIGN: THE EMERGENCE OF WELL-BEING DESIGN

Humanistic approaches to technology design have had a renaissance over the last 10 years. Among these humanistic perspectives, the terms often used in the computing disciplines, experience-centered design,

(Wright and McCarthy, 2010) value-sensitive design (Friedman, 1996) and positive computing (Calvo and Peters, 2014) are the most important for this chapter.

Experience-centered design is concerned with the "richness of human experience" (Wright and McCarthy, 2010), a focus that has become a mainstream aspect of software development (Hassenzahl and Tractinsky, 2006). Understanding the user serves as a balancing perspective to an absolute focus on the task to be accomplished using the technology. Furthermore, human experience can be seen as a construct incorporating emotion, cognition, and behavior, all of which can be interpreted and are influenced by our technological environments. As we will see later, this triad has received much attention among HCI and computer science researchers.

Technology designers are increasingly aware of the impact of their own personal views on the products they create. Value-sensitive design (Friedman, 1996) has been a movement emphasizing the impact that the designer's values have on the technologies being produced. Values relating to issues, such as privacy, sustainability, autonomy support, and well-being are seen from a techno-social perspective, using moral philosophy as an underpinning theoretical foundation. Value-sensitive design does not posit any one view as more important than another, but it makes the important point that whatever the values chosen (explicitly or not) by the designer, these values will have an impact on the final product. Therefore, designers have an obligation to take these values into account.

Improving Personal Experience—The Positive Technology Approach

Riva and colleagues (Botella et al., 2012; Riva et al., 2012) have suggested that it is possible to combine the objectives of positive psychology with technology design using the "Positive Technology" approach: the scientific and applied approach to the use of technology for improving the quality of our personal experience. In other words, personal experience is the dependent variable that may be manipulated through the technology. Specifically, they suggest that it is possible to reach this goal in three separate but related ways (Riva et al., 2012):

- *By structuring personal experience* using a goal, rules, and a feedback system (e.g., serious games): The goal provides subjects with a sense of purpose focusing attention and orienting his/her participation in the experience. The rules, by removing or limiting the obvious ways of getting to the goal, push subjects to see the experience in a different way. The feedback system tells players how close they are to achieving the goal and provides motivation to keep trying.
- *By augmenting personal experience* to achieve multimodal and mixed experiences. Technology allows multisensory experiences in which

content and its interaction is offered through more than one of the senses. It is even possible to use technology to overlay virtual objects onto real scenes.

- *By replacing personal experience* with a synthetic one. Using VR, it is possible to simulate physical presence in a synthetic world that reacts to the action of the subject as if he/she was really there. Moreover, the replacement possibilities offered by technology even extend to the induction of an illusion of ownership over a virtual arm or a virtual body.

Moreover, they identified three critical variables—Emotional Quality (affect regulation), Engagement/Actualization (presence and flow), and Connectedness (collective intentions and networked flow)—that can be controlled and assessed to guide the design and development of positive technologies (Inghilleri et al., 2015). This approach has been successfully used in cyberpsychology by Riva and colleagues (Carissoli et al., 2015; Gaggioli et al., 2014c; Munson and Resnick, 2012; Villani et al., 2013; Wiederhold and Riva, 2012). Positive technology can be considered an extension of cyberpsychology, a branch of psychology that aims at the understanding, forecasting, and induction of the different processes of change related to the use of new technologies (Riva et al., 2015a). Cybertherapy (or e-therapy) was the first area of cyberpsychology to have an impact on psychological treatments (Manhal-Baugus, 2001). Innovative e-therapy approaches are an opportunity for earlier and better care of the most common mental health problems (Christensen and Hickie, 2010). These e-therapy approaches allow patients to engage in treatment without having to accommodate to office appointments, often reducing the social anxiety of face-to-face treatment (Mair and Whitten, 2000). There is increasing evidence that Internet-based therapies are economically sound, effective at a low cost (Kadda, 2010). They can be used to reach people in isolated places, where mental health is often a problem (Hordern et al., 2011). Self-help interventions, a form of positive psychology, can be disseminated online (Schueller and Parks, 2012) as exercises aiming to increase positive thoughts, behaviors, and emotions. Some authors have suggested that these exercises may be an effective approach to reduce depressive symptoms. More recently, cybertherapy used technology for modifying the characteristics of personal experience using virtual reality (Riva et al., 2015b) and augmented reality (Chicchi Giglioli et al., 2015). In contrast to cybertherapy applications, positive technology is more focused on interventions that are specifically designed to support positive emotions and self-growth.

Universal Well-Being Design—The Positive Computing Approach

Well-being design builds on the factors that have been identified by psychology research as determinants of well-being—those that contribute directly to greater well-being and which can be used to identify individuals who are at the top of well-being scales.

In their description of positive computing, Calvo and Peters (2014) argue that design for well-being should go beyond psychological interventions and be integrated into the technology design cycle generally, thus informing the design of all software, including, for example, office productivity tools and social media. In this section, we focus on the research aiming to incorporate psychological well-being into everyday software design practices, while in the next section we will focus on dedicated applications with positive psychology interventions.

Software engineers have already incorporated cognitive science and psychology research into standard practices for usability and user experience methods. In the same way, positive computing aims to bring research findings from well-being psychology and neuroscience into the design of everyday technology interaction.

As we have identified earlier, there are multiple positive psychology theories and each identifies a set of determinant factors of well-being. PERMA, Self-Determination Theory, and the other theories described in Section "The Emergence of Positive Psychology as the New Science of Happiness" have enough research evidence on how those factors support well-being that they can directly inform design. Take for example, factors, such as self-awareness, empathy, and emotion regulation, which are at the core of emotional intelligence theories (Gallagher and Vella-Brodrick, 2008; Mayer et al., 1999). These have been validated as key to psychological well-being in dozens of studies. There are also many examples of empathy consciously incorporated into technology design, most notably in the area of role playing games. The Peacemaker game, the Frontier's game, and Shelter are just a few examples of digital environments that have been used to help people understand what it is like in someone else's shoes—sometimes experiencing the perspectives of two sides of a conflict. Research on empathy and conflict resolution can have even been incorporated into general applications like social media, (see Facebook's Compassion Research Day as an example). Empathy is, of course, just one determinant of optimal psychological functioning. There are many others described in the psychology literature.

Calvo and Peters (2015) have used a number of these well-being determinants (i.e., autonomy, competence, relatedness, compassion, engagement, and meaning) in workshops for interaction designers and HCI researchers in order to help them take into account the psychological well-being within the context of their design work. Within these workshops, designers are scaffolded and encouraged to consider the impact of various technology designs on various well-being determinants and how interface and interaction elements might be redesigned in light of well-being impact.

Positive emotions are possibly the first factor that interaction designers think about when considering well-being (Hassenzahl and Beu, 2001).

Design briefs often include terms, such as "pleasure" and "fun" because it is acknowledged that these attributes help bring and keep people to a product (e.g., a website or game). Yet, when designers consider positive emotions, the hedonic (Hassenzahl and Beu, 2001) aspect of a digital experience, they are generally not looking beyond ephemeral emotions. As we discussed earlier, positive psychologists instead see positive emotions as merely one, the hedonic, aspect of well-being scales. Well-being design argues that design for positive emotions is not sufficient to address support for overall psychological well-being. The other determinants identified in psychology literature must also be incorporated.

For example, autonomy has been identified as critical to well-being by various well-being theories. It has been defined as, "a state of being independent or self-governing" (Spear, 2001), "a capacity for thinking and acting independently" (Littlewood, 1996), and more narrowly, in the context of self-determination theory as "an internal perceived locus of causality" (Ryan and Deci, 2000). Within HCI, these theories have been used to describe the sense of agency required for enjoyable gameplay and the self-regulation users are seeking to enhance through the use of behavior change, healthcare and learning technologies. Calvo et al. (2014) have explored with the HCI community ways in which autonomy can be promoted in technology design.

Calvo and Peters argue that there are "inter- and intrapersonal" determinants like autonomy, competence, and empathy and there are a number of "extra-personal" factors. One of the most currently emerging extra-personal factors is compassion. Compassion has been shown by studies in both psychology and neuroscience as a determinant of resilience (Calvo and Peters, 2014), a combination particularly relevant to technologies designed for the helping professions, such as medicine, nursing, teaching, and emergency services. Although colloquially they are frequently lumped together, empathy and compassion are actually distinct (albeit connected). While empathy describes the mirroring of someone else's emotions (if you feel sad, I feel sad) right down to the neurophysiological level, compassion does not necessarily include a mirroring of the same emotion (Goetz et al., 2010), instead, it triggers concern and a motivation to help (caregiving behavior). Furthermore, empathy can have an inward focus, while compassion is definitively outward (approach-oriented). Empathic distress, for example, may cause someone to suffer when made aware of another's suffering while causing a feeling of helplessness or overwhelm leading to avoidance (as opposed to approach). This empathic distress is a common cause of burnout in healthcare and other professions. Compassion instead prepares the body for approach and caregiving and has been associated with positive affect patterns in the brain. The difference between these two emotions is evident in distinct facial expressions, physiological responses, and neurophysiological representations. Based on the literature for these two emotions, Peters and Calvo (2014) have proposed different strategies for the design of software that supports compassion and resilience: (1) addressing appraisals of deservedness, (2) supporting feelings of agency, (3) providing opportunities for the practice of altruism, (4) providing opportunities for the experience of elevation, and (5) supporting compassion-training practices. For some examples of how these might be applied in real-world technologies (Peters and Calvo, 2014).

Frameworks for Bringing Well-Being Design into New Interaction Design

A taxonomy of the different ways in which interaction designs take into account well-being can help designers better identify their approach. Some approaches are better suited for projects aimed at improving health while others for conventional software engineering projects with other aims, such as providing a service or entertainment.

Calvo and Peters (2014) describe four types of design approaches or ways in which well-being can be integrated into technology design. These authors come from technology background and so their positive computing framework was developed as a way of better understanding and communicating the various ways in which design for well-being can be incorporated into digital experiences in general.

Current State of Affairs—Productivity and User Experience

Most technologies today are not "positively" designed, that is, their design brief does not include well-being. It is not that these applications are negative, just that their designers have not taken well-being into account.

Generally, engineers take into account sensible things like productivity, speed, performance, and safety. Engineers' and computer scientists' collaboration with cognitive psychologists in particular has led to impressive developments on methodologies that can be used to improve usability and productivity and they have been at the center of the field of human factors or HCI. Without a doubt, increasing productivity has been the most important goal when developing technologies. A number of reasons can be used to explain this, including: productivity is easy to measure using output quantities or the time it takes to complete a task. Productivity is a culturally agreeable construct, few would question the benefits of increasing productivity (unless it has a negative impact on other variables). Furthermore, digital technology began as a tool for work, a context in which production is the goal.

Calvo and Peters (2014, 2015) have raised concerns with the computer–human interaction community that since technologies are now part of our everyday life, designers need to beware of the productivity mentality that is seeing us track, compare, and measure everything from miles run to hours slept. Although these measures may be useful as part of tools for personal reflection, as quantified-self advocates have suggested (Rivera-Pelayo et al., 2012), the boundary between personal improvement and giving to the tyranny of productivity can be subtle for designers but have opposing impact on the users' psychological well-being.

Dedicated Integration

Possibly, the most common way to introduce well-being into the design brief is by software that specifically supports the developing of one or more determinant factors. We have introduced this approach in the previous section.

Riva and colleagues (Riva, 2012; Riva et al., 2012) have focused on design of "positive technologies" that "manipulate and enhance the features of our personal experience for increasing wellness, and generating strengths and resilience in individuals, organizations and society" (Wiederhold and Riva, 2012). In their framework (Fig. 18.2), these technologies can be classified according to their effects on the above features of personal experience (Botella et al., 2012):

- Hedonic: technologies designed to induce certain emotions, generally positive, and pleasant.
- Eudaimonic: technologies created to support engaging and self-actualizing experiences.
- Social/interpersonal: technologies used to support and improve social integration and/or connectedness between individuals, groups, and organizations.

As concerns the first level (hedonic technologies), virtual environments, and mobile/wearable tools have shown an interesting potential as tools to induce positive emotional states and reduce distress. For example, in the "Green Valley" experience developed by Grassi et al. (2009), different audio narratives were presented together with a relaxing virtual environment showing a mountain landscape around a calm lake. After being immersed in the Green Valley, participants were asked to walk around the lake, to observe nature, and, after few minutes, to virtually sit on a comfortable deck chair and relax. Tested in a controlled trial, results showed a significant decrease in anxiety level and an increase in relaxation level (Grassi et al., 2009). The efficacy of virtual reality in supporting positive emotions may be further enhanced by combining the use of this technology with biofeedback training (Repetto et al., 2009). Biofeedback is a coaching and training technique that helps people learn how to change their physiological response patterns in order to improve their mental and

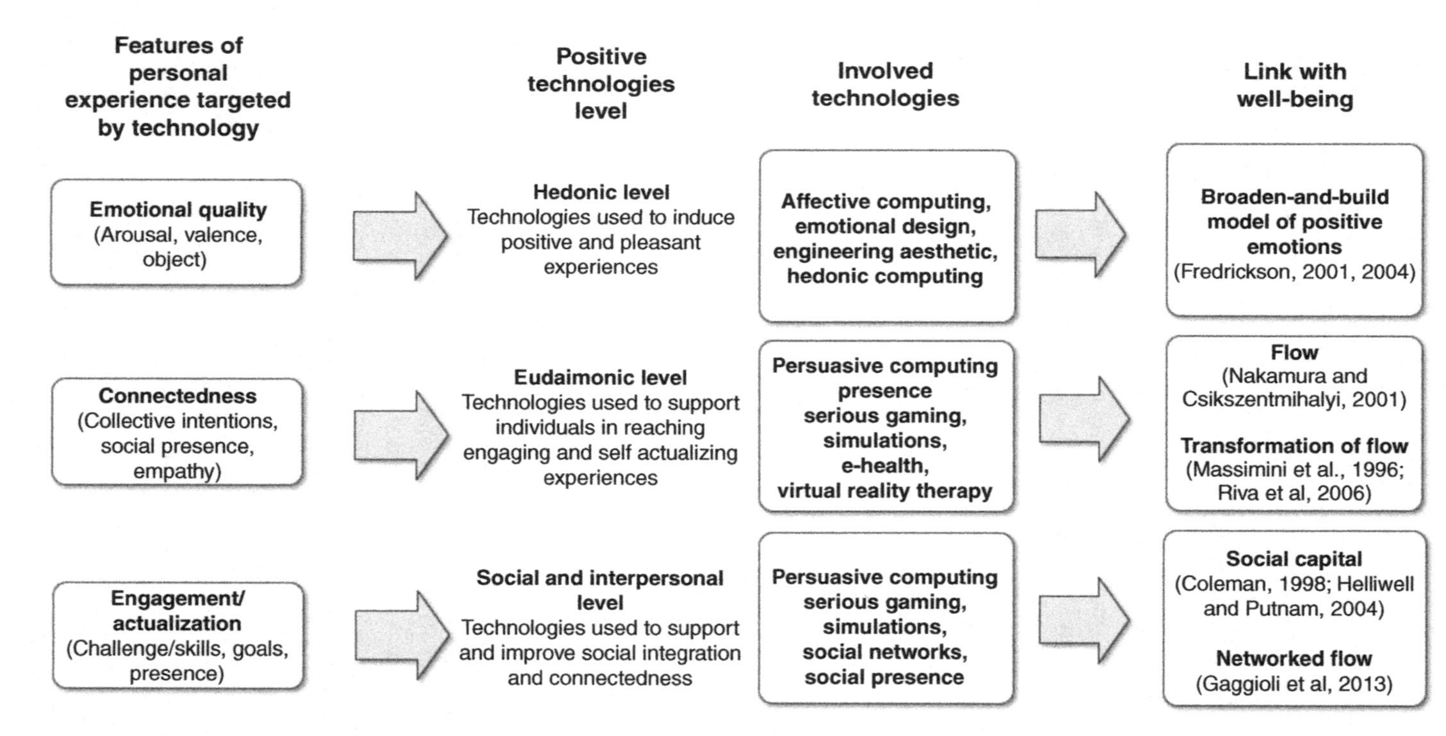

FIGURE 18.2 **Positive technology levels (Coleman, 1998; Massimini et al., 1996; Fredrickson, 2004; Gaggioli et al., 2013; Helliwell and Putnam, 2004; Nakamura and Csikszentmihalyi, 2001; Riva et al., 2012).**

emotional functioning. The person is connected to psychophysiological biosensors and uses the information provided as feedback to increase awareness or consciousness of the changes in the functioning of the body/mind (Riva et al., 2010). The combined VR-biofeedback approach has been also exploited by the "Positive App" (Gaggioli et al., 2014b), a free smartphone application for the self-management of psychological stress. The mobile platform features three main components: (1) 3D biofeedback, which helps the user learning to control his/her responses, by visualizing variations of heart rate in an engaging 3D environment; (2) guided relaxation, which provides the user with the opportunity of browsing a gallery of relaxation music and video-narrative resources for reducing stress; (3) stress tracking, by the recording of heart rate and self-reported levels of distress.

The second class of positive technology applications include tools that support the eudaimonic dimension of well-being, that is, the creation of engaging mediated experiences that support self-actualization and personal growth. One example is "transformation of flow"—the use of virtual reality to create artificial training environments that foster optimal experience by promoting unexpected psychological resources and sources of involvement (Riva et al., 2004, 2006). In this strategy, virtual reality can be used to engage the participant in challenging tasks that are matched to the user's personal skills and resources. This approach has shown promising results in the field of rehabilitation (Borghese et al., 2013; Shin et al., 2014; Robinson et al., 2015). For example, Robinson et al. (2015) examined the effects of "exergaming" (short for exercise video-games) against conventional training (balance training) on functional outcomes and flow experience in people with multiple sclerosis. These authors found that game-based treatment was as effective as conventional training in improving balance and gait, but was associated to higher levels of flow experience (thereby improving intrinsic patient's intrinsic motivation in doing the exercise).

A third category of positive technology applications include systems that are designed to support the social/interpersonal dimension of well-being, that is, by improving or maintaining social connectedness. An example of such applications is the use of smart tools, such as tailored internet programs, to reduce the feeling of isolation of older people living at home (Morris et al., 2013) or to foster communication between the elderly and the youth (Gaggioli et al., 2014a).

Preventative Integration

Digital environments sometimes contribute to people behaving in ways they would not in real life.

Certain antisocial behaviors have a negative impact on businesses. Different designs can promote or obstruct these types of behaviors (Phillips, 2015). In preventative integration approaches, obstacles or compromises to well-being are treated as errors. For example, when anonymity is found to increase the likelihood people abuse others online, anonymity can be eliminated.

Active Integration

Currently, consumers buy particular word processors and email systems, not because they will support any aspect of their well-being, but because these systems help achieve their goals, complete their tasks and work.

Calvo and Peters (2014) have argued that consumers will seek future technologies that support health and well-being. This is likely to occur in the same way they currently seek healthy foods not just for sustenance or even pleasure, but as a way to live healthy and meaningful lives.

Well-being can be actively integrated into technology by designing to actively support components of well-being in an application that has a different overall goal. We need techniques that allow designers to assess the impact that different choices have on the determinant factors of well-being. A number of fields within computing can contribute to measuring this impact. For example, affective computing, the discipline that studies how computers can detect and process human emotions is increasingly part of design considerations in health and education (Riva et al., 2015a). Currently, most approaches to measure psychological well-being require interrupting users to ask about their state of mind. These interruptions are needed for the sake of measuring but themselves can be disengaging and obtrusive. Affective computing techniques can be used to reduce the amount of questioning and self-reporting by automating some of the emotion detection. Furthermore, being able to detect emotions will allow computer interfaces to better adapt to users' states of mind and better engage, since emotional states are a most important aspect of psychological well-being.

A RESEARCH AGENDA

Interdisciplinary partnerships are necessary to solve complex problems, and few problems are as complex as the search for a path to happiness. It is clear to us that the meaning of the word, happiness is too convoluted and so we have referred to our end goal as psychological well-being and human flourishing. This description takes into account much more than ephemeral or only positive emotions.

Finding the right way of communicating ideas, in a way that resonates with distinct academic communities is not trivial and is maybe in itself the first important topic of any research agenda. Researchers and practitioners bring their disciplinary perspectives to any problem, and they need these to be able to tackle problems with the methods they know. Often their perspectives about certain issues—such as well-being —have never been made explicit, and they are founded on more personal rather than professional opinions.

Of course, the complexity of multidisciplinary perspectives is compounded by the difficulty that the general public often has with the well-being construct. It often leads to misunderstandings, such as when uninformed commentators believe—and tell others—it refers to "making people feel positive emotions". The same will happen with well-being design, particularly when we try to bring it into general software applications. Dedicated applications so far seem to fair better as they can be more easily associated with the self-help or a clinical category. And yet, a growing number of general software makers are already incorporating some well-being design without calling it that (such as the Facebook and Microsoft examples mentioned earlier).

In addition to the difficulty in accurately communicating the goals of well-being design, there are difficulties involved in understanding the design processes and roles of different stakeholders. Often psychologists see engineers exclusively as system builders, yet many engineers see themselves as scientists trying to expand human knowledge. Often engineers see psychology as a soft science that deals with subjective and inscrutable aspects of human experience, yet psychologists see themselves as empirical scientists with evidence-based, data driven approaches. Because of the history of their fields, HCI and cyberpsychology researchers are the best positioned to build the partnerships required between the different disciplines and find a common language for this new area.

A second topic for future research is in evaluation methods for well-being design practice—methods that can be used to determine whether particular design decisions are having a positive impact on well-being or not. The methods, such as experience sampling (Csikszentmihalyi and Larson, 1987) that requires constant interruptions to achieve ecological validity are often too intrusive for real-world applications. Although they are increasingly used in the software industry, new methods that combine automated emotion detection and behavioral analytics might provide more sustainable ways of measuring the impact of design.

A third area for future research involves a better awareness of how personalization and context-awareness can contribute to more "positive" designs. Although these are common in other forms of software design to our knowledge, they have not been systematically studied within well-being design approaches. In general, the studies described here have not considered (or have assumed as constant) the impact that the external environment or personal characteristics, have on how to design for well-being.

CONCLUSIONS

We have described different approaches to design technologies that support psychological well-being. These are grounded on different positive psychology theories, some of which have been reviewed. They can also be driven by the values and motivation of the designers, and their disciplinary background.

We used Calvo and Peter's (Calvo and Peters, 2014) framework to classify different ways for bringing well-being considerations into interaction design. Psychologists seem to be more inclined to use technology as a platform for supporting and sustaining the process of change (positive technology). Engineers instead look for ways of considering well-being in the design of any technology either as preventative or active integration.

This chapter is a collaboration between designers, engineers, and psychologists. Collaborations, such as this can be hard as each of us brings a different perspective. We hope that other multidisciplinary teams will find this framework useful to improve communication and outcomes.

References

Borghese, N.A., Pirovano, M., Lanzi, P.L., Wuest, S., de Bruin, E.D., 2013. Computational intelligence and game design for effective at-home stroke rehabilitation. Games Health J. 2 (2), 81–88.

Botella, C., Riva, G., Gaggioli, A., Wiederhold, B.K., Alcaniz, M., Banos, R.M., 2012. The present and future of positive technologies. Cyberpsychol Behav. Soc. Netw. 15 (2), 78–84.

Calvo, R.A., Peters, D., 2014. Positive Computing: Technology for Wellbeing and Human Potential. MIT Press, Cambridge, Massachusetts.

Calvo, R. A., Peters, D., 2015. Introduction to Positive Computing—Technology that Fosters Wellbeing. Paper presented at the ACM conference on Human Factors in Computing Systems (CHI 2015), Seoul, Korea.

Calvo, R. A., Peters, D., Johnson, D., Rogers, Y., 2014. Autonomy in Technology Design. Paper presented at the ACM CHI Conference on Human Factors in Computing Systems, Toronto.

Calvo, R. A., Dinakar, K., Picard, R., & Maes, P., 2016. Computing in Mental Health. In CHI '16 Extended Abstracts on Human Factors in Computing Systems. San Jose, California, USA: ACM.

Carissoli, C., Villani, D., Riva, G., 2015. Does a meditation protocol supported by a mobile application help people reduce stress? Suggestions from a controlled pragmatic trial. Cyberpsychol Behav. Soc. Netw. 18 (1), 46–53.

Chicchi Giglioli, I. A., Pallavicini, F., Pedroli, E., Serino, S., Riva, G., 2015. Augmented reality: a brand new challenge for the assessment and treatment of psychological disorders. Comput. Math. Methods Med. 2015, 862942.

Christensen, H., Hickie, I.B., 2010. E-mental health: a new era in delivery of mental health services. Med. J. Aust. 192, S2–S3.

Cohn, M.A., Fredrickson, B.L., Brown, S.L., Mikels, J.A., Conway, A.M., 2009. Happiness unpacked: positive emotions increase life satisfaction by building resilience. Emotion 9 (3), 361–368.

Coleman, J.S., 1998. Social capital in the creation of human capital. Am. J. Soc. 94, 95–120.

Csikszentmihalyi, M., 1975. Beyond Boredom and Anxiety: Experiencing Flow in Work and Play. Jossey-Bass, San Francisco.

Csikszentmihalyi, M., 1990. Flow. Harper & Row, New York.
Csikszentmihalyi, M., 1997. Finding Flow. The Psychology of Engagement With Everyday Life. Basic Books, New York.
Csikszentmihalyi, M., Beattie, O., 1979. Life themes: a theoretical and empirical exploration of their origins and effects. J. Humanist. Psychol. 19 (1), 45–63.
Csikszentmihalyi, M., Larson, R., 1987. Validity and reliability of the Experience-Sampling Method. J. Nerv. Ment. Dis. 175 (9), 526–536.
Csikszentmihalyi, M., Rathunde, K., Whalen, S., 1993. Talented Teenagers: The Roots of Success and Failure. Cambridge University Press, New York.
Danner, D.D., Snowdon, D.A., Friesen, W.V., 2001. Positive emotions in early life and longevity: findings from the nun study. J. Pers. Soc. Psychol. 80 (5), 804–813.
Deci, E.L., Ryan, R.M., 2000. The "what" and "why" of goal pursuits: human needs and the self-determination of behavior. Psychol. Inq. 11, 227–268.
Deci, E.L., Ryan, R.M., 2008. Self-determination theory: a macrotheory of human motivation, development and health. Can. Psychol. 49, 182–185.
Deci, E.L., Huta, V., Ryan, R.M., 2008. Living well: a self-determination theory perspective of eudaimonia. J. Happiness Stud. 9 (1), 139–170.
Delle Fave, A., Massimini, F., Bassi, M., 2011. Hedonism and eudaimonism in positive psychology. Delle Fave, A. (Ed.), Psychological Selection and Optimal Experience Across Cultures, Vol. 2, Springer, New York, pp. 3–18.
Diener, E., 1984. Subjective well-being. Psychol. Bull. 95, 542–575.
Diener, E., 1994. Assessing subjective well-being: progress and opportunities. Soc. Indic. Res. 31, 103–157.
Diener, E., 2000. Subjective well-being. The science of happiness and a proposal for a national index. Am. Psychol. 55 (1), 34–43.
Diener, E., Chan, M.Y., 2011. Happy people live longer: subjective well-being contributes to health and longevity. Appl. Psychol. Health Well Being 3 (1), 1–43.
Fredrickson, B.L., 2001. The role of positive emotions in positive psychology: the broaden-and-build theory of positive emotions. Am. Psychol. 56, 218–226.
Fredrickson, B.L., Branigan, C., 2005. Positive emotions broaden the scope of attention and thought—action repertoires. Cogn. Emot. 19, 313–332.
Fredrickson, B.L., Levenson, R.W., 1998. Positive emotions speed recovery from the cardiovascular sequelae of negative emotions. Cogn. Emot. 12 (2), 191–220.
Fredrickson, B.L., Mancuso, R.A., Branigan, C., Tugade, M.M., 2000. The undoing effect of positive emotions. Motiv. Emot. 24 (4), 237–258.
Fredrickson, B.L., Cohn, M.A., Coffey, K.A., Pek, J., Finkel, S.M., 2008. Open hearts build lives: positive emotions, induced through loving-kindness meditation, build consequential personal resources. J. Pers. Soc. Psychol. 95 (5), 1045–1062.
Friedman, B., 1996. Value-sensitive design. Interactions 3 (6), 16–23.
Fredrickson, B.L., 2004. The broaden-and-build theory of positive emotions. Philos. Trans. R. Soc. Lond. B. Biol. Sci. 359 (1449), 1367–1378.
Gable, S.L., Haidt, J., 2005. What (and why) is positive psychology. Rev. Gen. Psychol. 9, 103–110.
Gaggioli, A., Riva, G., Milani, L., Mazzoni, E., 2013. Networked Flow: Towards an Understanding of Creative Networks. Springer-Verlag, New York, NY.
Gaggioli, A., Morganti, L., Bonfiglio, S., Scaratti, C., Cipresso, P., Serino, S., Riva, G., 2014a. Intergenerational group reminiscence: a potentially effective intervention to enhance elderly psychosocial wellbeing and to improve children's perception of aging. Educ. Gerontol. 40 (7), 486–498.
Gaggioli, A., Cipresso, P., Serino, S., Campanaro, D.M., Pallavicini, F., Wiederhold, B., Riva, G., 2014b. Positive technology: a free mobile platform for the self-management of psychological stress. In: Wiederhold, B.K., Riva, G. (Eds.), Annual Review of Cybertherapy and Telemedicine 2014. IOS Press, Netherlands.

Gaggioli, A., Pallavicini, F., Morganti, L., Serino, S., Scaratti, C., Briguglio, M., et al., 2014c. Experiential virtual scenarios with real-time monitoring (interreality) for the management of psychological stress: a block randomized controlled trial. J. Med. Internet Res. 16 (7), e167.
Gallagher, E.N., Vella-Brodrick, D.A., 2008. Social support and emotional intelligence as predictors of subjective well-being. Pers. Individ. Dif. 44 (7), 1551–1561.
Goetz, J.L., Keltner, D., Simon-Thomas, E., 2010. Compassion: An evolutionary analysis and empirical review. Psychol. Bull. 136 (3), 351–374.
Grassi, A., Gaggioli, A., Riva, G., 2009. The green valley: the use of mobile narratives for reducing stress in commuters. Cyberpsychol. Behav. 12 (2), 155–161.
Harker, L., Keltner, D., 2001. Expressions of positive emotion in women's college yearbook pictures and their relationship to personality and life outcomes across adulthood. J. Pers. Soc. Psychol. 80 (1), 112–124.
Hassenzahl, M., Beu, A., 2001. Engineering Joy. IEEE Software 18, 70–76.
Hassenzahl, M., Tractinsky, N., 2006. User experience-a research agenda. Behav. Inf. Technol. 25 (2), 91–97.
Helliwell, J.F., Putnam, R.D., 2004. The social context of well-being. Philos. Trans. R. Soc. Lond. B. Biol. Sci. 359, 1435–1446.
Hordern, A., Georgiou, A., Whetton, S., Prgomet, M., 2011. Consumer e-health: an overview of research evidence and implications for future policy. HIM J. 40 (2), 6–14.
Inghilleri, P., Riva, G., Riva, E., 2015. Enabling Positive Change. De Gruyter Open, Berlin.
Kadda, A., 2010. Social utility of personalised e-health services: the study of home-based healthcare. Int. J. Electron. Healthc. 5 (4), 403–413.
Kahn, B.E., Isen, A.M., 1993. The influence of positive affect on variety seeking among safe, enjoyable products. J. Consum. Res. 20 (2), 257–270.
Kahneman, D., Diener, E., Schwarz, N., 1999. Well-Being: The Foundations of Hedonic Psychology. Russell Sage Foundation, New York.
Kasser, T., Dungan, N., Rosenblum, K.L., Sameroff, A.J., Deci, E.L., Niemiec, C.P., et al., 2014. Changes in materialism, changes in psychological well-being: evidence from three longitudinal studies and an intervention experiment. Motiv. Emot. 38, 1–22.
Kok, B.E., Fredrickson, B.L., 2010. Upward spirals of the heart: autonomic flexibility, as indexed by vagal tone, reciprocally and prospectively predicts positive emotions and social connectedness. Biol. Psychol. 85 (3), 432–436.
Littlewood, W., 1996. Autonomy: an anatomy and a framework. System 24 (4), 427–435.
Mair, F., Whitten, P., 2000. Systematic review of studies of patient satisfaction with telemedicine. Br. Med. J. 320 (7248), 1517–1520.
Manhal-Baugus, M., 2001. E-therapy: practical, ethical, and legal issues. Cyberpsychol. Behav. 4 (5), 551–563.
Maslow, A.H., 1954. Motivation and Personality. Harper, New York.
Massimini, F., Inghilleri, P., Delle Fave, A., 1996. La selezione psicologica umana. Cooperativa Libraria IULM, Milano.
Mayer, J.D., Caruso, D.R., Salovey, P., 1999. Emotional intelligence meets traditional standards for an intelligence. Intelligence 27 (4), 267–298.
McNulty, J.K., Fincham, F.D., 2012. Beyond positive psychology? Toward a contextual view of psychological processes and well-being. Am. Psychol. 67 (2), 101–110.
Morris, M.E., Adair, B., Ozanne, E., Kurowski, W., Miller, K.J., Pearce, A.J., Santamaria, N., Long, M., Ventura, C., Said, C.M., 2013. Smart technologies to enhance social connectedness in older people who live at home. Australas. J. Ageing 33 (3), 142–152.
Munson, S. A., Resnick, P., 2012. Learning from Positive Psychology to Promote Emotional Well-Being in Digital Environments. Paper presented at the ACM Conference on Human Factors in Computing Systems (CHI), Austin, Texas.
Nakamura, J., Csikszentmihalyi, M., 2001. Catalytic Creativity. American Psychologist 56 (4), 337–341.

Nakamura, J., Csikszentmihalyi, M., 2002. The concept of flow. In: Snyder, C., Lopez, S. (Eds.), Handbook of Positive Psychology. Oxford University Press, Oxford.
Niemiec, C.P., Ryan, R.M., Deci, E.L., 2009. The path taken: consequences of attaining intrinsic and extrinsic aspirations in post-college life. J. Res. Pers. 73 (3), 291–306.
Nussbaum, M., Sen, A. (Eds.), 1993. The Quality of Life. Clarendon Press, New York.
Park, N., Peterson, C., Seligman, M.E.P., 2004. Strengths of character and well-being. J. Soc. Clin. Psychol. 23, 603–619.
Peters, D., Calvo, R.A., 2014. Compassion vs. empathy: designing for resilience. ACM Interact. 21 (5), 48–53.
Peterson, C., Seligman, M.E.P., 2004. Character Strengths and Virtues: A Handbook and Classification. Am. Psychol. Assoc., Washington, DC.
Phillips, W., 2015. This Is Why We Can't Have Nice Things. MIT Press, Cambridge MA.
Raghunathan, R., Trope, Y., 2002. Walking the tightrope between feeling good and being accurate: mood as a resource in processing persuasive messages. J. Pers. Soc. Psychol. 83, 510–525.
Repetto, C., Gorini, A., Vigna, C., Algeri, D., Pallavicini, F., Riva, G., 2009. The use of biofeedback in clinical virtual reality: the INTREPID project. J. Vis. Exp. (33).
Riva, G., 2012. Personal experience in positive psychology may offer a new focus for a growing discipline. Am. Psychol. 67 (7), 574–575.
Riva, G., Mantovani, F., Gaggioli, A., 2004. Presence and rehabilitation: toward second-generation virtual reality applications in neuropsychology. J. Neuroeng. Rehabil. 1 (1), 9.
Riva, G., Castelnuovo, G., Mantovani, F., 2006. Transformation of flow in rehabilitation: the role of advanced communication technologies. Behav. Res. Methods 38 (2), 237–244.
Riva, G., Algeri, D., Pallavicini, F., Repetto, C., Gorini, A., Gaggioli, A., 2010. The use of advanced technologies in the treatment of psychological stress. J. Cyber Ther. Rehabil. 2, 169–171.
Riva, G., Banos, R.M., Botella, C., Wiederhold, B.K., Gaggioli, A., 2012. Positive technology: using interactive technologies to promote positive functioning. Cyberpsychol. Behav. Soc. Netw. 15 (2), 69–77.
Riva, G., Calvo, R., Lisetti, C., 2015a. CyberPsychology and Affective Computing. In: Calvo, R., D'Mello, S., Gratch, J., Kappas, A. (Eds.), The Oxford Handbook of Affective Computing. Oxford University Press, New York.
Riva, G., Dakanalis, A., Mantovani, F., 2015b. Leveraging psychology of virtual body for health and wellness. In: Sundar, S.S. (Ed.), The Handbook of the Psychology of Communication Technology. Wiley Blackwell, New York, pp. 528–547.
Rivera-Pelayo, V., Zacharias, V., Müller, L., Braun, S., 2012. Applying quantified self approaches to support reflective learning. Paper presented at the Second International Conference on Learning Analytics and Knowledge.
Robinson, J., Dixon, J., Macsween, A., van Schaik, P., Martin, D., 2015. The effects of exergaming on balance, gait, technology acceptance and flow experience in people with multiple sclerosis: a randomized controlled trial. BMC Sports Sci. Med. Rehabil. 7, 8.
Rogers, C., 1961. On Becoming a Person: A Therapist's View of Psychotherapy. Constable, London.
Rowe, G., Hirsh, J.B., Anderson, A.K., 2007. Positive affect increases the breadth of attentional selection. Proc. Natl. Acad. Sci. USA 104 (1), 383–388.
Ryan, R.M., Deci, E.L., 2000. Intrinsic and extrinsic motivations: classic definitions and new directions. Contemp. Educ. Psychol. 25 (1), 54–67.
Ryan, R.M., Deci, E.L., 2001. On happiness and human potentials: a review of research on hedonic and eudaimonic well-being. Ann. Rev. Psychol. 52, 141–166.
Ryff, C.D., Singer, B.H., 2008. Know thyself and become what you are: a eudaimonic approach to psychological well-being. J. Happiness Stud. 9, 13–39.
Schmitz, T.W., De Rosa, E., Anderson, A.K., 2009. Opposing influences of affective state valence on visual cortical encoding. J. Neurosci. 29 (22), 7199–7207.

Schueller, S.M., Parks, A.C., 2012. Disseminating self-help: positive psychology exercises in an online trial. J. Med. Internet Res. 14 (3), e63.
Seligman, M.E.P., 2002. Authentic Happiness: Using the New Positive Psychology to Realize Your Potential for Lasting Fulfillment. Free Press, New York.
Seligman, M.E.P., 2011. Flourish: A Visionary New Understanding of Happiness and Well-Being. Free Press, New York.
Seligman, M.E.P., Csikszentmihalyi, M., 2000. Positive psychology. An introduction. Am. Psychol. 55 (1), 5–14.
Seligman, M.E.P., Steen, T.A., Park, N., Peterson, C., 2005. Positive psychology progress: empirical validation of interventions. Am. Psychol. 60 (5), 410–421.
Shin, J.H., Ryu, H., Jang, S.H., 2014. A task-specific interactive game-based virtual reality rehabilitation system for patients with stroke: a usability test and two clinical experiments. J. Neuroeng. Rehabil. 11, 32.
Soto, D., Funes, M.J., Guzman-Garcia, A., Warbrick, T., Rotshtein, P., Humphreys, G.W., 2009. Pleasant music overcomes the loss of awareness in patients with visual neglect. Proc. Natl. Acad. Sci. USA 106 (14), 6011–6016.
Spear, H.J., 2001. Autonomy and adolescence: a concept analysis. Public Health Nurs. 21 (2), 144–152.
Tugade, M.M., Fredrickson, B.L., 2004. Resilient individuals use positive emotions to bounce back from negative emotional experiences. J. Pers. Soc. Psychol. 86 (2), 320–333.
Tugade, M.M., Fredrickson, B.L., Barrett, L.F., 2004. Psychological resilience and positive emotional granularity: examining the benefits of positive emotions on coping and health. J. Pers. 72 (6), 1161–1190.
Villani, D., Grassi, A., Cognetta, C., Toniolo, D., Cipresso, P., Riva, G., 2013. Self-help stress management training through mobile phones: an experience with oncology nurses. Psychol. Serv. 10 (3), 315–322.
Wadlinger, H.A., Isaacowitz, D.M., 2006. Positive mood broadens visual attention to positive stimuli. Motiv. Emot. 30 (1), 87–99.
Watson, D., Clark, L.A., Tellegen, A., 1988. Development and validation of brief measures of positive and negative affect: the PANAS scales. J. Pers. Soc. Psychol. 54 (6), 1063–1070.
Wiederhold, B.K., Riva, G., 2012. Positive technology supports shift to preventive, integrative health. Cyberpsychol. Behav. Soc. Netw. 15 (2), 67–68.
Wright, P., McCarthy, J., 2010. Experience-Centered Design: Designers, Users, and Communities in Dialogue. Synthesis Lectures on Human-Centered Informatics 3 (1), 1–123.

CHAPTER

19

Subliminal Perception or "Can We Perceive and Be Influenced by Stimuli That Do Not Reach Us on a Conscious Level?"

Andreas Riener

Johannes Kepler University Linz, Linz, Austria;
Ingolstadt University of Applied Sciences, Ingolstadt, Germany

HUMAN CAPACITY FOR INFORMATION PROCESSING

Information Processing in the Age of Ubiquitous, Disappearing Interfaces

The increasing amount of information to be processed and the rising quantity of interfaces to interact with in each point in time negatively affect the workload available for the primary task or activity. For example, for automotive user interfaces,[a] the attention of the driver is more and more disposed from the primary task of steering to driving independent tasks (Reynier and Hayward, 1993), for example, inspection of control elements, interaction with advanced driver assistance or information systems, or learning of new operational concepts. As a consequence, in recent times a great deal of vehicle accidents is caused by driver errors (actually, more than 90%) resulting from reduced situation awareness (SA) while driving at high workload levels (Ascone et al., 2009; Patten, 2007). Similar

[a]This chapter focuses on driver-vehicle interfaces as application domain.

Emotions and Affect in Human Factors and Human–Computer Interaction. http://dx.doi.org/10.1016/B978-0-12-801851-4.00019-7
Copyright © 2017 Elsevier Inc. All rights reserved.

examples can be found for other domains that build on human–machine interaction (HMI) for successful task completion, such as office work in general, operation rooms in hospitals, control rooms of power stations, or (with less dramatic consequences) home entertainment.

A research challenge in this regard is the development of novel user interfaces that make it easier to control and monitor technical systems, and this is mainly not a technical problem. Novel refers to the need that user interfaces should be understood intuitively and operated at low workload levels. Technological advances have made interfaces pervasive, ubiquitous, and finally disappearing. Using face tracking, gesture/posture recognition, conductive makeup/tattoos, metalized eyelashes (Vega, 2013), smart contact lenses, implants, etc. (i.e., "Natural user interfaces," Wigdor and Wixon, 2011, p. 14), nowadays the body itself is increasingly becoming the interface. This offers tremendous potential for recognition of input to systems based on context and current (mental) state, mimics and other forms of natural interaction. The hope is, that interaction in the near future would require only little immediate attention. Due to the omnipresence of interfaces and I/O capabilities, interaction is no longer happening only by intent or bound to specific locations as before, but may now occur everywhere and anytime and even without knowledge by the operator. The downside of this development progress is that people might want to selectively prevent certain input to the system (which is then a new source of cognitive demand) or feel anxious, which might induce distrust in technology or some sort of negative stress that could potentially result in false estimation by the system, an interaction performance drop, or reduced "willingness to interact." Ubiquitous interfaces might also require supplemental attention to keep track of their status and the variability of human decision-making (affected by momentary emotional state or mental condition) is said to be a major reason for unforeseen (re)actions. Unpredictable behavior of drivers is, for example, one of the most critical issues that might prevent the adoption of automated cars on large scale.

Technology adoption (e.g., prevalence of Smartphones) has made information and the capability for interaction available all the time. Continuous distraction and task switching between primary task, for example, working, studying/learning, leisure activities, and digital secondary tasks (media consumption or social media use on the phone) are one of the reasons why the attention span (= the ability to remain focused on a single task) has decreased from 12 s in 2000 to only 8 s in 2015 (Gausby, 2015). This means that on average we can maintain attention on the desired source and avoid the distracting influence of competing information sources for only 8 s. The process of task switching (= multitasking between Smartphone and primary activity) requires continual reallocation of both attentional foci (Wickens, 1992), which increases cognitive workload. The consequence is a declining task performance (as shown in

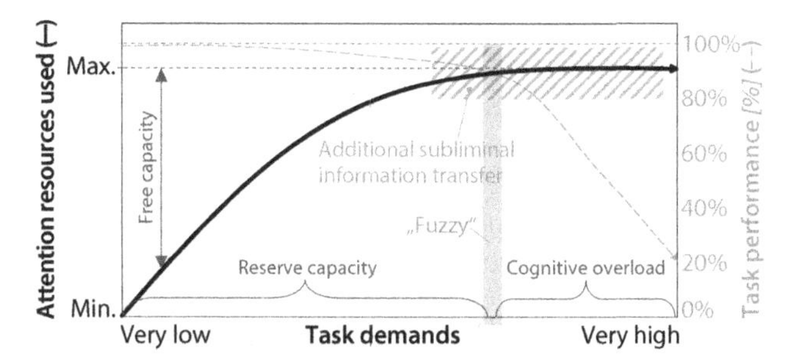

FIGURE 19.1 **Cognitive resources and their relationship to task performance.** The hatched area is critical in a sense that cognitive overload might appear. The application of subliminal techniques might help to reduce the effect of decreasing task performance. *Adapted from Patten, C. J., 2007. Cognitive Workload and the Driver: Understanding the Effects of Cognitive Workload on Driving from am Human Information Processing Perspective (PhD thesis). Department of Psychology, Stockholm University, Sweden, p. 59f.*

Fig. 19.1; dashed line) and the main reason for increased reaction times (Reimer et al., 2009). The same level of task demand consumes more attentional resources or, as converse argument, only lower demanding tasks could be completed.

This overview should have made it clear that it is necessary to look on the relationship between people and technology from a holistic point of view. Interaction performance and efficiency (error rate, etc.) can not only be explained and discussed through thought, experience, control of one's own cognitive processes, and environmental perception with our senses. In addition, remembrance, emotions and mental state, physiological conditions, and so on should be considered in future user interface research. One interesting yet controversially discussed aspect is information transfer below conscious awareness (and whether it can work or not). The hope discussed in this chapter is that information perceived unconsciously or "subliminally" can be used or adopted to compensate for the degradation of conscious cognition and awareness in real-world situations under certain circumstances.

State-of-the-Art and Limitations of UI Research

In the tradition of HMI, the user is an abstraction for every person (potentially) interacting with a system, being it a young child, an adult, or an elderly person, of female or male gender, and of any nationality or culture. With the advent of user centered design (UCD), an approach to UI design and development that involves, by definition, users in every phase throughout the design and development process (Stone et al., 2005, pp. 16), and related research fields, such as user interface design, which concentrates on the interface side (Opperman, 2002), human centered design, which is an (software engineering) approach focusing on the user at all stages (including the design process of the system) (Giacomin, 2012), participatory or interaction design (Greenbaum, 1993), which puts focus on a wider scope than just the computer and more emphasis on cognitive/experiential factors than traditional human factors and ergonomics, basic personal characteristics, such as age, gender, cultural background, or ethical heritage are now incorporated in the design of more effective (e.g., in terms of usability, interaction performance, perceived workload) user interfaces. These initiatives/research fields acknowledge that users are different, that is, might be more or less skilled, have differing degrees of background knowledge, varying willingness to use an interface, or variable familiarity with technology.

To further acknowledge human individuality, a huge number of sets of standards/rules have been proposed over time, and employed at different levels of design or different domains of application. TC 159/SC 4 Ergonomics of human–system interaction specifies the main principles and essential activities for human centered design to achieve a better usability of systems. ISO 9241[b] is one of the main standards under the responsibility of TC 159/SC 4. It is a multipart standard covering the ergonomics of human–system interaction including recent topics, such as tactile and haptic interaction (ISO 9241-910). Another standard in the field of human factors and ergonomics is ISO 10075[c] Ergonomic principles related to mental workload. Beside these and many other standards, interface designers also accept and follow guidelines or principles, such as *Shneiderman's 8 Golden Rules* (Shneiderman et al., 2009), *Norman's 7 Principles* (Norman, 1990), or *Nielen's 10 Usability Heuristics* (Nielsen, 1994) and Dix tried to combine and group all these principles to provide a superior system of rule (Dix et al., 2007, Chapter 3). Researchers in the field are finally supported by the Human Factors and Ergonomics Society (HFES),[d] a society whose "[..] mission is to promote the discovery and exchange of knowledge concerning the characteristics of human beings" (Human Factors and Ergonomics Society, 2016).

Unfortunately, following all the aforementioned procedures, standards, and guidelines, the individual user in its entirety is still "out of the loop" and the line of action unsatisfactory as it leaves out of consideration the personality and variability of subjects. Humans with some commonalities, such as age, gender, or cultural origin are within systems normally treated as one and the same subject, but are in fact individuals with different capabilities, behavior, attitudes toward or experience of life (= *personality*). Further on, situations and context (Dey, 2001) are perceived differently

[b]https://www.beuth.de/de/norm/din-en-iso-9241-110/110514174
[c]https://www.beuth.de/de/norm/din-en-iso-10075-1/34299836
[d]https://www.hfes.org/, Europe Charter: https://www.hfes-europe.org/

for all of us—as they depend a lot on our memories (Broek, 2013). Finally, it has to be substantiated that an individual might further behave differently based on his/her mental health and the spur of the moment (behavior may vary a lot depending on emotional condition or mood), and evolve over time (experience or "brain plasticity," age-related physical limitations, etc.). "Brain plasticity" is the ability of our brain to change over time, based on experiences and continuous development. It means that our brain has the ability to rewire and form new capabilities throughout the course of one's life and this finally allows us to adapt both to new or changing situations (Gausby, 2015).

Another problem is the representation of the sensory system. While human sensory processing is affected by various cross-channel interactions (Nishida, 2006), the whole sensory system is commonly modeled more simply as an aggregation of independent channels. The simplification also applies to this work—stimuli are considered as to excite only associated receptors, for example, light brightness innervates only retinal photoreceptor cells, odors are perceived through olfactory receptors in the nose, and vibrations are detected solely by mechanoreceptors in the skin.

Models of Human Memory

To better understand human memory operation, several models were developed and used to interpret how we make decisions, react on external stimuli, or why we refrain an action due to overlooking of information. Before discussing models more thoroughly, we will look at the basics of information processing.

Information processing in humans is measured in terms of time units (seconds) and information units (bits). The information unit "1 bit" is allocated to each decision in a dichotomous partition on a "yes/no" or "true/false" basis (Lehrl and Fischer, 1988). The grand information processing capacity of the human mind has the potential to handle about 11 million bits/s (Norretranders, 1997) (Fig. 19.2), but due to the inherent physical limitations of information processing, the number of bits processed *consciously* is much lower. In (Norretranders, 1997), an (optimistic) maximum of about 50 bits/s was indicated, but the exact number of bits to be processed explicitly (or consciously) actually depends on the task. Our explicit information processing speed can be reduced to as little as 45 bits/s when reading silently, 40 bits/s while using spoken speech, 30 bits/s when reading aloud, and 12 bits/s when executing calculations in our head (Hassin et al., 2005, p.82). Tactile information is processed at approximately 2–56 bits/s (Mandic et al., 2000). According to Overgaard and Timmermans (2010), "we may not even be able to experience all that we perceive." This coincides with our own observations in user studies, where we found that race car drivers (or other highly experienced drivers)

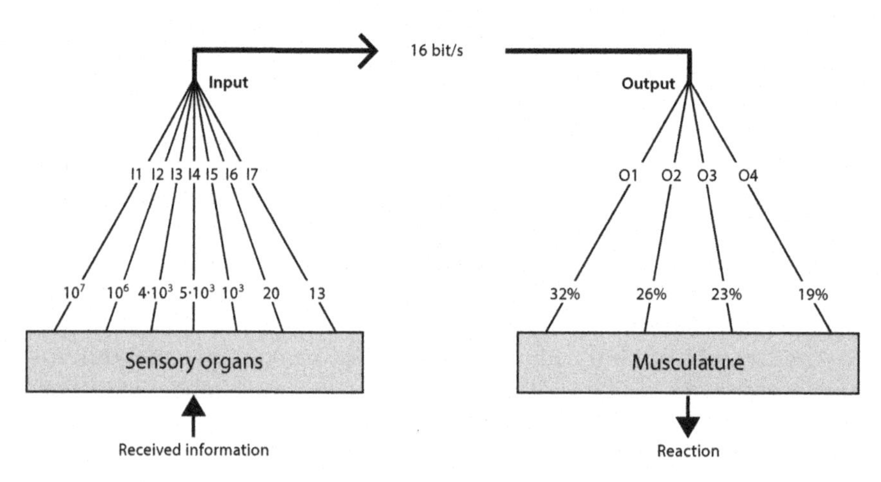

FIGURE 19.2 **Information flow in humans.** From a total of about 11 millions bits/s, only around 16 bits are perceived consciously. *I1*, Seeing, *I2*; Hearing; *I3*, Touching; *I4*, Temperature; *I5*, Muscles; *I6*, Smelling; *I7*, Sense of taste; *O1*, Skeleton; *O2*, Hand; *O3*, Speechmuscle; *O4*, Face.

are likely to be better equipped to process information and adapt steering behavior using a minimum of cognitive resources than less trained, regular drivers (Riener, 2010, pp.138).

Thus, compared to our total theoretical capacity, information processed consciously constitutes only a small fraction of all incoming information. The remainder is apparently processed without active awareness (or lost). The availability of such a large pool of processable information and the high speed at which information is processed further motivates us to explore making vital information accessible via subliminal routes in cognitive workload-sensitive settings.

Memory Models: Modal, HIP, and SEEV

One of the first and most influential is the modal model by (Atkinson and Shiffrin, 1968) that helps to better understand the processes that are part of our memory, starting with sensory input characterized by different modalities, such as seeing, hearing, or feeling (haptics). All the information perceived from the environment is registered in sensory stores, and after a short time of 1–2 s forwarded to the short-term memory (STM) store (if of particular relevance for us) or discarded. Atkinson and Shiffrin refer to memory capacity as we know from random access memory (RAM) in computers: We receive lot of information every second (Fig. 19.2) and if information would not be discarded after a short amount of time, we would run "out of memory." STM has limited capacity—Miller (1956) suggests that we can remember about 7 ± 2 information chunks, others, for example, Cowan (2001), found a more precise limit between 3 and 5 chunks. For a detailed discussion about memory limits, see Cowan (2001). To conserve information in STM, it needs to be (continuously) rehearsed (similarly to dynamic memory, where the information eventually fades unless the capacitor charge is refreshed periodically). For example, if a friend gives you his/her phone number, but you have nothing to write and your Smartphone not in the pocket, then you have to repeat the number to yourself over and over. If you rehearse the number too infrequently, it might get lost (i.e., discarded from short-term memory). If you rehearse it for a longer time, it might get stored in the permanent memory store (or long-term memory, LTM). According to Atkinson and Shiffrin, LTM works similarly than a hard disk drive in a computer. Information is stored in LTM for a long time (minutes up to years), but cannot be accessed directly. Instead, it needs to be moved to STM (analogy to RAM memory) before becoming accessible.

Another commonly used model is Wickens' human information processing (HIP) and memory model (Wickens, 1992). The model is much more detailed, including attention resources, decision-making, and response execution. A notable difference to the modal model by Atkinson and Shiffrin is that in the HIP model, STM is replaced by working memory (WM). This is to account that STM has a more active role in information processing (i.e., our "mental workbench"). Associations between external stimuli and memory store are basically similar (but described in much more depth) to the model of Atkinson and Shiffrin.

Wickens introduces attention resources and relates it to external stimuli received at the short-term sensory store (STSS). As we are continuously exposed to a great number of stimuli, we selectively focus on and attend to specific stimuli that are most relevant to our purpose (= conscious perception) and disregard all the rest. According to Wickens' model, the amount of stimuli that can be taken in by our STSS is considered to be unlimited, but our attention determines what information is finally transmitted to the working memory. The interpretation of sensory information requires retrieval of information from and interaction with LTM. Our prior experience and knowledge, emotional state, and value system (including prejudices) determine our perceptions of the external world and our response to these perceptions (= motor actions, feedback). The amount of information that can be held in working memory is, as before, limited [to four (Cowan, 2001) or 7 ± 2 information chunks (Miller, 1956)]. Working memory, therefore, creates a "bottleneck" for incoming information (Fig. 19.3). Feedback results in changed stimuli that are sensed again and thus "closes the loop." (To better understand the relationship of these memory components, in a classical computer architecture STSS would be

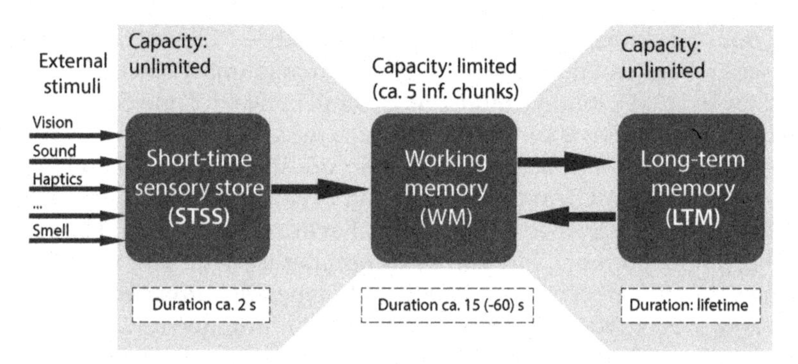

FIGURE 19.3 **Capacity and duration of memory components.** Working memory is a "bottleneck" with the purpose to filter out irrelevant stimuli. According to (Cowan, 2001), WM can hold up to 4, according to (Miller, 1956), 7 ± 2 information chunks. *Adapted from Wickens, C., 1992. Engineering psychology and human performance (second ed.). In: Yardley, L., (Ed.). HarperCollins Publishers Inc., New York, pp. 449–471.*

represented by a level 2/3 cache, WM corresponding to working memory and LTM similar to a hard disk storage.)

Relating back to the central question whether or not nonconsciously perceived information could actually induce a change in human mind or decision-making, based on the HIP model the answer must be no.

Humans are selective in choosing the information that should be processed. We consciously pay attention to particular information sources and easily get distracted by "more attractive" information. For example, a scholar sitting in class welcomes information from social network services (FaceBook, Twitter) to get distracted from education. The filtering process is carried out by mechanisms of human attention, represented by three modes. Selective attention characterizes "what" to process, focused attention is about efforts required to avoid distraction, and divided attention describes our multitasking ability, that is, to process more than one information source at a time (Wickens and Carswell, 2012, p.120).

The SEEV model of (visual) attention allocation (Wickens et al., 2003) can be used to describe processes related to selective attention. According to that model, the allocation of attention in dynamic environments is driven by bottom up attention capture of salient events, is inhibited by the effort required to move attention (as well as the effort imposed by concurrent cognitive activity), and is also driven by the expectancy of seeing valuable events at certain locations in the environment[e] or *P(Attend) = Salience – Effort + Expectancy + Value*. Good design should try to reduce the four components to only two by making valuable information sources salient and by minimizing the effort required to assess valuable and

[e]The first letter of each of the four terms defines the SEEV model.

frequently used expected sources (Wickens and Carswell, 2012). The SEEV model gives us the opportunity to predict what is or will be attended, but in contrast stands the problem of attentional or change blindness. For example, again for the automotive domain, several studies have shown that a head-up display (HUD) is a viable alternative for the delivery of (conscious) information (Charissis et al., 2009; Wittmann et al., 2006). The HUD opens up new possibilities, because it reduces the number and duration of drivers' sight deviations from the road [toward the dashboard; also known as head-down display (HDD)] and drivers can receive information without taking their eyes off the road. As a consequence, driver distraction is estimated to decrease since a driver is primarily focused on the traffic scene and not the dashboard instruments. Nakamura et al. (2005) did observe that information presentation via the head-up display resulted in reduced workload, decreased response times, more consistent speed, and increased driving comfort. At the same time, however, it has been reported that by overlaying additional information to the primary field of view, the driver might overlook important information on the road, for example, a child jumping onto the street (Levin and Baker, 2015). An interesting question in this regard is, if information shown on a HUD in a subliminal fashion, that is, not consciously perceivable, but still communicating some extra information, would also increase the risk of change blindness or improve situation awareness? (We are currently working on solutions for these problems.)

Suggestion for Improvements

The HMI discipline with its related guidelines and standards was improved over time to suit traditional interaction between human operators and their computers. Unfortunately, technological advances, such as the Internet, wireless connectivity, and the broad penetration with social media have recently outdated some of the existing concepts and made it necessary to rethink about HMI/HCI and its principles. This is increasingly a challenge in pervasive/ubiquitous computing as the computer disappears and is not longer visible, but the functionality, the interaction demand, and the attention required to interact is still there. Not surprisingly, humans are often overstrained from these (new) means of interaction, for example, they are unaware of the interaction channel, absent-minded or daydreaming or inattentive (fatigue, uninterested, or even cognitively overloaded). On the other side, however, the evolution of powerful information processing systems makes it now possible for the first time to measure and quantify the social nature of humans, for example, by tracking habits, detecting emotions and linking it to behavior, determining capabilities, etc. to the point of reverse engineering (parts of) the human brain or decoding the human genome (Vinciarelli et al., 2012). These innovations now facilitate the change over to cutting-edge, fine-grained interaction between man and machine, for example, by accounting all the cognitive/mental/emotional problems that have their roots in social and/or interpersonal communication and further incorporating aspects, such as trust, belief, friendship, or other social factors (Ferscha et al., 2012).

When humans communicate with humans, lot of information is exchanged on the unconscious level, derived from facial expressions (Rigato and Farroni, 2013), intonations of the voice, gestures, body posture or language, etc., and it is also agreed that in human social life actions are tightly linked with emotions (Ferri et al., 2013). In interpersonal conversation, humans are able to use implicit situational information, or context (Dey, 2001), to increase the conversational bandwidth. This sort of information is normally not accessible for computer mediated communication (Fig. 19.4), and the same holds true for other information transferred below threshold, mainly due to conflicting results of previous studies or

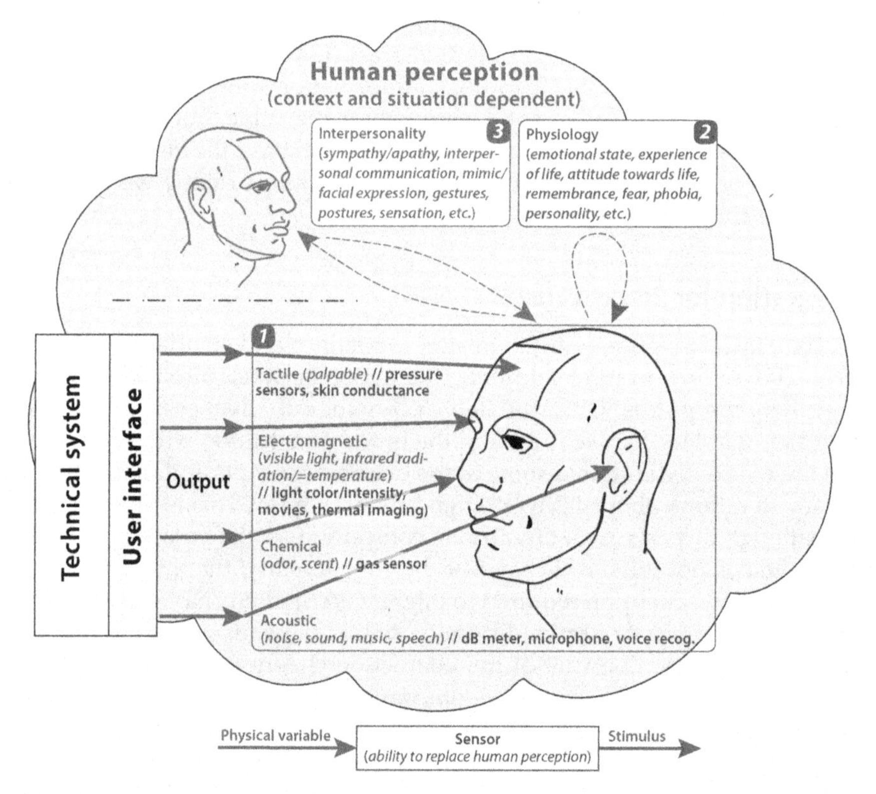

FIGURE 19.4 Traditional human–machine interaction only incorporates human perception from system output (1). Improved models (have to) include personal capabilities, individuality, and actual condition (2) as well as physiological processes involved in interpersonal communication (3) [after (Riener, 2014, p. 9)].

not available technology. With the availability of computation power, connectivity and more advanced sensors and measurement devices [e.g., ECG/heart rate variability (HRV), electroencephalography (EEG), event-related potentials (ERP), functional magnetic resonance imaging (fMRI), etc.], we now have everything at our disposal to start or continue research in this controversial area.

TERMINOLOGY AND TAXONOMY

Information Items and Reaction Time

Triggs and Harris (1982) (p.4) mentioned that the human reaction time depends almost linearly on the number of possible alternatives that can occur. As a result, cognitive activity is also limited to a small number of items at any one time. Testa and Dearie (1974) discovered that this number is between 5 and 9 and that a human being responds by grouping, sequencing, or neglecting items if more are present. More recently Cowan (2001) found the limit between 3 and 5 chunks. The limited capacity of information absorption for people was already identified earlier, for example, in George Miller's work "The magical number seven, plus or minus two" (Miller, 1956). He found that people can rapidly recognize approximately pieces of information at one time and hold them in STM. The evidenced correlation between information items to be perceived and the reaction time that is needed for it, together with our knowledge about the huge amount of information passing by when performing complex tasks, underpins the presumption that a lot of information must be gathered outside conscious awareness. Consider, for example, the process of steering a car through a busy city: a huge amount of information passes by the driver without being actively minded (but the driver might afterward have some abstract imagination about what was going on).

Attention and Situation Awareness

The level of attention (LOA) of a person executing a particular task is not constant. For example, driving on a winding road would require a higher LOA than on a straight street segment, driving at night a higher as compared to driving at daylight, heavy traffic a higher LOA than driving under light traffic condition (Dingus et al., 1989). Cognitive overload occurs when task difficulty exceeds resources available by the driver. In this case, driving performance and also safety starts to decline and it should be clear that such situations be avoided. The problem is that it is almost impossible to determine the exact point where cognitive overload is detected at the driver (the hatched area in Fig. 19.1). A possible reason is that the driver tends to alter his/her task management, for example, by delaying, excluding, or omitting certain elements. An example for "delaying" is, that drivers decrease their speed when engaged in a secondary task. This effect has been substantiated in several real driving and simulator studies (Young et al., 2008, p.338). A research carried out recently in a simulator also found that a driver's mean speed decreased while carrying out a naturalistic conversation on a cell phone (Rakauskas et al., 2004). Due to changing physiological, affective, emotional states of a person, as pointed out earlier, LOA and SA directed to a specific task might differ even for one and the same person.

Conscious, Unconscious (Subliminal), Preconscious Perception

The terminology as used in this chapter is grounded in neuroscience. Before explaining the neuroscience behind perception without conscious awareness (also called unconscious cognition) and the taxonomy used in this work, we need to address the general confusion surrounding the word subliminal.

Subliminal Perception

First, *subliminal perception* is not to be confused with *subliminal persuasion,* which is said to be unethical and aims to change the behavior or influence the decisions of individuals toward a goal without their knowledge. An example is audio tapes used to coerce people into spending to achieve wealth, fame, and self-esteem. The term *subliminal* was introduced by Johann Herbart (1776–1841) and is derived from the Latin words "sub," meaning under, and "limen" meaning threshold. Thus, subliminal perception, strictly speaking, is perception below the threshold of consciousness (Smith and McCulloch, 2012), without awareness (Egermann et al., 2006) or *"any situation in which unnoticed stimuli are perceived"* (Merikle, 2000, p.497). The common understanding of the term subliminal is *"beneath the level of perception"* or *"not strong enough to be recognized explicitly."* The notion that stimuli presented outside conscious awareness can involuntarily affect human beings has captivated the imagination and interest of a wide audience. One of the first experiments suggesting the existence of nonconscious influences on behavior can be traced to a landmark paper by Peirce and Jastrow (1884). The authors used an introspective approach on themselves to report that they could perceive small differences in pressure to the skin without conscious awareness of different sensations. A number of similar studies of perception without awareness were conducted, for example, Adams (1957) culminating in 1957 in Vicary's fallacious subliminal experiment. James Vicary, a marketing consultant, initiated a hoax on subliminal advertising and reported increased sales after subliminally flashing "Drink Coca-Cola" and "Eat popcorn" messages in a movie theater.

His claim that subliminal stimuli can not only influence behavior, but also have a long-term effect on consumer choices, was later withdrawn as a means of promotion. After that, the field of subliminal persuasion/perception lapsed into disarray until the 1990s, when the availability of more advanced sensor technology (EEG, ERP, fMRI, etc.) improved research methodologies. Two key papers reporting on robust subliminal priming methods showed that genuinely invisible primes could influence processing at a semantic level (Greenwald et al., 1996) and subliminal stimuli can elicit not only a behavior, but also a detectable neural activity in the motor cortex (Dehaene et al., 1998). Since then, one group of researchers has shown that invisible (subliminal) stimuli can elicit motor responses (Lleras and Enns, 2004; Schlaghecken and Eimer, 2006), while others have noted the importance of attention in modulating the *effect* of subliminal perception by suggesting that attention increases the processing of invisible stimuli at both perceptual and semantic levels (Kiefer and Brendel, 2006; Naccache et al., 2002; Summer et al., 2006).

Today, research on subliminal perception provides a way of testing whether conscious and unconscious processes are (as supposed) fundamentally different from one another. However, providing evidence for an impact of this type of stimuli is apparently not easy and has caused some confusion in the past. Does a particular stimulus, when presented subliminally, have different effects than when it is presented so that people are consciously aware of it (i.e., supraliminally)? In addition, and more relevant in the context of this chapter, presenting stimuli subliminally should allow us to rule out demand or other active strategies on the part of experimental subjects as alternative explanations for their effects (Smith and McCulloch, 2012).

In order to provide clear answers, a careful definition of the terminology is essential. Our terminology is based mainly on the theoretical taxonomy proposed by Dehaene et al. (2006) and the comparative analysis of the body of recent work by Hassin (2013). The taxonomy proposed by Dehaene et al. (2006) defines three distinct processing states of the brain when processing a stimulus: *conscious, unconscious (subliminal),* and *preconscious*. During *conscious processing*, the stimulus triggers intense neural activations in the brain—persons (e.g., drivers) are actively aware of the stimuli and can verbally report it. During *subliminal processing*, the stimulus is perceived by the brain, and the depth of the resulting neural activations depends on task, sensory channel, masking strength (the speed of projection of the stimuli in the case of visual stimuli) (Del Cul et al., 2007), and top-down attention. When top-down attention is present, the state is known as *subliminal attended* versus *subliminal unattended* when attention is absent (Dehaene et al., 2006).

In the context of this chapter, another important distinction is that between *subliminal priming* and *unconscious cognition*. Although subliminal priming is a subset of unconscious cognition in which the stimuli are not consciously perceived, the literature on this topic has developed somewhat independently. In the field of *subliminal priming*, scientists investigate the extent to which nonconsciously perceived stimuli can be processed and the effects they have on other processes. In the field of *unconscious cognition*, scientists examine unconscious processes without limiting themselves to subliminality. To them, it is critical that one is not aware of the relevant processes and/or their results. Each of these approaches has its unique set of strengths and shortcomings, and hence a comprehensive understanding of unconsciousness must rely on both. The main disadvantage of the literature on *unconscious cognition* is that it often relies on subjects' subjective reports regarding their internal states, which entails two major limitations. First, they cannot be observed from the outside. Second, psychological science has repeatedly demonstrated that our introspective capacity—crucial for many subjective reports—is limited (Hassin, 2013).

It is also important that subliminal processing is not independent of a driver's attention or working strategies. On the contrary, one cannot simply send invisible stimuli without consideration of task relevance and a driver's willingness to comply, as shown by numerous previous studies, for example, Kelly (2011), Kouider and Dehaene (2007), for an overview. The third and last brain state to be considered is *preconscious processing* and was incorporated into Dehaene's model to account for the fact that even strong stimuli can remain temporarily preconscious, which means that they are temporarily buffered in a "nonconscious store" and thus not consciously accessible (Dehaene et al., 2006).

Conscious Versus Subconscious Stimuli Perception

Ambiguity of the terminology continues also here, underpinned for instance by Rosalind Picards' statement that "the term consciousness refers to a complex morass of many things [..]" (Picard, 1997, p.42). Furthermore, Calvin and Ojemann (1994) remarked that different mental activities (which normally do not have a single origin) are all mapped to the single word consciousness—which makes it particularly tough to classify and understand what "conscious" should exactly be, see also Hudlicka and Fellous, 1996; Minsky, 1999, Chapter 4.

Indeed, the simple mention of the single word "subconscious" can lead to confusion and another problem is to precisely define what is meant by perception without awareness (Moore, 1988; Synodinos, 1988). In this chapter, we state that a person (driver) is consciously perceiving a stimulus if he/she is fully aware and able to identify the given stimulus. To give an example, the projection of a visual "turn right" symbol in the head-up display of a modern car would be consciously perceived by the driver. However, if this (or any other) visual symbol is projected very quickly, saying at a frequency of 100 Hz (10 ms), we can state that he/she has not perceived this symbol consciously (due to confirmed limitations of our visual spatial resolution, i.e., neurological properties of conscious perception). Del Cul et al. (2007) experimented with visual stimuli to cross the threshold to consciousness (visibility). It is generally accepted that exactly this threshold is the border between conscious and subconscious perception, thus the point at which awareness of a stimulus is reported or at which unreported awareness has a measurable effect on some subsequent behavior (Brannon and Brock, 1994, p.7). In their studies, Del Cul et al. (2007) substantiated the existence of such a threshold and that it is in the order of magnitude of 50 ms for the visual channel. The technique used to project a given stimulus below this threshold of awareness is called "subliminal priming" or "subliminal stimulation" (Chalfoun and Frasson, 2008). Both terms "subliminal" and "subconscious" are in general used interchangeable and in a equivalent manner.

However, according to the imprecise usage of the term subliminal in the past, for example, by Pratkanis and Greenwald (1989) and the fact that it carries a negative connotation (e.g., due to James Vicary's marketing joke), we will continue to use the term subconscious instead of subliminal for expressing information delivered below the level of conscious perception.

Subconscious Versus Unconscious Perception

The two terms *subconsciousness* and *unconsciousness* are again sometimes used similarly, for example, (Chalfoun and Frasson, 2008), but most frequently the terms are differentiated. The difference thereby lies in the fact that humans have deliberately virtually no access on our unconsciousness (for instance, our blood circulation runs totally unconsciously—we never have thought about that[f]).

According to Dunne and Jahn (2005, pp. 704), who have worked for more than 25 years in the field of physical anomalies in HMI, consciousness is defined as subsuming all categories of personal experience without presumptions of specific psychological or physiological mechanisms (it encompasses all dimensions of personal identity that can be distinguished from external circumstances and influences that consciousness perceives to be "not-self").

Subconscious (or subliminal) information presentation is considered a well-supported theory today, stating that perception can occur without conscious awareness and have a significant impact on later behavior and thought (Ramsøy and Overgaard, 2004). The timely manner of this field of research is confirmed by a rising attention given to the topic of subliminal perception, reported, for instance, by Ainsworth (1989), Merikle (2000), Reder et al. (1994), and Sandberg et al. (2010). One of the

[f]Ralf Sanftleben, "Das Unterbewusstsein—eine Einführung", http://www.zeitzuleben.de/artikel/denken/unterbewusstsein.html

first works presented in the scope of HCI/HMI was Wallace's text editor (Wallace, 1991). Their studies provided evidence for a reduced frequency of help requests when the requested information was projected subliminally. DeVaul et al. (2003) worked in his thesis "The Memory Glasses" toward the vision of zero attention user interfaces. As one example, he provided evidence (using wearable glasses that projected subliminal cues) that masked short-duration (=subliminal) visual cues may be effective in supporting memory recall in a just in time cuing applications. Furthermore, priming for memory support was elaborated by Schutte (2007). He assessed the effects of brief subliminal priming on memory retention during an interference task, examining in detail multiple parameters of subliminal projections, such as duration, relevance and contract. Latest achievements, for example, by Aarts et al. (2008) have demonstrated that subliminal priming of goal concepts in temporal proximity of the activation of affect alters people's readiness for goal pursuit. They have also shown that effects of nonconsciously shaped goals disappear if the ongoing mental operation aiding the nonconscious goal is taxed by conscious task goals. In contrast, research findings in this area have provided up to now very little evidence that stimuli below a person's subjective threshold actually have an influence on attitudes or decisions (Strayer, 2005).

Measuring Consciousness

Along with the discussion of the terminology goes the selection and definition of acceptable methods for detecting and measuring consciousness. There are two basic types of measures: *objective* and *subjective*. Their applicability was discussed, for instance, by Zehetleitner and Rausch (2013). Measures of consciousness are considered *objective* if a subject's state of awareness is determined on the basis of task performance (i.e., if a subject is able to discriminate a stimulus, it is assumed that he/she is conscious of, if, as opposed to a chance level performance on a discrimination task. Such a test is considered a reliable indicator of the absence of conscious awareness). The boundary between consciousness and absence of conscious awareness is typically identified by varying the stimulus onset asynchrony (SOA) between stimulus (prime) and mask. Objective measures stand in contrast with *subjective* measures (e.g., verbal reports), which are proposed to operationalize consciousness. These include confidence ratings by study participants, asking subjects about the reason for choosing a particular response or questioning the observers to judge their visual experiences directly, for instance, on a Likert scale (Del Cul et al., 2007). The validity of the measures used in the latter category has been questioned from an empirical science point of view. The debate whether these measures are valid for empirical science is ongoing, because they might be corrupted by uncontrolled changes of the response criterion. In our studies, we are applied both, objective and subjective measures.

REVIEW OF RELATED WORK

The main challenge in studies assessing the value of subliminal support and quantifying its effect lies in properly configuring and setting up the interfaces for providing information subliminally. Further, there is no guarantee that nonconsciously experienced stimuli will indeed be unconsciously perceived. This issue is compounded by the fact that subjects cannot be asked directly whether they perceived such a stimulus. In a recent study, we asked test subjects to report on a visual stimulus, but the answers were rather hard to interpret: about 10% of the control group (given no subliminal information) believed that they were exposed to subliminal stimuli, on the other hand, only 25% of the test group (given subliminal information) indicated the same. However, no subject had a concrete idea or could describe in detail what the subliminal stimulus or its aim was. Nevertheless, given that a (driver-vehicle) interface can offer information subliminally, it is to be expected that the subliminal information may replace other, consciously perceived information and thus help to reduce a driver's cognitive load (Table 19.1). This assumption is based on the result of cognitive and social psychologists showing that stimuli presented subliminally can have a considerable influence on a variety of cognitive processes and possibly even behavior (Moore, 1988).

Subliminal Visual Information

Due to the slowness of visual processing in human eyes, still images refreshed at about 25 fps (frames per second) are perceived as continuous course of movement. This is applied in motion pictures traditionally shown at 24 Hz. But the human eye is much more sensitive and capable to perceive flicker still at 50 Hz, causing optical illusions in motion picture. To reduce these effects and to make the path of motion even more precise and sharp, the film industry has adapted recording technology and films are shown today at a high-frame rate (HFR) of 48 Hz. The fact that visual processing (or more detailed, detection/perception of visual information in the eye rods) is slow, is utilized for subliminal visual information presentation (information perceived subconsciously). The underlying theory is that while the eye cannot detect single frames in a sequence updated at very high frequency (above 50 Hz, or below 20 ms), the brain might still have the capability to extract information from that single frame.

Driving is mainly a visual task (Broyce, 2008), but only very few approaches exist to support drivers in dealing with the increasing amount of

TABLE 19.1 Taxonomy of Mental Processing States Applied to Driving

Channel	Mental state	Examples relevant to driving
Visual	Conscious	Important information related to the primary task of driving, for example, speed indicator and arrows for lane changes in the dashboard.
Visual	*Subliminal*	*Secondary/tertiary information in a head-up display (HUD) (e.g., additional information of route navigation).*
Auditory	Conscious	Speech based guidance of the navigation system or auditory warning information.
Auditory	Subliminal	Subtle changes in engine, transmission, tire, and aerodynamic noise. Examples: quieter cars (i.e., passenger compartment) tend to encourage reduced headway and more risky gap acceptance (Horswill and Coster, 2002) or drivers for those auditory feedback of the car noise was diminished tend to drive faster compared to drivers who received louder car noises' (Horswill and McKenna, 1999).
Vibrotactile	Conscious	Vibrating car seat or waist belt; using strong tactile stimuli to perceive actions, such as for navigation (Boll et al., 2011).
Vibrotactile	Subliminal	Weak tactile feedback not consciously perceivable to improve on fuel efficiency (Riener et al., 2010).

Due to the nature of driving, subconscious perception of visual stimuli has most potential.

visual information in and around the car. Takahashi (2009) experimented with subliminal visual information on road safety hazards displayed in a driver's field of view and found that response times tend to be shorter with subliminal support than without added information. The majority of previous studies on visual subliminal priming were not conducted in a car driving setting. Nevertheless, in most cases the results can be translated to the car driving domain due to the similarity of the displays used. For example, head-up display (HUD) units in diverse forms have recently become a standard visual display device in vehicles and studies on visible content have substantiated that these interfaces have only a low impact on mental load compared to traditional (head-down) in-car displays (Weinberg et al., 2011).

Armstrong and Dienes (2013) presented a series of experiments that utilized subjective thresholds of subliminal priming to demonstrate a significant priming effect that cannot be attributed to partial conscious awareness or the retrieval of signal-response links. The results suggest that unconscious cognition is capable of processing the logical function of negation when instructed to choose between two nouns. This is an interesting finding that can be applied to priming studies in the car, for instance, using HUDs. Vorberg et al. (2003) experimented with different parameters (time course; stimulus onset asynchrony) of both the visual prime and the mask and finally proposed a model that provides a quantitative account of priming effects on response speed and accuracy. The findings (and the model) are not bound to a specific setting, and can thus also be used in the automotive domain to improve subliminal perception, ideally in harmony with the driver's current activity or workload. Another study investigating the potential of subliminal cues to support users in visual search tasks was presented in Pfleging et al. (2013). On one hand, the authors suggested that visible cues are effective, and on the other, they claimed that simple and nonblinking subliminal cues neither influence gaze behavior nor improve visual search performance. This contradicts a study by Bailey et al. (2009), where subtle cues resulted in an effect. These contradictory results highlight the challenge arising in priming studies: primes seem to work well in some but not in other settings. Kobilisek (2012) attempted to study subliminal primes in real life situations (defined by the author as a web browser window on the subject's personal computer). No significant difference was detected between the nonprimed control group and the primed test group, most likely due to the distributed setting, hardware differences (screen size, refresh rates, etc.) and unsupervised execution. Despite the study's shortcomings, the author raised a number of questions to be considered in future application of subliminal primes in vehicles. Larsen (2013) tackled the problem of reducing cognitive processing loads with subliminal cues for decision-making tasks using a game as a learning environment. Initial results indicated that a subset of participants had stronger propensities for subliminal perception. The first group (with fast responses and high skill/experience levels) showed better perception of subliminal cues in the experiment. Even though different results were attributed to different capabilities of users, the approach is interesting and has potential for application in the car, for instance, as an automated assistance system to support drivers in key tasks.

However, this result again raises the question of the general applicability of subliminal priming. In theory, priming should work for any driver, at least after proper interface configuration; failure to properly calibrate the interface might result in unsuccessful application of the approach, but how to validate? Another issue to be discussed in the parameterization of priming studies is the number of repeated presentations of a masked stimulus. Previous work has suggested that presenting a masked stimulus repeatedly improves priming without increasing perceptual awareness, but the neural theory of consciousness predicts the opposite (increased bottom-up strength should also result in increasing availability to awareness). Atas et al. (2013) tested these opposing views by manipulating the number of repetitions of a strongly masked small set of digits. They reported that successive repetitions of a strongly masked stimulus in a priming task increase both priming and perceptual awareness. They also suggested that the repeated masking method seems inefficient in improving stimulus signal without increasing visibility. This key finding must be considered in the setup of (visual) priming experiments.

In summary, previous studies on subliminal priming have shown its potential to provide additional information (to the driver) without increasing perceptual awareness, but have also raised a number of issues that need to be considered when designing and testing such interfaces. However, as our own studies have confirmed (see next section), subliminal visual priming is indeed a technique worth looking into to improve driver–vehicle interaction.

Auditory Driver Stimulation

Using purely auditory cues to "inject" subliminal information into the driver is not a commonly followed approach, but cross-modal links between different sensory modalities have been investigated for a while, for instance, the use of audition to improve spatial awareness. One example is the work by Ho and Spence (2005), where the authors looked at the potential benefits of using auditory warning signals (different car horn sounds and verbal cues) to capture visual attention. The results suggest that spatially predictive, semantically meaningful auditory warning signals may be a particularly effective means of capturing (visual) attention. Although the cues are not below conscious awareness, this work is to some extent of relevance here, because it discusses how auditory cues could be used to elicit a fast and automatic orientating reaction from a driver (Ho and Spence, 2005). In Nass et al. (2005), the authors investigated the potential impact of a car voice interface on a driver's performance and attitude. They concluded that drivers who interacted with a voice matching their own emotional state drove better while "speaking" more to the car than drivers interacting with mismatched voices. The question remains open whether this proves that emotions derived from the voice of an in-car information system carry subliminal information that influences the driver's behavior. There is no compelling evidence to support this claim, but it should not come as a surprise that drivers already use a variety of "hidden" sound or noise sources in and around the car to guide action, such as the best time to switch gears on the basis of engine speed or to assess and adapt driving speed based on ambient noise. Radel et al. (2013) investigated primes in a more general setting. The authors were interested in whether one's motivation can be influenced by unattended, subliminal audition. One study demonstrated that participants' "worked harder" when unattended speech contained highly motivational words than when it contained low-motivation words. Another study was conducted in a more real-life priming setting. A barely audible conversation was played while participants were focusing on a demanding task. Participants primed with a conversation that gave intrinsic motivation performed better and persisted longer in the subsequent task than those who were exposed only to a similar but unintelligible auditory signal. These findings suggest that the effect represents a real-life influence because most of the auditory inputs of one's usual environment (which could also be the car) are actually filtered out because of the limited capacity for higher-level processing.

Subliminal primes in auditory feedback are not "completely inaudible" (as opposed to visual priming) and are normally embedded imperceptibly in other sound sources (verbal demands in music, messages hidden in engine sound) or change the representation of existing sounds (variation of pitch, timbre in the voice of the navigation system). Thus, it has to be considered that auditory feedback interferes to a certain degree with driving capabilities. In addition, past experience has also shown that drivers do not like to receive driving instructions from a voice command system (Mauter and Katzki, 2003) and that speech modalities in native or known languages have very high saliency.

In conclusion, overuse of the auditory channel to provide information either consciously or subconsciously might distract the driver, particularly in high-workload situations. This holds true for information transmitted at the subconscious level, as the subliminal part is not completely inaudible, but embedded in a consciously perceived information carrier channel.

Olfactory Subliminal Perception

Subliminally perceived odors (taxonomically categorized under the mental processing state of *subliminally unattended* because of the absence of top-down attention and the weakness of the stimulus) connect directly to neuroscience because of the body of work showing strong links between smell and working memory, attention, reaction times, mood, and emotion (Brewster et al., 2006). More specifically, olfactory cues influence human perception and can induce behavioral responses (Parma et al., 2012). There is also evidence that smell can be used as a support measure, for example, to aid recall (Brewster et al., 2006). This effect was actually confirmed by a study showing that olfactory interfaces have a less disruptive effect on the primary task at hand than visual or auditory stimuli.

Few related publications on the application of smell-based subliminal interaction exist—for several reasons. First, the perception of odors is subtle, imprecise, and scents in particular will not work for everyone (this is known as *anosmia* or the inability of persons to smell one or more specific odors (Brewster et al., 2006)). Second, and more problematic, is the fact that the emotional state of healthy subjects has a clear effect on olfactory sensitivity. Since emotional states are likely to change quickly and in an uncontrolled manner in dynamic settings, such as vehicle operation, an efficient application of primes can be difficult to achieve. Pollatos et al. (2007), for example, reported that a negative emotional state reduces olfactory sensitivity. Lastly, olfactory interfaces normally saturate the entire space (i.e., passenger cabin), and information delivery is relatively slow (i.e., changes from one stimulus to another take time, as the former must first decay). Also, people are likely to adapt to olfactory stimuli (Li et al., 2007). This problem can be solved, at least partly, by using pulsed ejection of odors instead of continuous scent delivery (Kadowaki et al., 2007).

Despite all the limitations to a broad application of subliminal delivery through odor, there remains room for olfaction to deliver (on rare occasions) additional information to the driver. Its characteristics can be used, for example, to release a mild scent of burning oil in the passenger compartment to warn the driver in the case of oncoming motor defect. Alternatively, odors (e.g., lavender-based) can be used to calm down the driver to increase road safety. Lastly, strong odors, such as lemon or lime can be employed to shake off drowsiness or to avoid that the driver falls asleep. These examples are supported by concrete products on the market. For example, in 2009 Nissan presented an olfactory gratification system to enhance driving pleasure and Renault aims to "bring people health and wellbeing while on the road" by employing odors in the passenger compartment (therefore, the Renault ZOE includes an active fragrance diffuser with relaxing or stimulating properties along with a purifying and relaxing air ionizer[g]).

In conclusion, olfaction has the potential to be used as a channel for communicating information subliminally to the driver. However, since its effectiveness strongly depends on the individual and applications are limited, broad use of this channel is not anticipated in the near future.

Tactile Driver Feedback

Vibrotactile notification systems have been in production for several years now. For example, Citroen, Audi, and BMW are integrating simple warning systems based on vibrotactile notifications in pedals, seat, or steering wheel in their latest car models. More sophisticated tactile notification systems, for instance, to support the driver in route guidance activities, were also tested by several research groups.

Kern et al. (2009) investigated how vibrotactile representations of navigational information can be presented to the driver via the steering wheel to address the problem of cognitive overload and over-complexity. The results suggest that adding tactile information to audio feedback or particularly to visual representations (i.e., multimodal information presentation) can improve both driving performance and experience. In Asif and

[g]http://myrenaultzoe.com/index.php/zoe-description/comfort/

Boll (2010), the authors compared the presentation of spatial turn-by-turn information via a tactile belt with a conventional car navigation system in terms of cognitive workload, performance, and driver distraction. Their results show that drivers achieve better orientation performance with the tactile display than with the conventional navigation system, while cognitive workload, performance, and distraction remain unchanged. The study confirms that tactile interfaces are useful in presenting information in high-load driving conditions where drivers must focus on their visual and auditory senses. A similar study was presented by Kim et al. (2012), who evaluated the impact of multimodal feedback on driving performance and cognitive load (from a perspective of different perceptual and cognitive abilities). By measuring task performance, subjective workload, and objective workload (psychophysiological responses), it was shown that adding more modalities increased the already high workload of older drivers. In contrast, adding haptic feedback to traditional audio and visual feedback led to more attentive driving by younger drivers. More recently, Telpaz et al. (2015) used haptic feedback in the car seat to increase the situation awareness of a driver (and thus to facilitate drivers' preparedness for taking over in automated driving situations). Results of the study confirm that the haptic feedback system led to improved driving performance and increased driver's SA. In general, the benefit of interfaces integrating conscious and perceivable tactile feedback lies in the delivery of the information, in the fact that tactile stimuli can be presented very effectively while adding only little extra load to the cognitive channel (Varhelyi, 2002). (Taxonomically, this type of stimulus would be called *subliminally attended* because top-down attention is present and the stimulus is weak or subtle.)

Only little progress has so far been made in using the tactile channel for subliminal support. Virtual haptic feedback was discussed, for example, in the research project IMMERSENCE ([..] subjects did not recognize changes in the sensory cues, but behaved as if they did) (Hilsenrat and Reiner, 2009). The authors claim that "[..] this suggests that performance can be affected and possibly improved through subliminal cues by reducing the cognitive load, allowing the transfer of additional information otherwise impossible." It is not clear why these results did not motivate further research into employing the tactile channel to improve driver–vehicle interaction. Research focusing on subliminal driver support using tactile signals is even more uncommon and unexplored today. However, our own studies on assessing the capability of subliminal tactile techniques in vehicular interfaces have demonstrated the potential of subliminal driver support. We seek to share these results in the hopes of fostering continued efforts in developing applications for the car domain. Real driving experiments, for instance, those presented in Riener et al. (2010), have shown that drivers perceiving subliminal stimuli in the form of subtle tactile feedback tend to positively modify their driving behavior compared to the control group without any tactile feedback. This result is one example of the kind of nonconscious, positive behavioral change that our research aims to achieve.

EXPLORATORY STUDIES IN THE AUTOMOTIVE DOMAIN

Driving has become a very demanding activity so that safe vehicle operation on crowded highways and in complex traffic situations requires to continuously maintain high situation awareness. This conflicts with latest trends, such as communication technologies and Smartphones available in the car, as they have negative impact on attention span (Gausby, 2015) and further reduce attention resources available for driving. Supporting the driver to safely operate its vehicle is nowadays one of the greatest challenges (Jones, 2001). With the advent of automated driving, the problem seems to be solved, but is not. In order to achieve highest possible travel experience and comfort (an ultimate requirement), new concepts to get in touch with the (inactive) driver need to be developed.

One promising approach to reduce the (attentively perceived) quantity of information in the driver–vehicle interaction loop (Fig. 19.5) for non-critical situations, is the application of subliminal techniques, so to make information available on the subconscious level rather than providing it consciously.

In our research, we try to reach this goal by using various sensory modalities capable of providing additional information in a subtle, unnoticeable way without generating further load on the cognitive channel. Based on the taxonomy of nonconscious perception and in accordance with the

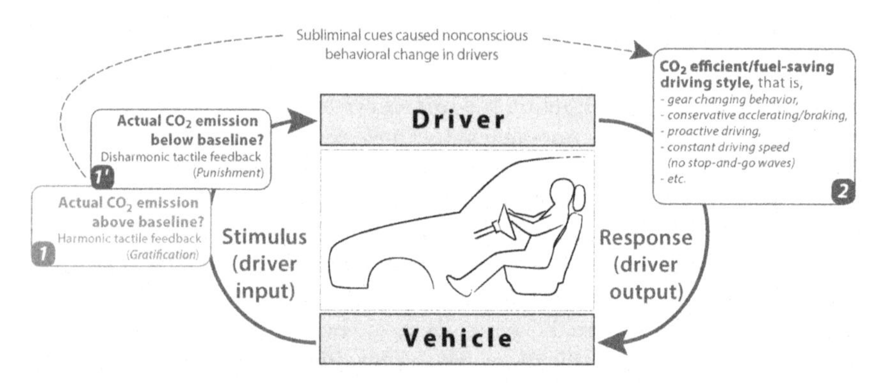

FIGURE 19.5 **Stimulus received (subliminally) by the driver and corresponding response.**

recommendations of related work, our research on subliminal perception focuses on the tactile and visual channels to deliver subliminal information to the driver. The tactile channel presently offers the highest potential for providing evidence for subliminal support, and the visual channel is used nearly all the time when driving and allows easy integration of subliminal information in current cars (e.g., as overlay in the head-up display).

The tactile channel is gradually finding its way into vehicular applications as a feedback modality offering additional information in a less demanding way. Tactile feedback has been underemployed and used mostly to present binary information (e.g., a lane departure warning system integrated into a seat or steering wheel). The somatic senses responsible for the perception of tactile stimuli operate all over the body and at all times (Driscoll, 2013). It should therefore be possible to integrate tactile systems into any operational control (steering wheel, driver seat, gearshift, safety belt, etc.). The purpose of such an interface is most-likely related to road safety, which could be easily achieved if part of the information currently transferred attentively could in the future be conveyed subliminally.

The visual modality has the highest capacity for information delivery. Since driving is essentially a visual task (Broyce, 2008), expectations of subliminal information transmission through this channel using HUDs or the front windshield itself (e.g., messages flashed so briefly that they cannot be perceived consciously) are high. As previously mentioned, numerous technologies are now in place to facilitate the projection of information into the driver's active field of view.

The aim of our studies presented subsequently was to condition drivers to change their driving behavior automatically based on hidden indication or cues—our experiments were designed accordingly.

Feasibility of Visual and Tactile Primes

In a first feasibility study, outlined in Fig. 19.6, we experimented with both different combinations of visible and invisible (i.e., primed) visual stimuli and tactile feedback perceived above/below the driver's awareness. The aim was to identify the general usability and potential of the front windshield or a HUD for visual cues and the seat and steering wheel as media for tactile feedback. In addition to different settings, combinations of different levels of workload were used in a dual-task setting to assess the influence of workload on subliminal perception.

Example 1: Subliminal Tactile Driver Notification

Based on the findings of the feasibility study, we chose to first test the effect of vibrotactile subliminal support by integrating a multitactor system into different controls of our test vehicle (driver seat and safety belt).

FIGURE 19.6 Driving simulator study (lane change task) with visible (consciously noticed) or invisible (subliminally perceived) information presented either on a head-up display, via speakers, or as tactile stimuli in the car seat.

Integration and parameterization (intensity, frequency, patterns, etc.) of the tactile elements followed comprehensive previous experience of its use for targeted, conscious driver stimulation. The field operational test (FOT) presented in Riener et al. (2010) was conducted according to the stimulus–response relation indicated in Fig. 19.5. Note that reaction types specified on the output side, for instance, gear changing behavior and conservative accelerating, are only examples; the effective behavioral change is fully controlled by the driver.

In the example study, embedded tactile actuators (Fig. 19.7) notified the driver of the current level of carbon dioxide emission/fuel consumption in comparison to an average value for the same section of the track. We used two types of tactile interfaces, one with tactors integrated into the driver seat and the other one with tactile elements embedded into the safety belt. The parameterization used to activate the tactors was the same for both types and the two interfaces were further installed in the test vehicle all the time [i.e., subjects tested with subliminal stimuli in the seat were also confronted with the (deactivated) interface in the safety belt]. A subtle harmonic tactile pattern was provided for efficient (better than average) driving, while a disharmonic pattern was transmitted for inefficient driving. Results compiled from about 500 km driven on two Saturdays with low traffic showed that drivers tend to drive more economically

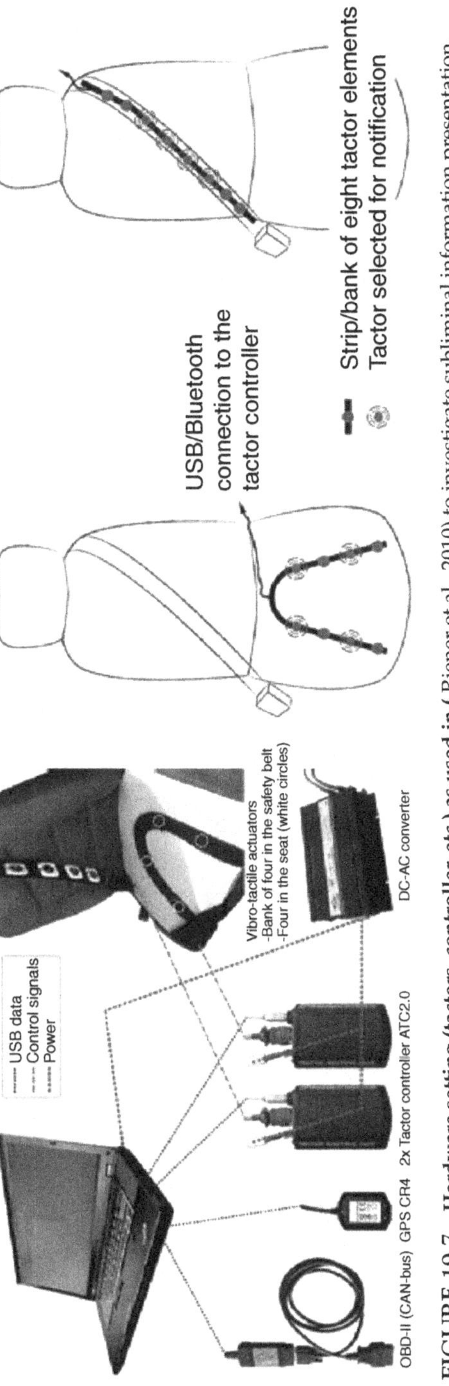

FIGURE 19.7 Hardware setting (tactors, controller, etc.) as used in (Riener et al., 2010) to investigate subliminal information presentation.

with harmonic/disharmonic tactile feedback on their current fuel consumption (test group) than without (control group). The potential to reduce CO_2 emissions (or fuel consumption) was in the range of about 2% (seat type interface) to 8% (belt type interface), which corroborates the unidirectional alternative hypothesis H_1 (subliminal perception reduces CO_2 emission). For details, we refer to Riener et al. (2010). In the postexperimental interview, none of the participants reported the experimental setting, in particular the tactile feedback, as being annoying or distracting, and all of them stated that they were not impaired by the modified safety belt or the foam mat placed on the seat. In conclusion, the results suggest that different tactile signals can be employed to provide ancillary driving-related information in a subtle, subliminal way.

Example 2: Visual Priming

In another priming experiment, the response–congruency effect (Elsner, 2008) was tested. It was assumed that response to a lane change request is faster for clearly visible targets if they are preceded by unconscious primes that are assigned to the same response (congruent trials) in contrast to trials where primes are assigned to another response (incongruent trials) or no primes at all (control). Our hypothesis was that drivers of the test group (visual prime presented) would change lanes earlier than control-group drivers confronted with no prime.

For the experiment, subjects with no previous knowledge about study setting and aim were recruited from the University campus. To minimize possible effects of age or gender, only male students aged 18–30 years were enrolled. In total, 20 people (10 test group, 10 control group) participated in the study. After being briefed in writing and verbally accepting the terms, each participant was asked to sit comfortably by adjusting the chair accordingly (Fig. 19.8). The subjects used a Logitech G27 steering wheel to control the driving simulator, and the driving scene was shown on a Samsung 46 in. monitor (8 ms response time) at a refresh rate of 60 Hz.

The driving experiment started with a warm-up phase in which no measures were taken. After a clearly visible start sign, recording of the steering behavior started. The task was to respond to lane change requests shown on overhead road signs as accurately as possible while maintaining a system-limited constant driving speed of 60 km/h (according to ISO 26022-2010). The overhead signs were continuously shown to the driver, displaying a "(x) (x) (x)" pattern while far away from the sign, and changing to a specified lane change request pattern when coming closer to the sign (display modes a and c in Fig. 19.8). The fact that drivers notice the overhead signs and their patterns is enough to assert that the drivers in both groups consciously perceived this information (taxonomically, we could say that the stimulus carries enough strength and obviously has

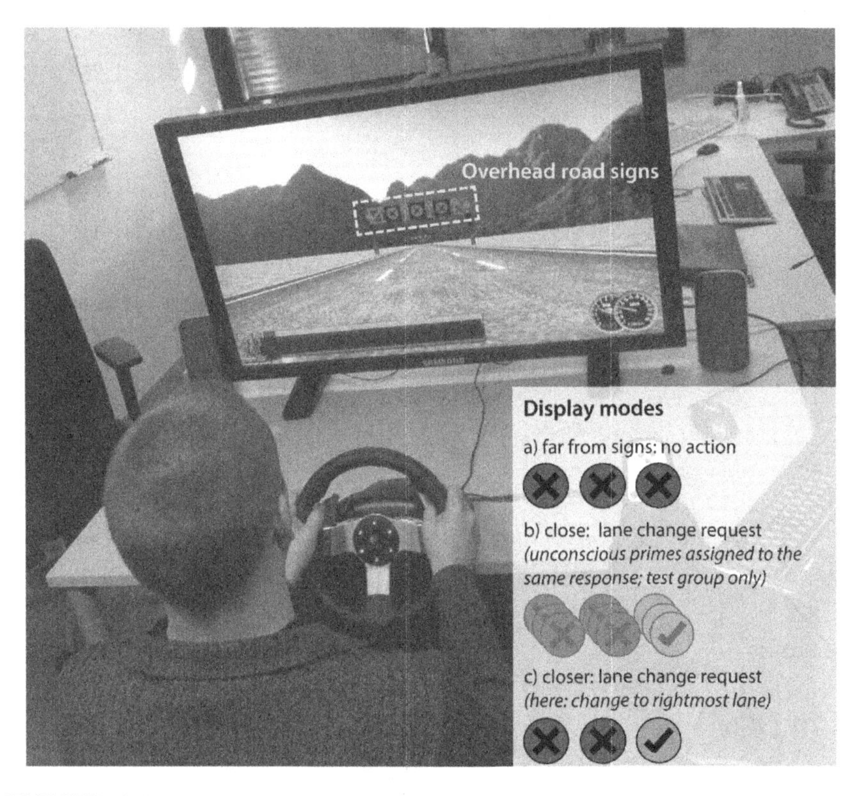

FIGURE 19.8 **Driving simulator experiment (lane change task) to assess the response-congruency effect and modes of priming (image overlay at bottom right).**

attention: a driver would be able to respond without hesitation when asked to which lane he is requested to change). Drivers in the test group received additional lane change information on the display 20 m ahead (or 1.2 s earlier) the panels. In a more realistic setting, this display would be most likely a HUD in the car. This additional information was shown only very briefly (in accordance with the monitor's refresh rate of 60 Hz, "primed" for just 16.66 ms; no external check performed; Fig. 19.9). We can state that drivers (of the test group) consciously perceived the visible target, but not the subliminally added information (display mode b) in Fig. 19.8). When asked what information could be read off the display or HUD, drivers would never indicate the briefly flashed lane change request information—this behavior was confirmed by the qualitative analysis of the study.

One test run took about 7 min and contained 33 lane changes on a straight, three-lane highway (ISO 26022-2010 conform). For the control group, all the lane changes were governed by the visible (consciously

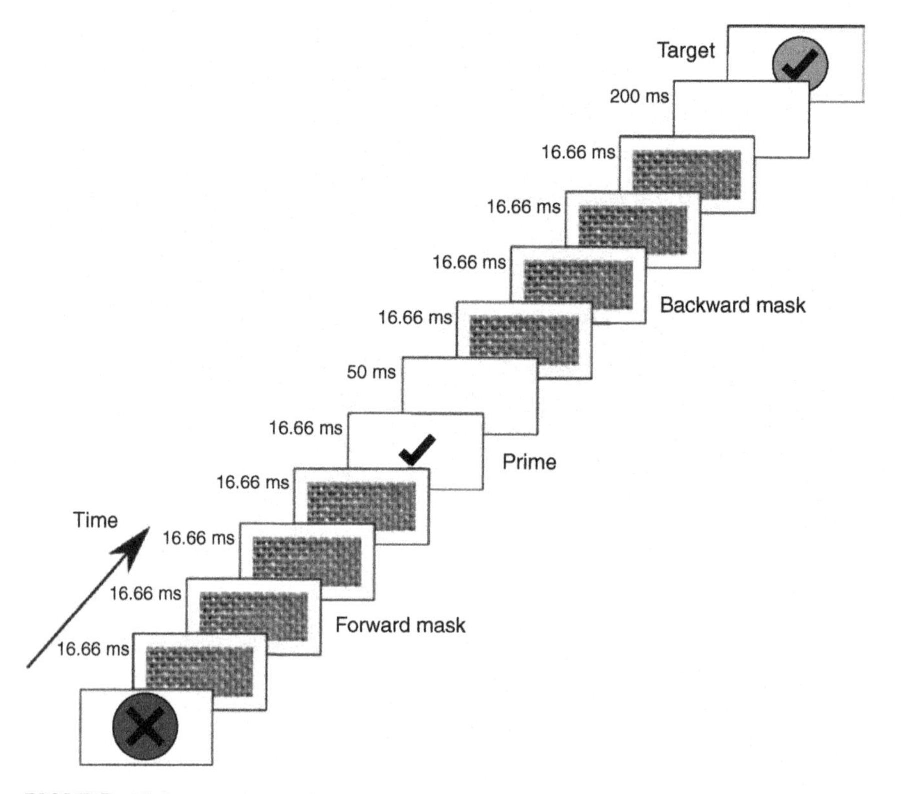

FIGURE 19.9 **Sequence of stimuli illustrated with congruent trials (unconscious primes that are assigned to the same response).** "Prime" is the information that was induced, "target" is the action that should be influenced by the prime, and the "masks" are stimuli in the form of a meaningless pattern, presented immediately following a critical stimulus, to eliminate its afterimage.

perceivable) overhead signs, while the test group received additionally "primed" ancillary information for half of the lane change requests (to assess whether there is a within-subject difference in primed/unprimed stimuli). After the driving experiment, subjects completed a NASA TLX questionnaire to rate their subjective workload and were then debriefed. At this stage, participants were not compensated for their effort.

Initial results revealed that while about 25% of the subjects from the test group believed that they were exposed to subliminal stimuli, not a single person had a concrete idea or could describe in detail what the subliminal stimulus or its aim was, and how it was presented. Our findings further suggest that there is a difference in the lane changing behavior between control and test groups.

CONCLUSIONS

With the advent of information and communication technology, Internet of Things and disappearing computers, information and communication expenditure increases continuously. As a consequence, new effective concepts to get in touch with the user while keeping attention resources low need to be developed.

In this chapter, we have discussed the potential and feasibility of information perceived on the subconscious (subliminal) level rather than presenting information actively, attentively, or in a conscious fashion. The history and progress of this field raised the question whether subliminal communication through nonconscious processing might in fact work. Recent empirical research suggests that subliminal interaction techniques improve HCI by reducing the workload of sensory channels. This assumption is still challenged by other researchers claiming that subliminal perception does not, or even cannot, work. Even though results from numerous experiments show different levels of performance with and without inattentively perceived information, it is difficult to judge whether or not the difference can be attributed to subliminal perception. While showing how this approach can be employed to avoid cognitive overload for the automotive domain, the method is generalizable and applicable to other fields of human–machine interaction.

We hope that this chapter has provided useful basic formalization with the approach of subliminal information presentation/perception and shown that it can be used to improve human–computer communication by reducing attention resources required for certain interactions. We further hope that this work will help to advance research in the field of subliminal perception and interaction. Even if such interfaces do not work for everyone and in every situation, it remains worth pursuing this exciting research.

References

Aarts, H., Custers, R., Veltkamp, M., 2008. Goal priming and the affective-motivational route to nonconscious goal pursuit. Soc. Cogn. 26 (5), 555–577.

Adams, J., 1957. Laboratory studies of behavior without awareness. Psychol. Bull. 54 (5), 383–405.

Ainsworth, L.L., 1989. Problems with subliminal perception. J. Bus. Psychol. 3 (3), 361–365.

Armstrong, A.-M., Dienes, Z., 2013. Subliminal understanding of negation: unconscious control by subliminal processing of word pairs. Conscious. Cogn. 22 (3), 1022–1040.

Ascone, D., Lindsey, T., Varghese, C., 2009. An Examination of Driver Distraction as Recorded in NHTSA Databases (Research note No. DOT HS 811 216). NHTSA's National Center for Statistics and Analysis, 1200 New Jersey Avenue SE., Washington, DC 20590: National Highway Traffic Safety Administration (NHTSA) (12).

Asif, A., Boll, S., 2010. Where to Turn My Car? Comparison of a Tactile Display and a Conventional Car Navigation System Under High Load Condition. In: Proceedings of, Automotive UI. Pittsburgh, PA (pp. 64–71). New York, NY, US.A. AC.M. Available from: http://doi.,acm.,org/10.1145/1969773.1969786

Atas, A., Vermeiren, A., Cleeremans, A., 2013. Repeating a strongly masked stimulus increases priming and awareness. Conscious. Cogn. 22 (4), 1422–1430.

Atkinson, R., Shiffrin, R., 1968. The psychology of learning and motivation. In: Spence, K. W., Spence, J. T., (Eds.), vol. 2. Academic Press, New York, pp. 89–195.

Bailey, R., McNamara, A., Sudarsanam, N., Grimm, C., 2009. Subtle gaze direction. ACM Trans. Graph 28 (4), 100:1–1100:.

Boll, S., Asif, A., Heuten, W., 2011. Feel your route: a tactile display for car navigation. IEEE Pervas. Comput. 10 (3), 35–42.

Brannon, L.A., Brock., T.C., 1994. Persuasion: Psychological Insights and Perspectives. In: Shavitt, S., Brock, T.C., (Eds.), (p. 15). Allyn & Bacon. Available from: http://courses.umass.edu/psyc392a/pdf/brannon&brock.1994.pdf

Brewster, S., McGookin, D., Miller, C., 2006. Olfoto: Designing a Smell-based Interaction. In: Proceedings of the Sigchi Conference on Human Factors in Computing Systems. New York, NY, USA: ACM, pp. 653–662. Available from: http://doi.acm.org/10.1145/1124772.1124869

Broek, E.L., 2013. Ubiquitous emotion-aware computing. Pers. Ubiquit. Comput. 17, 53–67.

Broyce, P.R., 2008. Lighting for Driving: Roads, Vehicles, Signs, and Signals. CRC Press, Taylor & Francis Group, Boca Raton, USA.

Calvin, W.H., Ojemann, G.A., 1994. Conversations with Neil's Brain—The Neural Nature of Thought & Language. Addison-Wesley, Boston.

Chalfoun, P., Frasson, C., 2008. Learning and Performance Increase Observed In a Distant, 3D., Virtual Intelligent Tutoring System When Using Efficient Subliminal, Priming. In: Bonk, C. J., Lee, M. M., Reynolds, T. (Eds.). Proceedings of World Conference on E-learning in Corporate, Government, Healthcare, and Higher Education 2008. Las Vegas, Nevada, USA: AACE, pp. 2543–2547. Available from: http://www.editlib.org/p/30027

Charissis, V., Papanastasiou, S., Vlachos, G., 2009. Interface development for early notification warning system: full windshield head-up display case study. Human-Computer Interaction. Interacting in Various Application Domains, Thirteenth International Conference, HCI International 2009, San Diego, CA, USA, July 19-24, Proceedings, Part IV. pp. 683–692. Available from: http://dx.doi.org/10.1007/978-3-642-02583-9_74

Cowan, N., 2001. The magical number 4 in short-term memory: a reconsideration of mental storage capacity. Behav. Brain Sci. 24 (1), 87–114.

Dehaene, S., Naccache, L., Le Clec'H, G., Koechlin, E., Mueller, M., Dehaene-Lambertz, G., Le Bihan, D., 1998. Imaging unconscious semantic priming. Nature 395 (6702), 597–600.

Dehaene, S., Changeux, J.-P., Naccache, L., Sackur, J., Sergent, C., 2006. Conscious, preconscious, and subliminal processing: a testable taxonomy. Trends Cogn. Sci. 10 (5), 204–211.

Del Cul, A., Baillet, S., Dehaene, S., 2007. Brain dynamics underlying the nonlinear threshold for access to consciousness. PLoS Biol. 5 (10), 2408–2423.

DeVaul, R.W., Pentland, A., Corey, V.R., 2003. The memory glasses: subliminal vs. overt memory support with imperfect information. IEEE International Symposium on Wearable Computers (ISWC'03), 146.

Dey, A.K., 2001. Understanding and using context. Pers. Ubiquit. Comput. 5 (1), 4–7.

Dingus, T., Hulse, M., Antin, J., Wierwille, W., 1989. Attentional demand requirements of an automobile moving-map navigation system. Transp. Res. Part A Policy Pract. 23 (4), 301–315.

Dix, A., Finlay, J.E., Abowd, G.D., Beale, R., 2007. Human computer interaction. Prentice Hall, New Jersey, USA.

Driscoll, R., 2013. By the light of the body: touch in the visual arts. Big Red Shiny (10), Available from: http://bigredandshiny.org/1653/by-the-light-of-the-body/, Article ID: 1653.

Dunne, B., Jahn, R., 2005. Consciousness, information, and living systems. Cell. Mol. Biol. 51, 703–714.
Egermann, H., Kopiez, R., Reuter, C., 2006. Is there an effect of subliminal messages in music on choice behavior? JASNH 4 (2), 29.
Elsner, K., 2008. Limited transfer of subliminal response priming to novel stimulus orientations and identities. Cons. Cogn. 17 (3), 657–671.
Ferri, F., Ebisch, S.J., Costantini, M., Salone, A., et al., 2013. Binding action and emotion in social understanding. PLoS ONE 8 (1), 1–9.
Ferscha, A., Farrahi, K.K., van den Hoven, J., Hales, D., Nowak, A., Lukowicz, P., Helbing, D., 2012. Socio-inspired ict. Eur. Phys. J. Spec. Top. 214 (1), 401–434.
Gausby, A., 2015. Attention spans (Tech. Rep.). Microsoft Canada Head Office, 1950 Meadowvale Blvd., Mississauga, ON, L5N 8L9: Consumer Insights, Microsoft Canada.
Giacomin, J., 2012. What is Human Centred Design? In: Tenth Conference on Design Research, Development (p/, d design 2012), Sao Luis, Maranhoa, Brazil, p. 14.
Greenbaum, J., 1993. Participatory design—principles and practices. In: Schuler, D., Namioka, A. (Eds.), pp. 27–37. Lawrence Erlbaum, New Jersey.
Greenwald, A., Draine, S., Abrams, R., 1996. Three cognitive markers of unconscious semantic activation. Science 273 (5282), 1699–1702.
Hassin, R.R., 2013. Yes it can: on the functional abilities of the human unconscious. Perspect. Psychol. Sci. 8 (2), 195–207.
Hassin, R.R., Uleman, J.S., Bargh, J.A. (Eds.), 2005. The New Unconsciousness. Oxford University Press, New York.
Hilsenrat, M., Reiner, M., 2009. The impact of unaware perception on bodily interaction in virtual reality environments. Presence Teleop. Virt. Environ. 18 (6), 413–420.
Ho, C., Spence, C., 2005. Assessing the effectiveness of various auditory cues in capturing a driver's visual attention. J Exp. Psychol. Appl. 11 (3), 157–174.
Horswill, M.S., Coster, M.E., 2002. The effect of vehicle characteristics on drivers' risk-taking behaviour. Ergonomics 45, 85–104.
Horswill, M.S., McKenna, F.P., 1999. The development, validation, and application of a video-based technique for measuring an everyday risk-taking behavior: drivers' speed choice. J. Appl. Psychol. 84 (6), 977–985.
Hudlicka, E., Fellous, J., 1996. Review of Computational Models of Emotion (Technical Report No. 9612). Psychometrix Associates, Inc, Arlington, MA.
Human Factors and Ergonomics Society, 2016. What Is Human Factors/Ergonomics? Available from: https://www.hfes.org//Web/AboutHFES/about.html
Jones, W.D., 2001. Keeping cars from crashing. IEEE Spectr. 09, 40–45.
Kadowaki, A., Sato, J., Bannai, Y., Okada, K., 2007. Presentation Technique of Scent to Avoid Olfactory Adaptation. In: Artificial Reality and Telexistence, Seventeenth International Conference.IEEEE CS Press, pp. 97–104.
Kelly, R., 2011. Subliminal computing: the support you don't see. XRDS 18 (1), 25–28.
Kern, D., Marshall, P., Hornecker, E., Rogers, Y., Schmidt, A., 2009. Enhancing navigation information with tactile output embedded into the steering wheel. Tokuda, H., Beigl, M., Friday, A., Brush, A., Tobe, Y. (Eds.), Pervasive Computing, vol. 5538, Springer, Berlin Heidelberg, pp. 42–58.
Kiefer, M., Brendel, D., 2006. Attentional modulation of unconscious "automatic" processes: evidence from event-related potentials in a masked priming paradigm. J. Cogn. Neurosci. 18 (2), 184–198.
Kim, S.J., Hong, J.-H., Li, K., Forlizzi, J., Dey, A.K., 2012. Route guidance modality for elder driver navigation. Kay, J., Lukowicz, P., Tokuda, H., Olivier, P., Krüger, A. (Eds.), Pervasive Computing, vol. 7319, Springer-Verlag Berlin Heidelberg, Heidelberg, Germany, pp. 179–196.
Kobilisek, K., 2012. Subliminal primes in real life situations (Master thesis psychology). University of Twente, Faculty of Behavioural Sciences.

Kouider, S., Dehaene, S., 2007. Levels of processing during non-conscious perception: a critical review of visual masking. Philos. Trans. R. Soc. B 362 (1481), 857–875.

Larsen, C. B., 2013. Subliminal Perception in 3D Computer Games—Towards an Invisible Tutorial. In: Proceedings of CHI'13, Paris, France (5). ACM.

Lehrl, S., Fischer, B., 1988. The basic parameters of human information processing: their role in the determination of intelligence. Pers. Individ. Dif. 9 (5), 883–896.

Levin, D., Baker, L., 2015. Change blindness and inattentional blindness. In: Fawcett, J., Risko, E.F., Kingstone, A. (Eds.), The Handbook of Attention. MIT Press, Cambridge, MA, pp. 199–232.

Li, W., Moallem, I., Paller, K.A., Gottfried, J.A., 2007. Subliminal smells can guide social preferences. Psychol. Sci. 18 (12), 1044–1053.

Lleras, A., Enns, J., 2004. Negative compatibility or object updating? A cautionary tale of mask-dependant priming. J. Exp. Psychol. Gen. 133 (4), 475–493.

Mandic, D., Harvey, R., Kolonic, D., 2000. On the choice of tactile code. In: IEEE International Conference On Multimedia And Expo (ICME 2000) (vol. 1, 195–198).

Mauter, G., Katzki, S., 2003. The Application of Operational Haptics in Automotive Engineering (Business Briefing: Global Automotive Manufacturing & Technology 2003 Nos. pp. 78–80). AUDI AG, D-85045 Ingolstadt, Germany: Team for Operational Haptics, Audi AG.

Merikle, P., 2000. Encyclopedia of psychology. In: Kazdin, A. E., (Ed.), (vol. 7). Oxford University Press, Oxford, pp. 497–499.

Miller, G.A., 1956. The magical number seven, plus or minus two: some limits on our capacity for processing information. Psychol. Rev. 63, 81–97.

Minsky, M., 1999. The emotion machine: from pain to suffering. In: Proceedings of the Third Conference on Creativity, Cognition (c, c '99), (7–13)., New York, N.Y., US.A., AC.M.

Moore, T., 1988. The case against subliminal manipulation. Psychol. Market. 46, 297–316.

Naccache, L., Blandin, E., Dehaene, S., 2002. Unconscious masked priming depends on temporal attention. Psychol. Sci. 13 (5), 416–424.

Nakamura, K., Inada, J., Kakizaki, M., Fujikawa, T., Kasiwada, S., Ando, H., Kawahara, N., 2005. Windshield display for safe and comfortable driving, SAE World Congress on Intelligent Vehicle Initiative (IVI) Technology 2005, Advanced Controls and Navigation Systems.

Nass, C., Jonsson, I.-M., Harris, H., Reaves, B., Endo, J., Brave, S., Takayama, L., 2005. Improving automotive safety by pairing driver emotion and car voice emotion. In: ACM., (Ed.), Chi 2005 Late Breaking Results: Short Papers.

Nielsen, J., 1994. Enhancing the explanatory power of usability heuristics. In: Proceedings of the Sigchi Conference on Human Factors in Computing Systems: Celebrating Interdependence. New York, NY, USA: ACM, pp. 152–158.

Nishida, S., 2006. Interactions and integrations of multiple sensory channels in human brain. In: 2006 IEEE International Conference on Multimedia and Expo (ICME 2006), pp. 509–512.

Norman, D., 1990. The Design of Everyday Things. Doubleday Business, New York.

Norretranders, T., 1997. Spüre die Welt. Rowohlt, Reinbek.

Opperman, R., 2002. Handbook on Information Technologies for Education and Training. In: Adelsberger, H., Collis, B., Pawlowski, J., (Eds.). Springer-Verlag, Berlin, pp. 233–248.

Overgaard, M., Timmermans, B., 2010. How unconscious is subliminal perception? In: Schmicking, D., Gallagher, S. (Eds.), Handbook of Phenomenology and Cognitive Science. Springer Science + Business Media B.V, The Netherlands, pp. 501–518.

Parma, V., Tirindelli, R., Bisazza, A., Massaccesi, S., Castiello, U., 2012. Subliminally perceived odours modulate female intrasexual competition: an eye movement study. PLoS ONE 7 (2), 8.

Patten, C. J., 2007. Cognitive Workload and the Driver: Understanding the Effects of Cognitive Workload on Driving from am Human Information Processing Perspective (PhD thesis). Department of Psychology, Stockholm University, Sweden. (ISBN: 978-91-7155-409-3).

Pfleging, B., Henze, N., Schmidt, A., Rau, D., Reitschuster, B., 2013. Influence of subliminal cueing on visual search tasks. CHI'13 Extended Abstracts On Human Factors In Computing Systems. ACM, New York, NY, USA, pp. 1269–1274.
Picard, R.W., 1997. Affective computing. The MIT Press, Cambridge, Massachusetts.
Pollatos, O., Kopietz, R., Linn, J., Albrecht, J., Sakar, V., Anzinger, A., Wiesmann, M., 2007. Emotional stimulation alters olfactory sensitivity and odor judgment. Chem. Senses 32 (6), 583–589.
Pratkanis, A., Greenwald, A., 1989. Advances in experimental social psychology. In: Berkowitz, L. (Ed.). vol. 22, Academic Press, San Diego, CA, pp. 245–285.
Radel, R., Sarrazin, P., Jehu, M., Pelletier, L., 2013. Priming motivation through unattended speech. Br. J. Soc. Psychol. 52 (4), 763–772.
Rakauskas, M.E., Gugerty, L.J., Ward, N.J., 2004. Effects of naturalistic cell phone conversations on driving performance. J. Safety Res. 35 (4), 453–464.
Ramsøy, T., Overgaard, M., 2004. Introspection and subliminal perception. Phenomenol. Cogn. Sci. 3 (1), 1–23.
Reder, L.M., Stoffolino, P., Ng, J., 1994. Word Frequency and Subliminal Perception (Tech. Rep.). 5000 Forbes Avenue Pittsburgh, PA 15213, US: Department of Psychology, Carnegie Mellon University, Pittsburgh.
Reimer, B., Mehler, B., Coughlin, J.F., Godfrey, K.M., Tan, C., 2009. An on-road assessment of the impact of cognitive workload on physiological arousal in young adult drivers. In: Proceedings of the First International Conference on Automotive User Interfaces and Interactive Vehicular Applications (Automotiveui '09). New York, NY, USA: ACM, pp.115–118.
Reynier, F., Hayward, V., 1993. Summary of the kinesthetic and tactile function of the human upper extremity (Technical Report No. CIM-93-4). 3480 University Street, Montral, Quebec, Canada, H3A 2A7: McGill Research Centre for Intelligent Machines, McGill University.
Riener, A., 2010. Sensor-Actuator Supported Implicit Interaction in Driver Assistance Systems (first ed.). Wiesbaden, Germany: Vieweg + Teubner Research (ISBN-13: 978-3-8348-0963-6).
Riener, A., 2014. Perceptual Computer Science—human-centric and reality-based human-machine interaction (Habilitation thesis). Johannes Kepler University Linz, Altenberger Strasse 69, 4040 Linz, Austria, pp. 279.
Riener, A., Ferscha, A., Frech, P., Hackl, M., Kaltenberger, M., 2010. Subliminal vibro-tactile based notification of CO_2 economy while driving. In: Proceedings of the Second International Conference on Automotive User Interfaces and Interactive Vehicular Applications (automotiveui 2010), Pittsburgh, Pa (p. 10). Pittsburgh, Pennsylvania, USA: ACM. (ISBN: 978-1-4503-0437-5).
Rigato, S., Farroni, T., 2013. The role of gaze in the processing of emotional facial expressions. Emot. Rev. 5 (1), 36–40.
Sandberg, K., Timmermans, B., Overgaard, M., Cleeremans, A., 2010. Measuring consciousness: is one measure better than the other? Consciou. Cogn. 19, 1069–1078.
Schlaghecken, F., Eimer, M., 2006. Active masks and active inhibition: a comment on Lleras and Enns 2004 and on Verleger, Jaskowski, Aydemir, van der Lubbe, and Groen 2004. J. Exp. Psychol. Gen. 135 (3), 484–494.
Schutte, P.C., 2007. Assessing the Effects of Momentary Priming On Memory Retention During An Interference Task (Technical Report No. NASA/TM-2007-214318). Virginia Commonwealth University, 907 Floyd Ave, Richmond, VA 23284, US: NASA Langley Research Center.
Shneiderman, B., Plaisant, C., Cohen, M., Jacobs, S., 2009. Designing the User Interface: Strategies for Effective Human-Computer Interaction, fifth ed. Addison-Wesley Computing, USA.
Smith, P., McCulloch, K., 2012. Subliminal perception. In: Ramachandran, V. (Ed.), Encyclopedia of Human Behavior. second ed. Elsevier, pp. 551–557.
Stone, D., Jarrett, C., Woodroffe, M., Minocha, S., 2005. User Interface Design and Evaluation, first ed. Morgan Kaufmann, Amsterdam.

Strayer, D., 2005. Lecture material for cognitive psychology (course psych 3120). Available from: http://www.psych.utah.edu/psych3120-classroom/12_01_05.pdf

Summer, P., Tsai, P., Yu, K., Nachev, P., 2006. Attentional modulation of sensorimotor processes in the absence of perceptual awareness. In: Proceedings of National Academy of, Science, pp. 520–530.

Synodinos, N.E., 1988. Subliminal stimulation: what does the public think about it? Curr. Issues Res. Advertising 11, 157–158.

Takahashi, H., 2009. A study on a method to call drivers' attention to hazard. Harris, D. (Ed.), Engineering Psychology and Cognitive Ergonomics, vol. 5639, Springer-Verlag Berlin Heidelberg, Heidelberg, Germany, pp. 441–450.

Telpaz, A., Rhindress, B., Zelman, I., Tsimhoni, O., 2015. Haptic seat for automated driving: Preparing the driver to take control effectively. In: Proceedings of the Seventh International Conference on Automotive User Interfaces and Interactive Vehicular Applications. New York, NY, USA: ACM, pp. 23–30.

Testa, C. J., Dearie, D. B., 1974. Human factors design criteria in man-computer interaction. In: ACM 74 Proceedings of the 1974 annual conference. New York, NY, USA: ACM, pp. 61–65.

Triggs, T. J., Harris, W. G., 1982. Reactions Time of Drivers to Road Stimuli (Human Factors Report No. HFR-12). Department of Psychology, Monash University, Victoria 3800, Australia: Human Factors Group, Monash University (Accident Re search Centre). Available from: http://www.monash.edu.au/muarc/reports/Other/hfr12.html

Varhelyi, A., 2002. Speed management via in-car devices: effects, implications, perspectives. Transportation 29 (3), 237–252.

Vega, K., 2013. Exploring the power of feedback loops in wearables computers. In: Proceedings of the Seventh International Conference on Tangible, Embedded and Embodied Interaction, tei '13. New York, NY, USA: ACM, pp. 371–372.

Vinciarelli, A., Pantic, M., Heylen, D., Pelachaud, C., Poggi, I., D'Errico, F., Schroeder, M., 2012. Bridging the gap between social animal and unsocial machine: a survey of social signal processing. IEEE Trans. Affect. Comput. 3 (1), 69–87.

Vorberg, D., Mattler, U., Heinecke, A., Schmidt, T., Schwarzbach, J., 2003. Different time courses for visual perception and action priming. Proc. Natl. Acad. Sci. USA 100 (10), 6275–6280.

Wallace, F., 1991. The effect of subliminal help presentations on learning a text editor. Inf. Process. Manag. 27, 211–218.

Weinberg, G., Harsham, B., Medenica, Z., 2011. Evalutating the usability of a head-up display for selection from choice lists in cars. In: Proceedings of AutomotiveUI 2011, Salzburg, Austria. ACM, New York, NY, USA, pp. 39–46.

Wickens, C., 1992. Engineering psychology and human performance (second ed.). In: Yardley, L., (Ed.). HarperCollins Publishers Inc., New York, pp. 449–471.

Wickens, C., Carswell, C., 2012. Information processing. Handbook of Human Factors and Ergonomics. John Wiley & Sons, Inc, Hoboken, New Jersey, pp. 117–161.

Wickens, C., Goh, J., Horrey, W., Helleberg, J., Talleur, D., 2003. Attentional models of multi-task pilot performance using advanced display technology. In: Human Factors. vol. 45, 360–380.

Wigdor, D., Wixon, D., 2011. Brave NUI World: Designing Natural User Interfaces for Touch and Gesture, first ed. Morgan Kaufmann Publishers Inc, San Francisco, CA.

Wittmann, M., Kiss, M., Gugg, P., Steffen, A., Fink, M., Pöppel, E., Kamiya, H., 2006. Effects of display position of a visual in-vehicle task on simulated driving. Appl. Ergonom. 37 (2), 187–199.

Young, K., Lee, J.D., Regan, M.A. (Eds.), 2008. Driver Distraction: Theory, Effects and Mitigation. CRC Press, Boca Raton, Fl.

Zehetleitner, M., Rausch, M., 2013. Being confident without seeing: what subjective measures of visual consciousness are about. Atten. Percept. Psychophys. 75 (7), 1406–1426.

CHAPTER

20

Physiological Computing and Intelligent Adaptation

Stephen H. Fairclough
Liverpool John Moores University, Liverpool, United Kingdom

INTRODUCTION

Physiological computing is characterized by a live connection between technology and the human nervous system. This act of monitoring renders the machine privy to a variety of data: electrochemical activity from the epidermis, fluctuations in muscular tension, the hemodynamics of the cardiovascular system and the electrocortical fluctuations of the brain. This connection between person and technology corresponds to an act of *digital embodiment*. By connecting to a computer, the human extends the boundaries of the central nervous system, communicating directly with technology via those physiological processes that underpin thoughts, emotions, and actions.

Digital embodiment has a number of implications for the way in which we interact with technology. Conventional human–computer interaction (HCI) is asymmetrical with respect to the flow of information. The user can interrogate a huge range of data concerning the internal processes within the computer (e.g., RAM use, disk space etc.) while the computer remains essentially blind to the psychological intentions and experience of the user. Continuous monitoring of the central nervous system is one way to facilitate a form of symmetrical HCI where information flows simultaneously from computer to user and vice versa (Hettinger et al., 2003). The implications of this innovation are potentially profound. For example, "smart technology" demonstrates a degree of intelligence by exhibiting sensitivity to task context and user intention without explicit information (Norman, 2007). Monitoring physiological data is a means of allowing a computer system to become aware of the user as a dynamic entity with the

Emotions and Affect in Human Factors and Human–Computer Interaction. http://dx.doi.org/10.1016/B978-0-12-801851-4.00020-3
Copyright © 2017 Elsevier Inc. All rights reserved.

major advantage of being continuously available, even in the absence of any overt forms of behavior (Byrne and Parasuraman, 1996).

This "wiretapping" approach shares a number of features with biofeedback technology. Both facilitate the process of self-regulation by providing feedback of physiological activity that can occur outside of conscious awareness. By interacting with data from the brain and body at an interface with technology, users can gain insight into states of cognition and emotion that can facilitate self-knowledge and associated strategies of self-regulation.

Categories of Physiological Computing

Physiological computing systems fall into one of two broad categories. The first are designed to extend the body schema, that is, the system of sensory-motor functions that we use every time we tap a key or move a joystick. These functions are guided by a sense of agency, that is, I am the one doing this. For example, the Brain-Computer Interface (BCI) offers an alternative mode of input control to extend the body schema (Allison et al., 2007). BCIs capture electrocortical activity at source (e.g., the intention that precedes movement or selection) and offer a highly novel form of hands-free interaction that is capable of communicating with standard screen-based technologies as well as specialized devices, such as prostheses. The same logic can be extended to monitoring muscle activity via electromyography (EMG). An EMG sensor on the forearm can detect patterns of gestures (Zhang et al., 2011) in a ubiquitous computing scenario. Similarly, muscle activity in the form of eye movements can be used for cursor control and other forms of input (Majaranta and Bulling, 2014).

The second category of physiological computing is relevant to perceptions of internal states related to psychological states. The body image has been defined as "a complex set of intentional states and dispositions… in which the intentional object is one's own body" (Gallagher, 2005). Physiological computing systems augment the body image by monitoring and responding in an adaptive fashion to spontaneous data originating from psycho-physiological interaction in the central nervous system. Biocybernetic adaptation covers a range of systems designed to capture psychological states relating to performance and wellbeing (Allanson and Fairclough, 2004; Fairclough, 2009). These states include psychophysiological signatures of emotions, such as anger, frustration, or fear, or changes in cognitive activity related to mental workload. For certain categories of software, such as games or autotutoring systems, we may be interested in changes that reflect elements of both cognition and emotion, that is, when someone is mentally overloaded (too much information, not enough time), they also may experience anxiety or anger. There are two important facets of biocybernetic adaptation that distinguish this category of control from those BCI systems designed to extend the body schema. The system is designed to adapt to *spontaneous* changes in the psychological state of the user. If the person is frustrated, the software may offer help; if the user is overloaded, the software may filter the flow of incoming information. This is an *implicit* mode of HCI with no requirement for the user to exhibit intentional behavior, which has also been termed passive BCI (Zander and Kothe, 2011).

Physiological computing systems are designed to extend the body schema or the body image by creating a dynamic quantification of those entities that may be accessed by a technological device. It is the quality of this dynamic quantification that will determine the efficacy of the interaction between user and system.

The Biocybernetic Loop

The biocybernetic loop (Pope et al., 1995) serves as a unifying concept for all physiological computing systems. This concept is derived from cybernetic model of control and communication within a closed loop (Wiener, 1948).

There are three generic stages of data processing within this feedback loop: collection, analysis, and translation. The first stage describes the collection of physiological data via sensor apparatus. A user at a desktop computer may wear a simple device to access heart rate or skin conductance level. The design of wearable, ambulatory sensors for data collection is a vital component of the data collection phase. The second stage of data analysis receives filtered data as an input and both quantifies the data in an appropriate way and identifies/corrects for the presence of artifacts. The analysis algorithm should be capable of both quantifying incoming data in real-time and identifying those periods that include "bad" data. It would be ideal if the analysis algorithm were capable of not only identifying sections of "bad" data but also subjecting these periods to a correction algorithm, in order to preserve the integrity of the data stream. The analysis stage should yield an appropriate and accurate quantification of physiological data but this is very loaded phrase; much depends on what particular aspect of psychology or behavior is the target or focus of the biocybernetic loop. The final process of the loop is translation. This stage describes how physiological units of measurement are converted into a computer command to be executed at the human-computer interface.

The three stages are realized in different ways depending on the category of physiological computing system. For EMG-based interfaces and some categories of BCI (e.g., those where motor functions are captured at the cortex), the function of the biocybernetic loop is to translate patterns of physiological activity into a specific command. This act of translation may be representative and functionally equivalent in some cases. A system that translates vertical and horizontal eye movement (monitored via EOG) into vertical and horizontal movements of a cursor on a screen is characterized by one-to-one correspondence as eye movements are scaled into x and y coordinates on the screen.

Other categories of physiological computing depend on the accurate identification of spontaneous psychological states to inform system adaptation. The obvious examples fall into the category of biocybernetic adaptation, such as affective computing technologies designed to capture changes in emotional states. In these cases, the link between physiological activity and psychological processes is analogous rather than strictly representative.

The translation of real-time physiological data into computer control is achieved by a component called the adaptive controller. This is an element within the biocybernetic loop that incorporates the translational rules of the system, for example, IF heart rate shows a rise of 30% THEN offer help, IF P300 amplitude is maximum for the "delete" command THEN activate delete function. For pattern-matching algorithms that translate physiology into input control at the interface, adaptive control is relatively straightforward. Detection of the upward movement of gaze can be translated into vertical movement of the cursor with an emphasis on low-level dynamics, such as the gain between EOG activity and sensitivity of cursor response. Controlling the movement of an avatar via BCI would require the adaptive controller to recognize and respond to template patterns of EEG activity that represented left/right and forward/backward. For those physiological computing systems that extend the body image, the purpose of the adaptive control is to translate physiological activity into an efficient (in terms of the rate of information transfer) and responsive mode of input control.

The controller serves a different function in the case of biocybernetic adaptation. These systems are designed to promote positive states and to prevent/ameliorate undesirable ones. Biocybernetic software serves as a dynamic mechanism for software to promote the same design goals. This is a very disruptive concept with respect to how people currently interact with technology. There is a shift of influence between user and system because biocybernetic control is designed to shape and manipulate the psychological state of the user. If an operator experiences a high level of mental workload, the system will intervene to reduce workload and preserve safety. If the player of a computer game is frustrated by the experience of repeated failure, the software may adapt to reduce challenge or offer help. The benign nature of these adaptations should not mask the fact that a working biocybernetic system is a machine with a prescribed agenda, a machine that deploys real-time adaptation as a means to achieve certain design goals, for example, to prevent accidents, to promote positive affect etc. This autonomy is achieved by incorporating a dynamic representation of the user, and by default the actual user, as an element in the control loop. The net result of the closed-loop design is an inevitable shift within the human-computer dyad toward the computer as a coworker or team-player that actively shares the goals of the user as opposed to the dumb slave system that I'm using to type these words.

The biocybernetic loop may function at different levels of the HCI. With respect to muscle interfaces and BCI, the biocybernetic loop functions within the HCI dyad and is designed to explicitly communicate commands to the interface. The loop mediates the intentions of the user to move the cursor down or to make the avatar move forward. Biocybernetic adaptation tends to function at the meta-level of the HCI, adjusting the parameters of the interaction (e.g., altering game difficulty) or making dynamic interventions (e.g., offering help, activating automation). This type of adaptation may also be achieved without any conscious intention on the part of the user. It is even possible for biocybernetic adaptation to occur in ways that are sufficiently subtle to escape the conscious perception of the user. The purpose of the biocybernetic loop is to adjust settings and make interventions in order to shape the interaction in a desirable way.

The biocybernetic loop functions both as a model for information flow and a unifying concept for physiological computing systems. It encapsulates the basic properties of sensor design and signal processing that underpin all categories of system.

MEASUREMENT AND CLASSIFICATION

The biocybernetic loop is the fundamental control unit of all physiological computing systems. The results of this data processing "pipeline" inform the mechanism of software adaptation. The loop is responsible for the interpretation of raw physiological data into a coherent response from computer software. The process encapsulated with the biocybernetic loop is associated with a string of important caveats:

1. physiological measures must be a valid measures of psychological concepts;
2. unobtrusive hardware must exist that is capable of capturing these measures in the field with sufficient fidelity;
3. data must be analyzed and categorized in near-real time in order to deliver a representation of the user to the system; and
4. changes in user representation must be translated into software control and adaptation that is both responsive and coherent.

These issues may be studied in isolation from one another (and often are), but from the perspective of an integrated system development, each part of the loop should be considered as mutually dependent on the others.

Inferring Psychological Meaning from Physiological Signals

The goal of measurement is the inference of the psychological or behavioral state of the user based on patterns of psychophysiological activity. One challenge for the designer of a physiological computing system is the identification of a distinct physiological pattern that is consistently associated with a target psychological concept. These patterns are used to trigger events at the interface and they must be salient, unique, and reliably detected in real-world conditions.

One way to increase the consistency of the psychophysiological inference is to capture the responses to a stimulus with known properties. An array of letters where each item flashes in sequential fashion can be presented in order to capture both the presence and magnitude of an evoked response potential (ERP) from EEG activity. It is known that the magnitude of certain components of the ERP (e.g., P300) respond to attentional processing, hence the letter one wishes to select from this array will generally deliver a greater magnitude of response. This "probe" strategy employs temporal coincidence in order to link a specific physiological pattern with a particular stimulus event. By contrast, the "wiretapping" approach that characterizes biocybernetic adaptation seeks to capture psychophysiological "signatures" of emotions and cognitive states against a background of spontaneous activity. This type of system must work with a low signal-to-noise ratio as the size of physiological response to an emotional event is relatively small compared to the magnitude of change due to physical movements and other confounding factors. It is a mistake for any designer to assume that physiological measures are capable of directly capturing psychological states in a "plug-and-play" fashion.

For those systems designed to extend the body schema, the accurate inference of intention is the critical issue. The "wiretapping" systems that fall into a broad category of biocybernetic adaptation are designed to classify spontaneous activity into pertinent categories that capture "target state" of user psychology. The process of inferring psychological events from physiological measures bears careful consideration in both cases. A penetrating analysis was provided by (Cacioppo and Tassinary, 1990), who scrutinized the specificity of association between concept and measure. The strongest category of psychophysiological inference for the development of a physiological computing system is the "one-to-one" relationship where a particular physiological measure operates as a unique marker of a specific psychological construct across all contexts of measurement. This kind of relationship is optimal but is also the rarest level of psychophysiological inference, particularly in the context of real-world testing.

Inference from psychophysiological measures can be a messy and inconclusive business, particularly in the context of everyday use outside of a laboratory. The most important starting point for the designer is a concrete understanding of the classification scheme that the act of measurement must deliver in order to make the system work, that is, how many categories must be distinguished for the system to work (Fairclough and Gilleade, 2012). In all cases of designing and developing physiological computing systems, it is important to select sensors/measures to the adaptive or input capabilities of the system. A concrete notion of what the measures are meant to achieve, in what environment and with whom provides an essential context within which to select, test, and validate the psychophysiological inference.

Wearable Sensors

The challenge for the design of sensors for physiological computing system is to create wearable sensors that maximize comfort, minimize intrusion and may be used in a public space without any embarrassment or self-consciousness—while maintaining a high fidelity of signal quality.

When designing sensors with low intrusiveness, there is a temptation to strip down the process of measurement. This simple strategy equates the number of sensors or measures directly to the comfort of the user (fewer sensors = greater comfort for the user) but this is a myopic approach. Ambulatory sensors are designed to be worn in the field, hence they must capture not only the signal of interest but also as many potential sources of noise as possible, in order to facilitate a process of artifact detection and correction. In the field, it is important to capture metabolic changes due to everyday activities using pedometers or accelerometers to quantify movement and to incorporate these variables into the diagnosis produced by the system. We may look to other data sources to provide additional context for our interpretation of spontaneous physiological activity, such as time of day or room temperature or background noise level. From the perspective of the user experience, the level of comfort associated with data capture is an overriding concern. As system designers, we need to move toward an invisible monitoring process wherein wearable sensors and the process of data capture, storage, and analysis are rendered as unobtrusive as possible.

Remote sensing offers the best chance to achieve an "invisible" process of physiological monitoring (Poh et al., 2011), but like all camera-based systems, the sensor requires a stationary user in order to work. Wearable devices, such as chest or wrist straps or earplugs, have the advantage of being ambulatory in the sense that the user can move around, albeit with the disadvantage that sensors are relatively intrusive as they involve contact with the skin.

The provision of sensor apparatus capable of comfort and signal fidelity is an essential development if we are to realize the potential of physiological computing systems. If these devices are not available, physiological computing will never reach the vast majority of users. It is also important for sensor technology to come equipped with Software Development Kit and the capability to use standard protocols for communication, such as Bluetooth. There is enormous potential for wearable sensors to interface with mobile devices via the specialized software niche of the app economy if the right combination of hardware and software tools can be brought to the market.

Signal Analysis and Classification

Once signals have been captured and filtered by the system, these data are subjected to a process designed to identify and classify significant patterns in the data. This combination of signal processing and diagnosis represents the operational crux of the biocybernetic loop whereby unique patterns of physiological activity are associated with psychological events.

There are several approaches to signal classification, which are applied to different types of categorization "problems" in physiological computing systems. The identification of a spontaneous psychological state falls in the domain of affective computing, where "target states," such as frustration or excitement are operationalized as a pattern of physiological activity distinguishable from spontaneous responses that are associated with other states. The successful operation of a BCI requires accurate identification of physiological features associated with the initiation of input control, such as activity in the somatosensory cortex or a positive deflection of electrical activity that occurs in the same temporal window as a particular stimulus. Signal classification must be both fast and accurate to facilitate real-time input control. The classification of physiological signals along a unidimensional continuum, such as anxiety or mental workload, represents a different category of assessment where estimates of magnitude (low, medium, high) on a unidimensional scale are the focus of signal classification.

The purpose of signal classification is to create a literal interface between the human nervous system and a repertoire of software responses. But the data processing "pipeline" that underpins this interface must be carefully constructed in a bottom-up fashion. To use an analogy, if we were to design a system for language translation, the first design question would concern the size of the vocabulary; in other words, exactly how many different words must be recognized and translated by the system in order to function with a degree of utility? This is also a good starting point for the process of data classification in a physiological computing system. All systems are associated with a *functional vocabulary* that describes how many commands, subjective states, or gradations of experience must be recognized in order to support successful operation. A system may function adequately on the basis of simple binary differentiation of two classes. Other systems may be equipped with repertoires of five or more adaptive responses, amplifying the challenge of accurate data classification. The number of items in the functional vocabulary defines the boundaries of the data classification problem inherent in the design of physiological computing systems, for example, 2 classes or 5 classes or 10 classes. Without this kind of operational context, the specification of an optimal process of signal classification remains an abstract proposition.

The degree of similarity or overlap between target states is also important for classification accuracy. Making a two-category distinction between happiness and anger would generally yield more accurate classification than an attempt to differentiate fear from anger. This is logical because similar emotional states contain greater overlap in terms of how they impact on psychophysiological reactivity.

The application of machine learning algorithms is a common strategy for classification within the biocybernetic loop. The general methodology for the construction of a classifier is to generate a training set, which accurately represents the dimensional space associated with the functional vocabulary of the system. This database will subsequently be used to train a classifier and represents the template for all subsequent acts of categorization conducted by the system, therefore it must provide a stable and well-defined mapping of physiological measures onto psychophysiological space. The first obstacle for classification is deriving and defining an optimal set of training data for the machine-learning algorithm. Data from the brain and body, particularly in the field, has poor signal-to-noise ratio, that is, contaminated by physical artifacts that may account for outliers in each "class" to be identified.

Signal classification is generally based around the creation of features (or vectors) that are derived from multiple measures of psychophysiology. There is a good deal of research literature where single data streams are analyzed in myriad ways; for example, measures of heart rate can be expressed in terms of descriptive statistics (mean, maximum, minimum, standard deviation) or subjected to further analysis to yield power in low and high bands, then further expressed as the ratio of both bands. The tendency to measure lot of variants from the same basic signal source creates the so-called "Curse of Dimensionality" where the amount of data required to describe different classes increase exponentially with the dimensionality of those features that are used as input. The practical implication of this "curse" is that the designer must acquire more training data when he or she adds new features as inputs to the classification process.

The act of classification may subsume both the process of discrimination and the implicit mapping onto psychological categories. It is hoped that the training set provides a good mapping in terms of a quantitative discrimination but whether it provides the best possible mapping remains an open question. If we consider which factors may contribute to classification errors, three main sources are most likely (Lotte et al., 2007):

1. The influence of noise from nonpsychological sources, as stated earlier, noise is a "fact of life" as far as ambulatory psychophysiology is concerned.
2. The degree of divergence between the estimated mapping provided by the training set and the best mapping possible. This second factor is determined purely by the representativeness of the training set, which is defined as the capacity of the training set to generalize across all instances of the target state or pattern. For instance, is the pattern of psychophysiological reactivity associated with a specific target state during the training set sensitive to all other instances of the target state that may be encountered by the system. The degree of this divergence between what is measured by the system now and what was measured during training is called Bias.
3. The degree of sensitivity of the classifier to the training set. It is important to note that different approaches to signal classification differ with respect to their susceptibility to specific and idiosyncratic qualities of the training dataset. The degree of sensitivity to the training set exhibited by the classifier is called variance.

This summary provides an overview of the challenges facing the design of a classification system using live data from brain or body as an input and producing real-time or near-real-time outputs. Noise is an irreducible component of the measurement process and may be dealt with by filtering the signal. If bias and variance are both low, we would expect classification errors to be minimal because the mapping is good and sensitivity to the training set is low. Obviously, a poor mapping can originate from several sources, such as multiple categories that are not clearly discriminated or a substantial overlap between the psychophysiological "signatures" of different target states. If the training set is situation-specific then classification errors will rise due to increased variance.

A good training set represents an essential prerequisite to enhance the accuracy of a classification algorithm. "Good" training data may be defined according to a number of different criteria. It is important to choose a psychophysiological measure (or collection of measures) that is sensitive to changes in psychological state or an intentional act. This measure must be sufficiently robust to exhibit sensitivity under the working conditions of the system. A training set that captures the range of physiological responses under realistic conditions will reduce the level of bias and variance in the classification system. Ecological validity is one key aspect of the training set that captures the ability of data to represent psychophysiological measurement as it would occur under real-world conditions.

A number of classification techniques have been utilized in the context of both BCI applications and the biocybernetic loop. The first category of classifiers may be described as *generative* and are designed to compute the likelihood of each class. This generative class includes Bayesian network approaches, which is a probablistic model that uses a "maximum a posterori" rule to assign a vector to the class with the highest posterori probability. *Static* classifiers, such as artificial neural nets (ANN), represent a sophisticated approach to network-based classification. ANNs are arranged in layers (e.g., multilayered perceptron) where each node or neuron receives a number of inputs in order to calculate the cumulative activation of that particular neuron, this output is relayed to the next layer of the network and so on. One disadvantage of the ANN approach is that the network is very sensitive to overtraining, particularly with noisy psychophysiological data as a set of inputs. The static approach to classification is contrasted with *dynamic* techniques, such as Hidden Markov Models (HMM), which incorporate temporal features into the process of discrimination. HMM have evolved to calculate the probability of observing a particular sequence of feature vectors; this approach has been adopted in some BCI systems but is rarely used for biocybernetic adaptation.

Nearest-neighbor approaches fall into the category of *discriminative* classifiers where the distance (e.g., Euclidean) between each input and a feature vector is calculated based upon the training space. Support Vector Machines (SVMs) represent another form of discriminative classifier and has been widely used in both BCI and biocybernetic systems. SVMs are capable of creating nonlinear decision boundaries and while they may be slow (computationally), they have good generalization and tend to be less sensitive to overtraining as well as the curse of dimensionality. *Stable* classifiers, such as linear discriminant analysis (LDA), are characterized by their relative simplicity and insensitivity to small variations in the training data. This is an advantage in the sense that the classification system is sensitive to gross rather nuanced trends in the training data; however, as the name suggests, LDA performs poorly where the classification is based on complex (i.e., nonlinear) boundaries in contrast to SVMs or ANNs.

A number of reviews (Lotte et al., 2007; Novak et al., 2012) have surveyed the prevalence of different classification techniques in the development of different categories of physiological computing systems. With respect to synchronous BCI, classification is mainly characterized by the use of SVMs, dynamic classification (HMM), and ensemble classification, particularly majority voting and boosting. The application of ANN to classification in BCI accounted for approximately a quarter of those systems in the review. Those systems designed for biocybernetic adaptation were characterized by a mixture of SVM, LDA, and Classification Trees, although a small number did use an ANN approach.

The process of measurement and classification is fundamental to the integrity of the biocybernetic loop. The challenge for designers is to create dynamic user representations that: (1) are scientifically sound, (2) rely on comfortable and nonintrusive sensors that are capable of delivering good signal quality in everyday settings, and (3) can be classified into different categories within an adaptive controller in order to drive adaptation at the interface.

INTELLIGENT ADAPTATION

The prioritization of current research on signal analysis and classification is understandable because the system cannot work without inputs and the act of classification is inextricably bound up with the quality of signal input. But the designer must make equally important decisions about how those categories are translated into a repertoire of software responses because action at the interface will ultimately determine user experience.

We must acknowledge the enormous scope for error that exists within the biocybernetic loop when data are collected in the field, or to phrase the statement in more precise terms, there is enormous scope for user perception of system error. A misclassification within the context of BCI interaction yields a response that is unintentional and obvious to the user. Spotting an error during interaction with a biocybernetic system is significantly more difficult for the user. Biocybernetic adaptation faces the significant hurdle of creating adaptation at the interface that resonates with the dynamic experience of the user. For systems designed to provide input control, the primary design issue is how to match the intentions with events at the interface within a small-time window. For biocybernetic systems, the capacity of the adaptive loop to synchronize with user experience is complicated by the complexity, spontaneity, and subjectivity of the latter.

The quality of adaptive response may be characterized by its accuracy, sensitivity, and intuitiveness from the perspective of the user. The *accuracy* of a physiological computing system is an obvious starting point for this discussion. It is reasonable to assume that an accurate response is desirable from the perspective of the user, but what does that mean? Accuracy is defined as matching a predefined pattern of physiological response with a specific command from the functional vocabulary. If the system selects a response based on an erroneous act of classification, it is an error of commission and will be perceived as invalid by the user. However, inaccurate systems are also guilty of errors of omission—when the user expects a response from the system but does not receive one. It is important that both categories are minimized, errors do not just annoy and inconvenience the user, but they fundamentally undermine the development of trust in the technology.

Consider the relationship between the size of the functional vocabulary (the repertoire of possible system responses) and the criterion of accuracy. It could be argued that increasing functional vocabulary is beneficial for user experience, making a greater range of adaptive options available to the system should lead to a nuanced response at the interface that increases the perceived "intelligence" of the system as a whole. However, probability dictates that the errors of commission become increasingly frequent as items are added to the functional vocabulary and the designer of the system faces a trade-off between accuracy and adaptive capacity. In order to resolve this dilemma, the designer ought to consider the minimum number of adaptive responses necessary for the system to meet its operational requirements. It is reasonable to assume that accuracy will be maximized for a system with minimal functionality. In some cases, a physiological computing system may be capable of operating quite effectively with only two or three types of adaptive responses; much depends on the type of interactive experience that the system is designed to deliver.

If accuracy is concerned with the design of adaptive systems where responses are tailored to the limits of psychophysiological classification, the *sensitivity* of the system response describes the temporal relationship between physiological activity and events at the interface. Many users develop an intuitive heuristic through interaction with input control devices so that x amount of mouse movement is equated with y amount of cursor control on the screen. The same logic applies to BCI and eye control of cursors. The temporal relationship between changes in psychophysiological activity and events at the interface is described as gradation sensitivity. There are two aspects of this relationship: proportion and dynamics. In its simplest form, a small change in electrocortical activity will cause a small movement in a desired direction, for example, an increase of EEG activity by one standard unit at a sensorimotor site causes an avatar to move forward 1 m in virtual space, and larger changes in the EEG cause greater movement and so on. However, the ratio between signal and its output at the interface may be designed in different proportions, hence a 2:1 relationship may mean that avatar moves forward 1 m in virtual space whenever sensorimotor activity increases by two standard units.

Gradation sensitivity can be weighted in different ways in order to create the largest amount of change at the interface for different levels of psychophysiological activity. The relationship between psychophysiological change and events at the interface may be conceived as an extension to Fitts Law (Fitts, 1954), where covert responses from the central nervous system are mapped onto the biocybernetic control loop. For the designer, creative adjustment of gradation sensitivity represents one mechanism to reinforce physiological self-regulation in order to improve the productivity and quality of the HCI.

The temporal dynamics between psychophysiology and software are perhaps less important in the case of biocybernetic adaptation. The sensitivity of biocybernetic adaptation is based mainly upon a perception of whether the system response is appropriate to a specific situation. The first criterion concerns the ability of the system to deliver the right response. If the user is extremely frustrated or very bored, there is an inherent expectation that a physiological computing system will intervene. If no response is forthcoming, the user perceives an error of omission. If the system makes an inappropriate response, for example, offering help when the user is calm and relaxed, the insensitivity of the system revolves around errors of commission. Both categories of error create unique forms of insensitivity, both of which are equally damaging to the user experience.

The perceived accuracy from the perspective of the user is an important determinant of the quality of the interaction. A system that is perceived to be accurate in the short-term will create a positive impression that encourages further use. However, perceived accuracy and trust can also be affected by bias or be linked to the cost of errors to the user, that is, errors may be more or less costly depending on a degree of annoyance or inconvenience experienced by the user as a direct result. The question of what is an acceptable level of accuracy for a physiological computing system has been addressed by several studies. Some simulated various levels of accuracy with respect to control of an input device and task difficulty respectively in order to explore levels of user acceptance and tolerance for system error (Novak et al., 2014; Van de Laar et al., 2013). One recent study assessed the ability of users to assess the classification accuracy of a physiological computing system designed to classify interest levels (high vs. low) in a series of movie trailers (Fairclough et al., 2015). The authors reported that participants tended to over-estimate a mathematical accuracy score of 0.82–0.91 by approximately 5%—hence participants may be able to accurately assess classification accuracy of a system if they are provided with overt feedback (as they were in this study).

One design option to deal with the inherent uncertainty of matching adaptive responses to dynamic user states is to vary the autonomy of the system. It is generally assumed that triggers from the biocybernetic loop will action events at the interface in an all-or-nothing fashion without any input from the user. But total automation is not the only option available to the design of a physiological computing system. The levels of automation analysis (LOA) (Sheridan and Parasuraman, 2006) represents system automation as existing on a continuum where responses vary from 100% manual to 100% automated. There are a number of hybrid forms on this continuum, such as automated prompts (e.g., would you like the game to increase difficulty now?) and automated cues (e.g., highlighting appropriate functions or areas of the screen). Adaptive technology based on physiological computing could adjust the autonomy of responses at the interface based upon the level of confidence underpinning the episode of classification. If the system is highly confident, the response occurs at the interface without any consultation with the user. If confidence is not as high, the system may prompt the user for confirmation before taking action. While biocybernetic adaptation is based upon correct classification, it is obvious that instances of greater physiological reactivity are easier to recognize than smaller ones. The physiological computing system would always identify a transition from low to high frustration but a change from low to medium frustration is harder to detect. Therefore, the system may again adopt a softer approach to automation, using prompts or cues, when the distance between categories is low and ambiguity of classification is high. This design option provides some insurance against both errors of commission and omission because the system makes a definite response but requires clarification from the user to resolve any uncertainty.

When a person interacts with a physiological computing system, the technology actions a series of adaptations in response to "live" changes in brain activity or psychophysiology. *Intuition* describes the intelligibility of those adaptations from the perspective of the user. This particular criterion focuses on the perceived meaning of adaptation at the interface as opposed to the process of measurement and classification supporting those adaptations. There is an analogy between adaptive interaction and a spoken conversation. Both activities are characterized in terms of turn-taking and mutual adaptation. Conversation is initiated in the biocybernetic loop when the user exhibits a particular pattern of psychophysiological activity that triggers an adaptive response. The interpretation of the system is communicated implicitly to the user via the specifics of adaptation at the interface: the presentation of calming music tacitly communicates an interpretation of stress or anxiety on behalf of the system, an assessment of the intention to move forward produces forward movement of an avatar in virtual space. This intuitive element informs users' understanding of how the system functions as an active agent working to a predetermined rationale. In order to engender user trust, it is important that the response from the system is understood with sufficient clarity to both inform and shape the HCI (Miller, 2005).

From the perspective of the user, intuition represents the extent to which an adaptation reflects feelings, thoughts, and intentions that are active in consciousness at that particular time. The informative content of system adaptation may simply reinforce what is already known, when an angry user experiences a calming response from the system, feedback is simply confirming the contents of awareness. Alternatively, an adaptive response may grant insight into unacknowledged feelings or thoughts and the presentation of a calming response can increase awareness of increased frustration (Picard et al., 2004). A second-order aspect of intuition concerns users' understanding or assessment of the rationale underlying the adaptive response from the system. The detection of frustration may prompt a number of adaptive responses: offer help, suggest a rest break, play calming music, recommend a number of breathing exercises. Each adaptive response is unique but the underlying rationale is consistent and it is important for this rationale (e.g., to counteract frustration) to be clearly communicated to the user. Systems associated with biocybernetic adaptation are designed with a distinct rationale in mind: to reduce mental workload, to promote safety, to help the user, to challenge the player etc. An understanding of this rationale will hopefully improve the intelligibility of system behavior.

The criterion of intuitiveness also concerns the quality of the adaptive response as assessed by the user. If the system makes a response that accurately reflects the current state of the user, correct classification will count for little if the actual adaptation has no utility from the perspective of the user. This aspect of intuition concerns the extent to which an adaptation meets the needs and desires of the user. If the user is frustrated because they are behind schedule, an automated request to take a rest break is unlikely to be welcomed or be perceived as particularly helpful. There is a degree of overlap between these aspects as an intuitive system response is accurate, comprehensible, timely, and useful from the perspective of the user.

The informative content of the adaptive response provides an explicit cue to the diagnosis of the user state. If the response from the system is opaque, the user may assess the system to be inert or unintelligible, both of which are undesirable. The informative element of adaptation is self-evident for input control systems where intention and feedback at the interface are closely coupled. For biocybernetic adaptation, the informative content of the adaptive response may be explicit and obvious to the user or implicit in the sense that the user may or may not notice a change at the interface. Explicit feedback concerns those categories of the adaptive response that are impossible for the user to miss: the appearance of an avatar offering help, an on-screen recommendation to take a break. This feedback represents an unambiguous statement of the design rationale underpinning the biocybernetic loop. There is a risk associated with explicit adaptation that errors of commission are obvious to the user and only a small number of high-profile errors are sufficient to damage trust in the system, but there are occasions, typically during extreme frustration or high mental workload, when the user might reasonably expect an explicit response from the system. Implicit adaptation may not be noticed by the user at the interface and have the advantage of making errors of commission without necessarily damaging trust because the user may not notice the error. The problem of implicit adaptation is that they may often exert a cumulative influence that takes time to achieve any tangible effect on the user.

For example, we constructed a biocybernetic version of the game Tetris (Fairclough and Gilleade, 2012) where the drop speed of the blocks were adapted to changes in real-time EEG using tiny adjustments that were barely noticeable to the player. The experiential effect of these implicit adaptations was double-edged; it took a period of several minutes for a clear trend of adjustment (to increase or decrease speed) to become apparent to the user and in the meantime the users generally perceived the system to be inert and creating errors of omission. The strengths and weaknesses of explicit/implicit adaptation may be designed into the interaction by using implicit adaptations for small magnitudes of change in the state of the user and relying on explicit adaptation when extremes are detected (Fairclough, 2009).

The repertoire of adaptive responses available to the system designer is similarly finite but the process of design is crucial. The designer must decide which adaptive options to include within the functional vocabulary of the system and which to omit. If a specific adaptation is deemed worthy of inclusion, the magnitude of response or the levels of response must be clearly defined. The whole purpose of this process is the production of a system capable of intelligible interaction.

CONCLUSIONS

Interaction with a physiological computing system represents one approach to the creation of a technology where control is achieved without touch and software responds to the psychological context of the user. The closed-loop logic of these systems describes how raw physiological data from the body and brain is translated into a series of dynamic control inputs and changes at the interface, which are conveyed directly to the user. This process of translation from raw physiology to input control contains a number of steps with significant hurdles, such as: the design of wearable sensors that deliver high quality data in an unobtrusive way, the process of inferring psychological states from physiological data in everyday life, the detection of artifacts, and classification of data in real-time. These challenges of measurement and signal processing in this field are substantial but the design of the adaptive controller is central to the user experience. The adaptive controller represents the rationale of the closed-loop, which describes the way in which data is translated into adaptations and responses at the interface with the user. This component remains relatively unexplored compared to signal processing and classification, but it is the efficacy of the adaptive controller that will largely determine the user experience and the degree of "intelligence" displayed by the system.

References

Allanson, J., Fairclough, S.H., 2004. A research agenda for physiological computing. Interact Comput 16, 857–878.

Allison, B.Z., Wolpaw, E.W., Wolpaw, J.R., 2007. Brain-computer interface systems: progress and prospects. Expert Rev. Med. Dev. 4 (4), 463–474.

Byrne, E., Parasuraman, R., 1996. Psychophysiology and adaptive automation. Biol. Psychol. (42), 249–268.

Cacioppo, J.T., Tassinary, L.G., 1990. Inferring psychological significance from physiological signals. Am. Psychol. 45 (1), 16–28.

Fairclough, S.H., 2009. Fundamentals of physiological computing. Interact. Comput. 21, 133–145.

Fairclough, S.H., Gilleade, K., 2012. Construction of the biocybernetic loop: a case study. Paper Presented at the Fourteenth ACM International Conference on Multimodal Interaction, Santa Monica.

Fairclough, S.H., Karran, A., Gilleade, K., 2015. Classification accuracy from the perspective of the user: real-time interaction with physiological computing. Paper Presented at the CHI '15 Proceedings of the Thirty-third Annual ACM Conference on Human Factors in Computing Systems, Seoul.

Fitts, P.M., 1954. The information capacity of the human motor system in controlling the amplitude of movement. J. Exp. Psychol. 47 (6), 381–391.

Gallagher, S., 2005. How The Body Shapes The Mind. Oxford University Press, Oxford, England.

Hettinger, L.J., Branco, P., Encarnaco, L.M., Bonato, P., 2003. Neuroadaptive technologies: applying neuroergonomics to the design of advanced interfaces. Theor. Issues Ergon. Sci. 4 (1–2), 220–237.

Lotte, F., Congedo, M., Lecuyer, A., Lamarche, F., Arnaldi, B., 2007. A review of classification algorithms for EEG-based brain-computer interfaces. J. Neural Eng. 4, 1–24.

Majaranta, P., Bulling, A., 2014. Eye tracking and eye-based human-computer interaction. In: Fairclough, S.H., Gilleade, K. (Eds.), Advances in Physiological Computing. Springer, London, pp. 39–66.

Miller, C.A., 2005. Trust in adaptive automation: the role of etiquette in tuning trust via analogic and affective methods. Paper presented at the First International Conference on Augmented Cognition, Las Vegas, NV.

Norman, D.A., 2007. The Design of Future Things. Basic Books, New York.

Novak, D., Mihelj, M., Munih, M., 2012. A survey of methods for data fusion and system adaptation using autonomic nervous system responses in physiological computing. Interact. Comput. 25, 154–172.

Novak, D., Nagle, A., Riener, R., 2014. Linking recognition accuracy and user experience in an affective feedback loop. IEEE Trans. Affect. Comput. 5 (2), 168–172.

Picard, R.W., Papert, S., Bender, W., Blumberg, B., Breazeal, C., Cavallo, D., Strohecker, C., 2004. Affective learning—a manifesto. BT Technol. J. 22 (4), 253–269.

Poh, M.-Z., McDuff, D., Picard, R., 2011. A medical mirror for non-contact health monitoring. Paper presented at the ACM SIGGRAPH Emerging Technologies, Vancouver.

Pope, A.T., Bogart, E.H., Bartolome, D.S., 1995. Biocybernetic system evaluates indices of operator engagement in automated task. Biol. Psychol. 40, 187–195.

Sheridan, T.B., Parasuraman, R., 2006. Human-automation interaction. Rev. Hum. Factors Ergon. 1, 89–129.

Van de Laar, B., Bos Plass-Oude, D., Reuderink, B., Poels, M., Nijholt, A., 2013. How much control is enough? Influence of unreliable input on user experience. IEEE Trans. Cybern. 43 (6), 1584–1592.

Wiener, N., 1948. Cybernetics: Control and Communication in the Animal and the Machine, second ed. M.I.T. Press, Cambridge, Mass.

Zander, T.O., Kothe, C., 2011. Towards passive brain-computer interfaces: applying brain-computer interface technology to human-machine systems in general. J. Neural Eng. 8, 1–5.

Zhang, X., Chen, X., Li, Y., Lantz, V., Wang, K., Yang, J., 2011. A framework for hand gesture recognition based on accelerometer and EMG sensors. IEEE Trans. Syst. Man Cybern. A Syst. Hum. 41 (1064–1074).

CHAPTER

21

Aesthetic Computing*

Paul A. Fishwick

University of Texas at Dallas, Richardson, TX, United States

OVERVIEW

The phrase "Aesthetic Computing" while taken literally applies the philosophical area of aesthetics to the field of computing, and work in the area is broadly defined as such; however, in my operational definition for the work we do in my research lab and in teaching, aesthetic computing is treated as *embodied formal language*. The purpose of aesthetic computing is to deliver knowledge and practice of formal languages using aesthetic products as a vehicle. Aesthetic Computing is founded on an increasing collection of literature on the role of the body in learning, specifically in mathematics. This foundation is then applied to the field of computing whose formal language elements are extensions of mathematics. There are two questions that this new area raises:

Q1: *How can embodied cognition be situated within formal languages?*
Q2: *How can embodied cognition result in novel computer interfaces for formal languages?*

Q1 surfaces a host of sub-questions revolving around theory, philosophy, and analysis. Asking this question raises issues of motivation: (1) Why am I interested in this topic? (2) How is the area of aesthetic computing built on top of embodied cognition and philosophy? (3) Who has worked in this area (e.g., the literature)? Q1 is not enough, however. It is one matter to analyze and develop theory, but another to ask oneself, "How can this theory be transformed into practice?" That is the essence of Q2. What should we be *doing, practicing, and creating* to take embodied cognition of mathematics and computing to the next level? We need to build a new generation of human-computer interfaces that are informed by embodied

*Portions of the chapter are from "Aesthetic Computing located at: https://www.interaction-design.org/encyclopedia/aesthetic_computing.html"

Emotions and Affect in Human Factors and Human–Computer Interaction. http://dx.doi.org/10.1016/B978-0-12-801851-4.00021-5
Copyright © 2017 Paul Fishwick. Published by Elsevier Inc. All rights reserved.

principles and use these principles as design elements for interacting with formal languages. A potential, and vital, third question would revolve around the effects on such computer interfaces on learning via assessment and scientifically-based research methods. This represents an area that aesthetic computing needs to investigate; however, most work to date is based on theory construction and engineering the novel interfaces.

The Aesthetic Computing Hypothesis is that given the embodied nature of cognition, we should realize this embodiment through novel human-computer interfaces for learning formal languages.

CONTEXT

I pose two questions as a means to provide context for the area of aesthetic computing: (1) Why is the term "Aesthetic Computing" being treated as "Embodied Formal Languages?" and (2) What are "Embodied Formal Languages"? For the first question, we must revisit the roots of the word "aesthetics." The original Greek definition of aesthetics, *α σθητικόζ (aisthetikos)*, stems from another Greek word aisthanomai, meaning "I perceive, feel, sense." At the core of aesthetics, then, lies the body, and its interactions in forming concepts and knowledge: aesthetics as embodiment. Aesthetics is, in breadth and depth, a much richer enterprise above this level (Kelly, 1998), yet we maintain a view of aesthetics that is body-based, even though Diffey (1995) notes that the term 'aesthetic' has largely lost its perceptual sense except in the word 'anaesthetic,' but retains its senses of "beautiful' and 'artistic."' As far as to why "Formal Languages" are used to characterize "Computing," we note that the bulk of theory of automata and computing is situated within linguistics—although a subset of general linguistics that requires a formally well-defined specification and treatment.

Let us now consider the definitions of embodiment and formal language. Embodiment suggests the perception/action feedback loop present when the body interacts with its environment. So, it seems clear that an embodied approach to anything would involve sensorimotor functions—using the mouse, keyboard, multi-touch displays as well as donning a head-mounted display or using a tactile feedback device. Human-Computer Interaction is chock-full of approaches that leverage such technologies. But, embodiment is a much deeper concept than sensory stimuli and physical manipulation. We have a sense of presence with certain advanced technologies such as multi-user virtual environments (i.e., achieving different types of presence, including social). We also have a sense of presence when reading a book since the book situates our "mind's body" within the narrative (ref. "narrative psychology" in Beck et al. 2011). Thus, embodiment can be measured objectively by hardware used to enable the senses, or subjectively through a presence instrument on the human subject. Embodiment should not be viewed as a rejection of abstraction, but rather as a complement to it (Devlin, 2006).

Formal languages define a category of language that is artificial, such as a programming language. These languages stem from formal grammars which can be based on text, shapes, or diagrams. FORTRAN, Java, and Perl are examples of formal languages, but so are the eXtensible Markup Language (XML), Unified Modeling Language (UML), data structures, Morse code, and dynamic model structures used for simulation (Fishwick, 1995, 2007a). Formal languages are frequently specified using grammars such as the Backus-Naur Form (BNF) and need not be text-based. For example, one can have formal audiovisual languages and also graph grammars. All formal language structures can be defined hierarchically using levels of abstraction (e.g., 3 finite state machine levels governing an underlying set of ordinary differential equations, which in turn are translated into the programming language Java, and then further into byte code). Languages, therefore, are frequently defined in long chains of specification and translation. Each language has its own target functionality, culture, and adherents. Ghezzi and Jazayeri (1997) provide general concepts of specification for programming languages.

PERSONAL EXPERIENCES AND INFLUENCES

Art

It is easy to take the idea of embodied cognition for granted since it seems like something so natural—that the body plays a central role in cognition. However, an adherence to embodiment tends to change your worldview when looking at objects. As an amateur artist, I collected many posters and prints of historically well-known artists. In middle school, I was strongly influenced by Thomas Gainsborough's work, in particular Fig. 21.1.

I imagined that with myself as an avatar, I could enter the painting, walk the wheat field, examine the trees, and engage in social discourse with Mr. and Mrs. Andrews. This led to a series of imaginary conversations and observations "in world." The key point here is the "reading" of this work *as a form of embodied experience*. The Gainsborough painting was not a remote object of study for me, but rather an example of virtual reality, a time machine—an illusion that allowed me to immerse myself within the world of 18th century England. This approach is an example of Dewey's *art as experience* (Dewey, 1934) and relates to Grau (2004) argument about artists as the first virtual reality creators. The approach stresses that when we approach an object, we can interpret it dynamically via a bodily simulation with all of the perceptual and motor-based actions that the body affords. This way of thinking and acting can be applied to all objects and media, including mathematics and computing.

FIGURE 21.1 Mr. and Mrs. Andrews, oil on canvas, Thomas Gainsborough, 1750. The National Gallery, London, UK.

Mathematics

In elementary school, like hoards of other students throughout the world, I was taught the elements of arithmetic—its methods and laws, with many examples that were exercised using rote memorization and intense practice. Doing mathematics was highly action-based, but the action was limited to solving multiple problems over extended periods of time. After the basic elements of arithmetic came algebra. Let's consider the following mathematical expression containing arithmetic with a sliver of algebra:

$$X = 2 * (3 + 4)$$

We have all been subject to such mathematical objects as they are critical to an educated public. Learning all components of this equation was not easy – one had to understand the concept of a variable, operations of multiplication and addition, followed by the concept of a parenthetically-delimited group. Order of operations is also critical, as suggested by the group. So, for example, I can add 3 to 4 and then multiply by 2 to obtain 14, which was then set to X as an equivalence. Certain laws of arithmetic were useful in transforming expressions such as this one. The Law of Distribution states that $x\ (y + z) = xy + xz$ where x, y, and z are numbers, and the multiplication is implicit rather than being defined explicitly using * as in the above equation. The teacher would define the law of distribution and give us many useful examples as a means to reinforce our understanding of the law and how it can be employed in symbolic manipulation. Such patterns of equivalence drove a static pattern-matching type of approach to mathematics.

However, during the ensuing lessons, I found it convenient to create an artificial method of solution that involved treating the numbers and symbols as physical objects. In mathematics education, this kind of process is termed reification (Sfard, 1994) and is related to constructivism (Piaget, 1950) and constructionism (Papert, 1980), where students create their own knowledge through a combination of ideas and life experiences. I used a virtual manipulation of the above expression by representing the distributive law through analogy and metaphor:

> Grab the "2" object, which when juxtaposed with the "*" operator, provides a biomechanical state where the "2" is pushed inward toward the group object defined within the parentheses "(...)". The "2" is pushed gradually and then when it reaches the edge of the spatial boundary denoted by "(", it moves through it to the other side and splits—in a biological fashion – into two clones that are attached to the "3" and to the "4," respectively. This cloning activity results in the expression (2 * 3 + 2 * 4). The sub-expressions 2 * 3 and 2 * 4 are evaluated through further bodily activity. Pushing the 2 and 3 into the *, for example, results in multiplications. Similar reactions occur to perform the + operation last, as dictated by the learned order of operations. The result is then placed manually in a box with an X printed on it.

Mathematics then, for me, had become akin to a full-body sport rather than simple operations requiring a collection of static text-based rules and patterns. The virtual manipulations might involve other embodied activities, where I might have "launched the 2" over a wall that bounds the parenthetical expression. While this is a personal experience, it is by no means unique, as Sfard observes in her dialogue with Thompson (Thompson and Sfard, 1994), where she notes the propensity for similar mental imagery: "My work with mathematicians brought lots of further evidence that, indeed, the inner world of a mathematizing person may look very much like a material, populated with objects which wait to be combined together, decomposed, moved and tossed around." Arzarello (2004) explains the difference between natural versus formal mathematical presentations, and surfaces the importance of gesture in using naturalistic explanations and interpretations in addition, or on the path, to the formal. The previous embodied description would be termed natural. Goldin and Kaput (1996) overview the effects of media on mathematical representation by noting "...changes in physical media that permit external representations to be action rather than display representations give these representations one characteristic of powerful internal representations." Hadamard (1996) studied mathematical thought which echoed similar cognitive processing. This action-based narrative on mathematical symbols was not limited to the distributive law for me. For example, in an expression such as $x + 2 = 4$, something interesting happens when moving numbers through the equals sign. There is a virtual line or plane that intersects at a right angle to the = when a number, such as 2 is dragged through this vertical plane, the number flips its sign on the other side with a mirror-like effect, resulting in $x = 4 - 2$. The laws of commutativity and associativity have similar pseudo-physical, material, behaviors that can be used to understand and process arithmetic expressions.

The problem with my early experiences with embodied sense of symbol manipulation is that none of the books (or teachers) explained mathematics in this way, and I, and likely many others, were forced to keep these somewhat peculiar cinematic episodes to ourselves. Whether this type of thinking is common requires more scientific studies and reflection upon the nature of mathematics. At the University of Florida, we have developed a web-based interactive tool that allows anyone to manipulate expressions in this fashion. We have also previously explored similar embodied representations involving a sense of presence in a virtual environment (Fishwick and Park, 2008a).

My purpose of relaying this experience is to emphasize the importance of the body in understanding formal languages such as mathematics. Lakoff and Nunez (2000) presented a landmark compilation of mathematical metaphors that build on top of the philosophy of embodied cognition (Johnson, 1987; Varela et al., 1992; Barsalou, 2010). In particular, Johnson's image schemata, such as containment, attraction, and equilibrium were integral aspects of my arithmetic experience. The literature in embodied thinking centers thought and knowledge on the body and is informed not only by areas, such as conceptual metaphor (Lakoff and Johnson, 1999, 2003), but also by subsequent empirical studies of the brain (Feldman and Narayanan, 2004; Feldman, 2008). Even more generally, language-based narratives appear to contain an embodied basis (Speer et al., 2007; Mar and Oatley, 2008) defining natural language in terms of simulation. Reading a story about grasping or running can result in a cognitive simulation of these events and activities, as if the reader had been physically active. Going back in time to when the Method of Loci flourished (Yates, 1966), we note that the act of memorizing a set of facts was turned into a rich, embodied process rather than viewed as mere associative retrieval. The area of situated learning and cognition (Brown et al., 1989; Lave and Wegner, 1991) meshes well with the embodied approach in terms of its goals and methods: learning by doing.

In closing the discussion of an embodied mathematics, we should note that the concepts of "action", "interaction", and "process" can be framed within standard mathematical notation containing explicit aspects of functional composition, dynamics, and procedure (i.e., embodied-types of thought). For example, the aesthetics of geometry and shape can be constructed generatively (Leyton, 2001, 2006) and dynamically via Blum's wave propagation-based medial axis (Leymarie, 2006). We can also use mathematics to create a formal representation of mathematical metaphors

(Guhe et al., 2009), thus making a loop: grounding metaphors on mathematical expressions, where the metaphors themselves are formally defined.

The embodied approach has profound implications for mathematics, and by extension for applied mathematics, and computing since computing is a direct outgrowth of mathematics, and formulas such as the one described earlier are common objects found in software "expressions." If our thought is embodied, then:

1. We should investigate the variety of metaphors used within mathematics and computing, and also their origins and cultural associations.
2. We should leverage the metaphorical, and embodied, substrate of language by creating new human-computer interfaces that reinforce and amplify this experience.
3. We should bring to bear other disciplines for whom "the body" is a natural component, such as the arts and humanities (Slingerland, 2008), thus forming new interdisciplinary collaborations that span the academy.

Programming

The embodied approach was extended from mathematics into learning programming and data structures. Programming, in particular, is known to be rich in metaphor. Loops are just that: patterns of cyclic behavior – small objects moving around a closed path as these objects perform other tasks. Sequential behavior is sometimes a movement along a spatial path, and functions are machines that take product inputs and produce outputs. Papert (1980) in his explanation of the LOGO language reinforces the importance of embodiment in a term he calls "syntonicity", where he notes "We have stressed the fact that using the Turtle as metaphorical carrier for the idea of angle connects it firmly to body geometry." Petre and Blackwell (1999) performed studies on programmers, and results indicate metaphorical reasoning involving objects, motion and general embodied interaction. Metaphors such as these are not only present in all programming languages, but also in the theory of computation on which the theory of computing is based. For example, the Turing machine is an excellent example: a machine envisioned by Alan Turing in the 1930s consisting of a tape read/write head and an infinite tape. This metaphor may have been because of the extensive use of magnetic tape at the time. In the previous century, Charles Babbage used a "mill" in his computing engine. Interestingly, in the vast history of computing where these historical concepts are discussed (Ifrah, 2001), most programming and computing was analog and embodied by definition and implementation. It is only relatively recently that the evolution from analog to digital has simultaneously sped up our computations, facilitated a computer revolution, but also disembodied our relations to computing.

Media

Media theorists have provided a host of approaches in understanding the evolution of media. McLuhan (1964) places importance, not only on the message created through a modulated medium, but on the medium itself which affects the message. McLuhan employs the example of a light bulb which he claims is a "medium without a message." However, the light bulb can host a binary digit, and perhaps more in the case of multi-way switch bulbs in a means not unlike Morse code manipulated through signal lamps. Bolter and Grusin (2000) present a theory of media forms undergoing gradual alteration, generally technology-driven, causing us to examine issues of immediacy (seeing beyond the medium to the target signified) and hypermediacy (being aware and reflecting on the medium). New media studies place specific importance on materiality, the medium, and embodiment. Manovich (2002, p. 317), when he considers the "loop as a narrative engine," with a loop being defined as a common programming structure enabling index-based iteration, asks "Can the loop be a new narrative form appropriate for the computer age?"

Popular media have significantly shaped my thought process underlying aesthetic computing. For example, Tron (Kallay, 2011), which debuted in 1982, is noteworthy because it was created based on a highly innovative screenplay which included a large piece of software, namely an "operating system," that could be experienced directly. Programs were bodies, and the operating system was composed of a city-like space with lighted, moving vehicles and interacting programs. Tron is fairly unique in this way within the science fiction/fantasy genre. Other more recent cinematic offerings, while impressive and engaging, tend to ignore the "program." For example, on Star Trek: The Next Generation, we were introduced to the Holodeck where one could experience an ultimate virtual reality with full sensory simulation. A user would stop at the outside of the Holodeck and say "Computer. Load Holodeck Program A-3" or some such phrase, and then the Holodeck would load this program and the user would enter. However, we never actually experienced the program itself—only its inputs and outputs. Similarly, in The Matrix, we have a rich embodied experience of human characters that, in reality, are stored inside of a network of fluid-filled pods.

Despite our familiarity and utility with text-based process descriptions, it is remarkable and ironic that a hyper-real environment such as the Matrix affording real-time synthetic interactions and simulacra would have to be programmed by strange-looking rivulets of green rain, which are not obvious to anyone, presumably except for the operator well trained in this postmodern descendant of cuneiform script. This semiotic condition presents a stark contrast: practically unlimited full-sensory simulation on one hand produced by the program, and what amounts to glorified typewriter symbols on the other defining the program itself. It is as if one provides you with a highly maneuverable hypersonic jet plane to fly with the caveat that you need to pilot the plane by tapping on a straight key to produce Morse code dots and dashes. One would expect that, just perhaps, the capabilities that form programs and data might avail themselves of the practically unlimited human-computer interface that the Matrix provides. Rotman (2000, p. 67) poses the question that forms this concern, "What if language is no longer confined to inscriptions on paper and chalkboards but becomes instead the creation of pixel arrangements on a computer screen?"

AESTHETIC COMPUTING: TURNING COMPUTERS INSIDE-OUT

Computers have shrunk in size, and increased in number, considerably over the past half-century. We are familiar with news stories about how ever smaller and thinner computers and software are now ubiquitous in our culture to the point where we carry or wear them in our daily routines. The decrease in size and increase in number creates a situation where computing effects most consumer products. For example, the digital video recorder enables time and place shifting for movies and television shows. What is just as interesting is exploring how computing affects us and our thinking. Turkle (2005) explains this psychological phenomenon and closes with the phrase "we are all computer people now."

Turkle's argument has significant ramifications for computing, and I would go one step further to suggest that the way in which our thinking is changing culturally surfaces deep abstract concepts in computing to us as we use these devices: from number, to information structure, to process. Digital watches and video recorders (DVRs) are good examples. Most digital watches are multi-function. These watches contain the ability to act as a way to tell time, set a stop watch, or wake up to an alarm. To use the watch, you have to learn how to navigate a menu by repeatedly pressing a mode button. In each mode, there are sub-functions refining that mode's interaction. This experience of mode-button pressing directly maps to a fundamental theoretical structure in computing called a *finite state machine* (Hopcroft et al. 2000). It is not just that the finite state machine is embedded within the watch's silicon, but also that the human wearing the watch becomes aware of this virtual machine's structure and its components through the experience of using the watch. The state machine *changes how the wearer thinks*, even though the wearer is probably unaware of the formal mathematical notation of a state machine. The watch's software internals become embedded within our psychology and culture. A similar process occurs within most other household appliances such as the DVR, however, the state machines in DVRs are more complex than in watches—yet to understand how to navigate the hierarchical menus, one has to become fully aware of a new type of thinking (Negroponte, 1996). Effects of computing on thought (e.g., neomillennial/digital native learning styles) have also been covered in the context of learning (Dieterle et al., 2007).

Experience with computing artifacts is a form of information representation, where the definition of "representation" is expanded as a form of interaction, rather than as a static object in the form of a sign. If the raw elements of computing—information, data, and software—are changing the way that we think and entering into our popular culture, it is natural to suggest that aesthetics of these raw elements can and should play a central role in computing. Aesthetics has evolved from the embodied, sensory, definition to a more comprehensive one offered by Kelly (1998), a "critical reflection on art, culture, and nature." Aesthetics within computing results in new interaction modalities for computing artifacts such as formal languages. Given the preponderance of new ways to connect human with computer, there are many opportunities for creative representation. We categorize and study these new ways using the phrase *aesthetic computing*.

WHY AESTHETIC COMPUTING?

Representation targets of aesthetic computing include terms such as data, information, software, and code. I use these terms somewhat interchangeably because of semantic overlaps. Data can be atomic or in the form of a structure. Code usually refers to software which encompasses both data as well as process. Information theory tells us that all of this is a form of information since information can be decoded as atomic, structural, or procedural. I prefer terms such as code, software, or information when referring to the "computing" part of aesthetic computing since these terms encompass broader categories of items that can be represented, whereas the term "data" in common parlance tends to denote non-procedural forms of information.

The argument for aesthetic computing involves emerging areas of computing which have changed:

- Our relationship to each other and to nature. These aspects include ubiquitous (Greenfield, 2006, Gershenfeld et al., 2004) and pervasive computing, customization and personalization of interfaces, and the new modalities for human-nature interaction as mediated through computing (e.g., the virtual reality continuum spanning physical, virtual, and augmented reality). Shared and customized interfaces

for information visualization (Viegas et al., 2007), code sharing (Reas and Fry, 2007), assisted with "remix culture" (Lessig, 2008) create a networked, customized (Pine, 1999) representational space.
- Our thought patterns, allowing computing artifacts such as information and software to permeate our experience. Salomon (1990) makes an argument for computing changing thought, resulting in *cognitive residues* from human-computer interaction. These studies are consistent with Turkle (2005).
- The importance of experience in computing in human-computer interaction (HCI). Cockton (2011) and Hassenzahl (2011) describe the shift in HCI from efficiency, alone, to experience, and Löwgren (2011) emphasizes the importance of interaction—a core aspect of experience. The emphasis on experience is related conceptually to embodiment (Lakoff and Nunez, 2000; Johnson, 2007) as a basis for cognition. The relevance of aesthetics in HCI is discussed by Tractinsky et al. (2000) and Norman (2004). Dourish (2004) lays out a philosophical foundation for embodiment in HCI through its beginnings in phenomenology.
- Our need as computer scientists to interact more frequently with artists and designers since they represent the creative component of aesthetic inquiry, and so experience-based representations for the diffusing computing artifacts need to be studied with the help of artist-scientist collaborations (Buxton 2008; Malina, 2011).

HISTORY OF THE AESTHETIC COMPUTING FIELD

I have been teaching a course in aesthetic computing for a decade (since 2000), and information on the most recent course can be found in (AC, 2012). A preliminary paper was published on the concept (Fishwick, 2002). A Dagstuhl seminar on Aesthetic Computing (Fishwick and Bertelsen, 2002) was co-organized in Germany (Dagstuhl, 2011) by myself, Roger Malina, and Christa Sommerer during the summer of 2002. This interaction resulted in several publications (Fishwick et al., 2005; Fishwick, 2006; Fishwick, 2007b; Fishwick, 2008b). Kelly et al. (2009) represents the most recent published workshop in the area. The use of the word "aesthetics" and "programs" can be found in several contexts, including Mohr (2011) and Nake (2009) who were early investigators in the aesthetics of interaction through the use of computer programs as a means of artistic expression. Knuth (1992) developed literate programming and made note of the importance of aesthetics in programming. Knuth's interest in aesthetics went beyond the purely cognitive, and included artistic forms of typography and layout design for programs. For Knuth, it would seem that computing was an embodied experience.

Aesthetic computing is unusual in that aesthetics is intended to be applied to computing rather than in the inverse direction: using computing to create artistic products. Examples of aesthetic computing, therefore, capture a kind of "boomerang effect" where elements of computer graphics, ubiquitous computing, and mixed reality interfaces can be used to interactively represent that which formed these technologies—namely the information and software.

In terms of academic curricula, Aesthetic Computing has been taught for a decade at the University of Florida in the form of two classes, which are usually combined: CAP 4403 (undergraduate) and CAP 6402 (graduate). The combined classes began as part of the Digital Arts & Sciences (DAS) programs (Fishwick, 2010) designed and developed to connect computing with the arts. The class has undergone several stages since 2000:

1. (2000–05) Representational alternatives to software artifacts – from numbers and expressions, to data structures and programs. There was one physical project, with the other two projects resulting in digital representations. The physical product was exhibited in several gallery areas on and off campus allowing passers-by to comment and explore.
2. (2006–09) Alternative representations for mass media and communications. This emphasis required students to employ representational creativity, but with the idea of starting with a contemporary news story and then mining this issue for the software artifacts in the story that were to be represented. The physical project was eventually dropped since many of the computer science students in the mid to later years had minimal design and art backgrounds.
3. (2010–11) Representation using web mining and APIs. This was an effort to create more automation in the representational process with students finding sources of information and then, mostly using APIs, to translate this information into creative representations. Most students use data as their information, but others used more complex web structures (e.g., XML) as sources.
4. (2012) A focus on representation of data structures, mathematical models, and dynamic models and programs. The end product is either an interactive game or video production whose goal is to facilitate education of computing concepts for early-age groups and non-computing specialists. This is the current incarnation of the class (AC 2012).

We use the term aesthetics in the spirit of Kelly's definition, but also extend the concept of "critical inquiry" to include the creative aspect of design and art. This is only natural, for engaging in critical inquiry presupposes and requires the creative act. Studies in aesthetics are numerous (Audi, 1999; Kivy, 2004) often with underlying attempts to find universal attributes of beauty (Scruton, 2011). My view on aesthetics is one that focuses on that which is generated as a result of cultural inquiry, which is to say the vast diversity of design and art forms. This "aesthetics as diversity" approach is similar in spirit to Hogarth (Burke, 1943) with the associated phrase, "unity in variety."

TOWARD SOFTWARE AS EMBODIED EXPERIENCE

Introduction

Partial justification for the use of embodiment as a form of representation is based on educational learning styles (Dede, 2005). Also, our ongoing research indicates a significant correlation between presence and memory in a virtual environment (Fishwick et al., 2010) with results currently in the journal submission phase. Recent mixed reality memory studies such as (Ikei and Ota, 2008) indicate positive effects on memory in an augmented environment. Instruments and studies on memory performance within virtual environments are being continually refined and investigated. Parsons and Rizzo (2008) introduce a test of validity for a virtual environment cognitive instrument called VRCPAT. Johnson and Adamo-Villani (2010) note significant effects of immersion on short term spatial memory. Embodied interaction with technology provides us with an understanding of internal logic, software, and process usually through pure experience. For example, we learn the state machine of a DVR through repetitive DVR use. While a large population may require this learning, not everyone may be required to take representation to the next step: from interaction to reflection and reification. The latter steps, however, have potential utility in entertainment (the arts, games) as well as in education.

Audiovisual Explorations: Steampunk Obesity Machine

Let's consider one such artifact, which is defined by a system dynamics model found in systems science and simulation. Figs. 21.2 and 21.3 are two different representations of a System Dynamics flow graph (Forrester, 1991) capturing the temporal nature of human metabolism.

The diagram in Fig. 21.2 represents a virtual machine based on the analogy of fluid flow. Fluid starts from source node (left-most "cloud" icon) and proceeds to flow through a system of levels separated by rates to a sink node (right-most cloud icon). More generally, the fluid flow can be construed as a kinetic energy flow since fluid velocity is the dominant flow variable. At the start of the machine, at the left, fluids pour into metabolism and food intake to suggest that the more energy, the higher the *Fitness Level*, but also the higher the *Weight*. The rate variable, *Metabolism*, is proportional to a functional combination of Fitness Level, Exercise, and Nutrition. The

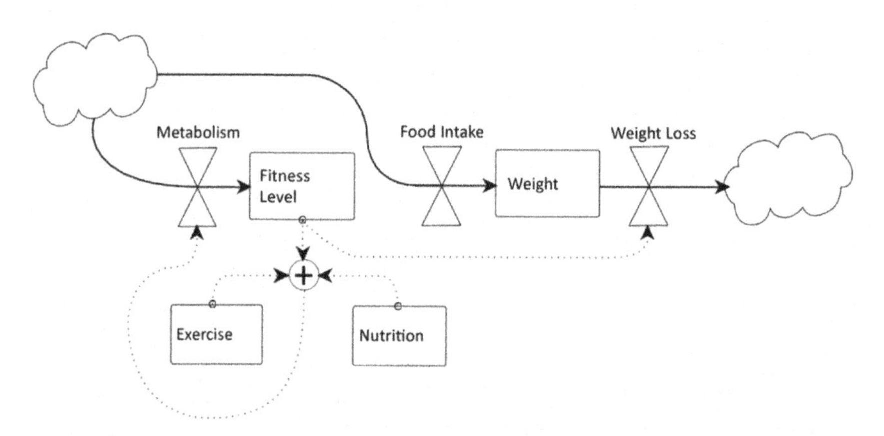

FIGURE 21.2 A System Dynamics flow graph with two levels (i.e., stocks) and three rates. *Source: Copyright 2008 Inderscience Publishing, Fishwick, P.A., 2008b. Software aesthetics: from text and diagrams to interactive spaces. Int. J. Arts Technol. 1 (1), 90–101.*

nature of this precise formula is not present in the model since the model is an abstract representation of the dynamics. The solid curve arrows reflect fluid flow through the system, and the dashed curve arrows reflect control settings to change the rates on the valves. Fig. 21.2 is a hypothetical example, and is not put forth as an accurate or valid simulation model of nutrition, but rather to demonstrate that similar diagrammatic models are widely used in science and engineering. These types of models were originally implemented as physical, analog computers although their more frequent existence today is as digital models with a diagrammatic front end authoring capability. The MONIAC, or "Phillips Machine," is one such example (Swade, 2000; Ryder, 2009) from the analog computing era.

Fig. 21.3 shows the same model which is a synthetic rendition of Fig. 21.2, reified using a "steampunk machine" since its structure is reminiscent of the cyberpunk aesthetic that continues to be popular since its inception in Gibson's work (Cavallaro, 2000). Steampunk culture has connotations of "reclaiming tech for the masses" (Grossman, 2009). Water is pumped using steam-power underneath the wooden floor. This water shoots out of two brass orifices that represent the two valve-icons in Fig. 21.2. Water filled glass containers represent the level quantities, and wood/brass control rods connect everything together as in Fig. 21.2. The human avatar on the left is demonstrating the machine in action to us, or we may become the avatar. The natural question is why anyone might want to construct such a machine when Fig. 21.2 might do. For the answer to this, we have additional questions to ask, with possible use-cases.

Fig. 21.2 and the equations that map to this diagram, are most often used by scientists familiar with the system dynamics method. It is unlikely that these scientists have any interest in structures, such as Fig. 21.3

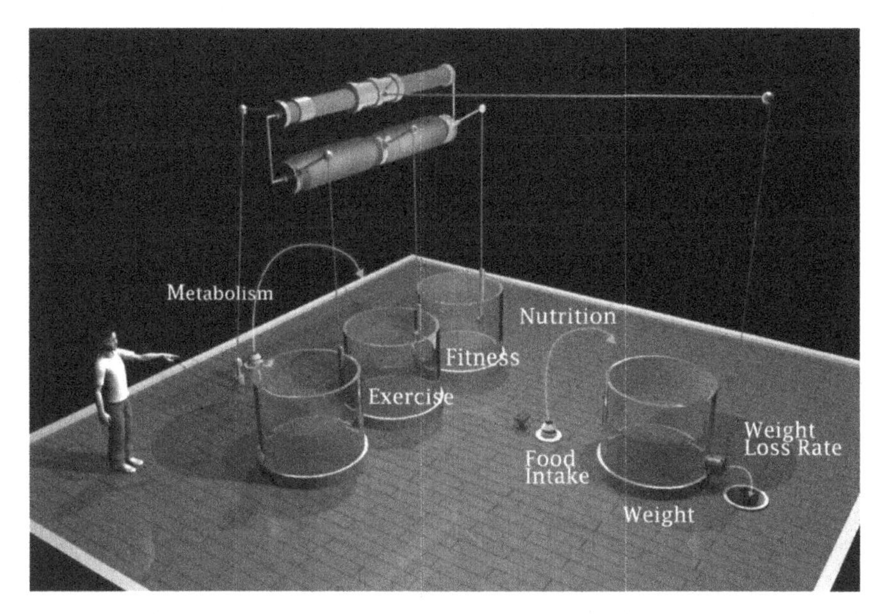

FIGURE 21.3 A steampunk obesity machine isomorphic to Fig. 21.2. *Source: Copyright 2008* Inderscience Publishing, *Fishwick, P.A., 2008b. Software aesthetics: from text and diagrams to interactive spaces. Int. J. Arts Technol. 1 (1), 90–101.*

mainly because they are comfortable and familiar with more formal representations. However, the vast majority of the population may require additional motivation if they are to understand, and be motivated or influenced by, the more formal representations. Therefore, the machine for Fig. 21.3 is appropriate for education and entertainment. It is easy to imagine the machine in Fig. 21.3 being engaging especially with game-like features that required certain goals such as stabilizing the water level in the Weight container.

Visual Representations of Data

There are numerous additional examples of artworks that, if used as guidance, can lead to aesthetic computing products useful for education. The vast majority of examples are the encoding and presentation of data rather than of program or model. It is logical given that data repositories and accessibility are expanding rapidly and that they represent the simplest and easiest to grasp forms of information. Consider the model of a single number shown in Fig. 21.4.

This encoding of number as a stack of one hundred dollar bills is given context by familiar objects whose size is known through pictures or experience (e.g., the Statue of Liberty, a football field, a truck). One might take this same approach to representing other analog representations of

FIGURE 21.4 The relative size of the U.S. debt if it reaches 15 trillion dollars. The large rectangular block represents stacks of one hundred dollar bills (Oto Godfrey, http://usdebt.kleptocracy.us)

monetary amounts through choosing different familiar objects. A participant's engagement can have both artistic and mathematical consequences. For example, we can imagine performing operations on numbers in this type of representation much as we have done manually in the past with quipus and abaci.

Consider Huff's prime number series (Huff, 2006) with two example encodings of prime factors shown in Fig. 21.5.

The two encodings in Fig. 21.5 are pieces of fine art, but could also be potentially used to motivate students to appreciate prime factorization

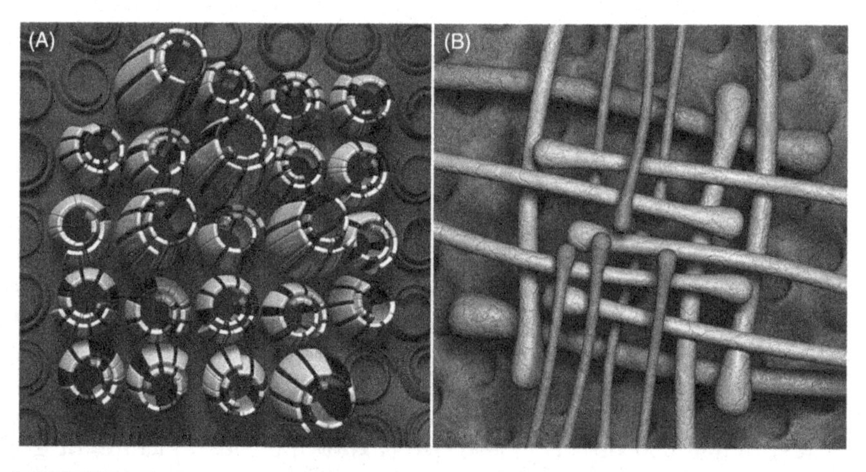

FIGURE 21.5 Prime number factorization encodings (EPF: 2003:V:A:997141 and 2000.24). (A) Infoviz graffiti and (B) living light. Kenneth A. Huff (http://www.kennethahuff.com)

FIGURE 21.6 Left (Infoviz Graffiti/Adjustable Pie-Chart Stencil, Golan Levin), Right (Living Light, David Benjamin and Soo-In Yang)

through puzzle-making. For example, consider where one might provide to someone a visual encoded integer and then ask that person to identify the number and factors. Fig. 21.6 shows two additional examples of information presence: Levin's infoviz graffiti for data, and Living Light. The graffiti is a deliberate mechanism for surfacing numbers of societal relevance in public places. Living Light is a permanent outdoor pavilion in Seoul, South Korea. The pavilion's purpose is to allow spectators to visualize environment levels such as air quality. As pervasive computing extends into the future, most flat surfaces become display surfaces opening up numerous possibilities for bringing information into our daily lives. Fig. 21.7 shows a model of a city which is turned into a computer program-like artifact, or automaton, whose output is a musical score.

FIGURE 21.7 Pianola city music (Akko Goldenbeld, http://akkogoldenbeld.com)

Textual Representations

The examples so far have been mainly visual; however, it is often desirable to use thinking similar to that described earlier for mathematical expressions, but to extend this to software, for example. Emerging areas in the humanities such as software studies (Fuller, 2008) and critical code studies (Marino, 2006) situate the need for studying formal languages using some dimension of hermeneutics. These areas also provide opportunity for creation of new human-computer interfaces. For example, we may treat software as a full hyper-mediated structure (Roth et al., 1994; Anderson et al., 2000). One can then, through an embodied approach facilitated by link interaction, treat formal language-based constructs as hypermedia.

ART AND DESIGN AS CREATIVE INFLUENCES FOR EMBODIED FORMAL LANGUAGES

The provided examples including Figs. 21.3—21.7 are related to aesthetic computing in different ways. Since aesthetic computing is embodied in formal languages with an educational goal as a final end product, I will overview how these examples might achieve that goal. Table 21.1 contains 5 columns: column 1 refers to a previously described image or product; column 2 is the original medium; column 3 is a hypothesized goal for the last 5 rows (i.e., since the original intention is not known but assumed); column 4 is an example repurposing of the original product for a formal language goal (column 5). Let's consider the 3rd row. The product has been designed to a highly compelling and attractive display of the national debt. This creative use can be recast as a new way to learn *number sense*. The formal language products are only examples and have not been constructed by anyone, however, the original art and designs are dually inspirational—for their original goal or purpose, and for a form that leverages their embodied characteristics for the purpose of formal language instruction.

Table 21.1 portrays aesthetic computing through repurposing existing art works, but this procedure is optional. Formal language-based products that capture the essence of embodied interaction can be designed directly from initial design, to detailed design, and onto an implementation. The Steampunk Obesity Machine (Table 21.1, Row 2) is a case in point. Even though a poster board image (Fig. 21.3) was part of a curated art exhibit (Harn, 2011), the image was meant as a preliminary design for a virtual machine to teach system dynamics concepts. The machine has not yet been constructed.

TABLE 21.1 An Aesthetic Transformation to Formal Language Learning Objectives

Example	Original product	Aesthetic goal (hypothesized)	Formal language product (example)	Formal language goal
Personal experience (arithmetic)	Typographic image	To illustrate elegance of the mathematical form	Game with moving operators and operands	To teach laws of arithmetic
Steampunk obesity machine (Fig. 21.2)	Raster image art work	To create steampunk genre-related imagery	Video illustrating functional mechanism and control	To teach system dynamics methodology
US national debt (Fig. 21.3)	Raster image art work	To illustrate the magnitude of the US debt using scale	A tactile set of blocks and objects	To teach number sense
Prime number factorization (Fig. 21.4)	Raster image art work	To celebrate organic forms using prime number encoding	An adventure game using encodings as 3D puzzles	To teach about prime numbers and factorization
Infoviz Graffiti (Fig. 21.5A)	Graffiti and template in Outdoor location	To present societal information to the public	An alternate reality game (hunting for graffiti)	To teach concept of percentage
Living Light (Fig. 21.5B)	Outdoor sculpture	To present environmental information to the public	A kinetic sculpture	To teach data structures
Pianola city music (Fig. 21.6)	Indoor kinetic art work	To explore an architecture-music interface	Indoor kinetic object	To teach concept of a data search via sound
Hyper-mediated software engineering	Web-based computational literature	To represent cultural knowledge	Hyper-mediated software/code	To encourage learning of how to code

FIGURE 21.8 Arithmetic logic unit built with "redstone" in an immersive play space using the Minecraft game engine.

EMBODIED COMPUTING USING SERIOUS GAMING

The discussion of aesthetic computing and the interpretation of it via embodied formal language would be incomplete without reference to video and console game cultures. An example is illustrative: logic circuits in the game Minecraft (2011). Minecraft is a "block game" where players move around a space and build blocks using a mining metaphor. Some of the procedural capabilities within the game have engaged members of the community to create basic circuits, leading up to full-fledged computers out of the logic circuitry. Since Minecraft is highly interactive, and invokes a sense of presence to boot, this type of hacking is consistent with the concepts in aesthetic computing: players are working together to form circuits through embodied interaction. Fig. 21.8 shows a Minecraft arithmetic logic unit (ALU) described by Ganapati (2010).

COLLABORATIVE ROLES, USABILITY, AND EXPERIENCE

Aesthetic Computing begins with a formal language construct such as a number, data, model, or software. Then the challenge is to represent this construct through embodiment. We noted that "embodiment" can be as simple as pure reification without representation of existing objects when we demonstrated the ability to grab hold of numbers and move them toward operators. However, reification can also suggest object representation as in Figs. 21.4–21.7. I need to address the "who" and "why" aspects of aesthetic computing.

First, who is going to be creating these representations? In the case of collaboration, I recommend teams of humanist scholars, artists, and computer scientists. Humanist scholars bring to bear different philosophies and theories which can help shape the resulting representation. The artist has the creative perspective and tools to create the representation, and the computer scientist can serve two roles: to help construct tools used by the humanist and artist in the extraction of information and in enabling the interaction that ensues through externalizing embodiment in the human-computer interface.

Second, who is going to use the representations? Students in my aesthetic computing class are often initially confused why one would construct anything but diagrams. This confusion is expected, but we must be careful when defining usability: usable for whom and for what purpose? We need to identify (1) the goal of the representation, and (2) the end target users. Goals for the embodied representations are education, arts, and entertainment (e.g., cinema, visual and performing arts, fiction). Target users may be any grade level in school or some segment of the general public. From a psychological perspective, a broad view of "usability" can encompass user goals including: increased valence, motivation, and attitudinal change, as well as improved short or long term memory. Mathematicians and computer scientists are not the target, as these populations are adept at using existing notations. Aesthetic computing is less stressed on information extraction and more on the use of entertainment, arts, and humanities on formal languages with the *largest practical effects being in education*. Thus the target users are formal and informal learners of all elements of formal language-based instruction (e.g., mathematics, computer science).

The roles of participants in aesthetic computing will likely be different given the interests of each party. For the computer scientist, for example, Fig. 21.5 serves as a design template for the creation of special effects and interactive games for the purpose of expressing elements of prime numbers and the factorization process into these numbers. The artist's work is a medium through which this aspect of formal language is creatively expressed. The goals of the artist and computer scientist are clearly different, but the means (i.e., representations of prime numbers) are common. This difference in ends, with similar means, plays out in the other examples. For instance, Perl poetry (i.e., poetry created using the programming language, Perl) may be an aesthetic product to the writer—a valid end in itself. To the computer scientist, this product represents a medium in which to express a different end—the formal language "message." Therefore, aesthetic computing by its arrangement of words comprising this phrase is focused on computing—the learning of formal languages. However, aesthetic products play a key role in this learning activity and allow for the artist, scholar, and computer scientist to collaborate with different intentions and goals.

Other areas related to aesthetic computing are information visualization (Card et al., 1999; Ward et al., 2010), and software visualization (Eades and Zhang, 1996; Stasko et al., 1998; Zhang, 2007; Diehl, 2010); however, the goals of these areas are generally quite different than for aesthetic computing. In information visualization, the goal is efficient communication of data and information, whereas for aesthetic computing, the goal is education through highly embodied, and interactive, aesthetic products in the forms of art and entertainment. As such, Aesthetic Computing fosters a deeper experience than building representations meant for immediate consumption (e.g., newspaper diagrams and maps). Readers will observe that the use of metaphor is rich within the high level interactions with computers. We are an interface culture (Johnson, 1997). However, the metaphors used on the "desktop," for instance, have not yet made their way into the core of mathematics and computing. Efforts such as computational thinking (Wing, 2006) are a move in the right direction.

Laurel (1993) presciently captures a prerequisite for aesthetic computing in her "Computing as Theatre." However, Laurel was mainly constructing a case for human-computer interaction as a complex theatrical production, involving many of the same elements found in theatre. The *use* of computing, and its associated interaction phenomena, are like theatre. However, what we find is that as we break open the lid of the black box containing the atomic elements of normally hidden data, formulas, code, and models is that computing is theatre *all the way down*.

TOWARD A METHOD OF AESTHETIC COMPUTING

While it is interesting to pose ideas and directions, a procedural method is something that can help to forge a discipline even if only as a general guide. Fishwick (2007b) was an initial attempt at this process with a small example of code that was represented as a collection of rooms in a building, complete with a partial narrative for context. Fig. 21.9 serves as a basis for describing the approach used in (AC, 2012).

We begin (in the top left of Fig. 21.9) with a formal language construct that is to be conveyed to non-specialists in mathematics and computing with the goal of broadening the exposure of computing concepts. The asterisks denote current emphases in (AC, 2012). Target users will depend on the type of formal language. If the goal is number sense, and the numbers are fairly simple, we may be looking at elementary school children. If the formal language is simple algebraic formulas, we may

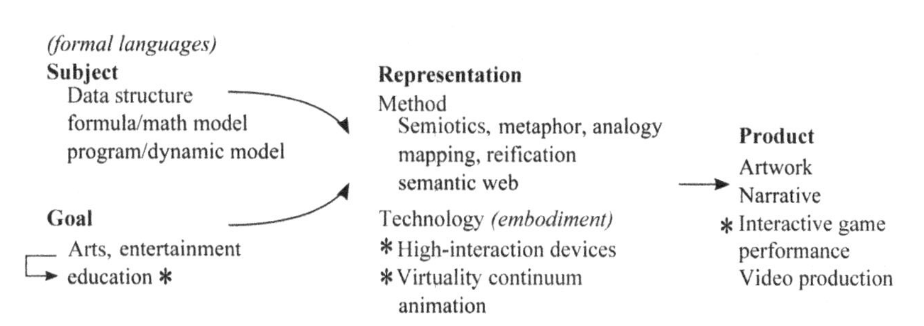

FIGURE 21.9 Aesthetic computing method.

be looking at 8th grade mathematics. More complex mathematical and computing structures may require higher grades, including universities and in postgraduate, informal learning contexts. One of the desirable outcomes of this approach to representation, though, is to expose very young children to seemingly complex data structures and programs by using games and video as motivational media. I expect that the approaches may serve as (1) scaffolding for later, more traditional, instruction and notations, and (2) secondary devices (e.g., puzzles) to reemphasize concepts that some learners find difficult using standard notations. The goal is not to eliminate standard notations as this would be counterproductive. Representation is divided, in Fig. 21.9, into two components: *methods* that achieve representation and *technologies* that support embodiment. End products that emphasize, or surface, embodiment can vary. A good piece of fiction can create a strong sense of presence and virtual embodiment, whereas a weak interactive game may be left ignored if not well designed.

NEW CONNECTIONS

A primary goal of mine in fostering aesthetic computing is to link disciplines—especially those in computing to the humanities and arts. As evidenced by designers and humanist scholars, artifacts such as "code" and "data" are now being interpreted and recreated. There are many reasons for this. Perhaps, the ubiquitous computing trend is the most significant driver—software is everywhere and so, by natural extension, cultural. I welcome the artists, designers, and humanists into the "formal languages" space and hope that through collaborations and interdisciplinary discussions and critique that we might re-humanize core elements of computing, and perhaps even mathematics.

DISCIPLINARY AND TECHNICAL CHALLENGES

The area of aesthetic computing is not without its challenges. The goal is to leverage embodiment theories toward building new computer-based interfaces for learning formal languages. Disciplines that I have covered have sub-areas that are all targeted toward this goal, but significant challenges remain for each area:

- *Mathematics*: the literature in mathematics education, and in the application of cognitive linguistics within mathematics learning, is well-founded and supports aesthetic computing. This body of knowledge, however, is more focused on analysis and theory construction rather than, through analogy, building new interfaces in mathematics education to take full advantage of the embodiment theories through realization. Some efforts in *virtual manipulatives* are a good start, but this work should expand to employ the next generation of interface capabilities that stress embodiment (e.g., multi-touch displays, body tracking, mixed/virtual reality technology).
- *Computing*: the literature in computing education provides fairly easy-to-use interfaces for seeing the results of executing programs; however, the programs are often limited to the canonical alphanumeric notation with all of the human interaction being in the program execution rather than inside the program. Efforts at software visualization move in the direction required by aesthetic computing, and yet, there is a much wider set of possibilities for representation if the goal is to teach non-specialists especially through immersion, situated learning, and interactive games. Diagrams are fine for communication, but if the goal is to explore deeply embodied approaches for learning, additional media and newer interfaces—as recommended for mathematics education—should be more thoroughly investigated.
- *Humanities:* the work in cognitive linguistics, and resulting embodiment theories, ground the work in aesthetic computing, but as with the work in the philosophy of mathematics learning, there is little corresponding effort in realizing these theories in a human-computer interface. Conversely, the work in cultural theory production is recently targeting "code" specifically as a new type of literacy (e.g., critical code studies). And yet, this production tends to avoid linguistic analysis and instead focuses on socio-historical analysis. New, embodied, interfaces for code can build off of the scholarly analysis, but these interfaces should also be informed by key facts of semiotics (e.g., analogy and metaphor) which lie at the foundation of formal languages. There appears to be a bias toward textual notation rather than exploring broader forms of "embodied literacy," which would include textual notation as one dimension.

- *Art & Design:* works of art have traditionally treated formal languages as "black boxes," tools needed to create art or designs. Unlike in the humanities, where code has become subject material, in art, code tends to be treated purely as a tool, whether embedded in package or programmed via a text-based development environment. The only exception to this observation would be in typography within graphic design, where the subject material is the text. More explorations are required so that formal languages become the active subjects of artwork.

Each one of these four areas has some common challenges. Observing that analogy is the engine of metaphor in scientific practice, aesthetic computing products can be created with an increased attention to analogy. Another observation is that with the exception of Art & Design, there is a classical focus on alphabetic notation. Such notation serves us well and has enriched our formal languages. However, there are other types of notations that exercise more of the body's sensorimotor functions. Diagrams are a good place to start in seeing this transition since with diagrams spatial metaphors for text-based notations abound, but we should not limit our embodied explorations to diagrams.

A primary aesthetic computing challenge is technological. It is still relatively expensive to build new interfaces based on the types of products described by the figures previously shown. "3D modeling" as a real-time technical interface capability is nowhere near the futuristic landscapes of Tron, the Matrix, and the Holodeck. Modeling and animating in three dimensions remains a major challenge compared with diagrammatic approaches, and even diagram-based software modeling (e.g., model-driven architecture) struggles for acceptance in the marketplace of software engineering solutions because of the relative ease of using textual symbols. Human-computer interaction solutions are expanding in scope and capability, but we still are a long way from being able to easily and inexpensively become embodied in our formal language constructs.

SUMMARY: THE ARGUMENT FOR EMBODIED FORMAL LANGUAGE

This chapter began with personal experiences in mathematics and then moved on to discussions of embodied cognition, along with some examples of where aesthetic computing could be applied. The area of aesthetic computing rests primarily on the foundation of embodiment—whether we believe that our bodily interactions form our thought. This assumption of embodiment runs deep in philosophy. We all recognize that we have body and mind, and most would agree that the latter is the effect of the former. It is only fairly recently, though, that literature has arisen to indicate a strong relationship to the extent that thought itself, even for abstract objects, is embodied. The theory that undergirds embodiment is compelling, but we have the nagging question about how this theory can change what we do and how we act. If I imagine that I am imagining grabbing and pushing a number through a pseudo-biological membrane during arithmetical operations, I want to build a human-computer interface that reinforces this mental sequence by infusing theory into practice. This perceived need matches the aesthetic computing hypothesis stated at the start of the chapter: *Given the embodied nature of cognition, we should realize this embodiment through novel human-computer interfaces for learning formal languages.*

Achieving this realization involves a more thorough understanding of the interplay among disciplines and how embodiment theories in those disciplines interact and connect. The realization also requires a host of newer "virtuality continuum" technologies that allow us to achieve what Biocca refers to as degrees of progressive embodiment (Biocca, 1997). The technologies and their characteristics are overviewed for virtual reality by Sherman and Craig (2002), and by Bowman et al. (2004), and for augmented reality by Bimber and Raskar (2005).

WHERE TO LEARN MORE ABOUT AESTHETIC COMPUTING

For a thorough understanding of computing as a discipline, and its artifacts which are represented in aesthetic computing, the 1998 ACM Computing Classification System (CCS, 1998) serves as a good starting point. Even though my treatment of aesthetics is based on its original, perceptual definition, Kelly (1998) collects that which erupted from this kernel in philosophy and the arts in four volumes. Even though information visualization is centered on efficient communication (e.g., reading the equivalent of a diagram in a newspaper), some archives such as *infosthetics* curated by Vande Moere (2011) are broader and contain a wide variety of potential use cases—from efficient communication to experience, education, and play. For text-based representations, HASTAC (2011) serves as a high level repository of bloggers and projects, many of which are associated with *digital humanities.* The reader is encouraged to review articles cited in this chapter.

Acknowledgments

I would like to first acknowledge all individuals who have participated in this journey from its inception, including my colleagues in the arts, natural and social sciences, computer science, and mathematics. Students in my Aesthetic Computing class have had to put up with these ideas, and they have produced wonderful products that I could never have imagined. Thanks to the following colleagues who took time to make very good critical remarks on earlier forms of this manuscript: Sophia Acord (University of Florida), Michael Kelly (University of North Carolina at Charlotte), Mads Søgaard (Interaction Design), and Kang Zhang (University of Texas at Dallas). I take responsibility for any errors and omissions.

References

AC (Aesthetic Computing) University of Florida Class. 2012. http://www.cise.ufl.edu/~fishwick/ac/2012.

Anderson, K.M., Taylor, R.N., Whitehead, E.J., 2000. Chimera: Hypermedia for Heterogeneous Software Development Environments. ACM Transactions on Information Systems 18 (3), 211–245.

Arzarello, F. 2004. Mathematical Landscapes and the Their Inhabitants: Perceptions, Languages, Theories, In Proceedings of the International Commission on Mathematical Instruction, (ICME., 10), Copenhagen, Denmark, 158–181.

Audi, R. (Ed.), 1999. The Cambridge Dictionary of Philosophy. Cambridge University Press, Cambridge.

Barsalou, L. W. 2010. Grounded Cognition: Past, Present, and Future. In *Topics in Cognitive Science*, Cognitive Science Society, Inc. 2, 716–724.

Beck, D., Fishwick, P. A., Kamhawi, R., Coffey, A. J., Henderson, J., 2011. Synthesizing Presence: A Multidisciplinary Review of the Literature, *Journal of Virtual Worlds Research*, 3(3), 3–35.

Biocca, F., 1997. The Cyborg's Dilemma: Progressive Embodiment in Virtual Environments. Journal of Computer-Mediated Communication 3 (2). http://jcmc.indiana.edu/vol3/issue2/biocca2.html#cyborg's.

Bimber, O., Raskar, R., 2005. Spatial Augmented Reality: Merging Real and Virtual Worlds. A. K. Peters.

Bolter, J.D., Grusin, R., 2000. Remediation: Understanding New Media. MIT Press.

Bowman, D., Kruijff, E., LaViola, J.J., Poupyrev, I., 2004. 3D User Interfaces: Theory and Practice. Addison-Wesley.

Brown, J.S., Collins, A., Duguid, S., 1989. Situated Cognition and the Culture of Learning. Educational Researcher 18 (1), 32–42.

Burke, J., 1943. Hogarth and Reynolds; a Contrast in English Art Theory. Oxford University Press, London, UK.

Buxton, B., 2008. The Role of the Artist in the Laboratory. Meisterwerke der Computer Kunst. Springer Verlag, Bremen, pp. 29–32, http://billbuxton.com/artistRole.html.

Card, S., Mackinlay, J., Schneiderman, B., 1999. Readings in Information Visualization: Using Vision to Think. Morgan Kaufmann Publishers.

Cavallaro, D., 2000. Cyberpunk and cyberculture: Science fiction and the work of William Gibson. Continuum Press, London, UK.

CCS. 1998. The ACM Computing Classification System. Association for Computing Machinery. http://www.acm.org/about/class/ccs98-html.

Cockton, G. 2011. Usability Evaluation. To appear in Encyclopedia of Human-Computer Interaction. http://www.interaction-design.org.

Dagstuhl. 2011. Schloss Dagstuhl, Leibniz-Zentrum für Informatik. http://www.dagstuhl.de.

Dede, C., 2005. Planning for Neomillennial Learning Styles. In Educause Quarterly 28 (1), http://www.educause.edu..

Devlin, K. 2006. The Useful and Reliable Illusion of Reality in Mathematics. In *Towards a New Epistemology of Mathematics*. In Workshop at GAP.6 Freie Universität Berlin. http://www.lib.uni-bonn.de/PhiMSAMP/GAP6/Talks/Devlin.pdf.

Dewey, J., 1934. Art as Experience. Penguin Group.

Diehl, S., 2010. Software Visualization: Visualizing the Structure, Behavior, and Evolution of Software. Springer.

Dieterle, E., Dede, C., Schrier, K., 2007. Neomillennial" Learning Styles Propagated by Wireless Handheld Devices. In: Lyras, M., Naeve, A. (Eds.), Ubiquitous and Pervasive Knowledge and Learning Management: Semantics, Social Networking and New Media to Their Full Potential,. IGI Global, pp. 35–66.

Diffey, T.J., 1995. A Note on the meanings of the term 'aesthetic. The British Journal of Aesthetics 35 (1), 61.

Dourish, P., 2004. Where the Action Is: The Foundations of Embodied Interaction. MIT Press.

Eades, P., Zhang, K., 1996. Software Visualisation. Series on Software Engineering and Knowledge EngineeringWorld Scientific Co, Singapore, 1996, ISBN: 981-02-2826-0, 268 pages..

Feldman, J.A., Narayanan, S., 2004. Embodied Meaning in a Neural Theory of Language. Brain and Language 89 (2), 385–392.

Feldman, J.A., 2008. From Molecule to Metaphor: A Neural Theory of Language. Bradford Books.

Fishwick, P.A., 1995. Simulation Model Design and Execution: Building Digital Worlds. Prentice Hall.

Fishwick P. A. 2002. "Aesthetic Programming: Crafting Personalized Software", Leonardo, MIT Press, 35(4): 383–390.

Fishwick, P.A. and Bertelsen, O W. 2002. Dagstuhl Seminar Report Number 348. http://www.dagstuhl.de/Reports/02/02291.pdf.

Fishwick, P., Diehl, S., Prophet, J. and Lowgren, J. 2005. Perspectives on Aesthetic Computing. *Leonardo* 38(2), 133–141, MIT Press, Cambridge. MA.

Fishwick, P.A. (Ed.), 2006. Aesthetic computing. MIT Press, Cambridge, MA.

Fishwick, P.A. (Ed.), 2007. CRC Handbook on Dynamic System Modeling. Chapman & Hall, CRC Computer & Information Science Series.

Fishwick, P.A., 2007b. Customized Visual Computing: The Aesthetic Computing Method. In: Ferri, F. (Ed.), Visual Languages for Interactive Computing. IGI Global, pp. 425–435.

Fishwick, P.A., Park, Y., 2008a. A 3D Environment for Exploring Algebraic Structure and Behavior, Ch. 30. In: Ferdig, R. (Ed.), Handbook of Research on Effective Gaming in Education,. Idea Group Inc, pp. 546–559.

Fishwick, P.A., 2008b. Software Aesthetics: From Text and Diagrams to Interactive Spaces. Int. J. of Arts and Technology. Inderscience 1 (1), 90–101.

Fishwick, P.A., 2010. A Decade of Digital Arts and Sciences. Leonardo. MIT Press, Cambridge, MA, Accepted for publication. http://cise.ufl.edu/academics/undergrad/das/LEO-DAS.pdf.

Fishwick, P. A. Kamhawi, R., Coffey. A. J., and Henderson, J. 2010. An Experimental Design and Preliminary Results for a Cultural Training System Simulation, In Proceedings of the 2010 Winter Simulation Conference, Eds. Johansson, B., Jain, S., Montoya-Torres, J., Hugan, J. and Yucesan, E. 799–810. http://www.informs-sim.org/wsc10papers/072.pdf.

Forrester, J.W., 1991. System Dynamics and the Lessons of 35 Years. The Systemic Basis of Policy Making in the 1990s 29.

Fuller, M. (Ed.), 2008. Software Studies: A Lexicon. MIT Press.

Ganapati, P. 2010. Geeky Gamers Build Working Computers out of Virtual Blocks. Wired Gadget Lab. http://www.wired.com/gadgetlab/tag/minecraft/.

Gershenfeld, N., Krikorian, R., and Cohen, D. 2004. The Internet of Things, Scientific American, October.

Ghezzi, C., Jazayeri, M., 1997. Programming Language Concepts. John Wiley and Sons.

Goldin, G.A., Kaput, J.J., 1996. A Joint Perspective on the Idea of Representation in Learning and Doing Mathematics. In: Steffe, L., Nesher, P. (Eds.), Theories of Mathematical Learning,. Lawrence Erlbaum Assoc, pp. 397–431, Chapter. 23.

Grau, O., 2004. Virtual Art: From Illusion to Immersion. MIT Press.

Greenfield, A., 2006. Everyware: The Dawning Age of Ubiquitous Computing. New Riders Publishing.
Grossman, L. 2009. Steampunk: Reclaiming Tech for the Masses. Time Magazine (Arts), Dec 14, 2009. http://www.time.com/time/magazine/article/0,9171,1945343,00.html.
Guhe, M., Smaill, A., Pease, A., 2009. A Formal Cognitive Model of Mathematical Metaphors. In: Mertsching, B., Hund, M., Aziz, Z. (Eds.), KI 2009: 32nd Annual Conference on Artificial Intelligence. Springer Verlag, pp. 323–330.
Hadamard, J., 1996. The Mathematician's Mind: The Psychology of Invention in the Mathematical Field. Princeton University Press.
Harn. 2011. Art in Engineering. Museum Nights at the Harn Museum of Art. Gainesville, Florida. October 13. http://www.harn.ufl.edu/octmuseumnight.pdf.
Hassenzahl, M. 2011. User Experience and Experience Design.. In Encyclopedia of Human-Computer Interaction. http://www.interaction-design.org/encyclopedia/user_experience_and_experience_design.html.
HASTAC. 2011. Humanities, Arts, Science, and Technology Advanced Collaboratory. http://hastac.org.
Hopcroft, J.E., Motwani, R., Ullman, J.D., 2000. Introduction to Automata Theory, Languages, and Computation, 2nd Ed., Addison Wesley, Reading, MA.
Huff, K.A., 2006. Visually Encoding Numbers using Prime Factors. In: Fishwick, P. (Ed.), Aesthetic Computing,. MIT Press, Cambridge, MA.
Ifrah, G., 2001. The Universal History of Computing: From the Abacus to the Quantum Computer. John Wiley & Sons, Inc.
Ikei, Y., Ota, H. 2008. Spatial Electronic Mnemonics for Augmentation of Human Memory. In Proceedings of, IEEE., Virtual Reality, 217–224.,
Johnson, E., Adamo-Villani, N., 2010. A Study of the Effects of Immersion on Short Term Spatial Memory. World Academy of Science, Engineering Technology, 71.
Johnson, M., 1987. The Body in the Mind. The Bodily Basis of Meaning, Imagination, and Reason. University of Chicago Press, Chicago, IL.
Johnson, M., 2007. The Meaning of the Body: Aesthetics of Human Understanding. University of Chicago Press, Chicago.
Johnson, S., 1997. Interface Culture: How New Technology Transforms the Way We Create and Communicate. Basic Books.
Kivy, P. (Ed.), 2004. The Blackwell Guide to Aesthetics. Blackwell Publishing, Malden, MA.
Kallay, W., 2011. The Making of Tron: How Tron Changed Visual Effects and Disney Forever. William Kallay: publisher.
Kelly, M. (Ed.), 1998. Encyclopedia of Aesthetics. Oxford University Press, New York, NY.
Kelly, M., Vesna, V., Fishwick, P., Vande Moere, A., and Huff, K. 2009. The State of Aesthetic Computing or Info-Aesthetics: curated panel discussion. In ACM SIGGRAPH '09, ACM, Panel held in New Orleans, LA. Published: New York, NY.
Knuth, D.E., 1992. Literate Programming. Stanford Computer Science Dept.
Lakoff, G., Johnson, M., 1999. Philosophy in the Flesh. Basic Books.
Lakoff, G., Nunez, R., 2000. Where Mathematics Comes From : How the Embodied Mind brings Mathematics into Being. Basic Books, New York, NY.
Lakoff, G., Johnson, M., 2003. Metaphors We Live By, 2nd Ed University of Chicago Press.
Laurel, B., 1993. Computers as Theatre. Addison-Wesley.
Lave, J., Wegner, E., 1991. Situated Learning: Legitimate Peripheral Participation. Cambridge University Press.
Lessig, L., 2008. Remix: Making Art and Commerce Thrive in the Hybrid Economy. Penguin Press.
Leymarie, F.F., 2006. Aesthetic Computing and Shape. In: Fishwick, P.A. (Ed.), Aesthetic Computing,. MIT Press, pp. 258–288.
Leyton, M., 2001. A Generative Theory of Shape. Lecture Notes in Computer Science Springer Verlag.

Leyton, M., 2006. The Foundations of Aesthetics. In: Fishwick, P.A. (Ed.), Aesthetic Computing,. MIT Press, pp. 289–313.
Löwgren, J. 2011. Interaction Design. In Encyclopedia of Human-Computer Interaction. http://www.interaction-design.org/encyclopedia/interaction_design.html.
Malina, R. 2011. The Strong Case for Art-Science Interaction. thoughtmesh: http://vectors.usc.edu/thoughtmesh/publish/120.php.
Manovich, L., 2002. The Language of New Media. MIT Press.
Mar, R.A., Oatley, K., 2008. The Function of Fiction is the Abstraction and Simulation of Social Experience. Psychological Science 3 (3), 173–192.
Marino, M. 2006. Critical Code Studies, Electronic Book Review, http://www.electronicbookreview.com/thread/electropoetics/codology.
McLuhan, M., 1964. Understanding Media: The Extensions of Man. MIT Press.
Minecraft Game. 2011. Minecraft Redstone Circuits. http://www.minecraftwiki.net/wiki/Redstone_Circuits.
Mohr, M. 1964-2011. Réflexions Sur Une Ésthetique Programmée, Sept. 9-Oct 15, 2011. Bitforms Gallery, New York, NY. http://www.bitforms.com/press-releases/405-2011-manfred-mohr.html.
Nake, F., 2009. The Semiotic Engine: Notes on the History of Algorithmic Images in Europe. Art J., 76–89.
Negroponte, N., 1996. Being Digital. MIT Press, Cambridge, MA.
Norman, D.A., 2004. Emotional Design: Why we Love (or Hate) Everyday Things. Basic Books, New York, NY.
Papert, S., 1980. Mindstorms: Children, Computers, and Powerful Ideas. Basic Books.
Parsons, T.D., Rizzo, A., 2008. Initial Validation of a Virtual Environment for Assessment of Memory Functioning: Virtual Reality Cognitive Performance Assessment Test. CyberPsychology & Behavior, 11 (1): 17–25.
Petre, M., Blackwell, A., 1999. Mental Imagery in Program Design and Visual Programming. International Journal of Human Computer Studies 51 (1), 7–30.
Piaget, J., 1950. The Psychology of Intelligence. Routledge, New York, NY.
Pine, B.J., 1999. The Experience Economy: Work is Theater & Every Business a Stage. Harvard Business Press, Cambridge, Mass.
Reas, C., Fry, B., 2007. Processing: A Programming Handbook for Visual Designers and Artists. MIT Press, Cambridge, MA.
Roth, T., Aiken, P., Hobbs, S., 1994. Hypermedia Support for Software Development: A Retrospective Assessment. Hypermedia 6 (3), 149–173.
Rotman, B., 2000. Mathematics as Sign: Writing, Imagining, Counting. Stanford University Press.
Ryder, W. 2009. A System Dynamics View of the Phillips Machine. In Proceedings of the 27 International Conference of the System Dynamics Society, http://systemdynamics.org/conferences/2009/proceed/papers/P1038.pdf.
Salomon, G., 1990. Cognitive Effects With and Of Computer Technology. Communication Research 17 (26), 26–44.
Sfard, A., 1994. Reification as the birth of metaphor. For the Learning of Mathematics 14 (1), 44–55.
Sherman, W.R., Craig, A.B., 2002. Understanding Virtual Reality: Interface, Application, and Design. Morgan Kaufmann.
Slingerland, E., 2008. What Science Offers the Humanities: Integrating Body and Culture. Cambridge University Press.
Speer, N.K., Zacks, J.M., Reynolds, J.R., 2007. Human Brain Activity Time-Locked to Narrative Event Boundaries. Psychological Science 18 (5), 449–455.
Stasko, J.T., Brown, M.H., Domingue, J.B., Price, B.A., 1998. Software Visualization. MIT Press.
Swade, D., 2000. The Phillips Machine and the History of the Computing. In: Leeson, R. (Ed.), A W. H. Phillips: Collected Works in Contemporary Perspective,. Cambridge University Press.

Thompson, P.W., Sfard, A., 1994. Problems of reification: Representation and Mathematical objects. Kirschner, D. (Ed.), Proceedings of the Annual Meeting of the International Group for the Psychology of Mathematics Education - North America, Plenary Sessions,, Vol. 1, Louisiana University, Baton Rouge, LA, pp. 1–32.
Tractinsky, N., Katz, A., Ikar, D., 2000. What is Beautiful is Usable. Interacting with Computers 13 (2), 127–145.
Turkle, S., 2005 January 30. How Computers Change the Way We Think. The Chronicle of Higher Education 50 (21), B26.
Scruton, R., 2011. Beauty: A Very Short Introduction. Oxford University Press, Oxford, UK.
Vande Moere, A. 2011. *Infosthetics Archive*. http://infosthetics.com/.
Varela, F.J., Thompson, E.T., Rosch, E., 1992. The Embodied Mind: Cognitive Science and Human Experience. MIT Press.
Viegas, F.B., Wattenberg, M., van Ham, F., Kriss, J., McKeon, M., 2007. Many Eyes: A Site for Visualization at Internet Scale. IEEE Trans. on Visualization and Computer Graphics 13 (6), 1121–1128.
Ward, C., Grinstein, G., Keim, D., 2010. Interactive Data Visualization: Foundations, Techniques, and Applications. A. K. Peters Ltd.
Wing, J.M., 2006. Computational Thinking. Communications of the ACM 49 (3), 33–35.
Yates, F.A., 1966. The Art of Memory. University of Chicago Press.
Zhang, K., 2007. Visual Languages and Applications. Springer.

Further Reading

Tufte, E., 1983. The Visual Display of Quantitative Information. Graphics Press, Cheshire, CT.

Index

D

E

G

H

Printed in the United States
By Bookmasters